Lecture Notes in Economic

and Mathematical Systems

Springer
Berlin
Heidelberg
New York
Barcelona
Hong Kong
London
Milan
Paris
Singapore
Tokyo

Nigel H. M. Wilson (Ed.)

Computer-Aided Transit Scheduling

Proceedings, Cambridge, MA, USA, August 1997

Editor

Prof. Nigel H. M. Wilson
Massachusetts Institute of Technology
Dept. of Civil and Environmental Engineering
77 Massachusetts Avenue
Cambridge, MA 02139, USA

Library of Congress Cataloging-in-Publication Data

International Workshop on Computer-Aided Scheduling of Public Transport (7th : 1997 : Massachusetts Institute of Technology)
Computer-aided transit scheduling : proceedings, Cambridge, MA, USA, August 1997 / Nigel H.M. Wilson (ed.).
p. cm. -- (Lecture notes in economics and mathematical systems ; 471)
Includes bibliographical references.
ISBN 3-540-65775-4
1. Local transit--Management--Data processing--Congresses. 2. Bus lines--Management--Data processing--Congresses. 3. Production scheduling--Data processing--Congresses. I. Wilson, Nigel H. M. II. Title. III. Series.
HE4301.I58 1997
388.4'1322'0285--dc21 99-14726
CIP

ISSN 0075-8442
ISBN 3-540-65775-4 Springer-Verlag Berlin Heidelberg New York

Typesetting: Camera ready by author
SPIN: 10699893 42/3143-543210 - Printed on acid-free paper

ERRATUM

Lecture Notes in Economics, Vol. 471
N.H.M. Wilson (Ed.)
Computer-Aided Transit Scheduling

Due to a regrettable error in compilation of this volume, the pagination in Part 2 is inaccurate. The correct pagination follows below.

X

Part 2: Vehicle Scheduling and Service Design

Part 3: Real-Time Control, Demand Responsive Systems and Information Systems

Foreword

This proceedings volume consists of papers presented at the Seventh International Workshop on Computer-Aided Scheduling of Public Transport, which was held at the Massachusetts Institute of Technology from August 5th to 8th, 1997.

In the tradition of alternating Workshops between North American and Europe – Chicago (1975), Leeds (1980), Montreal (1983), Hamburg (1986), Montreal (1990), and Lisbon (1993), Cambridge (Massachusetts, USA) was selected for the Workshop in 1997. As in earlier workshops, the central theme dealt with vehicle and crew scheduling problems and the development of software systems incorporating operations research techniques for operational planning in public transport.

However, following the trend that started in Hamburg in 1987, the scope of this Workshop was broadened to include topics in related fields. Two trends underlie this. First, the recognition that the core scheduling issues in public transport have important common elements with other application areas in which extensive work is also underway, and that it is vital to learn from these other initiatives. Second, while scheduling is indeed a core problem in public transport planning, and has shown the first and greatest benefits from computer application, it is intimately related to the preceding tasks in the planning hierarchy, such as service design, and the following tasks such as operations control and public information. Increasingly, computers are playing greater roles across this full spectrum of problems, building from the scheduling kernel. Therefore, the program of this Workshop included sections which dealt with scheduling problems, and approaches used in related scheduled systems such as airlines and long-distance rail service. There was also increased emphasis on the questions of service design, operations control and automated public information systems. With increasing emphasis on providing access to public transport for all, several papers were presented dealing with demand-responsive systems.

Many of the papers presented dealt with research and implementation of the latest state-of-the-art software systems. The capabilities in hardware, software, and operations research techniques are all continuing to advance at a rapid rate and it is clear that the state-of-the-art as reflected in this Workshop's papers will continue to advance dealing with the expanded, complex problems of planning and operational control in public transport, as they relate to scheduling.

This Workshop was organized by the Center for Transportation Studies at MIT under the direction of the Program Committee, which consists of the following members:

Nigel H.M. Wilson; Cambridge (MA) USA (Chairman)
Avishai Ceder; Haifa, Israel
Joachim Daduna; Berlin, Germany
Giorgio Gallo; Pisa, Italy
Jose Paixao; Lisboa, Portugal
Jean-Marc Rousseau; Montreal, Canada
Stefan Voss; Braunschweig, Germany
Anthony Wren; Leeds, United Kingdom

During the Workshop 36 papers were presented from various fields of research and application. As in the previous years, a parallel software exhibition accompanied the Workshop demonstrating state-of-the-art systems for computer-aided scheduling in public transport.

The Program Committee has selected twenty papers to be included in this proceedings volume of the Seventh International Workshop. These papers are organized into three sections, in accordance with the main topics of the Workshop. The papers of Part 1 deal with core problems of **crew scheduling,** a field which remains a focus of much public transport research and development, especially with regard to the application of operations research techniques. Part 2 contains papers dealing with the other classical field of research and application in computer-aided planning in public transport, the problems of **vehicle scheduling**. The main objective of papers in both of these sections is to present procedures for solving these complex combinatorial problems, especially by incorporating the latest research advances. Finally, Part 3 includes papers dealing with the newer aspects of **real-time operational control, demand-responsive systems and advanced public information systems**. All of these are critically dependent on the power of current computer systems (both hardware and software) and open up important new challenges in the development and application of operations research methods.

Looking ahead we hope to make considerable progress in this field in terms of both research and application by the next Workshop, the Eighth International Workshop on Computer-Aided Scheduling of Public Transport, to be organized by Joachim Daduna and Stefan Voss and held in Berlin in June 2000.

For the success of the Workshop we are greatly indebted to the MIT Center for Transportation Studies for its generous financial support. We also gratefully acknowledge the support of the following organizations who contributed to this Workshop:

International Union of Public Transport
Canadian Urban Transit Association
Massachusetts Bay Transportation Authority

Finally, I would like to express my great appreciation to several people who were key in completing this endeavor. First, the staff of the MIT Conference Services, Events, and Information Office, Anne Miracle (MIT), and Alan Castaline (MBTA Head of Service Planning and Scheduling) were instrumental in making the Workshop happen. Second, those who labored long and hard to produce this volume, including Anne Miracle (MIT), Martine Joseph, and most significantly, Ginny Siggia (MIT). Without them this volume would never have seen the light of day.

Nigel H.M. Wilson
Cambridge

Table of Contents

Part 2: Vehicle Scheduling and Service Design

Part 3: Real-Time Control, Demand Responsive Systems and Information Systems

Solution of Large-Scale Railway Crew Planning Problems: the Italian Experience

Alberto Caprara[1], Matteo Fischetti[2], Pier Luigi Guida[3], Paolo Toth[1] and Daniele Vigo[1]
[1] DEIS – University of Bologna – Italy
[2] DMI – University of Udine – Italy
[3] Ferrovie dello Stato SpA – Italy

Abstract: Crew scheduling is a very well known problem which has been historically associated with airlines and mass transit companies; recently railway applications have also come on the scene. This is especially true in Europe, where deregulation and privatization edicts are forcing a re-organization of the rail industry and better productivity and more efficient services are strongly required by the market and the public. Therefore, this sector is showing increasing interest in Operations Research and Management Science. Railway crew planning represents a hard problem due to both the dimensions and the operational/regulation constraints involved. This paper describes the development of a new crew planning system set up by Ferrovie dello Stato SpA (the Italian railway company) in co-operation with the University of Bologna.

1 European Railways: a New Scenario

The European railways are being reshaped by a new phase of their historical life cycle, where privatization and deregulation are the underlining themes of the sector. This also follows general European Union (EU) policy directives concerning the separation of the railway system into separate enterprises: infrastructure or track management companies, and transport or train operating companies. The infrastructure company should retain state control in the rail business through natural monopoly or government regulation and investments. On the other hand, train operations should give rise to a number of independent and more business-oriented companies, which have access to the infrastructure via a market mechanism such as payment of track fees. This represents a major change from the traditional status of the European railways as totally state-owned and vertically-integrated industries.

Although this scenario takes different forms in various European countries, it is nevertheless established as the current trend and long-term objective of the railway industry. (However, some protected segments, such as regional and local services with social interest, may not be affected by these objectives.) The new rail policy, which has close counterparts in other transport modes (e.g., new access procedures in air transportation and airport slot allocation mechanisms), has its roots in the general EU principles of "open markets" and free movement of people and goods between the Union's countries. In the railway sector, this also means improved exploitation of international and high-speed operations, technological inter-operability, and overall industry standards. This would also transform European railways into more productive, customer-oriented, cost-conscious and efficiently-managed companies.

In some cases the European railway "revolution" has already happened, as a result of parallel national policies. In Sweden, separation between infrastructure and transport operations took place in 1988, when two different companies were set up (Banverket and SJ, for infrastructure and transport operations, respectively). In the United Kingdom, track management now belongs to an independent, publicly traded company (Railtrack), and train services have been privatized (mostly franchised) to a number of transport operating companies. In other European countries the process is continuing towards the creation of independent *business units*, with their own accounting systems. This is the first step in the direction of further separation and privatization.

According to EU policy, railways should be given less state aid to avoid market distortions, and cross-subsidization across market segments and parts of the network should be eliminated. The *trunks* (i.e., the most heavily operated and financially sustainable rail lines) are providing the field where new market rules and concepts are introduced.

In this scenario the European railways are introducing new management styles in order to improve their business results. In particular, new tools are being sought in order to optimize their use of human and other production resources. As in other transport industries (air transportation and mass transit), where Operations Research and Management Science methods have been intensively exploited, railways are now exploring new paths to improve their system optimization.

In this paper we report the implementation of a new crew planning system at Ferrovie dello Stato SpA (FS), the Italian state railways. In the next section we give an overview of the crew planning problem to be solved. The solution of this problem is obtained by subdividing it into three phases, namely pairing generation, pairing optimization and rostering optimization, dealt with in Sections 3, 4, and 5, respectively. Preliminary computational testing of the proposed approach is illustrated in Section 6. Finally Section 7 draws some conclusions and describes future extensions.

2 Crew Planning at the Italian Railways

Various resource planning problems are peculiar to large railway companies. FS is still a traditionally-integrated enterprise, although it is now structured into *Strategic Business Units* (SBUs). In particular, the *Infrastructure* SBU provides train control and track/installations maintenance, whereas transport operations are managed by three SBUs: *Intercity*, *Metropolitan*, and *Freight/Logistics*. Locomotive and drivers are provided by the *Traction* SBU. Finally, *Engineering* and *Property* SBUs complete the basic organization. FS operates a network of 16,000 track kilometers, with a total manpower of about 125,000 people. The productivity level (i.e., trains·Km/employee) of FS is better than the Western European average. About 8,000 trains are run daily, approximately 800 of which are long/medium range passenger trains (Intercities), 1,500 are freight trains, and the remainder are regional, local and metropolitan services. The workforce includes about 40,000 crew members (25,000 drivers and 15,000 conductors) located in about 50 *depots* (crew home bases). These figures result in very large planning problems. In fact, one of the largest problems, limited to intercity and long-range passenger trains, involves about 2,000 trains and 5,000 crew working segments.

Following the re-organization in the early '90s, the company undertook the development of a new computer-based crew planning system aimed at:

a) looking at the process from an "integrated" (yet classical) point of view;

b) taking advantage of OR techniques and close co-operation between company staff and academics;

c) updating the state-of-the-art of railway crew planning methods.

The system, which also takes into account the locomotive scheduling problem (not considered here), was nicknamed ALPI ("Applicazione Locomotive e Personale Informatizzata", i.e., Automation of Locomotive and Personnel System).

In railways, crew planning is concerned with building the work schedules of train crews (drivers and conductors), given a planned timetable. This is also part of more general activities within the rail industry which can broadly be called "tactical planning" and must be normally redefined at every timetable update (i.e., annually or on a seasonal basis). Planning of locomotive schedules, designing train routes, station track assignment (platforming), maneuver, maintenance and other scheduling tasks must also be undertaken at this stage to guarantee a feasible and efficient use of the production resources. In this industry-wide planning effort, focusing on a "single" resource (e.g., crew) should also be considered a tactical way of approaching a very large and complex problem, which usually requires a lot of co-ordination, agreements and feed-back processes among the parties involved (functional or business unit managers). Furthermore, the assumption that the timetable is given should be carefully considered, as the timetable itself can be the result of an iterative decision process.

Notice that tactical planning is not concerned with:

i) long-term (i.e., multi-year) and resource acquisition problems, which are part of strategic planning;

ii) daily resource assignment tasks that must cope with unscheduled or contingent transport;

iii) resources, except by type; that is resources (crew or locomotives) are not identified at this stage by their personal name or identification code.

The crew planning problem examined in this paper may be defined as follows. We are given a planned timetable for *train services* (i.e., both the actual journeys with passengers or freight, and the movement of empty trains or equipment between different stations) to be performed every day of a certain time period. Each train service has first been split into a sequence of *trips*, defined as segments of train journeys which must be serviced by the same crew without rest. Each trip is characterized by a departure time, a departure station, an arrival time, an arrival station, and possibly by additional attributes (e.g., overnight). During the given time period each crew performs a *roster*, defined as a cyclic sequence of trips whose operational cost and feasibility depend on several rules laid down by union contracts and company regulations. To ensure that each daily occurrence of a trip is performed by one crew, for each roster whose length is, say, k days, k crews are needed. In fact, on a given calendar day each of these crews performs the activities of a different day of the roster. Moreover, on consecutive calendar days each crew performs the activities of (cyclically) consecutive days of the roster. The crew planning problem then consists of finding a set of rosters covering every trip once, so as to satisfy all the operational constraints with minimum cost. Analogous crew planning problems arising in different public transportation areas are examined in Wren (1981), Bodin/Golden/Assad/Ball (1983), Carraresi/Gallo (1984a), Rousseau (1985), Daduna/Wren (1988), Desrochers/Rousseau (1992), Barnhart/Johnson/Nemhauser/Savelsbergh/ Vance (1994), and Desrosiers/Dumas/Solomon/Soumis (1995).

In transportation jargon and literature, rosters that are produced within crew planning are sometimes called "blank", in the sense that they will be assigned to specific people during daily operations management, which also copes with contingencies, unscheduled traffic, service perturbations, etc.

The general approach used by FS for the solution of the above defined crew planning problem follows a methodology which is fairly traditional in the literature (see, for example, Bodin/Golden/Assad/Ball (1983)). The problem is decomposed into three phases as follows:

1. *Pairing generation*: a very large number of feasible *pairings* is generated. A pairing is defined as a sequence of trips which can be assigned to a crew in a short working period (1-2 days). A pairing starts and ends at the same depot and can have various characteristics (e.g., with/without intermediate rest period outside the depot, night working period, etc...) which determine its cost.

2. *Pairing optimization*: a selection is made of the best subset of all the generated pairings so as to guarantee that all the trips are covered at minimum cost. This phase follows quite a general approach, based on the solution of set-covering or set-partitioning problems (SCP/SPPs), possibly with additional constraints (e.g., lower and upper bounds on the number of pairings selected for each depot).

3. *Rostering optimization*: the pairings selected in the previous phase are sequenced into rosters, defining a periodic duty assignment to each crew which guarantees that all the trips are covered for a certain number of consecutive days (e.g., a month). Rosters are generated separately for each depot.

Phases 1 and 2 are often referred to as the *crew scheduling problem* (see, for example, Bodin/Golden/Assad/Ball (1983)).

It should be noted that an algorithmic approach to any of the previous phases has to comply with trade union and company operational rules. This adds a lot of complexity and binding constraints to the overall process, and usually makes this kind of planning problem very specialized and difficult. For instance, a number of constraints apply to: maximum daily, weekly and monthly working times (number of driving and other working hours), length of rest periods, number of night services, and so on. Moreover, the instances to be solved are often very large. As a consequence, the problems arising in Phases 2 and 3 are solved using heuristics.

Clearly the proposed sequential optimization method does not guarantee "global" optimality even if each of the phases is solved exactly. Nevertheless the solutions obtained may be improved by iterating and refining together some of the above phases, as the preliminary testing of Section 6 shows.

The above three-step approach was implemented due to its relatively low development risk, and also because it maps fairly well the process currently followed by the planners. The overall project had in fact to computerize the traditional, nation-wide organization, and this critical target was considered easier to achieve with the proposed approach.

To provide the algorithmic core of Phases 2 and 3 of the ALPI crew planning system, FS promoted, jointly with the Italian Operational Research Society (AIRO), two competitions among Italian Universities. These competitions aimed at both challenging classical and new solution methods with some real-life test cases, and stimulating the interest of the researchers in these large scale optimization problems.

The first competition, named FASTER (Ferrovie Airo Set-covering TendER), was on the design and implementation of effective heuristic algorithms for very large scale set covering problems arising in Phase 2, whereas the second one, named FARO (Ferrovie Airo Rostering Optimization), concerned the peculiar rostering optimization problem arising in Phase 3. A timetable pre-processor and a pairing generator were developed by FS to produce a set of relevant test problems to be used for the competitions and the successive validation of the algorithms. The competitions were held in 1994 and 1995, respectively, and the unit of DEIS, University of Bologna, was awarded the first prize in both of them. This resulted in a joint project between FS and the University of Bologna for the development of the actual codes to be included in the ALPI crew planning system.

In the following we give a more detailed description of the three phases above, presenting for each of them the most important constraints, and outlining the algorithmic approaches currently adopted within the ALPI system.

3 The Pairing Generation Phase

The pairing generation phase calls for the determination of a set of feasible pairings from the given timetabled trips. Each pairing is a trip sequence starting and ending at the same depot, and satisfying the constraints described below.

First of all, *sequencing rules* require that each pair of consecutive trips i and j in a pairing is *compatible*, namely:

- The arrival station of i coincides with the departure station of j.
- The time interval between the arrival of i and the departure of j is greater than a *technical time*, depending on i and j. This includes the times possibly required to change trains, to perform maneuver and other technical operations in the station.

An *external rest* of a pairing is a rest interval between two compatible consecutive trips, ending and starting at a station different from the depot, which exceeds the technical time by at least 7 hours. A pairing can contain at most one external rest. Moreover, the following characteristics are associated with each pairing:

- *spread time*, defined as the time between the departure of its first trip and the arrival of its last trip;
- *driving time*, defined as the sum of trip time plus all rest periods shorter than 30 minutes;

- *working time*, defined as the driving time plus all rest periods not shorter than 30 minutes, except the external rest;
- *paid time*, defined as the spread time plus possible additional transfer times within the depot.

A pairing is defined to be *overnight* if it requires some work between midnight and 5 am. The *operational constraints* require that for each pairing:

- The *spread time* must be smaller than 24 hours.
- If the pairing does not contain an external rest, the total working time must be smaller than 7 hours if it is overnight, and 8 hours and 45 minutes otherwise. If the pairing contains an external rest, the working time may not exceed 7 hours for each part before and after the rest.
- If the pairing does not contain an external rest, the total driving time must be smaller than 7 hours. Otherwise, the driving time may not exceed 4 hours and 30 minutes for each part before and after the external rest.
- In the intervals between 11 am and 3 pm, and between 6 pm and 10 pm, a *meal rest* of at least 30 minutes is required.
- A maximum total traveled distance may be imposed on a pairing depending on the set of trips included.

Each pairing has an associated cost based on its characteristics.

The typical sizes of the instances to be solved and the tightness of the above constraints make it practical to perform explicit generation of the feasible pairings. In addition, operational rules allow a crew to be transported at no extra cost as a passenger on a trip, hence the overall solution can cover a trip more than once. This implies that the pairing optimization phase actually requires the solution of a set covering rather than a set partitioning problem. As a result, only inclusion-maximal feasible pairings, among those with the same cost, need be considered in the pairing generation, thus considerably reducing their number.

When the number of generated pairings is too large, the best pairings are selected as input for the pairing optimization phase, based on a score reflecting the pairing cost and the trips covered.

FS implemented an enumerative algorithm which generates all the feasible pairings according to a depth-first scheme and backtracks as soon as infeasibilities are detected. Simple bounds on the additional spread, working and driving time required to complete a partial pairing are used to speed-up the enumeration.

4 The Pairing Optimization Phase

The pairing optimization phase requires the determination of a min-cost subset of the generated pairings covering all the trips and satisfying additional constraints. These constraints are known as *base constraints* and impose, for each depot:

- A lower and an upper bound on the number of selected pairings associated with the depot.
- A maximum percentage of selected overnight pairings over all those associated with the depot.
- A maximum percentage of selected pairings with external rest over all those associated with the depot.

Similar constraints are imposed with respect to the overall set of selected pairings.

Even without the above base constraints, set covering problems arising in Phase 2 appear rather difficult, mainly because of their size. Indeed, the largest instances at the Italian railways involve up to $5,000$ trips and $1,000,000$ pairings, i.e., they are 1-2 orders of magnitude larger than those arising in typical airline applications.

The pure *Set Covering Problem* (SCP) can formally be defined as follows. Let $I_1, \ldots, I_n$ be the given collection of pairings associated with the trip set $M = \{1, \ldots, m\}$. Each pairing I_j has an associated cost $c_j > 0$. For notational convenience, we define $N = \{1, \ldots, n\}$ and $J_i = \{j \in N : i \in I_j\}$ for each trip $i \in M$. SCP calls for

$$v(\text{SCP}) = \min \sum_{j \in N} c_j x_j \tag{1}$$

subject to

$$\sum_{j \in J_i} x_j \geq 1, \qquad i \in M \tag{2}$$

$$x_j \in \{0, 1\}, \qquad j \in N \tag{3}$$

where $x_j = 1$ if pairing j is selected in the optimal solution, $x_j = 0$ otherwise.

The exact SCP algorithms proposed in the literature can solve instances with up to several hundred trips and several thousand pairings, see Beasley (1987), Beasley/Jörnsten (1992), and Balas/Carrera (1996). When larger instances are tackled, one has to resort to heuristic algorithms. Classical *greedy* algorithms are very fast, but typically do not provide high quality solutions, as reported in Balas/Ho (1980) and Balas/Carrera (1996).

The most effective heuristic approaches to SCP are those based on *Lagrangian relaxation.* These approaches follow the seminal work by Balas/Ho (1980), and then the improvements by Beasley (1990), Fisher/Kedia (1990), Balas/Carrera (1996), Ceria/Nobili/Sassano (1995), and Wedelin (1995). Lorena/ Lopes (1994) propose an analogous approach based on *surrogate relaxation.* Recently, Beasley/Chu (1994) and Jacobs/Brusco (1995) proposed genetic and local search algorithms, respectively.

We next briefly outline the heuristic method proposed by Caprara/Fischetti/Toth (1995) which has been designed to attack very-large scale instances and is included in the ALPI system. The method is based on dual information associated with a Lagrangian relaxation of model (1)-(3). For every vector $u \in R^m_+$ of Lagrangian multipliers associated with constraints (2), the Lagrangian subproblem reads:

$$L(u) = \min \left\{ \sum_{j \in N} c_j(u) x_j + \sum_{i \in M} u_i : \; x_j \in \{0,1\}, j \in N \right\} \tag{4}$$

where $c_j(u) = c_j - \sum_{i \in I_j} u_i$ is the *Lagrangian cost* associated with pairing $j \in N$. Clearly, an optimal solution to (4) is given by $x_j(u) = 1$ if $c_j(u) < 0$, $x_j(u) = 0$ if $c_j(u) > 0$, and $x_j(u) \in \{0,1\}$ when $c_j(u) = 0$. The Lagrangian dual problem associated with (4) consists of finding a Lagrangian multiplier vector $u^* \in R^m_+$ which maximizes the lower bound $L(u)$. To solve this problem, a common approach uses the *subgradient vector* $s(u) \in R^m$ associated with a given u, defined by $s_i(u) = 1 - \sum_{j \in J_i} x_j(u)$ for $i \in M$. The approach generates a sequence $u^0, u^1, \ldots$ of non-negative Lagrangian multiplier vectors, where u^0 is defined arbitrarily.

For near optimal Lagrangian multipliers u_i, the Lagrangian cost $c_j(u)$ gives reliable information on the overall utility of selecting pairing j. Based on this property, we use Lagrangian (rather than original) costs to compute, for each $j \in N$, a *score* σ_j ranking the pairings according to their likelihood of being selected in an optimal solution. These scores are given as input to a simple heuristic procedure that finds a (hopefully) good SCP solution in a greedy way. Computational experience shows that almost equivalent near-optimal Lagrangian multipliers can produce SCP solutions of substantially different quality. In addition, no strict correlation exists between the lower bound value $L(u)$ and the quality of the SCP solution found. Therefore, it is worthwhile applying the heuristic procedure for several near-optimal Lagrangian multiplier vectors.

The approach consists of three main steps. The first one is referred to as the *subgradient step.* It is aimed at quickly finding a near-optimal Lagrangian multiplier vector. To this end, an aggressive policy is used for the updating of the step-size and the reduction of the subgradient norm. The second one is the *heuristic step,* in which a sequence of near-optimal Lagrangian vectors is determined. For each vector, the associated scores are given as input to a greedy heuristic procedure to attempt to update the incumbent best SCP solution. In the third step, called *fixing*, one selects a subset of pairings having a high probability of being in an optimal solution, and fixes to 1 the corresponding variables. In this way one obtains an SCP instance with a reduced number of pairings (and trips), on which the three-step procedure is iterated. After each application of the three-step procedure, an effective *refining procedure* is used to produce improved solutions.

When very large instances are tackled, the computing time spent on the first two steps becomes very large. To overcome this difficulty, one can define a *core problem* containing a suitable set of pairings, chosen among those having the lowest Lagrangian costs. The definition of the core problem is often very critical, since an optimal solution typically contains some pairings that, although individually worse than others, must be selected in order to produce a good overall solution. Hence it is better not to "freeze" the core problem, by using a *variable pricing* scheme to update the core problem iteratively in a vein similar to that used for solving large scale linear programs. The use of pricing within Lagrangian optimization drastically reduces computing time, and is one of the main ingredients in the success of the overall scheme.

5 The Rostering Optimization Phase

In the rostering optimization phase the previously selected pairings are sequenced into cyclic rosters.

Most of the published works on crew rostering refer to urban mass transit systems, where the minimum number of crews required to perform the pairings can easily be determined, and the objective is to distribute the workload evenly among the crews: see Jachnik (1981), Bodin/Golden/Assad/Ball (1983), Carraresi/Gallo (1984b), Hagberg (1985), and Bianco/Bielli/Mingozzi/Ricciardelli/Spadoni (1992). Set partitioning approaches for airline crew rostering are described in Ryan (1992), Gamache/Soumis (1993), Gamache/Soumis/Marquis/Desrosiers (1994), and Jarrah/Diamond (1995). Finally, related cyclic scheduling problems are dealt with in Tien/Kamiyama (1982), and Balakrishnan/Wong (1990).

We next give a description of the real-world crew rostering problem arising at the Italian railways. We are given a set of n pairings to be covered by a set of crew rosters. As mentioned in Section 3, each pairing is characterized by a *start time*, an *end time*, a *spread time*, a *driving time*, a *working time* and a *paid time*. Moreover, each pairing can have the following additional characteristics:

- *pairing with external rest*, if it includes a long rest away from the depot;
- *long pairing*, if it does not include an external rest and its working time is longer than 8 hours and 5 minutes;

- *overnight pairing*, if it requires some work between midnight and 5 am;
- *heavy overnight pairing*, if it is an overnight pairing without external rest requiring more than 1 hour and 30 minutes' work between midnight and 5 am.

A roster contains a subset of pairings and spans a cyclic sequence of groups of 6 consecutive days, conventionally called *weeks*. Hence the number of days in a roster is an integer multiple of 6. The length of a roster is typically 30 days (5 weeks) and does not exceed 60 days (10 weeks), although these requirements are not explicitly imposed as constraints.

The crew rostering problem consists of finding a feasible set of rosters covering all the pairings and spanning a minimum number of weeks. As already discussed in Section 2, the global number of crews required to cover all the pairings every day is equal to 6 times the total number of weeks in the solution. Thus, the minimization of the number of weeks implies the minimization of the global number of crews required.

We next list all the constraints of the crew rostering phase. For short, we call a *complete day* a time interval of 24 hours (i.e., 1440 minutes) starting at midnight. Moreover, a complete day is called *free* if no pairing or part of a pairing is performed during that day.

Each week can include at most the following number of pairings having particular characteristics: (i) 2 pairings with external rest; (ii) 1 long pairing; (iii) 2 overnight pairings. Furthermore, each week must be separated from the next one in the roster by a continuous rest, called *weekly rest*, which always spans the complete sixth day of the week. There are two types of weekly rests, conventionally called *simple* and *double* rests. Simple rests must be at least 48 hours long, whereas double rests must span at least two complete days, i.e., either the fifth and sixth day of a week or the sixth day of a week and the first day of the following one.

For each roster, the number of double rests must be at least 40% of the total number of weekly rests, and the average weekly rest time must be at least 58 hours. Moreover, for each (cyclic) group of 30 consecutive days within a roster, no more than 7 pairings with external rest can be included, and the total paid time of the included pairings cannot exceed 170 hours. Finally, for each (cyclic) group of 7 consecutive days within a roster the total working time of the included pairings cannot exceed 36 hours.

Two consecutive pairings of a roster, say i and j, can be sequenced either *directly* in the same week, or with a simple or double rest between them. The break between the end of a pairing and the start of the subsequent pairing within a week lasts at least 18 hours. If both pairings are overnight and at most one of them is a heavy overnight pairing, the minimum break lasts 22 hours, while if both are heavy overnight pairings the break must span at least one complete day. Moreover, after two consecutive overnight pairings in a week whose intermediate break does not span a complete day, the break before the start of any other pairing in the same week must last at least 22 hours.

When a simple rest is preceded by an overnight pairing, then either the first pairing in the next week starts after 6:30 am, or the rest must span two complete days. Finally, if the first pairing in a week following a double rest starts before 6 am, then the rest must span at least three complete days.

The solution approach implemented within the ALPI system is based on a heuristic algorithm proposed by Caprara/Fischetti/Toth/Vigo(1995), which is driven by the information obtained through the computation of lower bounds.

Simple lower bounds can easily be obtained by considering each of the operational constraints imposing a limit either on the total number of pairings with a given characteristic, or on the total working and paid time in a week and in a (cyclic) group of 30 consecutive days in a roster, respectively. A more sophisticated relaxation is proposed in Caprara/Fischetti/Toth/Vigo (1995) consecutive pairings within a roster. The relaxation also requires that the total number of weekly rests is equal to the total number of weeks making up the rosters, and that the total number of double rests is at least 40% of the total number of weekly rests. The resulting relaxed problem calls for the determination of a minimum-cost set of disjoint circuits of a suitably defined directed multigraph G, whose nodes represent the pairings and the arcs between each pair of nodes representing the different possible ways of sequencing the corresponding pairings, i.e., directly or with a simple or double weekly rest in between. Each of these circuits corresponds to a, possibly infeasible, roster. Moreover, the set of selected circuits must satisfy the following constraints:

(i) each node of G is covered by exactly one circuit;

(ii) the total number of simple- or double-rest arcs in the circuits has to be at least the total cost of the circuits, expressed in weeks;

(iii) the total number of double-rest arcs in the circuits has to be at least 0.4 times the total number of simple- or double-rest arcs.

The relaxed problem is further relaxed as a Lagrangian by dualizing constraints (ii) and (iii). The corresponding Lagrangian problem is solved as an Assignment Problem (AP) on a suitably defined cost matrix. Computational experience on railway instances showed that a tight lower bound may be obtained by simply considering two specific pairs of Lagrangian multipliers, i.e., by solving only two APs.

Finally, we describe the constructive heuristic of Caprara/Fischetti/Toth (1995), which extensively uses the information obtained from the solution of the relaxed problem mentioned above. The heuristic constructs one feasible roster at a time, choosing in turn the pairings to be sequenced consecutively in the roster. Once a roster has been completed, all the pairings it contains are removed from the problem. The process is iterated until all pairings have been sequenced.

The procedure we use to build each single roster, as it applies to the construction of the first roster, works as follows. One first computes the Lagrangian lower bound, and then starts building the roster by selecting its "initial pairing" which will be performed at the beginning of a week, i.e., preceded by a weekly rest. (The term "initial" is conventional as rosters are cyclic.) Once the initial pairing has been selected, a sequence of iterations is performed where:

a) the best pairing j to be sequenced after the current pairing i is chosen, based on a *score* taking into account the characteristics of pairing j and the possible lower bound increase occurring when j is sequenced after i;

b) the Lagrangian lower bound is parametrically updated;

c) the possibility of "closing" the roster is considered, possibly updating the best roster found.

The procedure is iterated until no better roster can be constructed, stopping anyway if the current roster spans more than 10 weeks.

When a complete solution to the problem is found, one can try to improve it by applying a *refining procedure*, which removes the last rosters constructed (which are typically worse than the others) from the solution, and re-applies the heuristic algorithm to the corresponding pairings. To this end, some parameters of the roster construction procedure are either changed with a random perturbation or tuned so as to take into account the constraints that made the construction of the last rosters difficult.

6 Preliminary Computational Testing

The heuristic algorithms implemented for Phases 2 and 3 of the ALPI system have been preliminarily validated through separate computational testing.

The set covering algorithm described in Section 4, hereafter called CFT, was tested both on instances from the literature and on real-life instances provided by FS. The technique outperforms previously published methods: in 92 out of the 94 literature instances the method found, within short computing time, the optimal (or the best known) solution. Moreover, among the 22 instances for which the optimum is not known, in 6 cases the solution is better than any other solution found by previous techniques. Table 1 reports the results for the FS instances. For each instance the table gives the instance name, the number of trips (m) and pairings (n), the instance density

Table 1: Results on real-life SCP instances from FS.

Name	$m \times n$	Density	LB	CFT Sol.	Others' Sol.
FASTER507	$507 \times 63,009$	1.2%	173	174	174
FASTER516	$516 \times 47,311$	1.3%	182	182	182
FASTER582	$582 \times 55,515$	1.2%	210	211	211
FASTER2536	$2,536 \times 1,081,841$	0.4%	685	691	692
FASTER2586	$2,586 \times 920,683$	0.4%	937	947	951
FASTER4284	$4,284 \times 1,092,610$	0.2%	1051	1065	1070
FASTER4872	$4,872 \times 968,672$	0.2%	1509	1534	1534

$\sum_{j \in N} |I_j|/(m \cdot n)$, the value of the lower bound LB computed by the subgradient procedure, the value of the heuristic solution found by code CFT, and the best solution obtained by other SCP algorithms tested by FS. The reported solutions were obtained within 3,000 CPU seconds on a PC 486/33 for the first three instances, and 10,000 CPU seconds on a HP 9000 735/125 for the remaining instances.

The table shows that algorithm CFT is capable of providing near-optimal solutions within limited computing time even for very-large size instances. The average percentage gap between the lower bound and the heuristic solution value is 0.9%.

The rostering optimization algorithm described in Section 5 was tested on real-life instances provided by FS, involving up to about 900 pairings. The results obtained are illustrated in Table 2. For each instance we report the instance name, the number of pairings, the best simple lower bound, the Lagrangian lower bound, the heuristic solution value, and the corresponding computing time, expressed in PC Pentium 90 CPU seconds. The table clearly shows the effectiveness of the approach, since 26 out of 36 instances have been solved to proven optimality within limited computing time.

The encouraging results obtained with the above experiments led to the implementation of the prototype version of the ALPI system. A first testing of the overall procedure was performed using the instance FASTER516, which includes all the pairings obtained from a set of 516 trips currently covered by crews from the Milan depot (one of the largest in Italy). The pairing set contains 47,311 pairings, 161 of which were selected in Phase 2. The rostering optimization phase determined within a few minutes a solution with 46 weeks. The information associated with this solution was used to better define the pairing costs for the pairing optimization phase, by taking into account the characteristics that made roster construction difficult. The new pairing optimization solution selected 151 pairings, and the corresponding rostering solution, again found in a few minutes, contained 45 weeks. The overall process was iterated, achieving a rostering solution of 44 weeks (significantly better than the initial one) at the fourth iteration. This gives an idea of the improvements that could be achieved by feedback between Phases 2 and 3. This is one of the main points to be investigated in the future.

Table 2: Results on real-life crew rostering instances from FS.

Name	n	Lower Bounds		Heuristic Solution	
		Simple	Lagrangian	weeks	time
FARO021	21	6	7	7	8
FARO033	33	9	11	11	17
FARO069	69	18	19	19	650
FARO134	134	34	39	39	365
FARO164	164	43	48	48	106
FARO360	360	108	108	111	342
FARO386	386	110	118	118	443
FARO525	525	154	164	164	1185
F21F33	54	15	17	17	29
F21F69	90	24	25	26	180
F21F134	155	40	45	45	645
F21F164	185	48	54	54	103
F21F360	381	114	114	118	1805
F21F386	407	116	124	124	447
F21F525	546	160	170	170	1026
F33F69	102	26	29	29	533
F33F134	167	43	50	50	117
F33F164	197	51	58	58	149
F33F360	393	117	118	121	638
F33F386	419	118	128	128	1587
F33F525	558	162	174	174	1936
F69F134	203	51	57	58	1135
F69F164	233	60	66	66	327
F69F360	429	125	126	130	1892
F69F386	455	127	136	136	7462
F69F525	594	171	182	182	11317
F134F164	298	76	87	87	595
F134F360	494	142	146	148	16160
F134F386	520	143	157	157	1396
F134F525	659	188	203	203	2107
F164F360	524	150	155	158	2134
F164F386	550	152	166	166	956
F164F525	689	196	211	211	7715
F360F386	746	217	225	228	4434
F360F525	885	261	271	274	6898
F386F525	911	263	282	282	3806

7 Conclusions

The ALPI system was scheduled to become operational in June 1997. However, crew planning for this year will not be done completely via the new system, due to some re-organization issues, that have delayed some training and testing activities. The "turmoil" in which European railways are operating these days does not usually help in making progress as scheduled;nevertheless it is inspiring a lot of innovative projects, including the one described here. The tests made so far with the new algorithms demonstrate the capability of ALPI in carrying out its future job: work timing and personnel savings are dramatic; the original planning workforce (about 50 people) may be substantially reduced, or better utilized, and the time needed to produce a new national planning worksheet will be incomparably shorter than before. In terms of driver resources, some test problems have identified reductions over hand made rosters corresponding to a saving of at least 1%.

Even better results are expected once the system is fully operational. Improved effectiveness will be used to increase output rather than to reduce the workforce size.

In Italy the market share of rail traffic has been increasing consistently over the last few years, although it is still small relative to road freight traffic. The new planning system will also help the railway staff to improve its market response and to propose new working rules and productivity/benefits trade-offs which can be more easily accepted by the trade unions. We also believe that this approach should represent a model for the introduction of MS methods into complex enterprises, like rail transportation. We might conclude that this "experiment" is rewarding for both the application and the progress in the state-of-the-art of this class of problems. Moreover, it confirms the high potential for OR techniques in the real world, and the benefits from open collaboration between industry and academia.

Acknowledgements: We would like to acknowledge Ferrovie dello Stato SpA, in particular the Infrastructure and Traction Business Units and the Computer Science Division for giving their support to the ALPI project and the related FASTER and FARO competitions. F. Olimpieri and A. Locatelli, from FS, have prepared the real-life instances; the former was a rail driver who graduated in engineering (no one else could have been able to do what he did in such a short time). Other people to be acknowledged for their support to OR development in FS are E. Maestrini, P. Vellucci, V. S. Achille and the third author. Finally, AIRO and its appointed members, who attended the computer evaluation tests, deserve special thanks: A. Colorni, G. Gallo, G. Improta, M. G. Speranza, and R. Tadei.

References

Balakrishnan, N./Wong, R.T. (1990): A network model for the rotating workforce scheduling problem. in: Networks 20, 25–42.

Balas, E./Carrera, M.C. (1996): A dynamic subgradient-based branch-and-bound procedure for set covering. in: Operations Research 44, 875–890.

Balas, E./Ho, A. (1980): Set covering algorithms using cutting planes, heuristics and subgradient optimization: a computational study. in: Mathematical Programming Study 12, 37–60.

Barnhart, C./Johnson, E.L./Nemhauser, G.L./ Savelsbergh, M.W.P./ Vance, P.H. (1994): Branch-and-price: column generation for solving huge integer programs. in: Birge, J.R./Murty, K.G. (eds.): Mathematical Programming: State of the Art 1994, 186–207. (The University of Michigan Press) Ann Arbor.

Beasley, J.E. (1987): An algorithm for set covering problems. in: European Journal of Operational Research 31, 85–93.

Beasley, J.E. (1990): A lagrangian heuristic for set covering problems. in: Naval Research Logistics 37, 151–164.

Beasley, J.E./Chu, P.C. (1996): A genetic algorithm for the set covering problem. in: European Journal of Operational Research 94, 392–404.

Beasley, J.E./Jörnsten, K. (1992): Enhancing an algorithm for set covering problems. in: European Journal of Operational Research 58, 293–300.

Bianco, L./Bielli, M./Mingozzi, A./Ricciardelli, S./Spadoni, M. (1992): A heuristic procedure for the crew rostering problem. in: European Journal of Operational Research 58, 272–283.

Bodin, L./Golden, B./ Assad, A./Ball, M. (1983): Routing and scheduling of vehicles and crews: the state of the art. in: Computers and Operations Research 10, 63–211.

Caprara, A./Fischetti, M./Toth, P. (1995): A heuristic method for the set covering problem. Technical report OR-95-8, DEIS University of Bologna. Extended abstract published in Cunningham, W.H./McCormick, S.T./Queyranne, M. (eds.): Proceedings of the Fifth IPCO Conference, Lecture Notes in Computer Science 1084, 72–84. (Springer) Berlin.

Caprara, A./Fischetti, M./Toth, P./Vigo, D. (1995): Modeling and solving the crew rostering problem. Technical report OR-95-6, DEIS University of Bologna. To appear in Operations Research.

Carraresi, P./Gallo, G. (1984a): Network models for vehicle and crew scheduling. in: European Journal of Operational Research 16, 139–151.

Carraresi, P./Gallo, G. (1984b): A multilevel bottleneck assignment approach to the bus drivers' rostering problem. in: European Journal of Operational Research 16, 163–173.

Ceria, S./Nobili, P./Sassano, A. (1995): A lagrangian-based heuristic for large-scale set covering problems. Technical report R.406, IASI-CNR, Rome. To appear in Mathematical Programming.

Daduna, J.R./Wren, A. (eds.) (1988): Computer-Aided Transit Scheduling. Lecture Notes in Economic and Mathematical Systems 308. (Springer) Berlin.

Desrochers, M./Rousseau, J.-M. (eds.) (1992): Computer-Aided Transit Scheduling. Lecture Notes in Economic and Mathematical Systems 386. (Springer) Berlin.

Desrosiers, J./Dumas, Y./Solomon, M.M./Soumis, F. (1995): Time constrained routing and scheduling. in: Ball, M.O./Magnanti, T.L./Monma, C.L./Nemhauser, G.L. (eds.): Handbooks in OR & MS, Vol. 8, 35–139. (Elsevier) Amsterdam.

Fisher, M.L./Kedia, P. (1990): Optimal solutions of set covering/ partitioning problems using dual heuristics. in: Management Science 36, 674–688.

Gamache, M./Soumis, F. (1993): A method for optimally solving the rostering problem. Les Cahiers du GERAD G-93-40. Montréal.

Gamache, M./Soumis, F./Marquis, G./Desrosiers, J. (1994): A column generation approach for large scale aircrew rostering problems. Les Cahiers du GERAD G-94-20. Montrèal.

Hagberg, B. (1985): An assignment approach to the rostering problem. in: Rousseau, J.-M. (ed.): Computer Scheduling of Public Transport 2. (North Holland) Amsterdam.

Jachnik, J.K. (1981): Attendance and rostering systems. in: Wren, A. (ed.): Computer Scheduling of Public Transport. (North Holland) Amsterdam.

Jacobs, L.W./Brusco, M.J. (1995): A local search heuristic for large set-covering problems. in: Naval Research Logistics 52, 1129–1140.

Jarrah, A.I.Z./Diamond, J.T. (1995): The crew bidline generation problem. Technical Report, SABRE Decision Technologies.

Lorena, L.A.N./Lopes, F.B. (1994): A surrogate heuristic for set covering problems. in: European Journal of Operational Research 79, 138–150.

Rousseau, J.-M. (ed.) (1985): Computer Scheduling of Public Transport 2. (North Holland) Amsterdam.

Ryan, D.M. (1992): The solution of massive generalized set partitioning problems in aircrew rostering. in: Journal of the Operational Research Society 43, 459–467.

Tien, J.M./Kamiyama, A. (1982): On manpower scheduling algorithms. in: SIAM Review 24, 275–287.

Wedelin, D. (1995): An algorithm for large scale 0-1 integer programming with application to airline crew scheduling. in: Annals of Operations Research 57, 283–301.

Wren, A. (ed.) (1981): Computer scheduling of public transport. (North Holland) Amsterdam.

Crew Pairing for a Regional Carrier

Guy Desaulniers[1], Jacques Desrosiers[2],
Arielle Lasry[3] and Marius M. Solomon[4]

[1] École Polytechnique and GERAD, 3000 Ch. de la Côte-Ste-Catherine, Montréal, Canada, H3T 2A7

[2] École des Hautes Études Commerciales and GERAD, 3000 Ch. de la Côte-Ste-Catherine, Montréal, Canada, H3T 2A7

[3] Numetrix Limited, 655 Bay St., Suite 1200, Toronto, Canada, M5G 2K4

[4] Northeastern University and GERAD, Boston, U.S.A., 02115

Abstract: This paper addresses the problem of generating valid crew pairings in the context of a regional air carrier. The classical column generation solution approach based on extensive enumeration of all valid duties is impractical in this context where duties comprise ten to twelve legs. In order to alleviate this difficulty, we propose an alternative approach which takes into account all work rules and air traffic regulations during the construction of valid crew pairings. Two network structures compatible with this approach are described. The first is a leg-on-node model while the second involves a leg-on-arc representation. Computational results obtained with the GENCOL optimizer on problems varying from 63 to 986 legs lead us to conclude that the leg-on-arc representation is substantially more efficient. In addition, we study the cost impact of three changes in the operating scenario. Finally, we illustrate how bounded perturbation variables virtually eliminate degeneracy, hence significantly decreasing CPU time.

1 Introduction

Airline companies are faced with several routing and scheduling problems, notably, flight scheduling, aircraft assignment, crew pairing and crew scheduling. This paper focuses on the third phase, *crew pairing*, which consists of determining a set of flight legs that can feasibly be flown by an airline crew. These pairings are round trips originating and terminating at the same crew home base composed of legal work days, called duties, separated by rest periods. In turn, each duty is composed of activities such as flight legs and ground time. The duration of a pairing in the operating environment of a regional air carrier varies between one and six days as opposed to up to three weeks for long haul flights. Each leg must be covered exactly once by a pairing, and all work rules and air traffic regulations must be satisfied.

The classical column generation solution approach consisting of enumerating all feasible duties and then combining them into valid pairings is impractical for regional air carriers. Typically, this approach is used in medium and long haul problems where the number of legs that a crew can cover per duty rarely exceeds five or six. In the context of a regional air carrier, this number can easily be as large as twelve. This results in millions of different valid combinations of activities that compose a legal work day. Hence, it becomes impossible to use the traditional approach to combine these duties into optimal pairings of several days. Nevertheless, this approach has the advantage of managing *a priori* the set of rules and regulations pertaining to a single duty, while leaving all constraints pertaining to the transition from one duty to the next and those globally restricting the admissibility of the set of pairings to the optimization problem.

The contribution of this paper consists of presenting a column generation approach that explicitly manages all rules and restrictions during the construction of feasible crew pairings. We do not eliminate *a priori* any combinations since this can increase the likelihood of infeasibility. We also examine the tradeoffs among two different network representations of the problem: a leg-on-node and a leg-on-arc network. In addition, we study the cost impact of three modifications to the operating scenario. Finally, we propose the use of bounded perturbation variables to handle degeneracy.

The first section of this paper reviews the literature on the airline crew pairing problem. Sect. 3 discusses specifics of the mathematical model for the application at hand while Sect. 3 presents the operating environment of the regional air carrier. Next, Sect. 5 describes the two network structures considered. Computational experiments conducted with the GENCOL optimizer are reported in Sect. 6. Finally, our conclusions are derived in Sect. 7.

2 Literature Review

The construction of crew pairings is one of the most studied problems in the airline industry. The classical formulation (see Salkin (1975)) used for this problem is the Set Partitioning model. Let F, indexed by f, be the set of flight legs to cover and P, indexed by p, the set of feasible crew pairings. Denote by c_p the cost of a pairing and by a_{fp} the binary parameter equal to 1 if pairing p covers leg f and 0 otherwise. Finally, let Y_p denote the binary decision variable related to pairing p; Y_p takes value 1 if pairing p is covered by a crew and 0 otherwise. The classical model is defined as:

$$\text{Minimize} \quad \sum_{p \in P} c_p Y_p \tag{1}$$

subject to:

$$\sum_{p \in P} a_{fp} Y_p = 1, \qquad \forall f \in F \tag{2}$$

$$Y_p \in \{0, 1\}, \qquad \forall p \in P. \tag{3}$$

The objective function (1) seeks to minimize the cost of covering the entire set of flight legs. Constraints (2) ensure that all legs are covered exactly once by a pairing. Integer constraints on the decision variables are imposed by (3).

This model may be enhanced by introducing restrictions on the bases, such as for example, on the maximum number of credited hours assigned to a crew base. Formally, let M, indexed by m, be the set of additional constraints. Denote by b_{mp} the coefficients of the Y_p variables and by b_m the right-hand side of such a constraint. These constraints would then take the form:

$$\sum_{p \in P} b_{mp} Y_p \leq b_m, \quad \forall m \in M. \tag{4}$$

The problem of generating crew pairings, with possible home base restrictions, is solved using branch-and-bound methods (see Hoffman/Padberg (1993)). The survey paper by Etschmaier/Mathaisel (1985) presents several heuristic and exact methods developed for problems of relatively small size. Due to the size of problems encountered by large air carriers, more recent methods have used column generation (see Dantzig/Wolfe (1960)). Several are based on an iterative process comprised of a heuristic generator of feasible pairings and an optimizer for the set partitioning model created by the generated pairings (see Ball/Roberts (1985), Crainic/Rousseau (1987), Gershkoff (1989), Anbil/Gelman/Patty/Tanga(1991), Graves/McBride/Gershkoff/Anderson/Mahidara (1993), Wedelin (1995) and Chu/Gelman/Johnson (1997)).

A different column generation approach consists of solving an optimization problem, also known as the subproblem, used to generate valid and potentially good quality pairings. These are added to the linear relaxation of the model (1)–(4) restricted to a subset of pairings, also known as the master problem. In this approach, the pairing generator is guided by the dual information associated with the current solution to the master problem to generate pairings that could improve the value of the objective function. The optimality of the current solution is proved when solving the subproblem no longer enables the generation of negative marginal cost pairings. This iterative process is embedded in a branch-and-bound method to derive integer solutions (see Lavoie/Minoux/Odier (1988), Barnhart/Johnson/Anbil/Hatay (1994) and Desaulniers/Desrosiers/Dumas/Marc/Rioux/Solomon/Soumis (1997).

The mathematical model formalizing the previous approach has been described in Desrosiers/Dumas/Solomon/Soumis (1995). It has been embedded in the optimizer GENCOL. This software package can be used to solve problems requiring the generation of paths on a network where vehicles or individuals travel to accomplish certain tasks. It allows, among others, the solution of vehicle routing (see Desrochers/Desrosiers/Solomon (1992)) and scheduling problems (see Desrochers/Soumis (1988)). Within the scope of this paper, a pairing corresponds to the path that will be assigned to a crew and the legs are the tasks to be accomplished. Next, we present the specifics of the subproblem for the application examined.

3 The Subproblem

The subproblem has a path structure on a network where resource variables to be defined next are added to model certain restrictions which do not appear in the underlying network structure. We define a shortest path problem with resource variables on a network $G = (V, A)$ where V is the set of nodes and A is the set of arcs. The set of nodes contains an origin node o, a destination node d, and a set N of intermediate nodes such that $V = N \cup \{o, d\}$. The set of arcs A is a subset of $V \times V$. The set of resource indices is denoted R. The problem variables are of two types: flow variables $X_{ij}, (i,j) \in A$ defined on the arcs and resource variables $T_i^r, i \in V$ and $r \in R$ defined on the nodes. The flow variable X_{ij} takes value 1 if the arc $(i,j) \in A$ is used by the path originating at o and terminating at d, and 0 otherwise. The resource variable $T_i^r, i \in V$ and $r \in R$ is restricted to the interval $[l_i^r, u_i^r]$; its values are integer if the functions that define it impose such values. The extension functions $f_{ij}^r(T_i^r)$, defined for each arc $(i,j) \in A$ and for each resource $r \in R$, allow the computation of resource variables at each node j while taking into account the information at node i. In most cases, resource variables are initialized to 0, i.e., $T_o^r = l_o^r = u_o^r = 0, r \in R$.

The marginal cost of a crew pairing is a function of the path followed, of the value of the resource variables and of the dual information provided by the restricted problem. Hence, the general formulation is $c(\boldsymbol{X}, \boldsymbol{T}, \boldsymbol{\alpha}, \boldsymbol{\beta})$, where $\boldsymbol{X} = (X_{ij} \mid (i,j) \in A)$ and $\boldsymbol{T} = (T_i^r \mid i \in V,\ r \in R)$, while $\boldsymbol{\alpha} = (\alpha_w \mid f \in F)$ and $\boldsymbol{\beta} = (\beta_m \mid m \in M)$ are the vectors of the dual variables associated with constraints (2) and (4), respectively. Then, the formulation of the subproblem is:

$$\text{Minimize} \quad c(\boldsymbol{X}, \boldsymbol{T}, \boldsymbol{\alpha}, \boldsymbol{\beta}) \tag{5}$$

subject to:

$$\sum_{j:(o,j)\in A} X_{o,j} = \sum_{i:(i,d)\in A} X_{i,d} = 1 \tag{6}$$

$$\sum_{j:(i,j)\in A} X_{ij} - \sum_{j:(j,i)\in A} X_{ji} = 0, \qquad \forall i \in N \tag{7}$$

$$X_{ij}(f_{ij}^r(T_i^r) - T_j^r) \leq 0, \qquad \forall r \in R,\ \forall (i,j) \in A \tag{8}$$

$$l_i^r \leq T_i^r \leq u_i^r, \qquad \forall r \in R,\ \forall i \in V \tag{9}$$

$$X_{ij} \in \{0,1\}, \qquad \forall (i,j) \in A. \tag{10}$$

The objective function (5) seeks the minimization of the cost function. Relations (6)-(7) are the classic path constraints, from the origin o to the destination d. Relations (8) describe the compatibility between the flow and resource variables. Finally, constraints (9)–(10) define the domain of the resource and flow variables.

Extension Functions: The extension functions f_{ij}^r, $(i,j) \in A$ and $r \in R$, allow the modeling of different situations and can take relatively complex forms depending on the application. The most commonly encountered form is that of an affine function:

$$f_{ij}^r(T_i^r) = T_i^r + t_{ij}^r, \tag{11}$$

where t_{ij}^r is the amount of resource r consumed on arc (i,j). They can also take the form of a constant:

$$f_{ij}^r(T_i^r) = const. \tag{12}$$

This form is typically used to reset resource variables. A more complex extension function is obtained by combining the two previous ones:

$$f_{ij}^r(T_i^r) = const \quad \text{if} \quad T_i^r + t_{ij}^r \leq u_j^r. \tag{13}$$

In this example, the cumulative resource r at node j must be less than the upper bound u_j^r in order to be valid and only then can it be reset to the value *const*. In the next section, different extension functions associated with the proposed network structures will be given.

Cost Function: The marginal cost function is considered as a resource and its value is computed at each node using the $\bar{Z}_i$, $i \in V$ variables and the extension functions $\bar{z}_{ij}(\bar{Z}_i)$ for $(i,j) \in A$ and $i \in V$. Once again, the most commonly used cost extension function is an affine function:

$$\bar{z}_{ij}(\bar{Z}_i) = \bar{Z}_i + \bar{c}_{ij}, \tag{14}$$

where, for each arc $(i,j) \in A$ of cost c_{ij}, is added the marginal cost $\bar{c}_{ij}$:

$$\bar{c}_{ij} = c_{ij} - \sum_{f \in F} a_{f,ij}\alpha_f - \sum_{m \in M} b_{m,ij}\beta_m. \tag{15}$$

The parameters $a_{f,ij}$ and $b_{m,ij}$ indicate the contributions of the arc (i,j) to the constraints $f \in F$ and $m \in M$, respectively. Depending on the network structure that is used, these contributions may differ. The marginal cost function is initially set at zero and the objective simply becomes:

$$\text{Minimize} \quad \bar{Z}_d. \tag{16}$$

More complex functions enable better modeling of different situations encountered in practice. In this paper, we consider a piecewise linear cost function in order to model credited hours to the crews. However, this is just one example of a non-decreasing extension function. Other linear, non-linear and even discontinuous functions of this type, instrumental in solving shortest path problems with constraints, are discussed at length in Desaulniers/Desrosiers/Ioachim/Solomon/Soumis/Villeneuve (1998).

4 A Regional Carrier's Operating Environment

Regional air carriers operate in a different environment from national or international carriers. Notably, legs and ground times are short, the number of passengers is small and there are generally no overnight flights. The large number of legs which can be covered by a crew during each duty renders the classical solution approaches impossible.

The application that we are considering involves a weekly scheduling problem comprised of 986 flight legs distributed across five types of aircraft. This regional carrier has three crew bases (Montréal, Québec and Sept-Iles) and services 30 cities across eastern Canada. All activities performed by crews are accounted for in terms of credited hours which serve as the basic unit for personnel salaries and permit computing the cost of a leg, a duty or a pairing.

Therefore the objective is to minimize the total number of credited hours. The following rules must be respected: the maximum credited hours per duty is 15, the number of landings per duty cannot exceed 12 and the number of duties per pairing is 1 or 2. All duties start with a 60-minute briefing and end with a 30-minute debriefing. The minimum and maximum connection times between two flight legs are respectively 15 minutes and 9.5 hours, regardless of the station where the connection occurs. There are two types of ground times, those lasting between 0 and 6 hours (which are credited proportionally) and those exceeding 6 hours (which are credited for a bonus b_1 in addition to their duration). Note that night time rest periods are not credited proportionally to their duration but receive a bonus b_2 which covers transportation and lodging expenses.

The previous scenario represents the basis for our computational experiments. We then proceed to make several modifications to the constraints in order to evaluate their impact. For example, we will consider a guaranteed minimum number of credited hours per duty and also pairings of three days. Moreover, we will introduce the option to start or end a pairing at a base which is not a crew home base by allowing deadheads at the beginning and end of pairings.

5 Network Structures

In this section, we present two types of space-time network representations, a leg-on-node and a leg-on-arc network.

5.1 Leg-on-Node Network

Fig. 5.1 illustrates a leg-on-node network which contains three types of nodes and four types of arcs. In the coordinate system chosen, the horizontal and vertical axes represent, respectively, the city and time of departure of the legs associated with the nodes. The arrival city is given by its IATA code inscribed within each node. The horizontal dotted line marks the passage from the first to the second day of the one-week horizon. In order to facilitate the viewing of the network, the YUL base, as an example, is represented by one starting node and two arrival nodes, one for each day.

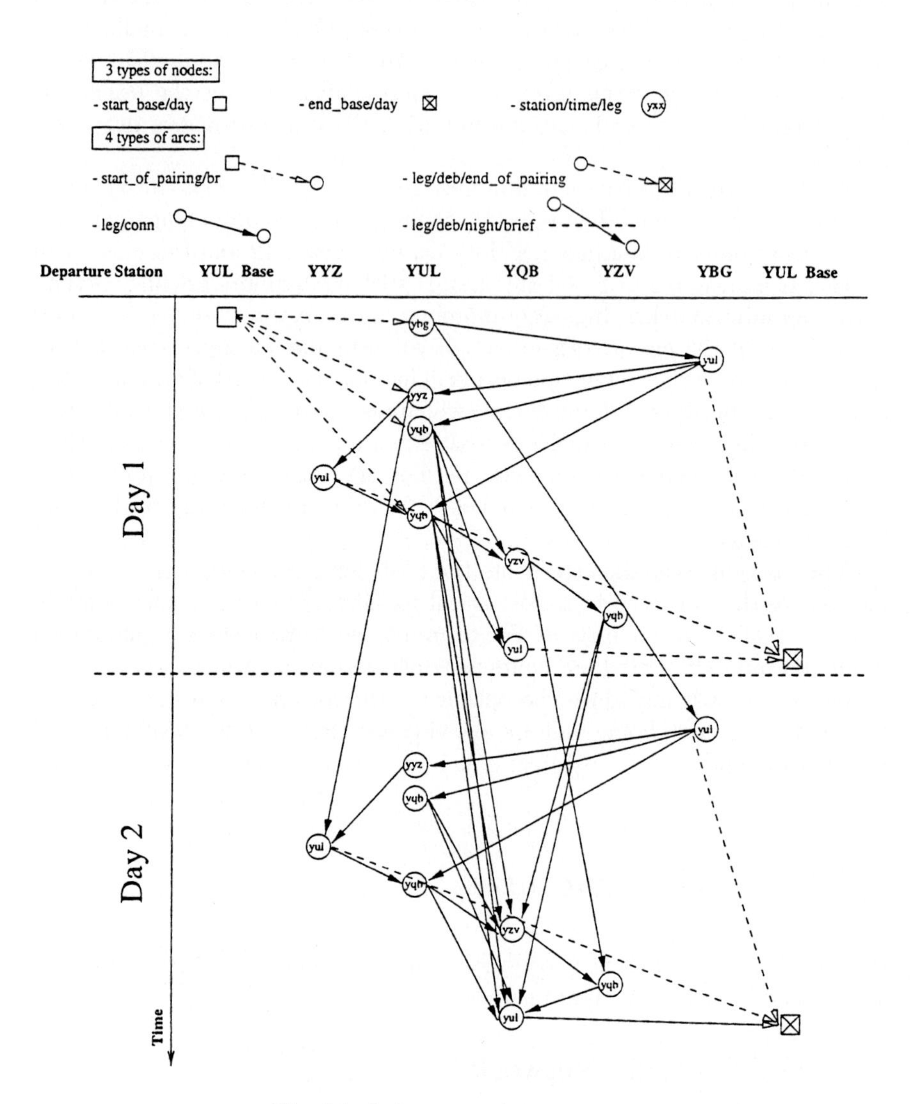

Fig. 5.1 A leg-on-node network

Nodes: A *start_base/day* node represents the departure from a base at the start of a pairing: there is one such node for each day of the week. Similarly, an *end_base/day* node represents the arrival at the same base at the end of each pairing: there is one such node for each day of the week. A *station/time/leg* node represents a leg that needs to be covered: its position on the horizontal axis indicates the departure city of the leg and its position on the vertical axis indicates the departure time of the leg, the arrival city being denoted within the node.

Arcs: Connections are represented by two types of arcs. In order for a connection to be valid, the arrival city of the first leg of the connection must be the same as the departure city of the second leg. In addition, the departure time of the second leg of the connection must be greater than the arrival time of the first leg offset by the minimum connection time and smaller than this arrival time offset by the maximum connection time. A *leg/conn* arc represents the connection between two legs, whose connection time is at least 15 minutes and at most 9.5 hours: this type of arc links two nodes of type *station/time/leg*. A *leg/deb/night/brief* arc corresponds to a connection between two legs separated by a nightly rest period: this type of arc links two nodes of type *station/time/leg*. A nightly rest period is preceded by a debriefing and followed by a briefing. This type of arc is the only kind that crosses the dotted line indicating the separation between two duties. Connections containing a nightly rest period at the YUL base are not permitted in the network of that base; however, they would be permitted on the network of the other bases.

Finally, an arc of type *start_of_pairing/briefing* corresponds to the start of a pairing and contains a briefing: it links a *start_base/day* node to all *station/time/leg* nodes where the departure city is the same as the base and the departure time is sometime during that day. An arc of type *leg/deb/end_of_pairing* corresponds to the end of a pairing and contains a debriefing after the flight leg: it links a *station/time/leg* node to an *end_base/day* node of the corresponding day. Since all pairings must end at a crew base, all legs whose destination is a crew base must allow for ending the pairing with a *leg/deb/end_of_pairing* arc.

Resources: In order to model certain constraints within the pairing construction process, the leg-on-node type of network requires two resources: the number of landings (L) and the amount of credited time per duty, also known as duty time (D). The reduced cost ($\bar{Z}$) can also be considered as a resource. Table 5.1 gives the resource extension functions for each of the four types of arc considered, while omitting the dual information as stated in (15). The cost resource (Z) is however included in Table 5.1. These functions ensure that a pairing is valid from node to node. Actually, they verify that the maximum number of landings and of credited hours is not violated. Note

Table 5.1: Extension functions for a leg-on-node network

	Cost (Z)	Number of Landings (L)	Duty Time (D)
- start_of_pairing/br	60	0	60
- leg/conn	$Z_i + t_{ij} + v_i + \delta b_1$	$L_i + 1$ if $L_i + 1 <= 12$	$D_i + t_{ij} + v_i$ if $D_i + t_{ij} + v_i <= 900$
- leg/deb/night/brief	$Z_i + v_i + 30 + b_2 + 60$	0 if $L_i + 1 <= 12$	60 if $D_i + v_i + 30 <= 900$
- leg/deb/end_of_pairing	$Z_i + v_i + 30$	$L_i + 1$ if $L_i + 1 <= 12$	$D_i + v_i + 30$ if $D_i + v_i + 30 <= 900$

Z_i : Cost accumulated up to node i
D_i : Duty time accumulated between the last night rest and node i
L_i: Number of landings accumulated between the last night rest and node i
t_{ij} : Connection time between flight legs i and j
v_i : Duration of the flight leg associated with node i
b_1 : Bonus for connections lasting at least 360 minutes
b_2 : Night rest bonus
δ : Parameter equal to 1 if the connection time exceeds 360 minutes, and 0 otherwise

that arcs of type *leg/deb/night/brief* force the solution process to reset the number of landings to 0 and the duty time to 60 minutes, corresponding to the time of briefing at the beginning of all duties. The cost of a pairing does not need to be reset nor does its value need to be compared to a bound at each node. The purpose of considering the cost as a resource is simply to keep track of the cost from node to node in a pairing.

The last rule stipulates that there should be no more than two duties per pairing. This as well as the minimum connection time constraint are dealt with by using the underlying structure of the network itself. For each base and for each day of the weekly planning period, a 2-day sub-network is constructed. Since there are three bases in the application, there will be a total of 21 sub-networks used, each containing one *start_base/day* node and two *end_base/day* nodes.

5.2 Leg-on-Arc Network

In this section, we describe a leg-on-arc type network. This network contains four types of nodes and five types of arcs (see Fig. 5.2). In this case, a third resource is necessary to validate certain rules during the construction of pairings.

Nodes: The nodes of type *start_base/day* and *end_base/day* represent the beginning and end of pairings at a crew base. An empty *station/time/leg/dep* node represents the departure of a leg and a solid *station/time/leg/arr* node represents the arrival of a leg.

Arcs: For each flight leg, an arc of type *leg* links a *station/time/leg/dep* node to its corresponding *station/time/leg/arr* node. The minimum connection time is included in the arrival time of all flight legs. Connections (or ground waiting periods) are represented by vertical arcs because these only move forward in time, not in space. These arcs link nodes of type *station/time/leg/arr* and type *station/time/leg/dep* associated with the same station and in chronological order of arrivals and departures. Fig. 5.2 presents the types of arcs which correspond to ground waiting (*ground*) and nightly rest periods (*deb/night/brief*). A nightly rest period arc crosses the dotted line which marks the transition between two days. Finally, arc types representing start and end of pairings are *start_of_pairing/briefing* and *deb/end_of_pairing*, respectively.

Resources: Three resources are necessary to validate pairings: number of landings (L), duty time (D) and connection time (C). Table 5.2 displays the extension functions of these resources as well as that of the cost function (without dual information). The resource corresponding to the connection time (C) serves several purposes. First, it computes the total connection time between two flight legs. Second, it permits limiting to 9.5 hours (or 570 minutes) the connection time between two flight legs. Third, it determines whether or not the connection time is greater than 6 hours (or 360 minutes) in order to credit (or not) the cost (Z) of bonus b_1 (i.e., when $\delta = 1$). This resource is reinitialized after each arc of type *leg* and reset to 0 between duties on arcs of type *deb/night/brief*. On *ground* arcs, the parameter γ prevents accumulation of connection time before the first leg of a duty. Note that connection time is added to the cost (Z) or the duty time (D) only when it lies between two legs of the same day.

As in the leg-on-node type network, the constraint restricting the number of duties per pairing to a maximum of two is dealt with using sub-networks spanning two days for each base.

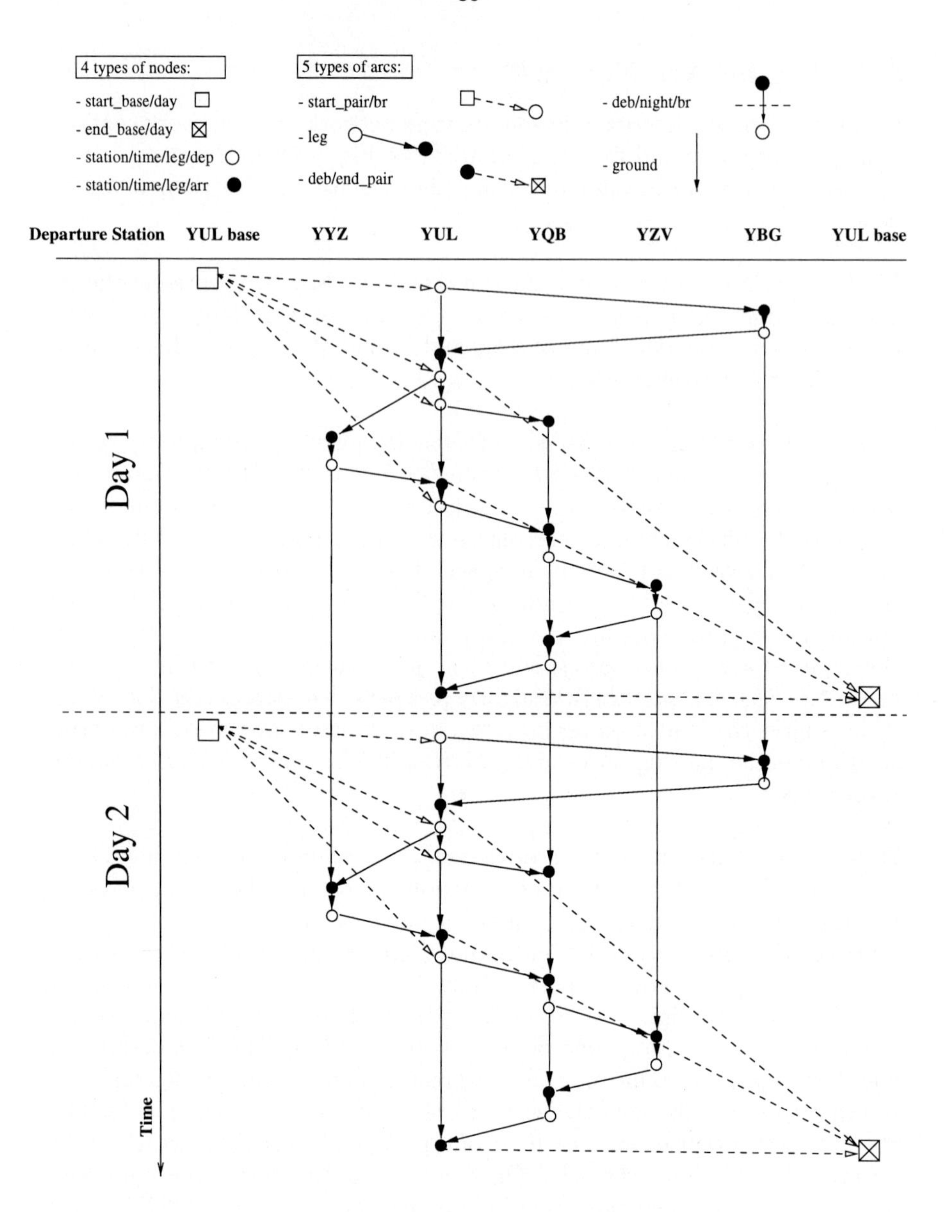

Figure 5.2: A leg-on-arc network

Table 5.2: Extension functions for a leg-on-arc network

	Cost (Z)	Number of Landings (L)	Duty Time (D)	Connection Time (C)
- start_of_pairing/br	60	0	60	0
- leg	$Z_i + v_i + C_i + \delta b_1$	$L_i + 1$ if $L_i + 1 <= 12$	$D_i + v_i + C_i$ if $D_i + v_i + C_i <= 900$	15 if $C_i <= 570$
- ground	Z_i	L_i	D_i	$C_i + \gamma t_{ij}$
- deb/night/brief	$Z_i + 30 + b_2 + 60$	0	60 if $D_i + 30 <= 900$	0
- deb/end_of_pairing	$Z_i + 30$	L_i	$D_i + 30$ if $D_i + 30 <= 900$	C_i

Z_i : Cost accumulated up to node i
D_i : Duty time accumulated between the last night rest and node i
L_i : Number of landings accumulated between the last night rest and node i
C_i : Connection time accumulated between the last flight leg and node i
t_{ij} : Ground waiting time between nodes i and j
v_i : Duration of the flight leg associated with node i
b_1 : Bonus for connections lasting at least 300 minutes
b_2 : Night rest bonus
δ : Parameter equal to 1 if $C_i >= 300$, and 0 otherwise
γ : Parameter equal to 1 if $L_i >= 1$, and 0 otherwise

5.3 Comparison of Network Structures

When comparing the above network structures, note that the leg-on-node structure contains many more arcs then the leg-on-arc structure. On the other hand, the leg-on-arc network involves one supplementary arc type, node type and resource. The connection time resource is unnecessary in the leg-on-node network because the connection arcs are created such that the minimum and maximum connection times are never violated. The requirement to add the bonus b_1 to the cost can also be easily evaluated using the difference between the departure time of the second leg of the connection and the arrival time of its first leg.

The leg-on-node network provides the simpler mathematical structure. However, the leg-on-arc network is much simpler to visualize because it contains many less arcs. Let n be the number of legs to cover, then the size of the leg-on-node network is $(6 \times 7) + n$ nodes and $O(n^2)$ arcs, and that of the leg-on-arc network is $(6 \times 7) + 2n$ nodes but only $O(n)$ arcs. The trade-off for this reduced number of arcs is the cost of dealing with an extra resource (C).

Deadheads: In both of these network types it is possible and likely essential to add deadhead arcs in order to obtain a feasible solution. These arcs are copies of all arcs containing a flight leg and are of equivalent cost in this application. Deadhead arcs allow the repositioning of a crew as passengers on a flight leg. In the leg-on-node type network, arcs of types *leg/conn*, *leg/deb/night/brief* and *leg/deb/end_of_pairing* are duplicated. In the leg-on-arc type network only the *leg* arcs are duplicated. Note once again the advantage of a leg-on-arc network structure.

6 Computational Experiments

The data sets used are the 986 flight legs covered by the company Intair between March 1st and 7th, 1991, inclusive. Intair was a regional aircraft carrier whose home bases were Montréal, Québec and Sept-Iles. For each flight leg, the departure and arrival times, dates and cities, as well as the type of aircraft used (F100, ATR, CVR, F28 or SWM) are specified. The duration of the legs varies from 15 minutes to 2 hours. All tests were run using GENCOL, version 4.1 on a HP 9000/735 workstation.

6.1 Results on the Fleet-Restricted Problems

Table 6.1 presents results obtained by solving the problem for each fleet separately. This table contains three sections. The first displays the types of aircraft and their corresponding number of legs to cover as well as the value of the optimal solution. The second and third display information and results on the five problems addressed using the leg-on-node and the leg-on-arc network, respectively.

The second and third sections of Table 6.1 are each divided into 3 subsections. The first of these contains information about the size of the networks, i.e., the number of nodes and the number of arcs. The second describes the solution process in terms of the number of column generation iterations performed and the number of columns generated. The last displays solution times. More specifically, the subproblem solution time, the master problem solution time and the total solution time. There is a slight discrepancy between the total solution time and the sum of the subproblem and master problem solution times due to overhead solution time required to load and pre-process data.

All problems were solved in under 7 minutes, regardless of the network structure used. Note as well that the branch-and-bound process was not needed as all solutions of the linear relaxations were integer. Table 6.1 shows that the total solution time for all leg-on-node problems was greater than

Table 6.1: The five fleet-restricted problems (basic scenario)

	Fleet				
	F100	**ATR**	**CVR**	**F28**	**SWM**
Number of Flight Legs	290	204	144	63	285
Cost (in minutes)	39315	21821	14195	7380	34160
Leg-on-Node Network (2 resources)					
Network					
Number of Nodes	332	246	186	105	327
Number of Arcs	**32096**	**31419**	**28063**	**27066**	**31410**
Solution Process					
Number of Iterations	146	90	231	43	192
Number of Columns	5841	3273	3563	819	5623
CPU (in seconds)					
Subproblems	243.7	110.1	212.8	37.6	306.8
Master Problem	47.4	12.6	21.4	0.6	43.6
Total	**299.9**	**131.0**	**242.0**	**45.2**	**359.5**
Leg-on-Arc Network (3 resources)					
Network					
Number of Nodes	622	450	330	168	612
Number of Arcs	**4209**	**4123**	**4063**	**3982**	**4204**
Solution Process					
Number of Iterations	206	168	199	82	140
Number of Columns	16914	6935	7434	1685	12310
CPU (in seconds)					
Subproblems	297.3	103.5	102.3	32.4	203.4
Master Problem	74.2	13.3	14.6	0.9	30.6
Total	**375.2**	**119.4**	**119.5**	**35.2**	**236.7**

that of the leg-on-arc problems, except for fleet F100. This average difference of 18% can be explained by the fact that the leg-on-node structure contains on average 7 times more arcs than the leg-on-arc network structure. Hence, the total solution time for the leg-on-arc network is faster than that of the leg-on-node type network, despite the extra resource necessary.

6.2 Scenario Analysis

This section analyzes the impact on cost and solution time of three modifications to the original parameters using the leg-on-arc network structure. The first scenario consists of introducing a guaranteed number of credited hours per duty in the cost function. The second allows pairings of up to three days. The last scenario consists of closing the Sept-Iles base and starting pairings only from the Montréal and Québec bases. Note that branching to obtain an integer solution was not needed as, once again, solutions to the linear relaxations were integer.

Guaranteed Credited Hours: In this scenario, four hours of duty time were guaranteed for all duties. In order to model this situation all costs associated with the arcs of the original scenario have been accumulated in a new resource, *ResCostGuarMin*. A check was imposed prior to each nightly rest period and end of pairing (i.e., before each arc of type *deb/night/brief* and *leg/deb/end_of_pairing*, respectively) to determine whether the *ResCostGuarMin* resource variable had attained the guaranteed duty time level of four hours. If it had, the value of the *ResCostGuarMin* resource variable was added to the cost function. Otherwise, four credited hours were added to the cost function of the pairing. In both cases, the *ResCostGuarMin* resource variable was reset to 0 for the upcoming duty.

Table 6.2: Minimum of 4 hours credited per duty (scenario 1)

	Fleet				
	F100	**ATR**	**CVR**	**F28**	**SWM**
Cost (in minutes)	**39801**	**22266**	**14195**	**7580**	**34385**
Cost Differential (w/r to Basic Scenario)	1.24%	2.04%	0.00%	2.71%	0.66%
Solution Process					
Number of Iterations	254	179	246	78	179
Number of Columns	26929	7075	10969	1638	14193
CPU (in seconds)					
Subproblems	883.5	219.4	424.1	41.0	501.0
Master Problem	218.3	16.6	19.5	1.0	48.5
Total	**1106.4**	**238.8**	**446.7**	**44.2**	**552.9**

Table 6.2 displays the computational results which show that the solution costs are slightly higher (by 1.33% on average) than the solution costs to the original problems. Furthermore, CPU time and the number of columns generated increase respectively by 170% and 37% on average. These increases are a result of the new resource introduced to handle this scenario.

The following two scenarios include minimum credited hours in their cost function. Comparisons were made with this scenario rather than with the original problem.

Pairing of up to three days: This scenario allows pairings which can last up to three days. This modification could decrease the cost of the solutions.

Table 6.3 shows that the cost was improved only for fleet F100 and SWM, by 0.82% and 2.53%, respectively. For the other types of aircraft, the opportunity to have pairings of up to three days did not decrease the cost of the

Table 6.3: Up to 3 duties per pairing (scenario 2)

	Fleet				
	F100	**ATR**	**CVR**	**F28**	**SWM**
Cost (in minutes)	**39474**	**22266**	**14195**	**7580**	**33515**
Cost Differential (w/r to Scenario1)	-0.82%	0.00%	0.00%	0.00%	-2.53%
Solution Process					
Number of Iterations	257	241	251	97	429
Number of Columns	30256	22226	14047	2462	66296
CPU (in seconds)					
Subproblems	1142.3	933.8	347.7	58.9	2211.3
Master Problem	1129.7	93.7	35.4	1.8	923.3
Total	**2277.2**	**1031.5**	**386.5**	**63.0**	**3143.4**

solutions. Although the number of iterations was slightly higher, the number of columns generated for this second scenario was 2.3 times greater than that of the first. This is due to the increased number of possibilities. As a result, the average CPU time increased threefold. The analysis of this scenario suggests that changing the union contract to allow 3-day pairings would not generate substantial savings.

Closing of the Sept-Iles base: This scenario evaluates the possibility of saving fixed costs related to the Sept-Iles base. It is important to note that the closing of a base does not imply that there will be fewer legs to cover; the city of Sept-Iles will continue to be serviced but there will no longer be pairings starting from that base. The networks will retain their size except for the start and end of pairing arcs associated with Sept-Iles. However, the number of sub-networks will decrease from 21 to 14.

Table 6.4: Closing of Sept-Iles crew home base (scenario 3)

	Fleet				
	F100	**ATR**	**CVR**	**F28**	**SWM**
Cost (in minutes)	**39801**	**22266**	**14950**	**7580**	**34890**
Cost Differential (w/r to Scenario 1)	0.00%	0.00%	5.32%	0.00%	1.47%
Solution Process					
Number of Iterations	243	102	253	52	229
Number of Columns	21362	6616	17942	1410	18607
CPU (in seconds)					
Subproblems	522.3	170.5	599.1	22.4	450.8
Master Problem	174.2	11.8	30.9	0.7	98.4
Total	**700.1**	**184.4**	**633.2**	**24.9**	**552.8**

Table 6.4 shows that closing the Sept-Iles base would be more costly only for the CVR and the SWM fleet where costs increased by 5.32% and 1.47%, respectively. For the other three types of aircraft, costs remained unchanged. Indeed, for any of these three fleets, pairings starting from this base were not present in the optimal solution of the first scenario. CPU time decreased by an average of 13% when compared with the first scenario. This can be explained by the fact that there were less sub-networks and the number of possibilities was more restricted.

Analyzing CPU times by fleet provides different conclusions for CVR and SWM. For these two types of aircraft, by closing Sept-Iles, the number of column generation iterations increased by 13% and 45%, respectively in comparison to the first scenario. Consequently, the associated CPU time increased by an average of 19%. Hence, closing Sept-Iles is a detriment to any fleet that used it as a base. To make a clear recommendation, this cost must be weighed against the savings derived from closing the base.

6.3 Results for the Global Problem

The following experiments were conducted by combining all five types of aircraft and solving one global problem consisting of about 1000 legs. The first set of results obtained for this problem are displayed in Table 6.5.

Table 6.5: Global problem (basic scenario)

	Leg-on-Node Network	**Leg-on-Arc Network**
Number of Flight Legs	986	986
Cost (in minutes)	108902	108902
Linear Relaxation Cost (in minutes)	108901.5	108901.5
Integrality Gap ($\times 10^{-3}$%)	0.45	0.45
Network		
Number of Nodes	1028	2014
Number of Arcs	49698	4905
Solution Process		
Number of Iterations	263	800
Number of Columns	149425	262861
CPU (in seconds)		
Subproblems	3249.6	4167.3
Master Problem	10073.9	25610.6
Total	**13345.6**	**29798.5**

The value of the optimal solution of the global problem depicted in this table (108,902) is smaller than the sum of the optimal solution values for the five constrained problems depicted in Table 6.2 (116,871). The solution of the global problem covers the same legs at a cost that is 7.3% lower. This percentage is significant in the context of air transport where personnel costs represent a major part of the company expenses.

Note that for both network models, an optimal integer solution was found before reaching the linear relaxation solution and hence no branching was required. That is because the integrality gap was negligible as the linear relaxation solution was 108901.5 and the optimal solution was 108902. With respect to CPU time, for the model representing legs by nodes, this was 3.7 hours, while for the leg-on-arc model it was 8.3 hours. These CPU times are abnormally long.

To analyze this behavior, we begin by observing that for both models, a large part of the CPU time is spent solving the master problem. Specifically, this represents 76% of the total time for the leg-on-node model and 86% for the leg-on-arc model. Seeing that the subproblem solution time is clearly not the source of the CPU time anomaly, the analysis of the network structure loses its importance, unless we are able to decrease the master problem CPU time to a level comparable to that of the subproblems.

The exceedingly long master problem CPU time is caused by degeneracy. To deal with this impediment, the linear relaxation of the problem has been solved as a set covering problem. Table 6.6 presents the results obtained for the global problem using this perturbation, for each of the two network models. The results presented in this table are related exclusively to the linear relaxation of the global problem.

Table 6.6: Global problem (over-covering scenario)

	Leg-on-Node Network	**Leg-on-Arc Network**
Number of Flight Legs	986	986
Linear Relaxation Cost (in minutes)	108901.5	108901.5
Network		
Number of Nodes	1028	2014
Number of Arcs	49698	4905
Solution Process		
Number of Iterations	102	277
Number of Columns	38374	26429
CPU (in seconds)		
Subproblems	1013.1	772.3
Master Problem	1347.9	1030.4
Total	**2373.8**	**1804.8**

By comparing the CPU times required to obtain the linear relaxation solution of the global problem with and without over-covering, we note a striking improvement provided by over-covering on each of the two network models. In absolute terms, the CPU times can be compared as follows: 39.2 minutes versus 3.7 hours for the model representing the flight legs by nodes and 30 minutes versus 8.3 hours for the model representing the flight legs by arcs. More exactly, the use of over-covering makes the CPU time 5.7 times faster for the former model and 16.7 times faster for the latter one.

Furthermore, we note that the master problem CPU time is still dominant. To be exact, for the leg-on-node model, the total time required for the master problem was 1347 seconds and 1013 seconds for the subproblems. For the leg-on-arc model, the master problem solution time was 1030 seconds and for the subproblems it was 772 seconds.

An analysis of the time required for the subproblems again confirms the efficiency of the network having legs represented by arcs over its node counterpart. By iteration, the total subproblem CPU time for the latter model was close to 4 times larger than that for the former model. Specifically, the respective times by iteration were 10 and 2.8 seconds.

The improvement in CPU times can be partly explained by the fact that by relaxing the Set Partitioning problem to a Set Covering one, we halve the dual variable space ($\alpha \geq 0$). To improve the solution time of the global problem linear relaxation even further, we introduce perturbation variables (i.e., bounded slack variables) that allow the sub-covering of the legs (see du Merle/Villeneuve/Desrosiers/Hansen (1997). These variables are randomly bounded between 0 and 0.0001. We encourage their utilization by associating with them a small unit cost. In this fashion, these variables will be present in the basis and will help reduce the degeneracy at the master problem level. The results obtained for the two network models are very good as can be seen in summary form in Table 6.7.

Table 6.7: Global problem (over- and under-covering scenario)

	Leg-on-Node Network	**Leg-on-Arc Network**
Number of Flight Legs	986	986
Linear Relaxation Cost (in minutes)	108899.3	108899.3
Network		
Number of Nodes	1028	2014
Number of Arcs	49698	4905
Solution Process		
Number of Iterations	47	39
Number of Columns	9283	16967
CPU (in seconds)		
Subproblems	386.4	259.1
Master Problem	209.2	194.5
Total	**608.9**	**455.6**

By comparing the results of this table with those of Table 6.6, we note a net improvement in total time produced by the solution with over- and under-covering. More specifically, the CPU time decreased from 39.2 minutes to 10 minutes for the leg-on-node model and from 30 minutes to 7.6 minutes for the arc based model. Compared to the total times without perturbation, the times obtained with over- and under-covering are respectively, 20 and 60 times faster for the former and latter network models. Therefore, the degeneracy of the master problem seems contained and the CPU times are divided equally between the master problem and the subproblems, for each of the two network models. In fact the CPU times for the subproblems are larger than those for the master problem. The minor differences between the costs of the solutions obtained by this method and all preceding ones is due to the presence of perturbation variables which permit the sub-covering at a small cost. The solutions are then optimal up to a negligible perturbation. Moreover, the average CPU time by iteration required by the subproblems was 8.2 seconds for the node based model and 6.6 seconds for its arc counterpart. This again indicates that the leg-on-arc networks are more efficient.

7 Conclusions

This research has analyzed crew pairing in the operating environment of a regional air carrier. As we have discussed, the classical solution approach for the medium and long haul problems based on enumeration of duties is not applicable in the regional context. Therefore, we have suggested a column generation approach which considers all work rules, including those related to duties, during the construction of feasible pairings.

We have also introduced two network structures compatible with this approach: the first represented legs by nodes while the second by arcs. The latter structure consists of substantially fewer arcs than the former, yet it requires the utilization of a supplementary resource in the model. The computational results indicate the superiority of this model over its node analogue. This supports similar findings in the school- and urban-bus areas.

We have also analyzed the impact on solution cost of three different operating scenarios. The proposed mathematical model has the flexibility to allow easy examination of these modifications. Moreover, we have shown that a stabilization method introducing bounded perturbation variables considerably decreases the master problem CPU time. This approach opens new horizons for column generation methods by practically eliminating the degeneracy phenomenon observed often.

Acknowledgments. This research was supported by the "Programme Synergie du Fonds de Développement Technologique du Québec" and by the Natural Sciences and Engineering Council of Canada. Marius M. Solomon was partially supported by the Patrick F. and Helen C. Walsh Research Professorship.

References

Anbil, R./Gelman, E./Patty, B./Tanga, R. (1991): Recent Advances in Crew Pairing Optimization at American Airlines. Interfaces 21, 62–74.

Ball, M./Roberts, A. (1985): A Graph Partitioning Approach to Airline Crew Scheduling. Transportation Science 19, 107–126.

Barnhart, C./Johnson, E.L./Anbil, R./Hatay, L. (1994): A Column Generation Technique for the Long-Haul Crew Assignment Problem. in: Ciriani, T.A./Leachman, R.C.(eds.): Optimization in Industry 2. (John Wiley) New York, 7–22.

Chu, H.D./Gelman, E./Johnson, E. (1997): Solving Large Scale Crew Scheduling Problems. European Journal of Operational Research 97, 260–268.

Crainic, T.G./Rousseau, J.-M. (1987): The Column Generation Principle and the Airline Crew Scheduling Problem. INFOR 25, 136–151.

Dantzig, G.B./Wolfe, P. (1960): Decomposition Principle for Linear Programming. Operations Research 8, 101–111.

Desaulniers, G./Desrosiers, J./Ioachim, I./Solomon, M.M./Soumis, F./Villeneuve, D. (1998): A Unified Framework for Deterministic Time Constrained Vehicle Routing and Crew Scheduling Problems. in: Crainic, T./Laporte, G (eds.): Fleet Management and Logistics.

Desaulniers, G./Desrosiers, J./Dumas, Y./Marc, S./Rioux, B./ Solomon, M.M./Soumis, F. (1997): Crew Pairing at Air France. European Journal of Operational Research 97, 245–259.

Desrochers, M./Desrosiers, J./Solomon, M.M. (1992): A New Optimization Algorithm for the Vehicle Routing Problem with Time Windows. Operations Research 40, 342–354.

Desrochers, M./Soumis, F. (1988): A Reoptimization Algorithm for the Shortest Path Problem with Time Windows. European Journal of Operational Research 35, 242–254.

Desrosiers, J./Dumas, Y/Solomon, M.M./Soumis, F. (1995): Time Constrained Routing and Scheduling. in: Ball, M.O./Magnanti, T.L./ Monma, C.L./Nemhauser, G.L. (eds.): Network Routing, Handbooks in O.R. & M.S. 8, (Elsevier) Amsterdam, 35–139.

du Merle, O./Villeneuve, D./Desrosiers, J./Hansen, P. (1997): Stabilized Column Generation. Les Cahiers du GERAD, G–98–06, École des Hautes Études Commerciales, Montréal, Canada, H3T 2A7.

Etschmaier, M.M./Mathaisel, D.F.X. (1985): Airline Scheduling: An Overview. Transportation Science 19, 127–138.

Gershkoff, L. (1989): Optimizing Flight Crew Schedules. Interfaces 19, 29–43.

Graves, G.W./McBride, R.D./Gershkoff, I./Anderson, D./ Mahidara, D. (1993): Flight Crew Scheduling. Management Science 39, 736–745.

Hoffman, K.L./Padberg, M. (1993): Solving Airline Crew Scheduling Problems by Branch-and-Cut. Management Science 39, 657–682.

Lavoie, S./Minoux, M./Odier, E. (1988): A New Approach for Crew Pairing Problems by Column Generation with an Application to Air Transportation. European Journal of Operational Research 35, 45–58.

Salkin, H.M. (1975): Integer Programming (Addison-Wesley) Reading, Mass.

Wedelin, D. (1995): An Algorithm for Large Scale 0-1 Integer Programming with Applications to Airline Crew Scheduling. Annals of Operations Research 57, 283–301.

An Improved ILP System for Driver Scheduling

Sarah Fores, Les Proll and Anthony Wren
Scheduling and Constraint Management Group, School of Computer Studies, University of Leeds, Leeds LS2 9JT UK

Abstract: Mathematical programming approaches to driver scheduling have been reported at many previous workshops and have become the dominant approach to the problem. However the problem frequently is too large for mathematical programming to be able to guarantee an optimal schedule. TRACS II, developed at the University of Leeds, is one such mathematical programming-based scheduling system. Several improvements and alternative solution methods have now been incorporated into the mathematical programming component of the TRACS II system, including a column generation technique which implicitly considers many more valid shifts than standard linear programming approaches. All improvements and alternative strategies have been implemented into the mathematical programming component of TRACS II to allow different solution methods to be used where necessary, and to solve larger problems in a single pass, as well as to produce better solutions. Comparative results on real-world problems are presented.

1 Introduction

The problem of scheduling public transport vehicles and their drivers has been the subject of six international workshops (see, for example, Desrochers/ Rousseau (1992), Daduna/Branco/Paixáo (1995)). With improvements in technology, mathematical programming solution methods are increasingly successful. One of the driver scheduling systems using a mathematical programming approach is TRACS II which was originally developed from the IMPACS system (see Smith/Wren (1988), Wren/Smith (1988)) and which uses a set covering formulation to ensure that all vehicle work is covered. This paper outlines several improvements incorporated into the mathematical programming component of the TRACS II system. Results on a selection of bus and train problems are reported.

2 The Driver Scheduling Problem

Typically a vehicle schedule is produced for which drivers need to be allocated. The vehicle schedule is represented graphically by a series of lines each of which depict the movements of a vehicle during the day, and at various stages the vehicle will pass a location which is convenient for driver changeovers. These location/time pairs are known as *relief opportunities*, and an indivisible period between any two relief opportunities is known as a *piece of work*. Each piece of work then needs to be allocated to a driver so that the minimum number of drivers is required and costs are minimized. The work of a driver in a day is known as a *shift*, which usually consists of two to four spells of work each of which covers several consecutive pieces of work on the same vehicle. The formation of shifts is governed by a set of labour agreement rules to ensure that there is adequate provision for mealbreaks and acceptable working hours etc.

It is possible to formulate the driver scheduling problem as a set covering or set partitioning problem which ensures that all of the vehicle work is covered. A suitable objective function can then be devised which ensures shift and cost minimization over a set of previously generated valid shifts. The difficulty with this approach is that the number of valid shifts is too large for this approach to be able to guarantee an optimal schedule in most cases. Thus mathematical programming is frequently combined with heuristic approaches to provide a viable solution method.

3 The TRACS II System

TRACS II is the driver scheduling system which was developed at the University of Leeds, and which originated from the commercially available IMPACS system (see Smith/Wren (1988), Wren/Smith (1988)). TRACS II has since been altered to incorporate features required in order to schedule train crews, (see Wren/Kwan/Parker (1994), Kwan/Kwan/Parker/Wren (1996)) and many of the algorithms contained in the individual components have been improved. In principle though, the TRACS II system exhibits a similar overall solution method to the IMPACS system:

- **Stage 1** *Generate a set of shifts which are valid according to labour agreement rules.*
  It should be noted that in a practical problem there are generally many million of such potential shifts. The shift generation process only produces a large subset of shifts, chosen heuristically in such a way that the most likely shifts are formed and that a good choice of shifts is available for each piece of work. These heuristics have proved effective in a wide range of practical applications.

- **Stage 2** *If necessary, reduce the size of the generated shift set.* Where a standard Integer Linear Programming (ILP) approach is used it optimises over the whole shift set, and this is limited in terms of data storage and is time-consuming, even with improvements in technology and algorithms. Heuristic methods have been developed which attempt to reduce the size of the set, while retaining the best possible subset of shifts.

- **Stage 3** *Select from the shift set a subset which covers the vehicle work.* The problem is one of minimizing the overall schedule cost, which includes a wage cost and sometimes a subjective cost reflecting penalties for shifts containing undesirable features. As the introduction of each driver incurs a large cost the primary objective is to minimize the number of shifts. Since the size of shift set originally generated is limited, it may not be possible to use a set partitioning approach which would imply that each piece of work should be covered by exactly one driver. A set covering model is used which guarantees that each piece of work is covered by at least one driver and the details of this method are discussed in the following sections.

4 Original ILP Solution Method (ZIP)

The solution method is referred to as ZIP (Zero-one Integer Programming). Given a problem with M pieces of work in the vehicle schedule, and a previously generated set of N shifts, we can define :

For $j = 1, .., N$

$$x_j = \begin{cases} 1 & \text{if shift } j \text{ is used in the solution} \\ 0 & \text{otherwise} \end{cases} \tag{4.1}$$

For $i = 1, .., M$

$$u_i = \begin{cases} 1 & \text{if workpiece } i \text{ is uncovered} \\ 0 & \text{otherwise} \end{cases}$$

$$o_i = \text{number of times that workpiece } i \text{ is overcovered}$$

The existence of the variables u_i and o_i allows workpieces to remain uncovered and have more than one shift covering them respectively. The desired situation would be where a piece of work is covered by exactly one shift so that the variables u_i and o_i would both have the value zero. Where overcover remains in the final schedule, the relevant shifts can be edited to form shorter shifts which were previously excluded at Stage 2.

4.1 The Objective Function

The complexity of requirements to produce an efficient driver schedule leads to several, sometimes conflicting, objectives being necessary. These arise out of the need to assign a driver to every piece of work and the desire to produce schedules that are satisfactory both from a management and a driver viewpoint. Normally these are addressed in the following decreasing order of importance.

- To minimize the number of uncovered pieces of work; the minimum will normally be zero.
- To minimize the number of shifts used in the schedule. This is given a high priority due to the large overheads associated with employing staff.
- To avoid shifts which contain undesirable features. It is possible to form a schedule consisting of many undesirable shifts with a relatively low wage cost but it is preferable to encourage more acceptable working conditions.
- To minimize wage costs. Wage costs can be affected by overall and spell durations and the wage cost of the shift combination should be reduced.
- To minimize the total duration of overcovered pieces of work.

This can be represented by an objective function of the form:

$$\text{Minimize} \quad \textstyle\sum_{j=1}^{N} C_j x_j + \sum_{i=1}^{M} D_i u_i + \sum_{i=1}^{M} E_i o_i \tag{4.2}$$

Setting the D_i coefficient to a large value will address the most important objective of reducing the number of uncovered pieces of work.

The E_i coefficient on the other hand should be lower to reflect a less important objective on the duration of overcovered pieces of work. E_i is proportional to the duration of workpiece i.

The remaining objectives require the C_j coefficient to reflect shift costs, penalty costs and wage costs. However, penalty costs would normally be added in later so as not to detract from the more important objective of minimizing the number of drivers. To prioritise the objectives a large constant is added to each shift cost so that minimization favours a schedule with fewer shifts.

Constants used in weighting the objectives are defined by the scheduler.

4.2 Constraints

As mentioned earlier the model is based on set covering, because the number of shifts generated is limited to those which are deemed to be 'efficient' and many shorter shifts are discarded. Although a final schedule cannot contain pieces of work which are covered by more than one driver the set covering approach potentially allows this. The following constraint ensures that every piece of work is covered by at least one driver:

$\sum_{j=1}^{N} A_{ij}x_j \geq 1$ for $i = 1, .., M$

Given that each piece of work has an associated undercover and overcover variable, we can rewrite this as:

$$\sum_{j=1}^{N} A_{ij}x_j + u_i - o_i = 1 \text{ for } i = 1, .., M \tag{4.3}$$

where the A_{ij} identify which pieces of work are covered by which shifts:

$$A_{ij} = \begin{cases} 1 & \text{if shift j covers workpiece i} \\ 0 & \text{otherwise.} \end{cases} \tag{4.4}$$

In practice, the introduction of overcovered pieces of work in the schedule is deterred by the minimization of overcover *and* wage costs. Although overcover is allowed, it can only be acceptable as an overlap in the middle of the day. Thus it is sensible not to allow overcover early or late in the day. Thus workpiece constraints corresponding to first pieces of work on early departing vehicles and to last pieces of work on late arriving vehicles are defined as equations. The inclusion of such constraints is, in fact, a requirement for the validity of the constraint branching strategy used within the branch and bound algorithm. The algorithm incorporated into the mathematical programming component automatically identifies the workpieces which shall be formulated as equality constraints, ensuring that the sets of shifts covering each of these workpieces are mutually exclusive.

Hence, constraint (4.3) can now be split into the following:

$$\begin{aligned} &\sum_{j=1}^{N} A_{ij}x_j + u_i = 1 && \text{for } i = 1, .., L \\ &\sum_{j=1}^{N} A_{ij}x_j + u_i - o_i = 1 && \text{for } i = L+1, .., M \end{aligned} \tag{4.5}$$

where L denotes the number of workpieces which can be identified as equality constraints.

In addition to the workpiece constraints, shifts can be categorised by an initially specified type and the user can impose constraints on any of these to ensure that the final schedule does not contain too many or too few shifts of any particular type. Constraints of this form can also be used to limit the total number of shifts in the final schedule and also eliminate undercover. The constraints can be imposed either initially or when reoptimising an existing solution.

The constraints can be expressed as follows:

$$\sum_{j=1}^{N} \delta_{kj} x_j \leq U_k, \quad \sum_{j=1}^{N} \delta_{kj} x_j \geq L_k \tag{4.6}$$

where U_k is an upper limit on the number of shifts of type k, and L_k is a lower limit on the number of shifts of type k.

The δ_{kj} are defined as :

$$\delta_{kj} = \begin{cases} 1 & \text{if } x_j \text{ is a shift of type k} \\ 0 & \text{otherwise} \end{cases} \tag{4.7}$$

4.3 Original (ZIP) Model

In summary, the model can be formulated as follows :

$$\text{Minimize} \quad \sum_{j=1}^{N} C_j X_j + \sum_{i=1}^{M} D_i u_i + \sum_{i=1}^{M} E_i o_i \tag{4.8}$$

$$\begin{aligned} \text{Subject to} \quad & \sum_{j=1}^{N} A_{ij} x_j + u_i = 1 && \text{for } i = 1,..,L \\ & \sum_{j=1}^{N} A_{ij} x_j + u_i - o_i = 1 && \text{for } i = L+1,..,M \\ & \text{Plus any user-defined side constraints} \\ & x_j = 0 \text{ or } 1, && \text{for } j = 1,..,N \\ & u_i \geq 0 \\ & o_i \geq 0 \end{aligned}$$

4.4 Method of Solution of the Model

It is possible to execute the model solver in a number of different ways, such as starting or stopping at different points in the solution strategy; however, the method will be described in full as if there were no user intervention in the process. The following is a summary of the stages involved in solving model (4.8).

Stage Z1

Shifts are ranked in groups in order of 'desirability' during the generation phase. The ranking is a crude measure of the ratio of work content to shift content. In the first stage of the solution all three-part shifts and shifts with a relatively low work content are temporarily excluded from the shift set. At this stage penalty costs are not included in shift costs. The integrality constraints on the shifts are relaxed so that a piece of work may be covered by fractions of several different shifts. The relaxed model is solved over the remaining shifts using a primal steepest edge (see Goldfarb/Reid (1977)) variant of the Revised Simplex Method, starting with an initial solution formed by a heuristic which considers all of the previously generated shifts.

The quality of the initial solution affects the number of subsequent iterations required to find the optimal continuous solution. It is developed heuristically and has to be transformed into a basic solution to (4.8) in order to provide an advanced start for the revised simplex method. The heuristic considers each piece of currently uncovered work in ascending order of the number of shifts available to cover it, and selects a shift to be included in the initial solution which minimizes the following cost function:

$$\text{Minimize} \quad \left(\frac{C_j}{NU_j}\right) + OC_j \tag{4.9}$$

where

C_j = cost of shift j
NU_j = number of currently uncovered workpieces covered by shift j
OC_j = increase in overcover costs caused by selecting shift j.

provided its inclusion does not violate a side constraint or an equality workpiece constraint. In order to satisfy a realistic upper limit on the number of shifts ofcertain types or on the total number of shifts it is possible that some workpieces will remain uncovered. For this reason the 'no uncovered work' constraint is not applied initially.

Stage Z2

The previously excluded shifts are restored and the LP relaxation is reoptimised with the complete set of shifts, again using the primal steepest edge algorithm.

Stage Z3

Having found the optimal Stage Z2 solution, the penalty costs are added to all appropriate shifts. These are not normally added initially because shifts displaying several undesirable features incur a very high cost, and although it is hoped that these shifts do not appear in a schedule it is actually preferable to include some of them rather than to exceed the minimum number of shifts. For this reason the solution strategy has historically been to devise two pre-emptively ordered objectives which are solved in stages. The first is to minimize a function combining the costs of the individual shifts with a large fixed cost per shift, so as to give a significant bias towards minimizing the number of shifts; the second is to minimize a cost function, incorporating penalty costs, with the added constraint that the total number of shifts does not exceed the number already ascertained at the end of the first stage. These objectives can be formulated as follows:

1) Minimize $\sum_{j=1}^{N} C_{1j}x_j + \sum_{i=1}^{M} D_i u_i + \sum_{i=1}^{M} E_i o_i$ (4.10)

2) Minimize $\sum_{j=1}^{N} (C_{1j} + C_{2j})x_j + \sum_{i=1}^{M} D_i u_i + \sum_{i=1}^{M} E_i o_i$

where
C_{1j} includes wage costs and the large constant shift cost

C_{2j} is the appropriate penalty cost.

In order to address the objective prioritisation the following two side constraints are now added:

$$\sum_{i=1}^{M} u_i \leq 0 \qquad (4.11)$$

$$\sum_{j=1}^{N} x_j \leq T.$$

The first ensures that there can be no uncovered work in the solution. The second constraint uses a target number of shifts to ensure that the minimization of the number of shifts in the solution is still addressed. If the number of shifts at the end of Stage Z2 is integral then T will take this value. However, it is more likely that the number of shifts at the end of Stage Z2 is fractional and, since an integer solution will be the final requirement, T will take this number rounded up to the next integer. This new model is now reoptimised by primal steepest edge.

Stage Z4
If the total number of shifts in the optimal Stage Z3 solution is non-integral and of the form $I.f(0 < f < 1)$, the side constraint

$$\sum_{j=1}^{N} x_j \geq I + 1 \qquad (4.12)$$

is added. This is because the current solution is minimal for a relaxed LP model, and so any integer solution must require at least the next highest integer number of shifts. The current solution will be infeasible for this new constraint and the new model must be re-solved using a Dual Simplex algorithm.

If the total number of shifts in the current solution is already integral then no side constraint is added and no reoptimisation performed.

Stage Z5
If the current solution is fractional then the shift set is first reduced and a branch and bound method is used to determine whether a particular shift should be included in the schedule or not.

Smith/Wren (1988) developed a technique to reduce the size of the shift set entering the branch and bound phase, which assumes that an acceptable integer solution can be found by restricting the choice of relief opportunities to the subset used in the LP optimum. In this way a shift which does not use any of these times can be excluded from the search. The *reduce* procedure uses the principle that the LP optimum gives a good indication of how the vehicle work would be covered in an integer solution, and this has been confirmed by extensive practical application. Application of *reduce* typically leads to a reduction in the number of shifts considered at the branch and bound stage of 50-80%, leading to a much reduced solution time. It is possible that the constraint limiting the search to find a schedule with an exact number of shifts may not be satisfied with the limited set of relief opportunities. However, in general there are many different integer solutions, some of which should be contained within the shift set available. The other possibility is that the reduced shift set produces a higher cost schedule than one which would be produced with the whole set, but the process is normally designed to terminate with the first good solution and its quality is largely dependent on the choice of path through the tree, rather than on the number of shifts which are considered.

The branch and bound method has been developed with an emphasis on finding a good integer solution quickly. The integer solution is found by developing a branch and bound tree, in which the lower bound on the objective cost is given by the optimal continuous solution. Once an integer solution has been found the nodes of the tree are fathomed if their cost is greater than or equal to a scheduler-specified percentage of the current best integer cost. In most cases the first integer solution found fathoms all of the remaining active nodes and hence the branch and bound process normally terminates with a possibly non-optimal integer solution. A maximum of 500 nodes can be created. The branch and bound process uses a specialised relief time branching strategy developed by Smith/Wren (1988).

5 Current ILP Solution Method (SCHEDULE)

5.1 Sherali Weighted Objective Function

Sherali (1982) noted that there are several disadvantages in using a sequential approach to find a solution which satisfies two objectives. Apart from the fact that two separate linear programs have to be solved which are essentially doing the same task but with a different cost coefficient, a high degree of degeneracy is likely to occur, with typically large numbers of iterations. Also the introduction of side constraints to maintain the subjective ordering of the objectives results in a more complex model to be solved which may increase execution times over a simple model for which efficient solution codes exist.

An alternative approach to solving the driver scheduling model has been explored in Willers/Proll/Wren (1993), Willers (1995). The approach combines the objectives of minimizing the number of shifts and the total shift cost, whilst retaining their preference ordering.

The two objective functions (4.11) can be simplified. The solution strategy adopted by ZIP involves adding a side constraint to prohibit any undercover and therefore it is unnecessary for the u_i variables to appear in the model. Overcover is unproductive and is discouraged by the minimization of wage costs and so zero costs can be attached to the o_i variables. Also since the more important objective is to minimize the number of shifts, all of the shift costs can be attached in the second phase. This removes the need to include a large prioritising constant to every shift. The two objective functions can now be expressed as:

1) Minimize $\sum_{j=1}^{N} x_j$ (5.1)

2) Minimize $\sum_{j=1}^{N} C_j x_j$

where
C_j combines wage and penalty costs for shift j.

Willers implemented a method of converting multi-objective models into single objective models. This involved weighting the objectives and combining them as follows:

$$\text{Minimize} \quad W_1 \sum_{j=1}^{N} x_j + W_2 \sum_{j=1}^{N} C_j x_j \tag{5.2}$$

The weights W_1 and W_2 must be selected so as to ensure that the objectives are correctly ordered. This produces:

$$W_1 = 1 + UB[\sum_{j=1}^{N} C_j x_j] - LB[\sum_{j=1}^{N} C_j x_j] \qquad (5.3)$$
$$W_2 = 1$$

where

$UB[\sum_{j=1}^{N} C_j x_j]$ is an upper bound value on the shift costs (5.4)

$LB[\sum_{j=1}^{N} C_j x_j]$ is a lower bound value on the shift costs.

An upper bound can be calculated by setting all shift variables to 1. This upper bound however corresponds to summing all shift costs for the N shifts generated, which would produce a very large weight, potentially leading to numerical difficulties in the solution method. A smaller weight can be calculated by adapting the upper bound so that:

$$W_1 = 1 + \text{ sum of } X \text{ largest } C_j \text{ values} \tag{5.5}$$

where X must be an upper bound on the number of shifts in the schedule. An appropriate weight can be ascertained by defining X to be the number of shifts in the initial solution plus the number of uncovered pieces, since the sum of the highest X cost values in this case must be an upper bound to any further schedule costs calculated. A potentially smaller weight would result from careful calculation of the lower bound. Willers (1995) investigated several possibilities and concluded that the simple lower bound of 0 was satisfactory.

5.2 The New Model

By incorporating the Sherali weight (5.5) into the objective function and defining :

$$D_j = W_1 + C_j \qquad \text{for } j = 1, .., N \tag{5.6}$$

as the new cost coefficient for every shift variable, the new model can be defined as follows:

$$\text{Minimize} \quad \sum_{j=1}^{N} D_j x_j \tag{5.7}$$

$$\text{Subject to} \quad \sum_{j=1}^{N} A_{ij} x_j = 1 \qquad \text{for } i = 1, .., L$$

$$\sum_{j=1}^{N} A_{ij} x_j \geq 1 \qquad \text{for } i = L+1, .., M$$

Plus any user-defined side constraints

$$x_j = 0 \text{ or } 1, \qquad \text{for } j = 1, .., N$$

It may be noted that this model, unlike (4.8), guarantees that the minimum number of shifts is obtained.

5.3 Method of Solution of the (SCHEDULE) Model

It was noted that the process of banning shifts in the first stage of the solution method did not contribute any improvement in solution time and is no longer used. The actual solution stages which remain depend upon the user's choice of solution strategy but a broad outline follows:

- Solve the LP relaxation over the whole shift set using either a conventional linear programming method or column generation. The single objective model is optimised using a primal steepest edge approach.

- Add a constraint which increases the shift total to the next highest integer and re-solve using a dual steepest edge approach due to Forrest/Goldfarb (1992).
- Reduce the number of constraints and variables and find an integer solution using the branch and bound technique used in the original model, but with reoptimisations performed by dual steepest edge.

5.3.1 Initial solution

Recognising that the overcover variables are no longer costed, a new cost function can be defined for the purpose of creating an initial solution:

$$\text{Minimize} \quad \frac{C_j}{NU_j} \tag{5.8}$$

where

C_j = combined wage and penalty costs for shift j
NU_j = number of currently uncovered workpieces covered by shift j.

The above method for constructing an initial solution is an adaptation of that used in ZIP in which the objective is cost minimization. In SCHEDULE, the primary objective is shift minimization and the following method addresses that more directly. If all pieces of work are to be covered by the minimum number of shifts, each newly selected shift must cover a lot of work not covered by shifts selected earlier. This suggests selecting shifts by:

$$\text{Maximize} \quad \sum_{i=1}^{M} \Delta_{ij} L_i \tag{5.9}$$

where

$$\Delta_{ij} = \begin{cases} 1 \text{ if shift j covers the currently uncovered piece of work i} \\ 0 \text{ otherwise,} \end{cases}$$
L_i = duration of workpiece i.

Willers' experiments (see Willers (1995)) show that while this method does not always outperform the earlier method in terms of number of shifts in the initial solution, it does so on average.

5.3.2 Column Generation

In order for a conventional mathematical programming approach to be used to solve the driver scheduling problem, heuristics are necessary to reduce the problem size. This restricts optimality to only those shifts remaining. The heuristics have been developed so that apparently inefficient shifts are removed and solutions have proved better than manual solutions. However it may be the case that some inefficient shifts link well with other shifts to produce the optimal solution. In order to guarantee optimality it must be possible to consider every valid shift.

Column generation is an approach which can be used to solve mathematical programming problems involving many variables and it is shown that column generation methods improve upon the solution and/or speed of the process. TRACS II uses a conventional mathematical programming method which requires all shifts to be available. The Revised Simplex Method is then used iteratively to improve a current solution by swapping a basic variable for a non-basic variable with favourable reduced cost, until no further improvement can be made. In the case of column generation only a subset of the columns is available at the outset. The Revised Simplex Method is used to find the solution which is optimal over the subset. One then generates new columns or searches through columns not previously considered and adds some or all shifts with favourable reduced costs as non-basic variables. The new subset is then reoptimised. For any LP relaxation which is optimal over its available subset the overall optimal solution is attained when no more columns which could improve the objective can be added to the set.

The HASTUS scheduling system (see Rousseau/Blais (1985)) contains a module, Crew-Opt (see Desrochers/Gilbert/Sauvé/Soumis (1992)), which uses column generation techniques to form bus driver schedules and has produced encouraging results (see Rousseau (1995)). A subset of valid shifts is identified and further shifts are added to the subset based upon utilising a shortest path algorithm to generate and identify shifts which will improve the current solution. Once no new shift can be found which will reduce the schedule cost the current solution is optimal over the relaxed model. A branch and bound technique which also incorporates column generation is then applied to find a good driver schedule. Crew-Opt allows all shifts to be available by generating them as constrained shortest paths through a network. In order to ensure optimality over *all* possible valid shifts it must be possible to generate any valid shift by using a method such as that used by HASTUS, or else all shifts must be generated initially and therefore available to the process. Since TRACS II already has a good shift generation process, and the shift costs include penalty costs which are more complicated to model using a shortest path technique, the option of generating complicated shift structures through potentially large networks was not considered. The technique which was implemented considers a previously generated shift superset which allows many more shifts to be available to the mathematical programming solver. Although the continuous solution is still not guaranteed to be optimal, better solutions lead to a branch and bound search for schedules with fewer shifts. A network formulation is not used, and so shifts from this larger set are selected to enter a working subset by means of a less sophisticated enumeration method.

Fores (1996) details the column generation implementation strategies. The solution method is outlined as follows:

- **Step 0 Generate a shift superset**
  This uses the TRACS II technique of generating shifts, possibly allowing more shifts to be produced by relaxing some of the conditions set previously to restrict the shift generation. No further heuristic reduction is then necessary.

- **Step 1 Create an initial solution and form an initial shift subset**
  An initial solution is generated from the shift superset, using one of the methods described. As shifts are being considered in forming the initial solution a shift subset is also being selected. This subset needs to be sufficiently large and varied to reduce the number of shift additions required.

- **Step 2 Solve the LP over the current shift subset**
  Use the Revised Simplex Method to solve the LP over the current shift subset.

- **Step 3 Add a set of shifts to the current subset which will improve the solution**

  Simplex multipliers produced at the end of Step 2 are used to calculate the reduced costs of shifts currently not selected from the superset. The reduced cost of a shift k is defined to be:

  $$\text{Minimize} \quad (c_k - \sum_{i=1}^{M} \pi_i a_{ik}) \tag{5.10}$$

  where:

  M is the number of constraints
  c_k is the cost of shift k
  π_i is the simplex multiplier for constraint i
  a_{ik} is the coefficient of shift k in constraint i

  Since the shift costs include the large Sherali weight they do not vary as significantly as the simplex multipliers. The simplex multipliers are therefore considered in decreasing order and shifts covering their corresponding pieces of work are added to the subset if they have a favourable reduced cost and if they are allowed to be added based upon parameters which control the shift addition.

- **Step 4 If no favourable shifts can be found then the LP solution is optimal, otherwise go to Step 2**

- **Step 5 Find an integer solution using branch and bound**
  Since we still cannot guarantee a relaxed solution optimal over *all* possible shifts, and the existence of subjective weightings allows ambiguity over the best cost solution, no alteration is made to the branch and bound strategy.

The column generation implementation was tested against a Sherali version of ZIP. This allowed the most accurate comparison of two single objective models. A summary of column generation results will be given in Sect. 6.2.

5.3.3 A Dual Approach

It has long been known that LPs of the form (5.7) are inherently degenerate (see Marsten (1974)) and that primal simplex approaches to their solution spend many iterations making no reduction in the value of the objective function. A number of degeneracy resolving procedures (e.g. see Wolfe (1963)) have been tried in SCHEDULE without success. An alternative is to solve the LP relaxation by a dual simplex approach to lessen the effect of degeneracy. In order to use a dual approach, the initial solution has to be transformed into a basic dual feasible solution, i.e. one in which all the reduced costs are non-favourable. Beale (1968) shows how this can be done. We then find the optimal solution to the LP relaxation using a dual steepest edge algorithm (see Forrest/Goldfarb (1992)) and use an improved *reduce* procedure and branch and bound to find a good schedule.

5.3.4 Improved REDUCE

The majority of the constraints in (4.8) are associated with pieces of work, each of which starts and ends at a relief opportunity. The vehicle block

where + denotes a relief opportunity and - denotes a vehicle movement, would generate four workpiece constraints. As any LP solution covers all the pieces of work, the first and last relief opportunity on any block must be used by at least one of the shifts in the LP solution. Thus any unused relief opportunity separates two pieces of work. Suppose the third relief opportunity in the above block is unused and that we eliminate it from the problem. The block would change to

and would generate only three workpiece constraints. Thus eliminating an unused relief opportunity allows one of the two workpiece constraints surrounding it to be removed. For technical reasons, it is not usually possible to remove one workpiece constraint for every unused relief opportunity and get a good starting solution to the reduced problem. Willers (1995) provides an algorithm for constructing an optimal LP basis to the reduced problem from the optimal LP basis to the original problem. This algorithm removed between 13% and 50% of the workpiece constraints over a set of twenty practical problems and also allowed the removal of additional shift variables beyond those removed by the earlier *reduce* procedure. The reduction in the number of constraints leads to reductions in the solution time of the branch and bound phase.

5.4 Other Improvements to SCHEDULE

Apart from the fact that the name of the mathematical programming component has been altered so that its function is more apparent to the users, the algorithms and code have been updated and the following two features have been added.

Since the earlier components of TRACS II have been modified to consider train crews as well as bus crews, SCHEDULE has also been altered to accept the train data generated. In particular it is now possible to define duty types rather than having specific names which differ depending on whether bus or train crews are being scheduled. Also it is necessary for SCHEDULE to handle four part shifts in train driver scheduling. In fact this has also been generalised so that shifts can be formed with any number of spells.

Since there are various different solution strategies which have been tested, those which have proved successful have been implemented in SCHEDULE. It is therefore possible for the user to select a solution strategy which would be more appropriate to a given problem. A default solution strategy can be adopted, along with various parameters which control the sensitivity of the algorithms.

6 Computational Results

One of the objectives in developing the SCHEDULE system was to allow several solution techniques to be available to the user. The viability of each method has been tested independently and it is infeasible to compare the quality and solution speed of each combination. Also, developments have taken place in parallel on different platforms and compared with versions in various stages of improvement. The justifications for including the dual approach and the column generation technique are given in the following two sections. Although these two lines of development have been pursued independently, there is some potential for integrating them which may lead to further improvement.

6.1 Dual Approach Results

Willers (1995) tested various implementations of the dual approach, including different methods of calculating the Sherali weights, different initial solution methods and different methods of determining an initial basic dual feasible solution. The best of these have been described above. On a sample of 20 bus crew scheduling problems, the dual approach reduced total solution time on a 33MHz 486 PC from 2525 minutes to 1243 minutes, an average reduction of 51% over the ZIP model.

6.2 Column Generation Results

Willers (1995) has shown that the Sherali weighted objective function gives improved results over the objectives used in the older version. The column generation implementation also uses the Sherali weighted objective function and was therefore tested against a version of ZIP using that objective in order to gauge any potential improvement made by this technique. Various improvements were made to both systems to allow the most accurate comparison.

For seven problems, two sizes of data set were used in order to test any improvements in solution or speed of solution for a problem which consisted of a larger set of previously generated shifts. The smaller set is that which would normally be run through the mathematical programming component of TRACS II. As the non column generation produces an LP solution, using the column generation method on the same data set will not improve the cost, as both will be optimal.

The following table compares the results obtained using the modified ZIP on a heuristically reduced shift set against those of a column generation system with a larger set of previously generated shifts. Timings are reported for a Silicon Graphics Iris Indigo Workstation with 33MHz R3000 MIPS Processor and are in minutes:

Table 1: Comparison of solutions and timings

Data Set	Modified ZIP			Column Generation		
	Shifts	Run Time		Shifts	Run Time	
		LP	Total		LP	Total
AUC	87	13	36	87	21	91
CTJ	88	10	16	87	9	19
CTR	–	–	–	88	18	54
GMB	34	0.6	0.9	33	0.7	0.9
RI2	45	0.9	2.2	44	1.0	1.2
STK	61	5.2	16	59	21	24
SYD	56	5.4	5.8	56	15	18

It can be seen from Table 1 that the number of shifts required was reduced in 5 out of the 7 sets. This includes the one problem where no solution could previously be found. It is difficult to compare timings through a branch and bound tree because the route through it will be different. However, it is useful to note that where there is an increase in the overall execution times, the times themselves would still be acceptable to users. Where the same size of data set were compared, results showed an average reduction in execution time of 41% using column generation (see Fores (1996)).

7 Conclusion

The original TRACS II system consistently achieves better results than those obtained by conventional methods and users are happy with the quality of the solution obtained. Improvements to the system, including the mathematical component, will therefore benefit the user in terms of ease of use and of the potential improvements in solutions. The expansion to a more generalised system allows different types and sizes of problems to be addressed. Also, the improvements in algorithms and the introduction of the Sherali weighted objective function and the column generation technique allow better solutions to be found more quickly. A parameter driven approach gives the user more flexibility in solving problems using different solution techniques. Since the original draft of this paper, the column generation approach has been used successfully in many scheduling exercises for clients. This has given us the confidence to tackle the solution of considerably larger problems than previously.

References

Beale, E.M.L. (1968): Mathematical Programming in Practice. (Pitman) London.

Daduna, J.R./Branco, I./Paixáo, J.M.P. (eds), (1995): Proceedings of the Sixth International Workshop on Computer-Aided Scheduling of Public Transport, Computer-Aided Transit Scheduling. (Springer-Verlag) Berlin, Heidelberg, New York.

Desrochers, M./Gilbert, J./Sauvé, M./Soumis, F. (1992): CREW-OPT: subproblem modelling in a column generation approach to urban crew scheduling. in: Desrochers, M./Rousseau, J.-M. (eds.): Computer-Aided Transit Scheduling. (Springer-Verlag) Berlin, Heidelberg, New York.

Desrochers, M./Rousseau, J.-M. (eds), (1992): Proceedings of the Fifth International Workshop on Computer-Aided Scheduling of Public Transport, Computer-Aided Transit Scheduling. (Springer-Verlag) Berlin, Heidelberg, New York.

Fores, S. (1996): Column Generation Approaches to Bus Driver Scheduling. PhD Thesis, University of Leeds.

Forrest, J.J./Goldfarb, D. (1992): Steepest edge simplex algorithms for linear programming. in: Mathematical Programming 57, 341 - 374.

Goldfarb, D./Reid, J.K. (1977): A practicable steepest-edge simplex algorithm. in: Mathematical Programming 12, 774 - 787.

Kwan, A.S.K./Kwan, R.S.K./Parker, M.E./Wren, A. (1996): Producing train driver shifts by computer. in: Allan, J./Brebbia, C.A./Hill, R.J./Sciutto, G./Sone, S. (eds.): Computers in Railways V, Vol. 1: Railway Systems and Management, (Computational Mechanics Publications) Southampton, Boston.

Marsten, R.E. (1974): An algorithm for large set partitioning problems. in: Management Science 20, 774 - 787.

Rousseau, J.-M. (1995): Results obtained with Crew-Opt, a column generation method for transit crew scheduling. in: Daduna, J.R./Branco, I./Paixáo, J.M.P. (eds.): Computer-Aided Transit Scheduling. (Springer-Verlag) Berlin, Heidelberg, New York.

Rousseau, J.-M./Blais, J.-Y. (1985): HASTUS: an interactive system for buses and crew scheduling. in: Rousseau, J.-M. (ed.): Computer Scheduling of Public Transport 2. (North-Holland) Amsterdam, New York, Oxford.

Sherali, H.D. (1982): Equivalent weights for lexicographic multi-objective programs: characterizations and computations. in: European Journal of Operations Research 18, 57 - 61.

Smith, B.M./Wren, A. (1988): A bus crew scheduling system using a set covering formulation. in: Transportation Research 22A 97 - 108.

Willers, W.P. (1995): Improved Algorithms for Bus Crew Scheduling, PhD Thesis, University of Leeds.

Willers, W.P./Proll, L.G./Wren, A. (1993): A dual strategy for solving the linear programming relaxation of a driver scheduling system. in: Annals of Operations Research 58, 519 - 531.

Wolfe, P. (1963): A technique for resolving degeneracy in linear programming. in: SIAM Journal 11, 205 - 211.

Wren, A./Kwan, R.S.K./Parker, M.E. (1994): Scheduling of rail driver duties. in: Murthy, T.K.S./Mellitt, B./Brebbia, C.A./Sciutto, G./Sone, S. (eds.): Computers in Railways IV, Vol. 2: Railway Operations. (Computational Mechanics Publications) Southampton, Boston.

Wren, A./Smith, B.M. (1988): Experiences with a crew scheduling system based on set covering. in: Daduna, J.R./Wren, A. (eds.): Computer-Aided Transit Scheduling. (Springer-Verlag) Berlin, Heidelberg, New York.

An Exact Branch and Cut Algorithm for the Vehicle and Crew Scheduling Problem

Christian Friberg[1] and Knut Haase[2]
[1]Institut für Informatik und Praktische Mathematik, Christian-Albrechts-Universität zu Kiel, Preußerstraße 1-9, 24105 Kiel, Germany
[2]Lehrstuhl für Produktion und Logistik, Institut für Betriebswirtschaftslehre, Christian–Albrechts–Universität zu Kiel, Olshausenstraße 40, 24118 Kiel, Germany

Abstract: We present a new model for the vehicle and crew scheduling problem in urban public transport systems by combining models for vehicle and crew scheduling that cover a great variety of real world aspects, including constraints for crews resulting from wage agreements and company regulations. The main part of the model consists of a set partitioning formulation to cover each trip. A column generation algorithm is implemented to calculate the continuous relaxation which is embedded in a branch and bound approach to generate an exact solution for the problem. To improve the lower bounds, polyhedral cuts basing on clique detection and a variant of the column generation algorithm that suits the cuts were tested.

1 Introduction

A central problem in urban public transport is the scheduling of vehicles and crews to serve trips defined by a timetable at minimum cost.

In this paper we introduce a formulation for the vehicle and crew scheduling problem (VCSP) that combines the approaches of Desrochers/Soumis (1989) and Ribeiro/Soumis (1994). That is the VCSP is formulated as a set partitioning problem where the columns represent different workdays of crews or vehicles with additional constraints to combine both schedules. For the optimal solution we present a branch and bound algorithm in which column generation is applied to solve the continuous relaxation in the nodes of a branch and bound scheme. To improve the lower bounds, an approach basing on polyhedral cuts for the set partitioning problem as described by Hoffman/Padberg (1993) is taken. As these cuts do not easily match with the column generation algorithm, an additional branch and bound algorithm to generate the new pivot column is presented. Several heuristics to schedule vehicles and crews simultaneously have been proposed (see, for example, Ball/Bodin/Dial (1983), Tosini/Vercellis (1988), Patrikalakis/Xerocostas

(1992), and Falkner/Ryan (1992)). Freling/Boender/Paixão (1995) gives the only mathematical formulation for exact solutions of the VCSP we are aware of. The proposed model uses a quasi-assignment problem to describe the vehicle scheduling and a set partitioning formulation for the crew scheduling. To calculate a lower bound two algorithms are presented. One approach uses a two step column generation method to solve the crew scheduling part while the other one is based on a Lagrangian relaxation. Both are tested with an example composed of 7 trips without embedding it in a branch and bound scheme. In Freling (1997) the Lagrangian heuristic is applied to derive upper and lower bounds on a Pentium 90MHz PC with 32MB memory for randomly generated instances with up to 120 trips and some practical instances with up to 171 trips. The solution gap for the random instances ranges up to 2.9% and for the practical instances up to 30.43%.

The remainder of the paper is organized as follows: We formalize the VCSP in terms of a set partitioning problem in Sect. 2. Sect. 3 presents an overview of the optimal solution algorithm. In Sect. 4 the solution of the continuous relaxation of the VCSP by a column generation algorithm with suitable subproblems is shown. An approach for combining column generation and polyhedral cuts is given in Sect. 5. Sect. 6 provides computational results. Finally, a summary is given in Sect. 7.

2 The Vehicle and Crew Scheduling Problem

In this section we present the set partitioning model of the VCSP. First the binary linear program is given, using sets of blocks and workdays as parameters. Then we show how these sets can be generated using vehicle and crew scheduling graphs.

2.1 The Model

As the first step we will define the formulation of the optimization problem to get a minimum cost vehicle and crew schedule.

A vehicle and crew schedule has to ensure that every *trip* defined in the timetable is served by a bus and a driver. One vehicle can serve a number of trips per day. A collection of such trips is called a *block*. Therefore the number of blocks is equal to the number of vehicles. The driver of a bus may leave it during a trip at a *relief point*, for example to stop work and deliver the bus to another driver. Thus the trips are divided into parts — called *dtrips* — which can be served by different crews. The collection of dtrips served by one crew per day is called a *workday*.

The covering of trips and dtrips is not sufficient for a feasible schedule. Every bus has to be driven to the beginning and from the end of each trip by a driver. This leads to a conjoint covering of pull in and pull out trips and deadheading trips — called *links*. A link is covered by a vehicle *if and only if* it is covered by a crew.

For the moment let us assume, that all feasible blocks and workdays are collected in the sets B and W, respectively.

Defining the parameters

$vtrip_{ik}$ $= 1$, if block b_i serves trip k, 0 otherwise

$cdtrip_{jk}$ $= 1$, if workday w_j serves dtrip k, 0 otherwise

$varc_{ik}$ $= 1$, if block b_i serves link k, 0 otherwise

$carc_{jk}$ $= 1$, if workday w_j serves link k, 0 otherwise

$c(b_i)$ the cost of block b_i

$c(w_j)$ the cost of workday w_j

and the binary variables

$x_i = 1$, if block b_i is scheduled, 0 otherwise

$y_j = 1$, if workday w_j is scheduled, 0 otherwise

the VCSP can be formalized as follows:

Definition 2.1 (vehicle and crew scheduling problem) Let B and W be the sets of all feasible blocks and workdays, respectively.

$$\text{Minimize} \sum_{b_i \in B} c(b_i)x_i + \sum_{w_j \in W} c(w_j)y_j \tag{1}$$

$$\sum_{b_i \in B} vtrip_{ik}x_i = 1 \quad \forall \text{ trips } k \tag{2}$$

$$\sum_{w_j \in W} cdtrip_{jk}y_j = 1 \quad \forall \text{ dtrips } k \tag{3}$$

$$\sum_{b_i \in B} varc_{ik}x_i - \sum_{w_j \in W} carc_{jk}y_j = 0 \quad \forall \text{ links } k \tag{4}$$

$$x_i, y_j \in \{0,1\} \quad \forall\, b_i \in B, w_j \in W \tag{5}$$

The objective function (1) minimizes the total cost for vehicle and crew schedules. The constraints (2) and (3) ensure that all trips and dtrips are covered, while restriction (4) links blocks and workdays together so that every vehicle is driven by exactly one driver. The number of blocks and workdays are not fixed and not restricted.

Blocks and workdays can be defined as paths through graphs. In the following subsections we will present the exact definition of the vehicle and crew scheduling graphs and their relationship.

2.2 The Vehicle Scheduling Graph

Assuming a homogeneous fleet of vehicles at a single depot with no extra constraints on their use (e.g. refueling) we can easily define a network for the vehicle scheduling. The nodes are the beginnings and ends of trips, arrivals or departures at the depot and the start and end of the day. The arcs are annotated with the costs to serve them. These costs include driving and sign on/sign off costs and therefore cover both variable and fixed costs of a bus.

Definition 2.2 (vehicle scheduling graph) Let $VG = (VN, VA)$ be a graph with nodes

$$VN = \{\text{start, end}\} \cup \{\text{bt}_i, \text{et}_i, \text{bdepot}_i, \text{edepot}_i \mid \forall\ \text{trip}_i\}$$

where start and end are the source and the sink of the network, and bt_i and et_i are the beginning and end of trip i, respectively. The nodes bdepot_i and edepot_i denote the departure from and the arrival at the depot, respectively.

The set of arcs VA includes the start of the day (start,bdepot_i); the end of the day (edepot_i,end); the trips (bt_i,et_i); pull out trips from the depot (bdepot_i,bt_i) and pull in trips to the depot (et_i,edepot_i); deadheading trips (et_i,bt_j), if the time difference between the trips $teb_{ij} \in [0, vwmax]$; and breaks at the depot (edepot_i,bdepot_j), if the time needed to go from trip i to the depot and back to trip j $(ted_i + tdb_j) \in [0, vwmax]$.

For every arc a its cost will be denoted $c(a)$. □

The depot nodes corresponding to the beginning and the end of each trip are used to model a vehicle starting from or coming back to its depot at a certain point of time. The importance of this approach will be seen in the combination with the crew scheduling.

In the graph in Fig. 1, a vehicle can start the day with trip 1, return to the depot, and then serve trip 3 and finally return again to the depot to end the day. Other possibilities are to go from the end of trip 1 **directly** to the beginning of trip 2 or 3. After serving trip 2 the vehicle can go directly to trip 4 or end the day by driving back to the depot. Note that there is no arc from the end of trip 1 to the beginning of trip 4 even though both can be served by one vehicle if trip 2 is taken between them. This can be explained by the waiting time $vwmax$ between those trips.

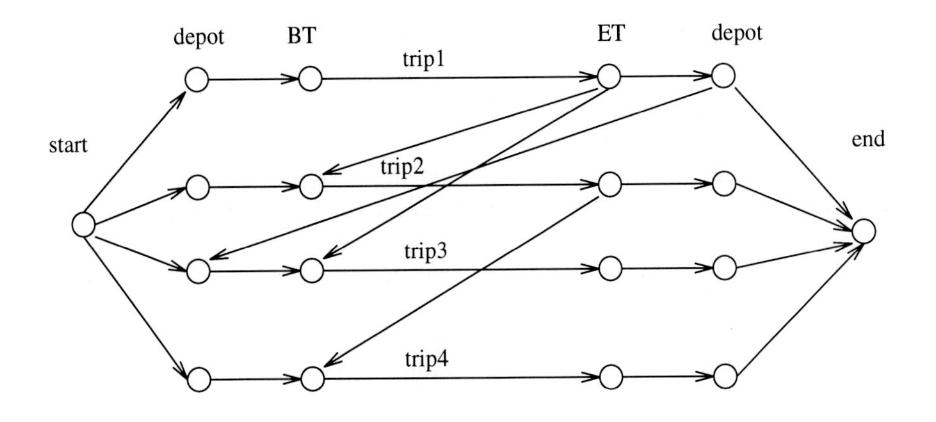

Figure 1: Example of a vehicle scheduling graph

For the application of the column generation algorithm in Sect. 4 we also need to define a *minimum cost block.*

Definition 2.3 (block) A *block* b in the vehicle scheduling graph VG is defined as a finite sequence $node_0, \ldots, node_{r_b}$ of nodes $\in VN$, where $node_0 =$ start, $node_{r_b} =$ end and $(node_i, node_{i+1}) \in VA$ for all i. Its *cost* is defined as

$$c(b) = \sum_{i=0}^{r_b-1} c((node_i, node_{i+1})).$$

A *minimum cost block* is a block b such that $c(b) \leq c(b')$ holds for every other block b'. □

2.3 The Crew Scheduling Graph

The crew scheduling graph can be described similarly to the vehicle scheduling graph (see, for example Desrochers/Soumis (1989) and Desrochers/Gilbert/Sauvé/Soumis (1992)). The only extension necessary is the consideration of the relief points where the crew can change. At this point the crew can stay with the same vehicle, switch to another vehicle, return to the depot, or end the workday. A graph containing such information will be called a *simple crew scheduling graph.*

Definition 2.4 (simple crew scheduling graph) Let $SCG = (CN, SCA)$ be a graph with nodes

$$CN = \{\text{start, end}\} \cup$$

$$\{\text{rp}_{ij}, \text{bt}_i, \text{et}_i, \text{bdepot}_i, \text{edepot}_i, \text{bdepot}_{ij}, \text{edepot}_{ij} \mid \forall\ \text{trip}_i,\ j = 0, \ldots, n_i\}$$

where start and end are the source and the sink of the network, respectively. The different types of depot nodes are for the start and end of trips (single index) and travel to/from relief points (double index).

SCA includes the following arcs: the start of the day on foot (start,bdepot_{ij}), and by vehicle (start,bdepot_i); similar but for end of the day (edepot_{ij},end) and (edepot_i,end); travel on foot from the depot to a relief point (bdepot_{ij}, rp_{ij}) and vice versa (rp_{ij},edepot_{ij}); pull out trips from the depot (bdepot_i,bt_i) and pull in trips to the depot (et_i,edepot_i); dummy arcs to distinguish between driving and on foot (bt_i,rp_{i0}) and (rp_{in_i},et_i) with cost 0; the dtrips (rp_{ij-1},rp_{ij}); travel on foot between relief points (rp_{ij},rp_{kl}), if $trp_{ijkl} \in [0, cwmax]$; deadheading trips ($\text{et}_i$,$\text{bt}_j$), if $teb_{ij} \in [0, cwmax]$; and breaks (waiting time) at the depot (edepot_{ij},bdepot_{kl}), if $(tdrp_{ij} + trpd_{kl}) \in [0, cwmax]$, and ($\text{edepot}_i$,$\text{bdepot}_j$) with cost $cwcd_{ij}$ if $(tdb_j + ted_i)$ is in $[0, cwmax]$.

For arc a its cost is $c(a)$. □

The arcs associated with links, i.e. the pull out and pull in trips and the deadheading trips for combining vehicle and crew scheduling are called *linking arcs*. Definition 2.1 requires that these arcs are taken if and only if the same stage is served by a vehicle, i.e. a crew has to **transfer** this vehicle.

Even in a simple example we can see how many arcs are required for *SCG*.

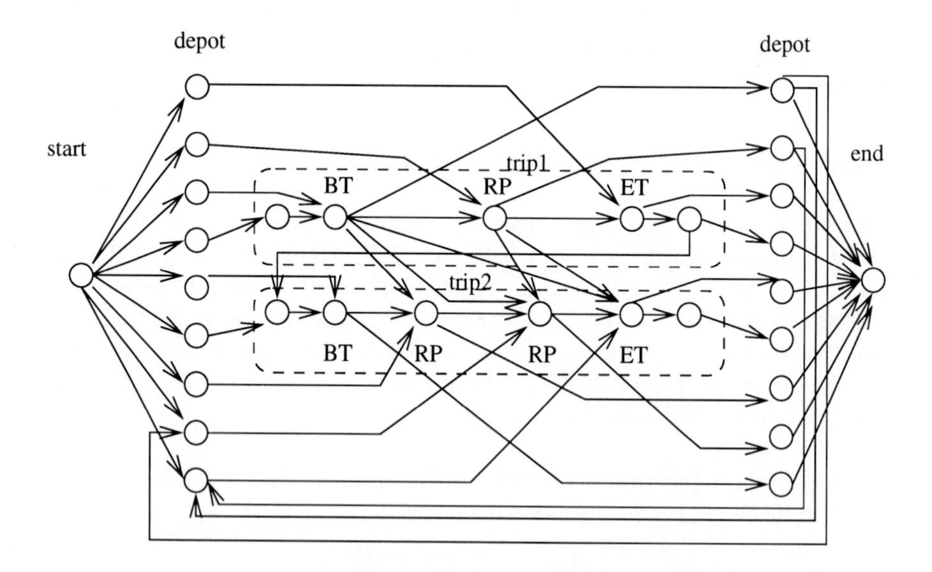

Figure 2: A simple crew scheduling graph

In Fig. 2 two trips (dotted boxes) with one or two additional relief points, respectively, are connected with the arcs allowed. Every relief point (i.e. BT, RP and ET) is associated with two depot nodes that model travel to the depot without a vehicle. The beginning and the end of each trip is connected with a second depot node to describe the stage to the depot with a vehicle. Note that a driver is allowed to take a bus from the end of trip 1 to the beginning of trip 2 but may not go there **alone**, even if an arbitrary relief point is chosen. This can be used to model time differences between driving the vehicle or walking from one relief point to another.

In crew scheduling we have to consider restrictions on the length of a workday, the length of work without a break, and other regulations. Following Desrochers/Soumis (1989), there are two ways to model these restrictions in a graph: with path feasibility constraints and in the definition of the arcs.

The first idea used here is the introduction of a *piece of work*. A piece of work is a sequence of dtrips and links (further called *task*) operated by a driver on the **same** bus. So every block can be split into several pieces of work. The wage agreements usually restrict the number of pieces of work in a workday and the number of tasks in a piece, so it is useful to form the graph mainly with such arcs.

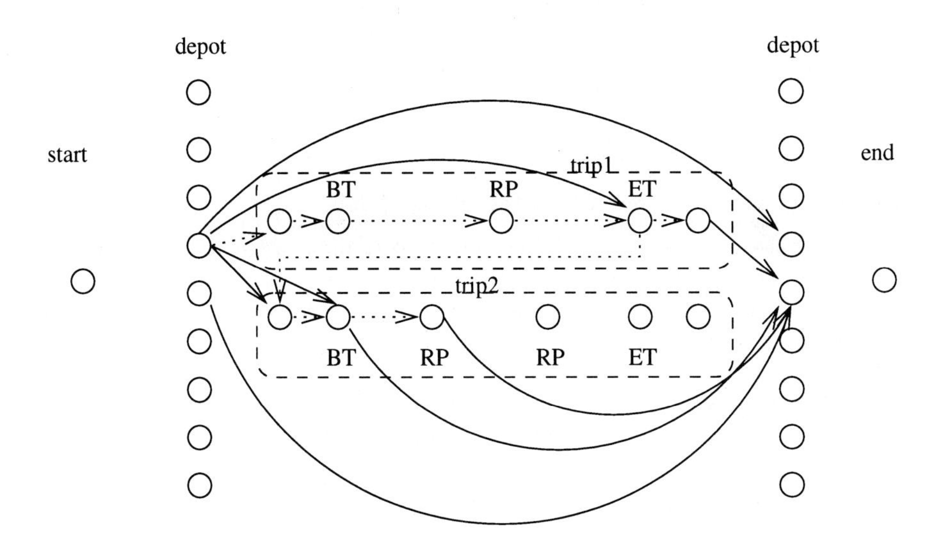

Figure 3: Pieces of work

In Fig. 3 we see an example of pieces of work derived from Fig. 2, where one piece may consist of 3 to 5 tasks. All pieces that arrive at nodes where no arc departs (and vice versa) can be eliminated. Note that for example the arc from the beginning depot of trip 1 to the first relief point of trip 2 symbolizes a piece of work that covers the complete first trip, the stage to, and the first dtrip of trip 2, all served by one vehicle. The corresponding arcs of the simple scheduling graph are represented in a dotted version. In practical situations also the amount of time available for some piece of work will be restricted, so even more arcs can be eliminated.

The remaining arcs are used to model sign on, sign off (connected with start and end activities, respectively), and breaks. In practice, most of these arcs can be eliminated because of restrictions such as the length of breaks.

Global constraints such as the total working time or the number of pieces of work on a day can be modeled as resource constraints at each node and resource consumption on each arc. Resource constraints are formulated as resource constraint windows that restrict consumption on the arcs. For an example see Desrochers/Gilbert/Sauvé/Soumis (1992).

We can now formulate the crew scheduling graph that will be used in the remainder of the work.

Definition 2.5 (crew scheduling graph) Let $CG = (CN, CA)$ be a graph. CG is called a *crew scheduling graph* if there exists a simple crew scheduling graph $SCG = (CN, SCA)$ with a mapping Φ from CA to the set of **sequences** of SCA, such that for every arc a in CA the sequence $\Phi(a)$ is a **path** in SCG.
The costs of each arc a of CA are determined by

$$c(a) = \sum_{(node_i, node_{i+1}) \in \Phi(a)} c((node_i, node_{i+1}))$$

as the sum of the covered arcs of the simple graph.

CG is called a *resource restricted crew scheduling graph* if there exists a number $r_{CG} \geq 1$ of resources so that for every node we have the resource constraint windows $[a^l_{node}, b^l_{node}]$ and for every arc k we have resource consumption d^l_k with $1 \leq l \leq r_{CG}$. □

Further we will use only resource restricted crew scheduling graphs and so will call them crew scheduling graphs to simplify the notation.

For the application of the column generation algorithm in Sect. 4 we additionally need to define a *minimum cost workday*.

Definition 2.6 (workday) A *workday* w in the crew scheduling graph CG is defined as a finite sequence $node_0, \ldots, node_{s_w}$ of nodes $\in CN$ where $node_0$ = start, $node_{s_w}$ = end and $(node_i, node_{i+1}) \in CA$ for all i.
A workday is *feasible* if for all $t \in \{0, \ldots, s_w\}$ and all $l \in \{1, \ldots, r_{CG}\}$

$$\sum_{i=1}^{t} d^l_{(node_{i-1}, node_i)} \in [a^l_{node_i}, b^l_{node_i}]$$

holds.

Its *cost* is defined as

$$c(w) \;=\; \sum_{i=0}^{r_w-1} c((node_i, node_{i+1})).$$

A *minimum cost workday* is a workday w such that $c(w) \leq c(w')$ holds for every other feasible workday w'. □

3 The Branch and Bound Algorithm

In this section we introduce an approach that solves the VCSP with a branch and bound algorithm. Lower bounds are calculated by dropping the integral constraints of the set partitioning formulation and solving the continuous linear program with column generation. The application of this special approach is explained in detail in the following sections, so we will now concentrate on the branching. For reasons of simplicity, we will show the concept for vehicles. The branching for crews requires only slight and obvious modifications.

Definition 3.1 (branching in the VCSP) Let S be the set of feasible integral solutions of the VCSP as in Definition 2.1. Define for every index set I of columns the set

$$S^I \;=\; \{x \in S \mid i \in I \;\Rightarrow\; x_i = 0\}$$

Obviously we have $S^I \subseteq S$ for all I and $S^\emptyset = S$. A solution for the continuous relaxation can be found easily by solving the relaxation of S with the simplex algorithm without considering the columns in I.

Now let x be an optimal feasible solution for the relaxation of S^I and P_j a column representing a block or a workday. Then calculate for every linking arc $link_k$ its violation

$$v_k \;=\; 0.5 - \mid 0.5 - \sum_{j=1,\; P_j\; block}^{n} varc_{jk} x_j \mid$$

and detect the maximum violation $v_{max} \;=\; \max v_k$ with a corresponding linking arc $link_{k_{max}}$. If $v_{max} \neq 0$, then x is not integral, i.e. the problem is not leveled for this branch, and there is at least one column P_j, $j \notin I$ and $x_j \neq 0$ that covers $link_{k_{max}}$. If $v_{max} = 0$, then x may still be fractional and one has to calculate similarly the violation v_k^b for all break arcs $break_k$ and their maximum violation v^b_{max} with the corresponding break arc $break_{k_{max}}$ to find the column P_j.

Let I' be $I \cup \{j \notin I \mid P_j \text{ covers } link_{k_{max}}\}$ and I'' be $I \cup \{j \notin I \mid P_j \text{ covers link } l, l \neq k_{max}, link_l \in Ray(link_{k_{max}})\}$, where $Ray(a)$ is the set of arcs that have the same starting point as the arc a. Define

$$\rho(S^I) = \{S^{I'}, S^{I''}\}$$

as the (binary) branching function. □

It can be shown, that if $v_{max} = 0$ and $v^b_{max} = 0$, the solution x is integral in every component x_j, with column P_j representing a **block** (see Friberg/Haase (1997)).

To get an integral solution for all columns including the workdays, the violation of links associated with **travel on foot** have to be calculated. The branching mechanism follows the same lines as for vehicles. Afterwards, a branching function ρ is defined that obviously defines a finite tree whose leaves are the integral solutions of S.

So far we have described how columns that have already been generated can be eliminated. Now we have to consider the explicit columns that are coded into the feasible paths of the scheduling graphs. As the index set I is derived from sets of linking arcs, this can be done by removing arcs from the graph. In the implementation the arcs are simply marked to reuse them in other branches.

4 The Column Generation Approach

The complexity of the linear relaxation of the VCSP is determined by the number of rows of the constraint matrix A which is equivalent to the number of **feasible paths**, increasing exponentially with the number of arcs. Moreover, if the number of columns is too big (more than 5 million columns for 30 trips!) they can not be stored in main memory which affects the computational performance. So a column generation algorithm (see Lasdon (1970)) is used to calculate the actual pivot column instead of enumerating all columns in advance. Therefore the simplex multipliers of the optimal solution of the restricted linear program are subtracted from the costs of the corresponding trips, dtrips, and links. The pivot column of the complete matrix is then equivalent to a **shortest path** in one of the scheduling graphs.

In Definition 2.1 the constraint matrix was defined such that every column corresponds to a feasible path of the vehicle scheduling graph, i.e. a block, or of the crew scheduling graph, i.e. a workday. This can be formalized by introducing ϕ_v and ϕ_c that map the set of blocks B to the set of vehicle columns $VC = \{P_1, \dots, P_{n_b}\}$ and the set of workdays W to the set of crew columns $CC = \{P_{n_b+1}, \dots, P_{n_b+n_w}\}$, respectively, where n_b and n_w are the number of blocks and workdays, respectively. Define further the number of trips n_t, dtrips n_d, and linking arcs n_l. The vehicle mapping is defined as

$$\phi_v : B \mapsto VC,$$

$$\phi_v(b) = P_j : a_{ij} = \begin{cases} 1 & \text{if } 1 \leq i \leq n_t \text{ and trip}_i \in b \\ & \text{or } n_t + n_d < i \leq n_t + n_d + n_l \\ & \text{and } link_i \in b \\ 0 & \text{otherwise} \end{cases}$$

while the crew mapping is

$$\phi_c : W \mapsto CC,$$

$$\phi_c(b) = P_j : a_{ij} = \begin{cases} 1 & \text{if } n_t < i \leq n_t + n_d \text{ and} \\ & \text{dtrip}_i \in b \\ -1 & \text{if } n_t + n_d < i \leq n_t + n_d + n_l \\ & \text{and } link_i \in b \\ 0 & \text{otherwise} \end{cases}$$

where only the entry for the linking arcs is different from ϕ_v to couple vehicles and crews.

The column generation algorithm now proceeds by first solving a smaller linear program with a submatrix A' of A, using a simplex solver, and then generating a column P_j not in A' with

$$\bar{c}_i = \min_k \bar{c}_k = \min_k (c_k - \pi P_k)$$

This generation is done by solving a *subproblem.* As every column corresponds to a feasible path of a scheduling graph, the generation process is a **shortest path problem** where the simplex multipliers π_i are added to the costs of the arcs.

Definition 4.1 (marginal cost graphs) Let $VG = (VN, VA, c_v)$ be a vehicle scheduling graph and $CG = (CN, CA, c_c)$ be a crew scheduling graph with arc costs c_v and c_c as described in Sect. 2. Further let $\pi_1, \dots, \pi_m$ be simplex multipliers of the linear program.
Now define the marginal cost graphs $VG' = (VN, VA, c'_v)$ with

$$c'_v(a) = \begin{cases} c_v(a) - \pi_i & \text{if } a = trip_i \\ c_v(a) - \pi_{i+n_t+n_d} & \text{if } a = link_i \\ c_v(a) & \text{otherwise} \end{cases}$$

and $CG' = (CN, CA, c'_c)$ with the costs

$$c'_c(a) = c_c(a) - \sum_{dtrip_i \in \Phi(a)} \pi_{i+n_t} + \sum_{link_i \in \Phi(a)} \pi_{i+n_t+n_d}$$

calculated with the **simple** crew scheduling graph (see Definition 2.5 in Subsection 2.3). □

With these definitions we can now formulate the column generation algorithm.

Definition 4.2 (column generation) Let A' be a **nonsingular** submatrix of A and $P(A')$ be the linear program for the VCSP restricted to A' and the corresponding subvector c' of c.

1. Solve the linear program $P(A')$
2. Calculate the marginal cost graphs VG' and CG' as in Definition 4.1. Let b be a shortest feasible path (minimum cost block) in VG' and w be a shortest feasible path (minimum cost workday) in CG'.
3. If both $c'_v(b)$ and $c'_c(w)$ are nonnegative, stop. The current solution is optimal for $P(A)$.
4. If $c'_v(b) < 0$, redefine A' as

   $$A' \leftarrow (A', \phi_v(b))$$

   and c' as $(c', c_v(b))$. If $c'_c(w) < 0$, redefine A' as

   $$A' \leftarrow (A', \phi_c(w))$$

   and c' as $(c', c_c(w))$.
5. Continue with Step 1.

In Step 4 the matrix A' can be expanded by more than one column to accelerate the algorithm (multiple pricing). □

Following Desrochers (1988) we assume that all possible resource consumptions are **discrete**, for example minutes of working time or number of breaks. A shortest path can then be calculated as for the vehicle scheduling graph but for every **resource combination** instead of just every node which accelerates the computation.

5 Polyhedral Cuts and Column Generation

The linear relaxation of the set partitioning problem already provides very good lower bounds for the branch and bound algorithm. Nevertheless, methods that use polyhedral cuts produce lower bounds that are higher, thereby allowing more branches to be cut in the search tree. In this section we will describe an approach to combine polyhedral cut generation and column generation.

For the detection of clique cuts Hoffman/Padberg (1993) present several algorithms. The one which we have implemented proceeds by concentrating on a subset of the column indices $F = \{j \in \{1, \ldots, n\} \mid 0 < x_j < 1\}$ that correspond to the fractionally valued variables of the current solution vector x.

Now one possibility to combine the polyhedral cut approach with our column generation approach is to ignore the information provided by the simplex multipliers of the new rows (clique cuts) and generate new columns with the set partitioning and linking constraints. Obviously, for all implicit columns not being generated by the algorithm, their entry in all polyhedral cuts is 0. So if the simplex algorithm terminates no column with an entry other than 0 in a cutting row is the pivot column.

The main idea is to generate a shortest block or workday that is **not represented by a column already existing** and take it as a candidate for the pivot column. This approach leads to an exact algorithm that combines both polyhedral cuts and column generation. This is achieved by a branch and bound scheme similar to a kth shortest path problem as in Lawler (1972). It calculates the current shortest path and if the corresponding column already exists, the branch step is performed as follows:

Definition 5.1 (branching on paths) Let $p = (n_1, \ldots, n_r)$ be the actual shortest path. Let

$$M = (m_1, \ldots, m_s), 1 \leq m_1 < m_s \leq r$$

be the set of indices of nodes in p that have more than one successor (usually the end of a trip with a number of possible links or a relief point with several possibilities to continue on foot).
If $M = \emptyset$ no branching is possible and hence no feasible shortest path exists and the current branching node is leveled. Otherwise, s subproblems are generated (instead of two in the binary global branch and bound algorithm) as in Lawler (1972). Subproblem i $(1 \leq i \leq s)$ is defined by:

1. forbid arc (n_{m_i}, n_{m_i+1})
2. for all $j < i$ force arc (n_{m_j}, n_{m_j+1}) by forbidding all other arcs that start on node n_{m_j}.

The algorithm is extended by taking the value of the current shortest path as a lower bound for the desired path not coded in the LP matrix. □

To detect if the current shortest path corresponds to an existing column, all paths are coded into a set of binary variables each denoting a trip or a link, respectively. This can be easily incorporated into the shortest path and column generation algorithms. The sets are hashed by adding the numbers of the arc and let the final sum modulo 512 be the index for the hash queue. So a test of the actual path results in just a few operations in the respective hash queue and therefore is very cheap.

In summary, we obtain a branch and bound algorithm, that enumerates *all* possible paths in the branching tree, and additionally checks in the bounding part, if the actual path already exist in the LP using a hash queue for the search, before the lower bound is tested. Doing this, the algorithm clearly finds the optimal solution. Unfortunately, the branching algorithm – which, as should be emphasized here, is used to solve the *subproblem* of the VCSP – has exponential complexity and so shows rather poor performance for large examples.

6 Computational Analysis

The algorithms – including the revised simplex algorithm – have been coded in C and implemented on a SUN Sparc-Station 10/40.

To analyze the performance of the algorithms we have derived some preliminary computational results. All considered instances have two resources, the working time and the number of breaks. They are generated randomly and are varied by the main characteristics as shown in Table 1. (Note, for the 7-th example we have defined $c(b_i) = c(w_j) = 1$ for all blocks b_i and workdays w_j.)

Table 2 provides the compution times for solving the first linear program to optimality (second column), to derive the first integer solution (fourth column), and to compute the optimal solution (sixth column). Furthermore, the percentage, compared with the optimal objective function value, of the objective function values of the first optimal linear programming solution (third column) and of the first integer solution (fifth column) are given. All instances are solved without applying polyhedral cuts. Only the third instance is also solved by applying polyhedral cuts based on clique detection (last column in Table 2).

Table 3 shows some additional results: the percentage of the overall time spent for the simplex algorithm in comparison with the final integer solution

Table 1: Characteristics of the Instances

No.	trips	dtrips	links	breaks	on foot	g[1]	minimize
1	10	16	38	51	95	1	cost
2	10	16	38	51	95	5	cost
3	10	16	38	51	47	1	cost
4	20	34	92	300	229	1	cost
5	20	34	92	300	229	5	cost
6	20	34	92	190	112	1	cost
7	20	34	92	300	229	1	busses + crews
8	20	40	86	352	243	1	cost
9	30	54	142	616	374	1	cost

[1] granuality of resource windows in minutes

Table 2: Results

No.	first LP		first integer		optimal integer	cuts
	time	%	time	%		
1	38	100.0	38	100.0	38	no
2	26	100.0	26	100.0	26	no
3	9	98.1	14	100.0	64	no
4	1433	98.0	2564	100.0	5693	no
5	844	98.8	1712	100.3	16541	no
6	1198	-	3784	-	-	no
7	1335	-	5246	-	-	no
8	2555	100.0	4001	100	5066	no
9	21112	-	21112	-	-	no
3	21	98.6	45	100.6	134	yes

Table 3: Additional Results

No.	simplex	iterations	crew subp.	cols.	rows	depth
1	38%	883	60%	321	64	0
2	27%	660	72%	289	63	0
3	68%	3278	20%	457	60	10
4	82%	36187	15%	2268	145	3
5	90%	142412	8%	3790	135	9
6	-	-	-	-	-	> 40
7	-	-	-	-	-	> 40
3	82.9%	4326	11%	443	98	3

(second column), the number of iterations of the revised simplex algorithm (third column), the percentage of the overall time spent for the generation of shortest paths in the crew scheduling graph in comparison with the final integer solution (fourth column), the number of columns at the end of the calculation (fifth column), number of rows at the end of the calculation (sixth column), and the length of a maximum path from the root to a leave in the branch and bound tree (last column).

As the results in Table 2 and Table 3 show, the limit of the current algorithm turns out to be around 20 trips. If an optimal solution is found, the lower bound by the first continuous solution is excellent (over 98 per cent of the optimal solution) and the first integer solution comes close to or even is the final solution.

To calculate the first continuous solution the time used by the simplex algorithm is significantly shorter than the shortest path time. If some branch steps are needed, this ratio changes to the disadvantage of the simplex algorithm because all generated columns are reused and so only a few new columns have to be calculated on deeper nodes of the branch and bound tree. As all problems with depth greater than 0 show, this approach leads to a much faster generation of the lower bounds on the nodes as if all columns are erased and must be calculated again. Nevertheless, the algorithm to solve the linear programs is just a simple revised simplex, and for the larger problems 80 to 90 per cent of the time are consumed by it, a faster algorithm could improve the implementation greatly.

The simultaneous application of polyhedral cuts and column generation reduces the depth of the branch and bound tree from 10 to 3 for the small instance $\mathrm{p1}_{rw}$ (Table 3, Rows 3 and 8). At first sight this is a very promising result. However, for larger instances the computation time for the generation of shortest paths in the crew scheduling graph increases drastically, and no solution was obtained in reasonable time.

Another problem that may arise if the crews are scheduled first seems to be that the branching rule is not as appropriate for the CSP as for the VCSP. In almost all cases that could not be solved the solver finds a first feasible integer solution on a step deeper than 40 in the branch and bound tree. So the tree becomes too large to be leveled in a reasonable time.

7 Summary

In this paper we presented a mathematical formulation of the vehicle and crew scheduling problem (VCSP). It is defined as a combination of the set partitioning formulations for the vehicle scheduling problem and the crew scheduling problem with resource windows on the scheduling graph as proposed in Desrochers/Soumis (1989). By using this technique to model constraints for crew schedules, a great variety of different real world problems can be tackled.

The VCSP is solved by a column generation approach for both the vehicle and the crew scheduling part. The simplex multipliers of the set partitioning constraints for trips, dtrips, and links are subtracted from the costs of the corresponding arcs of the vehicle or crew scheduling graph, respectively. To generate the new pivot column, shortest path algorithms for both types of arcs are presented.

The column generation approach provides a good lower bound for the branching algorithm. For 10 trips the solution time is below one minute, for 20 trips above one hour, and sometimes no solution is found in an acceptable range of time. For 30 trips only the continuous relaxation could be solved.

To obtain lower bounds that are higher and closer to an integer solution, polyhedral cuts for the set partitioning polyhedron as described by Hoffman/Padberg (1993) are incorporated in the branch and bound scheme and combined with the column generation approach. To ensure that no already existing column is generated, a branch and bound algorithm is developed. This algorithm increases the overall computation so much that only very small instances are solvable in reasonable time.

References

Ball, M./Bodin, L./Dial, R. (1983): A matching based heuristic for scheduling mass transit crews and vehicles. in: Transportation Science 17, 4–31.

Dell'Allmico, M./Fischetti, M./Toth, P. (1993): Heuristic algorithms for the multiple depot vehicle scheduling problem. in: Management Science 39, 115–125.

Desrochers, M. (1988): An algorithm for the shortest path problem with resource constraints. Cahiers du GÉRAD G-88-27, École des H.E.C., Montreal, Canada.

Desrochers, M./Desrosiers, J./Solomon, M. (1992): A new optimization algorithm for the vehicle routing problem with time windows. in: Operations Research 40, 342–354.

Desrochers, M./Gilbert, J./Sauvé, M./Soumis, F. (1992): CREW-OPT: Subproblem modeling in a column generation approach to urban crew scheduling. in: Desrochers, M./Rousseau, J.M. (eds.): Computer-Aided Transit Scheduling: Proceedings of the Fifth International Workshop. (Springer) Berlin, 395–406.

Desrochers, M./Soumis, F. (1989): A column generation approach to the urban transit crew scheduling problem. in: Transportation Science 23, 1–13.

Falkner, J.C./Ryan, D.M. (1992): Express: Set partitioning for bus crew scheduling in Christchurch. in: Desrochers, M./Rousseau, J.M. (eds.): Computer-Aided Transit Scheduling: Proceedings of the Fifth International Workshop. (Springer) Berlin, 359–378.

Freling, R./Boender, G./Paixão, A. (1995): An integrated approach to vehicle and crew scheduling, Report 9503/A, Erasmus University Rotterdam.

Freling, R. (1997): Models and techniques for integrating vehicle and crew scheduling, Tinbergen Institute Research Series 157, (Thesis Publishers), Amsterdam.

Friberg, C./Haase, K. (1997): An exact branch and bound algorithm for the vehicle and crew scheduling problem, Manuskripte aus den Instituten für Betriebswirtschaftslehre der Universität Kiel, 416, University of Kiel.

Hoffman, K.L./Padberg, M. (1993): Solving airline crew scheduling problems by branch-and-cut. in: Management Science 39, 657–682.

Lasdon, L.S. (1970): Optimization Theory for Large Systems. (MacMillan) New York.

Lawler, E.L. (1972): A procedure for computing the k best solutions to discrete optimization problems and its application to the shortest path problem. in: Management Science 18, 401–405.

Nemhauser, G.L./Wolsey, L.A. (1988): Integer and combinatorial optimization. (Wiley) New York.

Patrikalakis, I./Xerocostas, D. (1992): A new decomposition scheme of the urban public transport scheduling problem. in: Desrochers, M./Rousseau, J.M. (eds.): Computer-Aided Transit Scheduling: Proceedings of the Fifth International Workshop. (Springer) Berlin, 407–425.

Ribeiro, C.C./Soumis, F. (1994): A column generation approach to the multiple-depot vehicle scheduling problem. in: Operations Research 42, 41–52.

Tosini, E./ Vercellis, C. (1988): An interactive system for extra-urban vehicle and crew scheduling problems. in: J.R. Daduna and A. Wren (eds.), Computer-Aided Transit Scheduling: Proceedings of the Fourth International Workshop, (Springer) Berlin, 41–53.

Driver Scheduling Using Genetic Algorithms with Embedded Combinatorial Traits

Ann S.K. Kwan, Raymond S.K. Kwan and Anthony Wren
Scheduling and Constraint Management Group
School of Computer Studies, University of Leeds, Leeds LS2 9JT, UK

Abstract: The integer linear programming (ILP) based optimization approaches to driver scheduling have had most success. However there is scope for a Genetic Algorithm (GA) approach, which is described in this paper, to make improvements in terms of computational efficiency, robustness, and capability to tackle large data sets. The question "What makes a good fit amongst potential shifts in forming a schedule?" is pursued to identify combinatorial traits associated with the data set. Such combinatorial traits are embedded into the genetic structure, so that they would play some role in the evolutionary process. They could be effective in narrowing down the solution space and they could assist in evaluating the fitness of individuals in the population.

The first stage of research uses as a starting point the continuous solution resulting from relaxing the integer requirement of an ILP model for driver scheduling. The continuous solution consists of a relatively small set of shifts, which usually contains a high proportion of the shifts in the integer solution obtained. The aim is to derive a GA to evolve from the non-integer solution to yield some elite schedules for further exploitation of combinatorial traits. This first stage is already very effective, yielding in some test cases schedules as good as those found by ILP.

The second stage of research is still ongoing. The aim is to extract from the fittest individuals in the population various forms of combinatorial traits. The genetic structures are then dynamically transformed to make use of the traits in future generations.

1 The Driver Scheduling Problem

Driver scheduling is to construct 'a set of legal shifts which together cover all the work in a vehicle schedule, that is, in a schedule for several vehicles which may reflect the whole operation of an organization, or a self-contained part of that operation' (see review by Wren/Rousseau (1995)).

Driver shifts have to conform to certain rules determined by the union agreements. Drivers can only be changed when a vehicle passes one of a number of designated *relief points*; the times at which the vehicles passes these points are called *relief opportunities*. The work of every vehicle may be considered as a number of indivisible *work pieces* joined at the relief opportunities. The work of a vehicle and its relief opportunities for a day may be represented diagrammatically in the form of a *vehicle graph* (Fig. 1.1).

0732	1017	0843	1145	1302	1435	1550	1718	1920	2031
+---	--+--	--+--	--+--	--+--	--+--	--+--	--+--	--+--	--+--
C	A	B	C	B	C	B	C	A	A

Fig. 1.1 An example vehicle graph

The horizontal line represents the time that the vehicle needs a driver and each '+' represents a relief opportunity at the time and location indicated above and below it respectively. Contiguous work pieces covered by the same shift are together called a *spell*. A shift consists of a number of spells usually drawn from a number of vehicles. The quality of a schedule is dependent on how the vehicle graphs are partitioned into spells and how the spells are combined into shifts. We shall call any such combinatorial patterns that may lead to the formation of an efficient schedule *combinatorial traits*.

If there exist n different valid shifts that could be used to cover a given vehicle schedule and m work piece constraints, the problem may be formulated as follows:

$$\text{Minimize} \quad \sum_{j=1}^{n} c_j x_j \tag{1.1}$$

Subject to the constraints:

$$\sum_{j=1}^{n} a_{ij} x_j = 1, \quad \text{for } i = 1, 2, ..., m \tag{1.2}$$

where x_j are 0-1 variables ;

a_{ij} are constants whose values are either 0 or 1 ;

c_j are positive constants

The variable x_j represents whether a shift j is selected, and each constraint is associated with a work piece to be covered; a_{ij} is 1 if shift j covers work piece i and is 0 otherwise. The cost of shift j is c_j.

This is a *set partitioning* problem in which every work piece i must have exactly one shift assigned to it. The *set covering* problem is a relaxation of the set partitioning problem in which (1.2) is replaced by the condition that every work piece i must have at least one shift assigned to it:

$$\sum_{j=1}^{n} a_{ij}x_j \geq 1, \quad \text{for } i = 1, 2, ..., m \qquad (1.3)$$

2 Motives of Current Research

Research in driver scheduling and advances in computer technologies have made automatic driver scheduling ever more feasible. The mathematical programming based optimization approaches have had most success (see review by Wren/Rousseau (1995)), including the IMPACS and TRACS II (see another paper at this workshop by Kwan/Kwan/Parker/Wren (1997)) systems developed by the authors and their associates. Both IMPACS and TRACS II are based on the set covering ILP model. However the nature of the mathematical approach is such that each computer run may take a long time, larger problems have to be sub-divided, and there is no guarantee that an integer solution can be found within practical computational limits. This research aims to use a Genetic Algorithm (GA) approach to tackle these shortcomings.

Earlier work on heuristic approaches to this problem did not yield very satisfactory results. However, metaheuristics have rekindled hope in recent years. The GA approach is one of the major paradigms under metaheuristics. It mimics the genetic evolution process to derive good, hopefully optimal or near optimal solutions very quickly. GAs are based on the idea of a chromosome as a string of genes defining the characteristics of a particular member of a species, or of a particular solution to a problem. The chromosome may be represented as illustrated in Fig. 2.1.

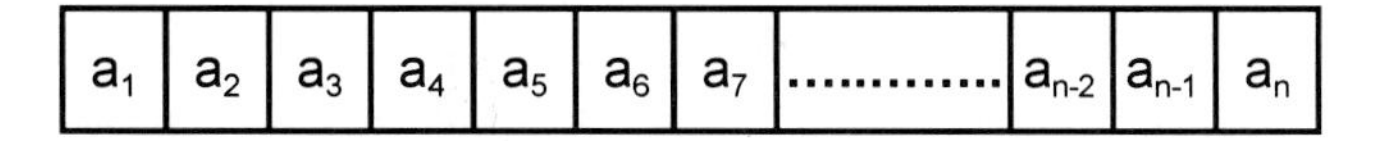

Fig. 2.1 A chromosome representation

In a particular member of a natural species the genes, a_i, take on values, or *alleles*, each of which defines some set of characteristics, such as color of hair, etc. In a GA, the gene values point to some feature of a solution.

A GA starts by generating a population of perhaps a hundred potential solutions (chromosomes) chosen by a random process. In a conventional GA, each population member is evaluated to determine its fitness, and pairs of parents are chosen from the population, usually by some process based on relative fitness. Potential offspring are formed by selecting and recombining genes from the parents, and evaluated; less fit population members are allowed to die and are replaced by fit offspring. In some GAs, a whole new population is generated before any of the older generation is replaced, while in others a continuous regeneration takes place, with suitable offspring replacing weak population members as soon as they are formed.

Attempts have been made using the GA approach to form an efficient schedule out of a very large set of valid shifts already generated, i.e. to replace the ILP component of the TRACS II system. Research shows that GAs can converge very quickly, but previous GA approaches often converge to a sub-optimal solution. Thus, research into the new GA approach described in this paper was started.

In the first phase of the ILP component of TRACS II, the integer requirement on shift variables is relaxed, yielding a "continuous" solution consisting of possibly fractional values for some shift variables. This process is then followed by a branch-and-bound method to find an integer solution. Normally the branch tree will not be exhaustively searched. However as an experiment, the branch tree was exhaustively searched for a small set of data from a train operating company. About twenty thousand potential shifts were fed into the mathematical process, and the best schedule has 14 shifts. The exhaustive search surprisingly yielded only five 14-shift solutions. There would be an enormous number of ways to replace one or more of the shifts in the 14-shift solutions to give rise to solutions with 15 or more shifts. It is strongly suspected that for this data set, there are thousands of 15-shift solutions and millions of 16-shift solutions. The implication is that it would be relatively easy for a search process to yield a near optimal solution, but much more difficult to yield schedules using the optimal number of shifts. The new GA approach therefore attempts to derive effective means of controlling the search space.

The first stage of research aims at deriving a new GA to replace the branch-and-bound process. Research shows that the non-integer solution already found at this stage would have contained on average about three-quarters of the shifts found in the best integer solution. Some of the shifts involved in the non-integer solution, which is a relatively small set, therefore would combinatorially fit well together.

Initially, the GA uses a chromosome structure such that each gene position corresponds to the selection or not of one of the shifts involved in the non-integer solution. Using "repair" heuristics, a schedule is completed giving rise to an evaluation of the chromosome. Applying some genetic operators, the population of chromosomes will evolve through generations to yield some elite schedules. This paper will describe the development of this initial GA scheme and some test results, which in some cases are already as good as those obtained by the mathematical programming process.

The initial GA has resulted in even smaller subsets of shifts, found in the elite chromosomes, that fit well, which can be regarded as combinatorial traits. Furthermore, the schedules yielded by the repair heuristics and associated with the elite chromosomes would contain more combinatorial traits. On-going research in recognizing and utilizing such combinatorial traits in the evolutionary process will be discussed in this paper.

3 Review of GAs Related to Driver Scheduling

Genetic algorithms (GAs) and other evolutionary processes have been proposed for the solution of a wide range of problems for over thirty years, and over the past ten years their use has become extensive. They have been particularly interesting to the scheduling community in the widest sense. Standard texts by Goldberg (1989) and Davis (1991) establish the theoretical basis of GAs and present some applications, while Michalewicz (1994) shows how careful consideration of data structures can be important in influencing the success of a GA or evolutionary program.

In many scheduling problems there is a high probability that after a single point crossover, neither offspring will represent a feasible schedule. For example, Wren/Wren (1995) represented a bus driver schedule formed from a large set of potential shifts by a long chromosome with as many gene positions as there were work pieces in the schedule. The value of the gene was the index of the shift from the large set chosen to cover it. The first parts of two chromosomes might have been as illustrated in Fig. 3.1.

431	431	2584	2584	2584	164	164	431	431	431	431	5472

1826	1826	519	519	519	3324	3324	1826	1826	931	931	931

Fig. 3.1 Representation of two example chromosomes

The first chromosome represents a solution in which shift 431 covers the first two work pieces, together with pieces 8 to 11, while shift 2584 covers pieces 3 to 5, and probably other pieces later in the chromosome.

A problem with this representation is that most offspring produced by a standard crossover would not represent a legal schedule. For example, a crossover between the above two parents after the fifth position would have resulted in one child containing the first two pieces of shift 431, but not the next four. The other child would contain the first two pieces of shift 1826, but not the next two. In some scheduling problems it is possible to repair illegal offspring, but this is often difficult in driver scheduling.

Wren/Wren (1995) overcame this problem by defining a new type of mating which they called *balls in the air*. A *fertilized cover* of the schedule was obtained by combining the shifts used to cover the work pieces in each of the two parents, and this was *distilled* by discarding redundant shifts at random until no further shift could be discarded without leaving some work pieces uncovered. At this stage the resultant schedule may cover some work pieces more than once (indeed this was allowed in the initial population members); however, overlapping shifts can be cut back to provide a legal schedule.

Although this work started by considering chromosomes as defined above, in practice the chromosome can be represented by a list of the shifts contained in it. This is more concise and lends itself easily to forming the union of parents.

Wren/Wren (1995) discovered that on average the children were fitter than the parents, and they obtained best results by mating three parents in this way, rather than two. Good, possibly optimal, schedules were consistently obtained using randomly generated starting solutions for a small problem requiring sixteen shifts. However, subsequent work on significantly larger problems failed to find good solutions.

Further research both in our group and elsewhere has been devoted to finding better representations, better types of mating and better measures of fitness. For example, Hernandez/Corne (1996) have developed an algorithm for a general set covering problem and applied it to a driver scheduling problem provided by our group. They use a variant of the above chromosome representation, in which it is not necessary for any shift to be indicated in all the gene positions which it covers, and thus the schedule may contain overcovered work pieces. For example, gene positions 1 to 5 are holding shift numbers 22, 22, 22, 87, 87 respectively when in fact shift 87 also covers work piece number 3. which is overcovered. Offspring may be generated using simple crossover; the solution is always legal, and the corresponding schedule may be obtained by distillation. There are of course generally many possible corresponding schedules, depending on the random nature of the distillation process, and fitness was defined as the mean of several distilled schedules.

Beasley/Chu (1996) have solved many set covering problems using a chromosome in which each gene position represents one of the possible covering variables (driver shifts in our problem), and has a value of 1 or 0 depending on whether the variable is or is not present in the solution. Crossover operators lead to an infeasible solution which they then repair. This process has been successfully applied to air crew scheduling problems among others, but it is believed that these are easier to solve by this approach than bus driver scheduling problems, because the normal air crew duty in their problems has fewer separate pieces than the average bus driver shift, and there are not as many possible distinct shifts.

In our group, experiments have been undertaken using each of the above representations, and some variations of them, with, until now, limited success. Issues that have been considered include:

- **Fitness measures.** Simply measuring the quality of the schedule represented by a chromosome may not be sufficient. A bad driver schedule may have good features which would be desirable to carry forward. For example, a schedule may have an efficient set of shifts covering the morning peak, but poor coverage of the other peak. It may be reasonable to judge a schedule by the number of shifts in operation at each of several critical times of day, and to classify a schedule as very fit if it has the minimum number of shifts yet found at any one of these times, irrespective of the quality at other times.
- **Parent selection.** Various considerations have been observed in parent selection. Bearing in mind the above comment on fitness measures, it may be sensible to choose one parent who is fit at one time of day and another parent who is fit at a different time, in the hope that an offspring will be fit at both times.
- **Mating operators.** There are many different types of crossover operators described in the GA literature, including single- or multi-point crossover, and uniform crossover, in which genes are chosen from one or other parent according to a randomly generated mask. Other crossovers may take account of chromosome structure, endeavoring not to break good strings of genetic material.
- **Chromosome structure.** Choice of structure is very important in a GA. Possible structures include work pieces sorted in chronological order; this would allow a single-point crossover to take place at a particular time of day and would assist the combination of a good morning schedule with a good afternoon schedule. Other useful structures may require rearrangement of the chromosome according to the shifts included in the solution; an ordering of the shifts in such a way that successive shifts fit well together may allow a crossover operator to preserve good strings of genetic material.
- **Mutation.** Although mutation in a GA is usually carried out at infrequent intervals it can be very important for breaking out of a local optimum. Mutation may take the form of replacing one group of shifts in a solution by another group chosen from a previously defined super set or generated to meet the occasion. It might also involve slight adjustments to shifts in the solution.

- **Other issues :**
  - regeneration of parts of the population from scratch;
  - elitism in which one ensures that chromosomes which are believed to contain good genetic material are preserved through all stages;
  - random repetitions of the whole process;
  - evolution of populations in parallel, with some cross-mating;
  - artificially preserving some population members that differ strongly from others.

  These procedures may be used to assist a general GA to evolve towards a good population. There are many ways in which these may be adapted to the specific problems of bus driver scheduling.

Despite our having had as yet little success in the above work, we are encouraged to believe that further research above will lead to fruitful results. In the meantime this paper presents a different approach.

4 Development of a GA that Exploits the Relaxed LP Solution

The TRACS II system first builds a large set of valid and potentially good shifts according to some parameters and some hard coded rules, which reflect the union agreement. A subset of the shifts will then be selected by solving a set covering ILP model in the second stage.

The ILP method first relaxes the integer requirement on the shift variables, and solves the resulting model by a variant of the Simplex method (Ryan (1980)) to obtain a *continuous solution*. It then proceeds with the branch-and-bound (B&B) strategy to find an integer solution, using the continuous solution as the first node of the B&B tree.

Previous experience in driver scheduling research suggests that a good solution method would need to exploit specific knowledge in the problem domain, and be able to cut down the search space as much as possible. At the start, we therefore reviewed the mathematical component which we aim to replace by a GA. The study led us to embark, as a starting point, on developing a GA that would replace the B&B process after a relaxed LP has been solved. The large set of valid shifts that are input to the mathematical process is also input to the GA process.

The continuous solution provides two pieces of useful information. First, the sum of all the shift variables is a very accurate estimation of the minimum number of shifts required in the integer solution. Since the objective function of the ILP is heavily weighted towards minimizing the total number of shifts, the integer solution very rarely has a number of shifts less than the sum of the shift variables in the optimum solution of the relaxed LP. The estimate is set equal to the sum of the shift variables, rounded up unless the sum's fractional part is very close to, or equal to, zero. In practice, having solved many thousands of practical problems, this estimate is exact in almost all cases. Second, it is a general property of LPs that the continuous solution will involve a (relatively small) number of shifts that is no more than the total number of work pieces. The corresponding variables for some of these shifts may have values of one, but most of them will have fractional values.

The minimum number of shifts needed as estimated from the continuous solution has been used as the target number of shifts in the B&B process. This will also be useful as one of the termination criteria for a GA.

The relationship between the continuous solution and the integer solution obtained by the subsequent B&B process has been analyzed for 21 data sets from bus and train operators. In the results shown in Table 4.1, at least 50% of the shifts in the final integer solution can be found in the continuous solution, on average the percentage is 74% and the maximum is 97%. Since there may be many different integer solutions using the minimum number of shifts, those shifts in the continuous solution but not in the best integer solution returned by solving the relaxed LP may still be interesting in terms of contributing to a minimum shift schedule.

The results indicate that there may be subsets of well fitting shifts in the continuous solution, and each such subset could constitute a combinatorial trait. We therefore proceeded to develop a GA to derive such subsets. Each subset will be supplemented with shifts selected from the original large pool of potential shifts to form a schedule. Thus the GA will also be yielding indirectly a population of complete schedules, many of which are expected to be at least near optimal; these are called *elite schedules*. From the current elite schedules, it is hoped that other traits will be identified and utilized by the GA to converge ultimately to solutions using the minimum targeted number of shifts. Thus, the GA will be a replacement for the B&B process.

The B&B process sometimes may be terminated before any integer solution is found because it is generally only computationally practical to search the B&B tree up to a certain extent. A GA may be more robust in the sense that even if it cannot find a solution with the target number of shifts, it will always have available some (sub-optimal) solutions. GAs are also generally much faster than mathematically based methods. In our experience some data sets, particularly those from train operators, may involve searching through several hundred nodes in the B&B process and take a very long time to run. Moreover, the development of the GA may suggest moving towards replacing the entire mathematical component by a much more superior GA.

Let us call the shifts in the continuous solution the *preferred shifts*, and the full set of potential shifts at the start of the mathematical process the *potential shifts*.

Table 4.1 Proportion of shifts in the integer solution that are also present in the continuous solution

Data set	Number of shifts *(a)* Integer solution	*(b)* Continuous solution	*(c)* Common in *(a)* & *(b)*	*(c)* / *(a)* x 100%	Total shift variables	Total work pieces
TEST	14	113	9	**64**	5000	165
G34A	106	476	70	**66**	30701	502
G14A	61	268	41	**67**	31913	372
G149	65	307	43	**66**	30510	372
EXNR	74	212	69	**93**	11950	236
ALL	87	358	51	**59**	29743	454
G309	113	425	76	**67**	27973	502
W7	61	281	52	**85**	29995	297
W3	52	237	41	**79**	29913	256
WAKH	106	428	88	**83**	30000	446
SWAX	48	321	32	**67**	30574	339
SWBX	132	474	104	**79**	30423	632
WKH	109	420	98	**90**	30000	468
MON	49	220	33	**67**	20761	222
TL96	112	368	85	**76**	17430	537
TL97	112	377	99	**88**	22917	538
G209	95	361	69	**73**	22000	454
GMB	34	126	23	**68**	12869	154
G5X	42	192	41	**97**	27973	196
R207	40	225	31	**78**	11451	244
NEUR	60	269	31	**52**	25000	340
			Average =	**74**		

4.1 Chromosome Representation

In designing a GA for driver scheduling formulated as a set covering problem, it is intuitive to choose a chromosome representation in which each potential shift occupies one gene having a value of either one for inclusion or zero for exclusion in the schedule. However, the very large number of potential shifts renders such a representation impractical with respect to the usual genetic operators. The very long length of the chromosomes makes them expensive to process. A simple mating process is unlikely to yield a valid schedule as offspring. Mutation of one or a few genes would be largely ineffective. As reviewed earlier, some other representation schemes have been used. Nevertheless, we shall use a slightly modified version of this simple representation scheme.

Instead of one gene for each potential shift, we shall use one gene for each of the preferred shifts. Hence the chromosomes will be short, e.g. 100 genes in a small problem. Fig. 4.1 illustrates a chromosome whose genes are labelled with their corresponding indices in the original pool of potential shifts and randomly chosen to have the value '1'.

83	101	1132	1157	2198	3056	3214	......	8796	9654	9982
0	1	0	1	1	0	0	......	0	1	0

Fig. 4.1 A chromosome of preferred shifts

4.2 The Repair Heuristic

In almost all cases, the shifts selected by a chromosome do not represent valid schedules. A *repair heuristic* is therefore used to derive a *valid schedule* for each newly formed chromosome. A valid schedule is a schedule in which every work piece is covered by at least one shift and *overcover* (i.e. work pieces being covered by more than one shift) is allowed. The derived schedule is evaluated to determine the fitness of its associated chromosome.

The repair heuristic first uses all the preferred shifts selected by the chromosome, which is followed by two processes called FILL and DISCARD. The heuristic updates information about the coverage of each work piece as shifts are inserted or deleted from the schedule being formed. The information includes which shift, if any, is currently assigned to cover the piece. For a currently uncovered work piece, there is a list (called the *coverage list*) of all the potential shifts that can cover it. The FILL process takes a yet uncovered work piece with the shortest coverage list and selects a shift from the coverage list for the schedule. The newly selected shift may at the same time cover some other work pieces. FILL takes a greedy approach and chooses from the coverage list a shift that covers the most amount of yet uncovered work. This process is repeated until there are no more uncovered work pieces left. At the end of the FILL process, some of the shifts selected for the schedule may be redundant and the DISCARD process will systematically remove them. It leaves the preferred shifts to be the last to be considered for discarding.

4.3 Measurement of Fitness

The fitness of a chromosome is expressed in terms of the weighted cost of the schedule derived from it by the repair heuristic. The weighted cost of a schedule is defined as the sum of the costs of its shifts plus a heavy weight (5000) per shift in the schedule, i.e.

$$WeightedCost(S) = \sum_{j=1}^{n} C_j + n \times 5000$$

where n = number of shifts in the schedule S, and C_j is the cost of shift j. The heavy weight puts stress on reducing the total number of shifts. The lower the weighted cost of the schedule the fitter the chromosome is. However, the above costing formula lacks an assessment of how many good combinatorial characteristics the schedule contains. Further research for improving the measurement of fitness is suggested later in the paper.

4.4 Crossover and Mutation Operators

The crossover operation combines individuals in the population and exchanges information to produce a new offspring. Two types of crossover operators have been investigated in our GA process. The first type is *uniform crossover* and the other, which we have adopted, is *single point crossover*. Uniform crossover uses a binary mask to indicate from which parent each gene is to be chosen. The mask is generated at random with equal probability, at each gene position, of choosing any one of the parents for contributing its gene value to the child. Experiments have shown that the quality of offspring produced by uniform crossover is poor compared with those produced by single point crossover, even when good quality parents were used. The explanation could be that uniform crossover tends to fragment the bit strings of the parent chromosomes, and some useful characteristics conveyed through groups of neighboring bits in the original bit strings of the parents might be lost. In order to have some retention of patterns of gene values from the parents, crossover in just a small number of locations is normally more advantageous. A single point crossover operator has therefore been adopted as illustrated in Fig. 4.2.

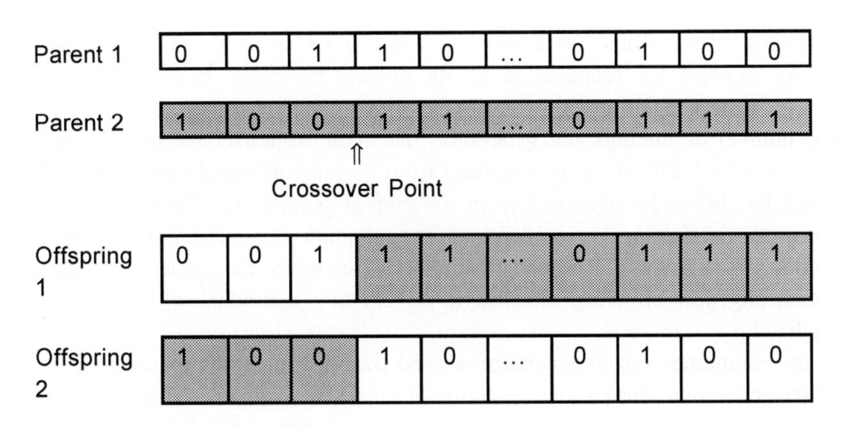

Fig.4.2 Single point crossover

The mutation operator is used for two purposes. It reintroduces 'lost' information into the population, and it ensures that the species develop in unexpected ways. Two types of mutation have been used. The first type is a *standard mutation,* which occurs infrequently throughout the GA process. Mutation on offspring occurs after a constant number of generations and the first gene found to be of the same value throughout the population will be mutated to a different value. Standard mutation was found to have negligible effect on our GA, and did not help prevent premature convergence.

The other mutation operator used is called *aggressive mutation.* Aggressive mutation uses a variable rate of M mutations per generation and a variable number G of genes to be mutated. When the GA process starts, both M and G are kept to small values so that the population will evolve progressively and the role of mutation is minimal. The values of M and G are progressively increased as the GA gradually converges. When near convergence, mutation is made very intensive. The mutation is described as aggressive also because the criterion for it to mutate is when the gene has the same value for both parents rather than for the entire population. The genes to be considered for mutation are randomly selected. Aggressive mutation has the effect of enlarging the search space when the GA process starts to converge. Some experiments have shown that by using aggressive mutation, the population size could be significantly reduced without affecting the rate of convergence. Because of smaller population sizes, the GA runs a lot faster.

4.5 Selection of Parents and Population Replacement

The method used for the selection of parents for reproduction is probabilistic. The probability of any individual of the population being selected is proportional to the individual's fitness. Thus even a very "bad" individual has a chance, though small, of being selected for propagation of its genetic materials, helping to sustain diversity in the population.

The number of matings per generation has been arbitrarily set to be half the population size. Offspring are separated from the current population so that they will not be chosen for reproduction in the current generation. Since each mating produces two offspring, at the end of a generation the maximum number of individuals would be twice the population size when the generation started. However any new offspring identical to an existing individual will be discarded immediately.

At the beginning of each generation a fixed size, n, population is maintained. In the first generation, the entire population is created at random. Thereafter before starting a new generation, up to some maximum number of the fittest individuals, from the old population plus the offspring, equal to a percentage of n will be allowed to survive. The percentage is a pre-specified parameter called the *survival rate*. The other individuals will be discarded. New individuals are then created randomly to bring the total population size back to n before the next generation of reproduction starts again.

Suppose there are a total of P genes in the chromosome, which means there are P preferred shifts. When a chromosome is to be created at random, the GA first decides a number m of the P genes to be set to the value one. Experiments have shown (see Table 4.1) that if there are T shifts in the final integer solution, at least 50% of them will be present in the set of preferred shifts. Therefore, m is randomly chosen between $T/2$ and T. Then, m positions on the chromosome will be selected at random and set to the value one to complete the creation of the chromosome. The repair heuristic already described is then used to yield a feasible schedule to be associated with the chromosome.

There is no conclusive theory in deciding the best population size. Small population sizes may have the danger of not covering the solution space satisfactorily. Experience from various experiments indicates that, in general, the population size should be proportional to the size of the problem. The size of a problem is measured principally proportional to the total number of work pieces, but it is also dependent on the total number of potential shifts. However, it is believed that by using aggressive mutation, the population size could be significantly smaller than otherwise. In our experiments, population sizes ranging from 50 to 800 have been used.

The GA stops after a pre-specified number of generations.

4.6 Results

The GA described has been implemented using Borland C++ (version 5) and tested on a 200MHz Pentium-Pro PC with 32 MB RAM. All the test data sets are from real life problems in the UK that have been studied by our group in recent years.

They represent various types of driver scheduling problem with five for train and two for bus operations. The data sets contain from around 10,000 to almost 30,000 potential shifts, and from around 150 to slightly over 500 work pieces. The GA was first tested on a small train problem for which a solution of 14 shifts was found by a mathematical programming based system. The GA also managed to find a 14 shift solution after a few trials on different parameter settings. It was then applied to six other data sets and the results are shown in Table 4.2.

Table 4.2 GA results

Data Set			Integer Solution				Continuous Solution of relaxed ILP		GA Parameters	
			Number of shifts		Total cost (unweighted)					
Name	Type	No. of Potential shifts (Work pieces)	*(a)* **ILP**	**GA**	ILP	GA	*(b)* No. of preferred shifts	% of shifts in *(a)* found in *(b)*	Population size	No. of generations
TEST	train	22041 (165)	**14**	**14**	7403	7375⁺	113	64	50	50
EXNR*	train	11950 (236)	**74**	**74**	33873	33909	216	96	450	100
GMB	bus	12869 (154)	**34**	**35**	17261	17571	126	68	50	50
G5X	train	27973 (196)	**42**	**42**	18650	19022	192	97	200	50
NEUR*	train	25000 (340)	**60**	**62**	29828	30971	269	52	400	100
TL96*	train	14818 (537)	**112**	**113**	53878	54497	392	80	800	50
R207*	bus	11451 (244)	**40**	**41**	19317	19658	225	78	800	50

⁺ The GA was able to produce a lower cost solution because the ILP process does not search the B&B tree exhaustively for practical reasons.

Aggressive mutation was used for all the data sets, but for those marked with an asterisk, large population sizes are still needed to achieve the results shown. However, the current aggressive mutation method is still quite crude and there is scope for further research to refine it. In terms of the number of shifts in the solution, the GA has matched ILP results in three cases, is worse off by one shift in three cases and is worse off by two shifts in one case.

5 Combinatorial Traits in "Elite" Schedules

Elite schedules refer to a proportion of the current GA population that has been evaluated to be among the fittest. The elite schedules are expected to be closer to optimum the more generations the population has evolved. Current research is aiming at identifying within the elite schedules key patterns of work piece combinations that have enabled the schedules to fall into place like jigsaw puzzles. We call such patterns and characteristics combinatorial traits. Various driver scheduling systems may already be using heuristic rules concerning combinatorial characteristics, but they are typically engaged in sequential searches involving only one schedule, or partial schedule, at a time. GAs work with many schedules in each generation and therefore can gather a lot more information.

There has been no attempt in GAs for driver scheduling to manipulate combinatorial traits explicitly, which are left to chance emergence within population members. Since combinatorial traits are undetected, they cannot be used to influence the evolutionary process. The current GA research seeks dynamically to revise the chromosome structure to incorporate combinatorial traits as genetic materials so that they will also be subject to the survival-of-the-fittest principle. Inherited traits will directly affect the way the repair heuristic constructs new schedules. Traits may be accumulated depending on the line of inheritance through generations, and thus the evolutionary process may be guided towards a progressively confined solution space. The evolutionary approach has the advantage that "bogus" combinatorial traits arising from imperfect diagnosis will not strike a fatal blow because they will lead to unfit schedules, which would be discarded in new generations.

In the following, we shall briefly describe four types of combinatorial trait that have been investigated. They are namely *well-fitted shifts*, *relief chain*, *handover relief* and *overcover link*. Exactly how they will feature in the GA is still under research.

5.1 Well-Fitted Shifts

The set of *preferred shifts* fit optimally when fractions of the shifts are allowed in forming a schedule. When only integral shifts are allowed, there will be subsets of the preferred shifts that will fit well together. The number of such subsets is likely to be more than one because there is usually more than one different schedule using the minimum number of shifts. At the end of the initial GA described in the previous section, the elite chromosomes will each represent a selection from the preferred shifts. The selected preferred shifts constitute a set of well-fitted shifts.

This trait refers to a set of just a few shifts that simply fit very well together. This trait describes characteristics exhibited at a macro level. The other three traits all involve finer grain analysis of interactions between sub-structures in the shifts.

5.2 Relief Chain

An important feature in driver scheduling is that driver shifts have one or more breaks in them. After a break, the driver is usually assigned to work on another vehicle, often taking over from another driver who is due for a break. Thus a chain of events of drivers relieving each other is formed, and we call the relief opportunities involved a "*relief chain*". For example, Fig.5.1 shows a relief chain (R1 - R2 - R3). Shift 1 starts its break at R1, relieves shift 2 when it resumes work at R2. Shift 2 relieves shift 3 when it resumes work at R3.

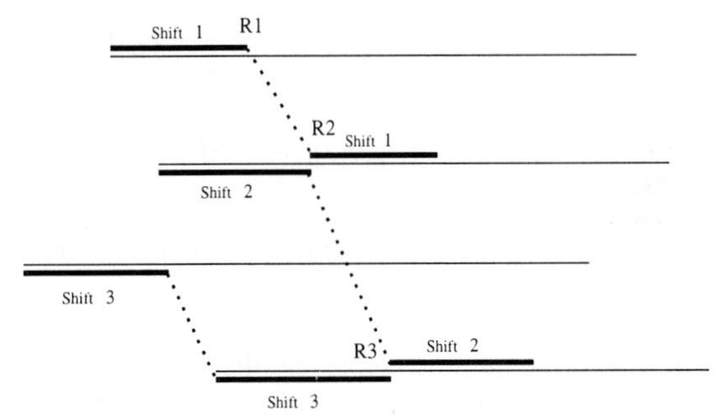

Fig. **5.1** A relief chain R1 - R2 - R3

Relief chains have been the subject of much investigation in previous research into driver scheduling heuristics. Experience suggests that well formed relief chains, especially critical during peak periods, can avoid using extra drivers unnecessarily. For the current GA research, relief chains might be used in two ways. First, some important relief chains might be identified from the current elite chromosomes and incorporated into the genetic structure to influence subsequent evolution. It should be noted that in some cases although two relief chains may be different, one of the chains may be just shifted from the other by roughly the same amount of time along the chain as shown in Fig. 5.2.

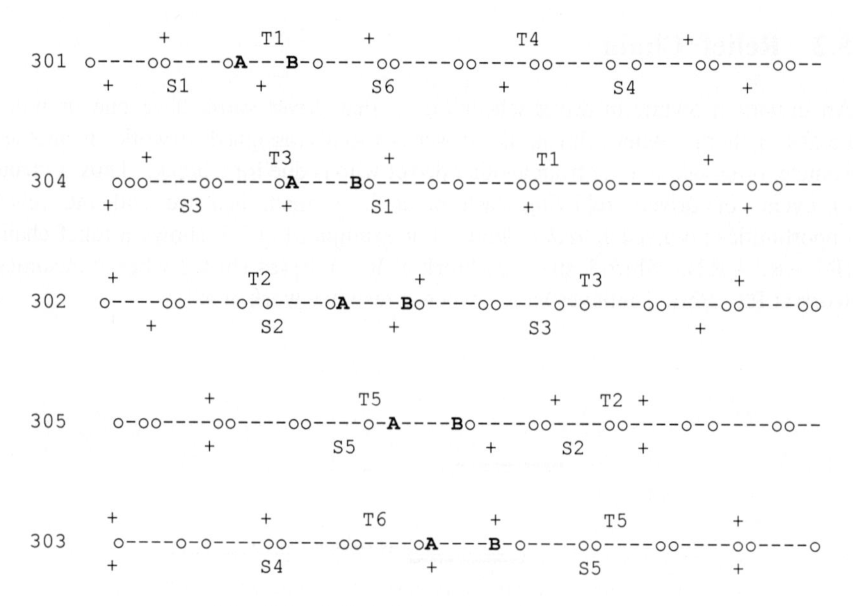

Fig. 5.2 Relief chains of similar patterns

In Fig. 5.2, the horizontal lines represent the work of vehicles 301 to 305 drawn to timescale and are truncated at the right hand side. The relief opportunities are marked with the letter "o". Two schedules, whose shifts are labelled by numbers with prefixes S and T respectively, are shown one above and the other below the horizontal line. The "+" signs delimit the work covered by driver shifts. The relief opportunities forming relief chains are marked by "A" and "B" for each of the schedules respectively. Both schedules consist of 14 shifts.

The relief chains present in an offspring schedule may be assessed as part of the fitness measurement. For example, the longer relief chains are during a peak period, the better the schedule will be.

5.3 Handover Relief

When a shift finishes on a vehicle that still has work to be covered, the relieving driver may be either starting a new shift or resuming work after a break. In the former case, we call the relief opportunity used a "*handover relief*.' In the latter case, the relief opportunity terminates a relief chain as described earlier.

Some handover reliefs found in the elite schedules might be critical to the formation of good schedules. When passed on to an offspring, handover reliefs could be used by the repair heuristic to give priority to selecting shifts covering the work pieces adjacent to the handover reliefs. Shifts covering both work pieces adjacent to a handover relief, i.e. not using it for driver relief, would be banned.

5.4 Overcover Link

One frequent and frustrating phenomenon in this area of research is that a GA may get “stuck” at solutions using more shifts than the known minimum required. Since more shifts than minimum are used, some overcover is expected. It is interesting that although they are inefficient, the pieces of overcover have persisted in the evolutionary process. One explanation could be that there is something critical with one or more work pieces linked to the overcovered piece, and the GA has not been able to yield an optimal combination of shifts to accommodate the critical feature. It is likely that only one of the pair of shifts involved in an overcover is obstructing the formation of an optimal combination of shifts, and it might be possible to avoid the linkages leading to the overcover. However, the remedy would need more reformation than simply replacing the shift at fault. What could be done is for the subsequent evolutionary steps to enforce the critical features identified and to ban the linkages suspected to be at fault.

The GMB data set, included in Table 4.1 and Table 4.2 before, was studied. So far, the best solutions the GA could find use 35 shifts, which are one more than the ILP solution. Three different 35 shift solutions yielded by the GA were analyzed. In the example extracted from one of the schedules shown graphically below, the work piece (bus 14, 1311-1429) is fully contained within another one covered by a different shift as shown in Fig. 5.3.

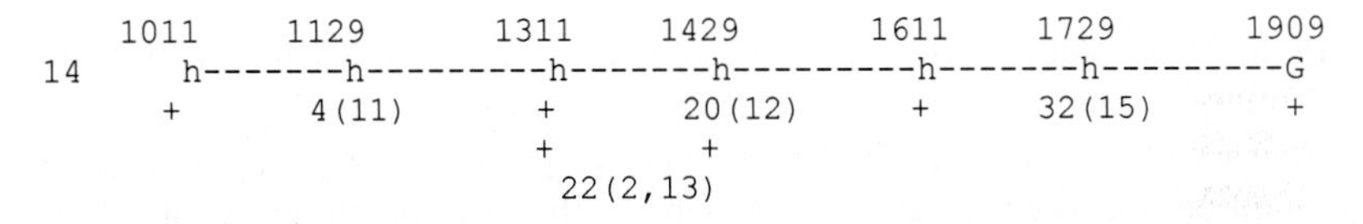

Fig. 5.3 Work fully covered by another shift

Underneath the timescale line, the “+” signs delimit the work on bus 14 covered by specific shifts, e.g. “+ 20(12) +” denotes a work piece covered by shift number 20 which also covers another work piece on bus 12, and it is necessary to refer to other parts of the schedule document to find out the complete details of shift number 20.

The fully contained piece is the second part of a 3-part shift (no. 22):

Shift 22:	bus 2,	1152 - 1255
	bus 14,	1311 - 1429
	bus 13,	1511 - 1927

Since it has a fully contained work piece, we assume that shift 22 is not a good shift, but it may possess some critical characteristics. The relief opportunities (bus 2, 1255) and (bus 13, 1511) are suspected to be critical, i.e. they might be needed to achieve a good schedule, but they probably should be linked differently from the above. We shall call the links before and after the overcover '*overcover links*.' In the above example, the overcover links are: (bus 2, 1255) to (bus 14, 1311), and (bus 14, 1429) to (bus 13, 1511).

The analysis of the three 35 shift solutions has identified 16 critical relief opportunities and 20 overcover links (four of them are end of shift links). 160 different 34 shift solutions were obtained using a slightly adapted ILP process. None of the 34 shift solutions contains any overcover. The 16 critical relief opportunities are used in all the 34 shift solutions.

Table 5.1 Distribution of 20 overcover links in 160 minimum shift schedules

No. of 34-shift schedules	Distribution of overcover links
0	9
1 - 10	5
11 - 20	3
21 - 50	2
51 - 160	1

In the above study, there were two types of overcover other than that shown in the example. One type is when two shifts cover the same work piece exactly, i.e. complete overlap, on a bus. The other type is a partial overlap. In these cases, no assumption has been made as to which of the pair of shifts involved in the overcover should be replaced, and therefore links from both shifts were extracted to be overcover links. In fact often only one of the pair of shifts is at fault. This explains why some of the overcover links might not be bad as indicated in Table 5.1. Nevertheless, Table 5.1 shows that nine out of 20 overcover links are most likely to be bad, and the banning of another eight of them probably would not be detrimental. The remaining three overcover links if banned would probably lead to inferior offspring that would not survive.

6 Further Research and Conclusions

Driver scheduling is a very difficult problem. Where x is the optimal minimum number of shifts needed, to derive solutions with $x + 2$ or more shifts probably would not be difficult. Thereafter, each move to a solution with one fewer shift represents a quantum leap. We have taken an approach that supplements the genetic algorithm metaheuristic with rich domain knowledge. The main advantages of the approach are:

- For a given data set, the GA provides a diversity of schedules for analysis and discovery of data specific knowledge (traits)
- The mechanism of inheritance and evolution of traits is flexible and fault tolerant

The research has already identified four types of traits, and methods for extracting them from the currently elite schedules have already been developed. Some experiments of utilising the traits identified have already begun, and we hope to report our results in the near future. Apart from using the traits for guiding the creation of schedules, we are also investigating possible roles of traits in improving the measurement of fitness of chromosomes. The current weighted cost formula used is good for comparing the fitness of schedules using different numbers of shifts, or for comparing schedules all using the minimum number of shifts. For comparison of schedules all using a equal number of shifts greater than the minimum, we would ideally want a cost element to reflect the extent useful traits the schedule possesses.

Mutation is generally used at constant and infrequent rates in GAs. Our experiments indicated that *aggressive mutation* could be very useful for a driver scheduling GA. There will be further research to refine the method and to perform thorough testing on more data sets.

Acknowledgement Thanks are due to the Engineering and Physical Sciences Research Council for providing funding support for the work described in this paper.

References

Beasley, J.E./Chu, P.C. (1996): A genetic algorithm for the set covering problem. in: European J.O.R., 94, 392 - 404.

Davis, L. (1991): Handbook of genetic algorithms. (Van Nostrand Reinhold).

Fores, S./Proll, L./Wren, A. (1997): An improved ILP system for driver scheduling. in: Preprints of the 7th International workshop on computer-aided scheduling of public transport, Boston, US.

Goldberg, D.E. (1989): Genetic algorithms in search, optimization and machine learning. (Addison-Wesley).

Hernandez, L.F.G./Corne, D.W. (1996): Evolutionary divide and conquer of the set covering problem. in: T.C. Fogarty (ed.): Evolutionary Computing, (Springer) 198 - 208.

Kwan, A.S.K./Kwan, R.S.K./Parker, M.E./Wren, A. (1997): Producing train driver schedules under different operating strategies. in: Preprints of the 7th International workshop on computer-aided scheduling of public transport, Boston, US.

Michalewicz, Z. (1994): Genetic algorithms + data structures = evolution programs. Second, extended edition. (Springer-Verlag).

Ryan, D.M. (1980): ZIP - a zero-one integer programming package for scheduling, Computer Science and Systems Division, Atomic Energy Research Establishment, Harwell, Oxfordshire.

Wren, A./Rousseau, J.-M. (1995): Bus driver scheduling - An overview. in: J.R. Daduna /I. Branco/J.M.P. Paixao (editors) Computer-aided transit scheduling. (Springer-Verlag) 173 - 187.

Wren, A./Wren, D.O. (1995): A genetic algorithm for public transport driver scheduling. in: Computers Ops Res, 22, 101 - 110.

An Integrated Approach to Ex-Urban Crew and Vehicle Scheduling

Andrea Gaffi[1] **and Maddalena Nonato**[2]
[1] MAIOR srl, Viale san Concordio 639 Lucca, Italy maiorlu@tin.it
[2] Computer Science Department, University of Pisa, Italy nonato@di.unipi.it

Abstract: Scheduling of vehicles and crews, traditionally performed sequentially by scheduling vehicles prior to crews, has to be carried out simultaneously in particular settings such as the ex-urban mass transit, where crews are tightly dependent on vehicles' activity or crews' dead-headings are highly constrained. In this paper we propose an integrated approach to vehicle and`crew scheduling which exploits the network structure of the problem. A heuristic method based on Lagrangean relaxation is presented, which determines a set of pieces of work suitable for both vehicle activities as well as for crew duties. Crew duties are fixed step by step, while vehicles are scheduled once all the trips have been partitioned into pieces. Extended use is made of Bundle methods for polyhedral functions and algorithms for constrained shortest paths and assignment within a dual greedy heuristic procedure for the set partitioning problem. Computational results are provided for Italian public transit operators, which show some improvements over the results of the sequential approach.

1 Introduction

In spite of a variety of very efficient methods for the sequential scheduling of vehicles and crews, few proposals have been made in the literature for the solution of the joint problem. Scheduling vehicles prior to crews does not affect the overall solution quality except for a few particular cases, including ex-urban public transport with multiple depots. Since in most industrialized countries crew-related costs are large relative to costs associated with vehicles, it is reasonable to tackle the joint problem in those cases where a very efficient vehicle schedule may lead to a poor crew schedule or even to no feasible solution at all. The problem dimension, which is a critical aspect of this class of combinatorial problems, dramatically increases when the two problems are handled together. In fact, most models describing the joint problem are solved in practice by heuristics in which the problem is indeed decomposed: either crew constraints are taken into account at vehicle scheduling time, as in Scott (1985), or crews are scheduled before vehicles, as in the pioneering paper Ball/Bodin/Dial (1983); other attempts made to solve the joint problem deal with the single depot case only (see Freling/Paixao (1994)) which does not fit into many ex-urban settings where several depots are scattered over a large geographical area.

In order to delve into the drawbacks of the sequential process when applied in the ex-urban setting, let us briefly review the basic terminology. Transit service is defined by a set of *lines* and *timetables*; each traversal of a line is a *trip* which is described by both starting and ending times and locations. A trip is the basic unit of service for the purpose of vehicle scheduling since a trip must be operated by a single vehicle. A vehicle *block* is a sequence of compatible trips starting and ending at any pair of depots, while a *vehicle duty* is a feasible sequence of blocks, each block ending at the depot from where the next block starts, and such that the last block ends at the starting depot of the first block of the sequence, namely the *depot of residence* of the vehicle. A *relief point* along a block provides a time and place for a possible crew relief. A *piece of work* is defined as a continuous crew working period on a single vehicle between two relief points on a block. In *crew dead-headings* drivers move from one relief point to the next. A *crew duty*, often called a *run*, is a feasible sequence of pieces that starts and ends at the driver's residence depot. Bus driver duties will be hereafter referred to as *runs* while vehicle duties will be referred to as *duties*.

As in other European countries, three kinds of public bus transport service arise in Italy at the district level: urban service, providing transport within the city; sub-urban service, providing connections from the suburbs to the city center, and ex-urban service, inter-connecting minor towns of the province and to the main city.

While urban and sub-urban services are often scheduled together and share the same fleet and crews, ex-urban service is dealt with separately. This kind of transport is characterised by trips over longer distances and by many depots distributed over a large area. When the service cannot be split among depots, allowing many single depot scheduling problems to be solved independently, we claim that specific tools should be devised to schedule vehicles and crews together. Since vehicle scheduling packages usually minimize fleet size and operational cost, when applied to ex-urban service they are likely to yield pieces of work which are too long to meet union regulations as well as situations in which drivers have no means of returning to their depot of residence at the end of their duty. Since the first issue affects individual blocks, it can sometimes be dealt with by considering some of the union constraints, such as spreadtime or maximum driving time, at vehicle scheduling time. On the other hand, issues related to the latter question cannot be solved without knowledge of the crew schedule. Moreover, while in the urban case drivers can move from a relief point to another regardless of the vehicle schedule, thus providing many ways to combine pieces of work within a crew duty, in the ex-urban setting distances are such that drivers are tied to their vehicle for the purpose of reaching relief points, so that pieces and blocks collapse into the same entity. For this reason, it would sometimes be convenient in terms of the crew schedule to split a block into two and force a vehicle to detour in order to pass by a depot or any other relief place, thus providing a relief opportunity to the crew.

In conclusion, we feel that since in the ex-urban setting the vehicle schedule can heavily affect crew schedule, crews must be scheduled at the same time as vehicles in order to achieve efficiency in the use of vehicles without spoiling the crew schedule.

In our approach we focus on the intermediate stage of the solution process where trips are partitioned into pieces of work out of which both runs and duties are constructed: the Lagrangean relaxation of the constraints enforcing consistency between the pieces of runs and the pieces of duties provides a guideline on how to partition the trips into pieces.

In Sect. 2 a review of the literature on both vehicle and crew scheduling is provided, focusing on graph theory models; in the following section our model is introduced, and a heuristic solution approach via Lagrangean relaxation is proposed in Sect. 4; the computation of a vehicle schedule is discussed in Sect. 5, and computational results provided for real life cases are presented in Sect. 6; finally, conclusions and perspectives for further research are discussed in Sect. 7.

2 Network Models for Crew and Vehicle Scheduling Problems

Both vehicle and crew scheduling problems (VSP and CSP, respectively) consist of covering a set of service units by a set of duties; since these units are fully characterised in terms of time and place, the compatibility relation between units that can be operated in sequence is often exploited through the *compatibility graph*; here service units are modelled by nodes, and arcs connect compatible units, i.e. units that can be operated in sequence by the same subject. On compatibility graphs duties are modelled as constrained paths, which inspired several network-based solution approaches as described in Carraresi/Gallo1984.

When feasibility constraints are overwhelming, such as when crew duties must fulfil all requirements of the collective agreements, Set Partitioning/Covering models are preferred: here one variable for each crew duty is explicitly introduced in the model, while the network structure is eventually exploited using column generation mechanisms.

For a more general discussion on routing and scheduling problems, we refer the interested reader to Desaulnier/Desrosiers/Ioachim/Solomon/Soumis/Villeneuve (1997), where similarities and differences between the structure of these problems are discussed and a unified framework is proposed for their solution.

2.1 The Vehicle Scheduling Problem

The VSP consists of finding a set of vehicle duties such that: each trip is covered by exactly one vehicle; each vehicle returns to its depot of residence at the end of its duty; the number of duties operated from each depot does not exceed the number of available vehicles. When dealing with a single depot and homogeneous fleet, the VSP can be formulated and solved polynomially as a network flow problem, but it turns *NP*-Hard when multiple depots are introduced (see Bertossi/Carraresi/Gallo (1987)): both heuristic and exact approaches have been proposed for the latter case (see Daduna/Paixão (1995) for a review), to which most real life problems in both urban and ex-urban mass transportation systems belong.

Multiple Depot Vehicle Scheduling Problem (MD-VSP) can be formalized as follows: let $V = \{v\}$ be the set of trips to be operated, each $v = 1,..,n$, starts from location s_v at time st_v and arrives at location e_v at time et_v; travel time τ_{uv} from e_u to s_v is provided for each pair $u,v \in V$; trips u and v are said to be *compatible* if and only if $et_u+\tau_{uv} \leq st_u$, which is to say that a vehicle can operate them in sequence. Let $D=\{d^1,..,d^K\}$ be a set of depots, each d^k housing at most C^k vehicles, $k=1,..,K$. Vehicle transfers without passenger service connecting compatible trips either to each other or to a depot are called *dead-headings* (*dh-trips* in the following). Between consecutive blocks vehicles wait idle at the depots, where no crew attendance is required.

MD-VSP is usually modelled as a multicommodity flow problem on the compatibility graph extended with depot connections, with one commodity for each depot. A vehicle duty corresponds to a cycle through exactly one depot node; at most C^k cycles through depot d^k are allowed, for each $k=1,..,K$. Ribeiro and Soumis (see Ribeiro/Soumis (1991)) describe a very efficient exact method where the column generation subproblem produced by the Dantzig Wolfe decomposition applied to the linear relaxation of the multicommmodity flow formulation reduces to K shortest path problems on an acyclic graph.

2.2 The Crew Scheduling Problem

In the CSP blocks are first partitioned into pieces by breaking blocks at relief points; pieces are then sequenced to yield a set of feasible runs. This approach has a nice network representation in terms of (constrained) path covering problems on acyclic graphs (see Carraresi/Gallo (1984)), but the most popular approach is the one based on Set Partitioning/Covering (SP/C), as we shall discuss more in detail in Sect. 2.2.1. In SP/C based models, each *task* (a piece between two consecutive relief points) must be covered by (at least) one run, and the sum of the costs of all selected runs must be minimized. Because of the large number of variables involved, column generation techniques are often applied in order to solve the SP/C continuous relaxation, and the process is embedded in a Branch&Bound framework to produce an integer solution. The master problem is solved by means of packages for linear programming while the network structure is exploited in the column generation phase. The related subproblem, where feasible runs are generated, must deal with union regulations which are rather cumbersome and vary locally. They usually concern *local* constraints as well as *global* constraints. The firsts involve issues such as total working time, driving time, and spread time of the single run, plus, for each individual run, number of pieces and duration of each piece. The latters are imposed on the overall crew schedule and concern features of the final solution, such as the average working time of the schedule or the maximum percentage of runs of a specific type.

Dual heuristics for the SP/C based on Lagrangean relaxation and column generation have also received attention recently (see Carraresi/Girardi/Nonato (1995)), thanks to improved algorithms to solve the Lagrangean dual, as described for example in Carraresi/Frangioni/Nonato (1995).

Finally, the use of Genetic Algorithms (GA) for the CSP has been investigated in Clement/Wren (1995) and more recently at this workshop in Kwan/Kwan/Wren (1997), although it appears that GA - which occasionally prove competitive with classical set covering algorithms - need to be specially tailored to deal with the peculiarities of CSP.

2.2.1 Set Partitioning and Covering Approaches

Due to their generality and flexibility, SP/C based approaches are rather popular to solve CSP in most transportation sectors. Set Partitioning (SPP) is more popular in the airline context where the cost of dead-heading is high, since it corresponds to the loss of revenue from seats occupied by crew members flying off-duty. Set Covering (SCP) models, which are more appealing since computing a feasible solution for SCP is much easier than for SPP, are preferred when tackling bus driver scheduling and in crew scheduling for railway applications. We shall briefly review some of the most recent contributions in the area to place our approach in the proper context. We shall use the term *duty* indiscriminately to denote a run, a pairing (duty in airline crew scheduling) or a railway crew duty.

Two critical issues often arise in real life applications, namely a high number of duties and a unicost objective function, since generally the aim is to minimize the number of crews' working days; this frequently leads to instances characterized by massive primal degeneracy, which makes the solution of the continuous relaxation very slow by way of traditional pivoting; moreover, it prevents a straightforward use of the sophisticated techniques that have been devised to solve SP/C problems to optimality unless only a small subset of all feasible duties is considered.

As far as SPP is concerned, for small instances the primal continuous solution is often integer (see Marsten/Shepardson (1981)) and the problem can be easily solved to optimality via Branch & Bound within a few iterations. Nevertheless, this nice feature vanishes with increasing size as well as with the introduction of side constraints. In order to deal with the first issue, two approaches are rather popular in airline crew scheduling. According to the first, a procedure for the optimal solution of small SPP instances is embedded within an iterative procedure; this procedure, at each iteration, randomly selects a subset of columns until a certain number of rows is covered, exhaustively generates all feasible duties defined on these rows, and solves the associated SPP to optimality; if the objective function has been improved the optimal columns replace the selected columns and the process is iterated. According to the second strategy, at first a subset of duties defined on the whole set of rows is heuristically generated, and then the corresponding SPP is solved once; this second strategy often goes under the name of *once and for all* approach. Note that these approaches do not provide an optimal solution and do not even provide a lower bound on this value. Moreover, in airline crew scheduling, other approximations are often introduced in order to work on a manageable size problem; for example it is quite popular to solve the daily problem first, where each flight is supposed to be flown every day, and then adjustments are made to deal with weekly exceptions.

Polyhedral theory has provided some tools to improve the solution techniques for SPP. Hoffman and Padberg exploited polyhedral cuts for SPP within a branch and cut approach in Hoffman/Padberg (1993), allowing for the exact solution of instances up to 825 rows and 8627 columns and up to 145 rows and over 1 million columns, able to cope with side-constraints for much smaller instances. Nevertheless, this method by itself is still far from handling real life airline crew scheduling problems of average size, since it provides the optimum only when all columns are given, and it is well known that for real life cases the feasible duties can number in the billions. Therefore the approach can be used only within one of the previous strategies, together with a heuristic duty generator.

Regarding bus driver scheduling, the polyhedral structure of the problem is investigated in Ryan/Falkner (1988), where it is shown that the constraint matrix of the SPP defined on a special subset of duties built using the so-called *next available* principle is totally unimodular, and thus enjoys the integrality property. Nevertheless, this subset is not representative of the whole set of feasible duties since the next available principle does not honor major union constraints such as meal break requirements.

Column generation seems to be the only way to deal effectively with the entire set of variables and to compute a real lower bound. Its successful use for SP/C is reported in Lavoie/Minoux/Odier (1988) and in Desaulniers/Desrosiers/Gamache/ Soumis (1997) for airline crew scheduling, and in Desrochers/Soumis (1989) for urban bus driver scheduling, just to mention a few. The network structure of the problem is exploited in the column generation phase.

Several networks can be adopted to model the pricing subproblem. In *time-space oriented* networks, nodes represent locations at different times, while arcs model crew activities, including breaks as well as on-duty and off-duty transfers. Another family of networks (hereafter called *compatibility oriented* networks) is derived from compatibility graphs; a node is associated with a crew activity which may be either a task or a piece of work, or even a daily service activity in airline crew scheduling, while arcs connect compatible activities. Selecting the most suitable network for the purpose of modelling a given problem is a question related to the trade off between network size and feasibility constraints embedded in the network topology. For example, consider a compatibility oriented network with nodes corresponding to pieces of work: the feasibility constraints concerning the individual pieces are satisfied by default for any duty generated as a path on this network and do not need to be checked during column generation. On the other hand, the number of pieces is higher than the number of tasks, although the former is usually a linear function of the latter provided that vehicle switching is forbidden during a piece. In Desaulniers/Desrosiers/Lasry/Solomon (1997), presented at this workshop, time-space oriented and compatibility oriented networks have been experimentally compared with respect to a specific airline crew scheduling problem for a regional carrier; with the first resulting in more appropriate solutions for that specific context.

Moreover, the choice between time-space oriented and compatibility oriented networks is affected by the number of different locations involved in the problem with respect to the number of trips. In fact, time-space networks are mostly used in the airline setting rather than in bus driver scheduling. Whatever the network, the search for a column with minimum reduced cost is modelled as a resource constrained shortest path, whose complexity is pseudo-polynomial in the number of nodes in the paths.

Regarding the solution of the master problem, the most recent approaches apply perturbation strategies to reduce degeneracy and increase the convergence rate (see DuMerle/Villeneuve/Desrosiers/Hansen (1997)).

On the other hand, much effort has been devoted to the development of cheap and quick heuristics for the SCP; these are either applied within a column generation scheme in order to compute a feasible solution at each iteration with respect to the current set of variables, or are just repeatedly re-run on (randomly) perturbed instances in order to provide a few alternative feasible solutions among which the best is selected. Greedy heuristics typically meet these requirements, where the variable with the best *score* is selected at each iteration; many such score rules have been suggested in the literature, most of which are based on the variable cost and the number of uncovered items covered by that variable. After the pioneering work of Balas and Ho in 1980, computing lower bounds for the SCP by way of a dual heuristic instead of solving the primal continuous relaxation and using reduced costs to score variables within greedy heuristics has recently become more and more popular (see Beasley (1990)). In Fisher/Kedia (1990) such tools are deeply exploited to address the mixed problem, involving both SCP and SPP, and this is the strategy behind the best performing algorithms intended to solve SCP for crew scheduling at the Italian Railway Company, as documented in Ceria/Nobili/Sassano (1995) and in Caprara/Fischetti/Toth (1997). Note that, in all these procedures, the computation of a primal feasible solution is carried out with respect to a feasible dual solution that does not change along the computation. On the other hand, in the procedure we shall describe in Sect.3.2, dual multipliers vary along the computation of a primal feasible solution for SPP, intertwining with the updating of the set of explicit primal variables within a column generation context.

2.3 The Joint Case

One of the most significant works on this subject, Ball/Bodin/Dial (1983), is almost 15 years old but the network models proposed there still inspire the most recent contributions (see for example Friberg/Haase (1997) presented at this workshop). The paper deals with the single-depot urban case, where crews are not tied to vehicles. The problem structure is exploited through two different graphs sharing the same set of nodes which model trips or portions of trips that must be operated by a single driver and a single vehicle. On each graph the arc set models vehicle/crew deadheadings, with different kinds of arcs modelling different kinds of deadheadings. For the first time the joint problem is modelled as a feasible path cover (here a *path cover* of a graph refers to a set of node disjoint paths covering all the nodes) on both graphs, although two drawbacks prevent this model from being practical: first, network dimensions are prohibitive; second, coherence between the two covers must be guaranteed. The authors resort to a problem decomposition, emphasising the crew scheduling phase where the service is initially partitioned into a set of crew pieces by heuristically solving a constrained path cover by a sequence of matching problems. This approach, suitable for the urban setting, would not straightforwardly apply to an ex-urban setting where costs of pieces vary depending on the depot of residence of the driving crew.

Since then, few proposals follow, such as Patrikalakis/Xerocostas (1992) and Tosini/Vercellis (1988), which reverse the sequential approach by scheduling crews directly on the trip set, disregarding vehicles completely, and solve a VSP in a later step.

In Patrikalakis/Xerocostas (1992) a set of runs is generated by sequencing trips using the *next-available* rule (a sequence of trips is built by repeatedly adding the closest-in-time among compatible trips). Then two set partitioning problems are solved - each one for a half day activity; the VSP is solved on the basis of the selected runs, and finally the CSP is solved again on the basis of the vehicle schedule. Note that the first selection of runs is performed with only approximate knowledge of the real features of these runs, since vehicle activity is not already defined. Moreover, according to the authors, their approach performs better when vehicle dependent constraints are less important, while an integrated approach would yield its potential benefits in just the opposite situation.

In Tosini/Vercellis (1988), a set of runs is computed according to Ball's idea extended to deal with multiple depots and introducing a stochastic flavor by searching for the M-best matching, where M is a random variable. An SCP with additional constraints is heuristically solved on this set of runs by a greedy procedure, and then a VSP is solved as a minimum cost network flow. The whole procedure is embedded within an interactive system allowing the operator to interact directly with the solver in order to modify the proposed schedule.

The second attempt after Ball/Bodin/Dial (1983) to solve VSP and CSP simultaneously is described in Freling (1997) and has been presented at this workshop. Vehicles and crews are explicitly represented in the same mathematical model; both crew dead-headings with and without vehicles are included in the model since the urban transit problem is being addressed. However, the author tackles the single depot case, exploiting the fact that the VSP with a single depot is solvable polynomially within the solution of a Lagrangean dual. Moreover, the problem of scheduling crews first has also been investigated under the name of *independent crew scheduling problem*, but the author reports that it compares unfavorably to the other approaches.

In conclusion, no practical models and approaches have been proposed so far to our knowledge to tackle *the integrated vehicle and crew scheduling problem in an ex-urban setting with multiple depots*, which is where the sequential approach most often fails. This challenge will be specifically addressed in this paper.

3 An Integer Linear Programming Model for the Joint Problem

Let us introduce our model by describing the joint problem in terms of networks. Consider the standard compatibility graph where nodes represent trips and are linked by compatibility arcs; connections with the depots are explicitly represented by introducing depot source and sink nodes connected to all the trip nodes. A block corresponds to a path from a source to a sink depot; the idea is to find a set of blocks covering the service such that this set can be partitioned into feasible vehicle duties as well as into feasible crew duties, all complying with some global constraints.

Note that, since in this particular context vehicle blocks correspond to pieces of work, we shall hereafter refer to both simply as pieces.

Two sets of variables are explicitly introduced in order to model duties and runs while blocks are implicitly described as paths from a source depot to a sink depot in terms of the arc variables modelling dh-trips.

Hereafter, vehicles are assumed to be homogeneous but our approach can easily be generalized to deal with different types of vehicles.

3.1 Definitions and Notations

Let x_j be the Boolean variable associated with run j, with cost c_j, $j \in J$, where J is the set of feasible runs; z_i be the Boolean variable associated with duty i with cost a_i, $i \in I$ where I is the set of feasible duties; $D=\{d_k\}$, $k=1,..,K$, be the set of depots with capacity C^k; $V=\{v\}$ be the set of trips to be covered. Let $J(v) \subseteq J$ ($I(v) \subseteq I$) be the set of runs (duties) covering trip v and, conversely, let $V(j)$ ($V(i)$) be the set of trips covered by run j (duty i); we shall say that run j and run j' intersect if they share a common trip, i.e. $V(j) \cap V(j') \neq \emptyset$; likewise, pieces p and p' intersect if $V(p) \cap V(p') \neq \emptyset$.

Moreover, let y_{uv} be the variable modelling the dh-trip from u to v indicating whether trips u and v are operated in sequence by the same vehicle and the same driver, and b_{uv} be its cost; for $u=d$ ($v=d$), $d \in D$, y_{dv} (y_{ud}) models the dh-trip from (to) depot d. Let DH indicate the set of all dh-trips. Finally, let $J(u,v)$ ($I(u,v)$) be the subset of runs (duties) covering the dh-trip from e_u to s_v.

The following Integer Linear Programming model fully describes our problem:

$$\min \sum_{j \in J} c_j x_j + \sum_{i \in I} a_i z_i + \sum_{(u,v) \in DH} b_{uv} y_{uv} \qquad \textbf{(P1)}$$

$$\sum_{u \in V \cup D} y_{vu} = 1 \qquad \forall v \in V \qquad (3.1)$$

$$\sum_{u \in V \cup D} y_{uv} = 1 \qquad \forall v \in V \qquad (3.2)$$

$$\sum_{j \in J(u,v)} x_j = y_{uv} \qquad \forall (u,v) \in DH \qquad (3.3)$$

$$\sum_{i \in I(u,v)} z_i = y_{uv} \qquad \forall (u,v) \in DH \qquad (3.4)$$

$$\sum_{i \in I_k} z_i \leq C^k \qquad \forall d_k \in D \qquad (3.5)$$

$$\sum_{j \in J(t)} x_j \leq R^t \qquad \forall t \in T \qquad (3.6)$$

$$x_j, z_i, y_{uv} \in \{0,1\} \qquad \forall j \in J, \forall i \in I, \forall (u, v) \in DH$$

Constraints (3.1)-(3.2) require that for each trip $v \in V$ a driver drive a vehicle to s_v within st_v, leaving from e_v at time et_v; this set of constraints can be reduced to an assignment problem. Constraints (3.3)-(3.4) require that exactly one run and one duty/no runs and no duties among the ones including the dh-trip from u to v belong to the schedule if and only if arc (u,v) does/does not belong to the matching. Therefore, set partitioning constraints such as:

$$\sum_{j \in J(v)} x_j = 1 \quad \text{and} \quad \sum_{i \in I(v)} z_i = 1 \qquad \forall\, v \in V$$

that are usually included in order to guarantee that the service is covered, are here redundant being surrogate constraints of (3.1)-(3.4), thanks to the fact that drivers are tied to vehicles.

Constraint (3.5) models capacity for each depot, requiring that no more than C^k vehicle duties can be assigned to depot k.

Finally, as mentioned in Sect. 2.2, the model must ensure that at most a fixed percentage of runs in the solution belong to certain classes of runs. At least three types of runs are usually present in a regular crew schedule, i.e. *split-shifts*, *straight-runs* and *trippers*. Split-shifts and straight-runs mainly differ in the duration of breaks and spread time: split-shifts have a long break in the middle of the run, while straight-runs have many short breaks and a shorter spread time. For this reason, split shifts are not appreciated by drivers, although it is usually necessary to include them in any efficient schedule, because of the two peak periods that characterize the service at the beginning and at the end of the work day. Therefore, union constraints tend to limit the maximum number of split shifts in the schedule. Trippers are very short runs which are often operated as overtime work; due to their cost, the company requires that they do not exceed a certain number.

Let $T=\{t\}$ be the set of different types of run; constraints (3.6) model the restriction on the maximum number of runs of type t allowed for each t in T. As long as the different types of run considered are disjoint over J, as is the case when split-shifts, straight-runs or trippers are concerned, constraints (3.6) can be implicitly handled as described in Carraresi/Girardi/Nonato (1995) and will be disregarded hereafter. The same consideration holds with respect of constraints (3.5) since set I can be partitioned into K disjoint subsets $I_1,..,I_K$, on the basis of the depot of residence d_k of each duty i.

We shall refer to (**P1**) when disregarding constraints (3.5)-(3.6) as problem (**P**).

3.2 A Network Description of the Model

Model (**P**) allows a nice network representation: consider the usual compatibility graph $G = (N, A)$ where N is isomorphic to V and $A = \{(u,v) : u \text{ is compatible with } v\}$ where compatibility now is intended for both vehicles and drivers.

Depot connections are modelled by adding additional node sets D' and D'', with $d'_k=n+k \in D'$ and $d''_k=n+\mathrm{K}+k \in D''$ $\forall\ d_k \in D$, and by adding additional arcs $A^D=\bigcup_{k=1..\mathrm{K}}\bigcup_{v\in V} \{(n+k,v), (v,n+\mathrm{K}+k)\}$ connecting each d'_k to each trip v and each v

to each d_k'', respectively. Let this new graph be $G' = (N', A')$ with $N' = N \cup D' \cup D''$ and $A' = A \cup A^D$; set A' is isomorphic to DH, the set of dh-trips.

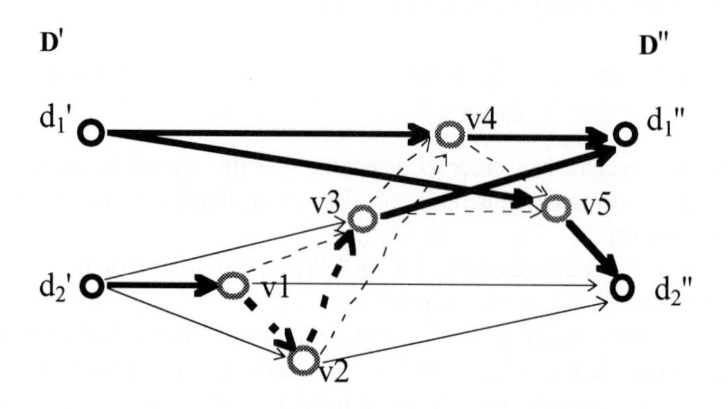

Fig. 3.1: Graph G' is depicted for 5 trips and 2 depots

In Fig. 3.1 graph G' is depicted for an example with five trips and two depots: the arcs in A are represented as dotted lines and a solution to constraints (3.1)-(3.2) is shown with thicker lines; this solution identifies the following three pieces $(d_2, v_1, v_2, v_3, d_2)$, (d_1, v_4, d_1), and (d_1, v_5, d_2).

In order to limit the number of possible connections, it is a common practice to set a threshold value on layover time, which is the time spent by the idle vehicle waiting for the beginning of the next service, and on the distance of the dh-trip. The values of these thresholds vary depending on the kind of service considered, in order to allow a reasonable set of options on how to sequence the trips. As a consequence, not all depots are considered eligible for connections with a trip, only those that are located in the neighborhood of the locations visited during the trip. Moreover, since trips are distributed over a large area, G' tends to be sparser than compatibility graphs for MD-VSP in the urban setting.

Let P be the set of paths on graph G' going from D' to D'' that correspond to feasible pieces. Two pieces $p, q \in P$ can be operated in sequence by the same vehicle only if p ends at depot d_k'' at time t_p and q starts at depot d_h' at time t_q with $h=k$ and $t_p<t_q$. This defines a compatibility relation among pieces in terms of vehicles. Let us represent such *vehicle-compatibility* relation with a compatibility graph $G^V = (P^V, A^V)$, where P^V is isomorphic to P (there is a node $p \in P^V$ for each piece in P) and there is an arc for each pair of vehicle-compatible pieces. Now consider the *crew-compatibility* relations among pieces, defined by the pairs of pieces that can be operated in sequence by the same crew. Again we can represent this relation by means of a crew-compatibility graph $G^C = (P^C, A^C)$, where P^C is isomorphic to P (there is a node $p \in P^V$ for each feasible piece in P), and there is an arc for each pair of pieces that can be operated by the same crew.

Note that the two graphs G^V and G^C have the same node set, that is $P^V=P^C$. On the other hand, the two arc sets are different: A^C is a subset of A^V since crew-compatibility among pieces is restricted by union regulations, such as the minimum duration of a break.

Each solution $(\boldsymbol{y},\boldsymbol{x},\boldsymbol{z})$ that is feasible for (**P**) induces:

- $P(\boldsymbol{y}) \subseteq P$ a set of paths on G' from D' to D'' corresponding to the partition of the trip set into pieces given by $\boldsymbol{y}$;
- $G^V(P^V(\boldsymbol{y}))$ the subgraph of G^V induced by the subset of nodes $P^V(\boldsymbol{y}) \subseteq P^V$, where $P^V(\boldsymbol{y})$ is isomorphic to $P(\boldsymbol{y})$;
- $G^C(P^C(\boldsymbol{y}))$ the subgraph of G^C induced by the subset of nodes $P^C(\boldsymbol{y}) \subseteq P^C$, where $P^C(\boldsymbol{y})$ is isomorphic to $P(\boldsymbol{y})$;
- a partition of $P^V(\boldsymbol{y})$ into a set of feasible paths on $G^V(P^V(\boldsymbol{y}))$ according to the value of $\boldsymbol{z}$;
- a partition of $P^C(\boldsymbol{y})$ into a set of feasible paths on $G^C(P^C(\boldsymbol{y}))$ according to the value of $\boldsymbol{x}$.

Note that, once $\boldsymbol{y}$ is given, (**P**) reduces to two independent set partitioning problems.

In Fig. 3.2 subgraph $G^V(P^V(\boldsymbol{y}))$ is depicted where $\boldsymbol{y}$ is the solution shown in bold in Fig. 3.1; subgraph $G^C(P^C(\boldsymbol{y}))$ is shown in Fig. 3.3: note that because of union restrictions the piece corresponding to (d_1,v_4,d_1) can not be operated by the same crew staffing piece (d_1,v_5,d_2).

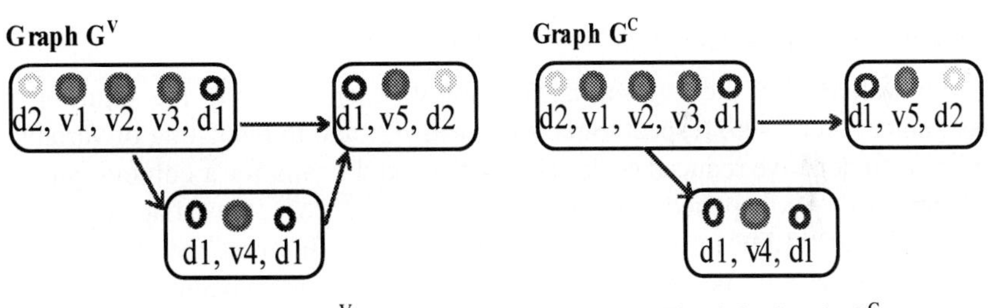

Fig. 3.2: Graph G^V **Fig. 3.3:** Graph G^C

A few remarks on these two graphs: first note that G^V and G^C are compatibility oriented networks, where nodes model pieces of work, corresponding to crew activities in G^C and vehicle activities in G^V. As in many approaches for crew scheduling, this network is more suitable for the purpose of generating columns than the trip compatibility graph, where nodes correspond to single trips. In fact, as we discussed before, local constraints affecting pieces of work are already resolved at this level, and the algorithms for constrained shortest path perform quite efficiently, since feasible paths turn out to be made by very few nodes. On the other hand, the overall number of nodes is higher than in the trip compatibility graph; this is particularly true in our case, because any feasible sequence of trips is eligible to be a piece of work, since vehicle blocks have not been previously determined. In fact, only subgraphs of G^V and G^C will be made explicit in our approach.

The Lagrangean Relaxation of constraints (3.3) and (3.4) in **(P)** leads to the following linear programming problem:

$$\Phi(\lambda,\mu) = \qquad \textbf{(LRP)}$$

$$min \sum_{j\in J} x_j(c_j - \sum_{uw\in A'(j)} \lambda_{uw}) + \sum_{i\in I} z_i(a_i - \sum_{uw\in A'(i)} \mu_{uw}) + \sum_{uw\in A'} y_{uw}(b_{uw} + \lambda_{uw} + \mu_{uw})$$

$$\text{s.t.} \quad \sum_{w\in N'} y_{vw} = 1 \qquad \sum_{u\in N'} y_{uv} = 1 \qquad \forall\, v \in V$$

$$x_j,\, z_i,\, y_{uv} \in [0,1] \qquad \forall\, j \in J,\ \forall\, i \in I,\ \forall\, (u,v) \in A'$$

where $A'(j)$ *and* $A'(i)$ denote the set of dh-trips in run j and in duty i, respectively.

In order to simplify the notation let us define the following "reduced costs" as

$$\bar{c}_j = (c_j - \sum_{(u,w)\in A'(j)} \lambda_{uw}) \qquad \forall\, j \in J$$

$$\bar{a}_i = (a_i - \sum_{(u,w)\in A'(i)} \mu_{uw}) \qquad \forall\, i \in I$$

$$\bar{b}_{uw} = (b_{uw} + \lambda_{uw} + \mu_{uw}) \qquad \forall\, (u,w) \in A'$$

The evaluation of the Lagrangean function $\Phi(\lambda,\mu)$ reduces to the solution of an assignment on G' with respect to reduced costs $\bar{b}_{uw}$, and to the listing of variables x_j and z_i with negative reduced costs. The size of J and I calls for a column generation scheme to be applied within the solution of the Lagrangean dual $max_{\lambda,\mu}\, \Phi(\lambda,\mu)$ (**LDP** in the following) but, since in this case the size of set P is also critical, P will be dynamically handled. $\Phi(\lambda,\mu)$ is maximized via a bundle algorithm for which we refer to Carraresi/Frangioni/Nonato (1995). For each value of λ,μ let $\boldsymbol{y^*},\boldsymbol{x^*},\boldsymbol{z^*}$ be any optimal solution to $\Phi(\lambda,\mu)$: a vector $[g_\lambda, g_\mu] \in R^{2A'}$ such that

$$g_{\lambda_{uw}} = y^*_{uw} - \sum_{j\in J(uw)} x^*_j \qquad \text{and} \qquad g_{\mu_{uw}} = y^*_{uw} - \sum_{i\in I(uw)} z^*_i$$

is a subgradient of the function in (λ,μ). Because of the integrality property of the relaxed problem, the optimal multipliers (λ^*,μ^*) - for which a null subgradient exists - define a set of pieces (the solution to the assignment problem with respect to (λ^*,μ^*)) and a set of duties and runs with zero reduced costs that belong to the optimal solution of the continuous relaxation of **(P)**.

4 A Heuristic Procedure Based on Lagrangean Relaxation

In this section we shall describe the outline of the heuristic approach that has been applied to solve VSP and CSP in an integrated manner, under the hypothesis that a driver staffs the same vehicle during its service from depot to depot; as we claimed, this is quite realistic in the ex-urban setting and it requires that pieces of work and blocks coincide. Therefore, the problem reduces to finding a set of pieces P^* which represents a partition of the trip set, such that P^* can be partitioned in to a set of duties as well as into a set of runs, and such that the overall cost is minimum.

In our approach, a heuristic procedure computes many alternative feasible crew schedules; each one is obtained by fixing runs in a greedy fashion guided by the values of the Lagrangean multipliers. During this process, vehicle duties are also made explicit; their contribution in the evaluation of the Lagrangean function influences the selection of the runs by way of the Lagrangean multipliers. A feasible vehicle schedule is then devised on top of the pieces of the best computed crew schedule.

Although we cannot certify the (sub)optimality of such a solution other than by considering the gap with respect to the value of the Lagrangean function, which in turn provides a lower bound for the current subproblem, it is widely believed that for this class of SP/C problems many almost-equivalent and near-optimal solutions exist and tailored dual heuristics pay off because of the large duality gap (see for example Caprara/Fischetti/Toth (1997) and Fisher/Kedia (1990)).

We first describe a greedy procedure based on reduced costs which yields a single crew schedule, under the hypothesis of being able to handle the entire piece set; then we describe how a sequence of different crew schedules are obtained, and finally how to generalize this method in order to cope with the cardinality of the piece set.

4.1 Computing a Single Crew Schedule

Suppose we are able to handle the whole set P. Let us apply a column generation scheme in a dual context. At iteration h, only a subset $J_h \subseteq J$ of the runs and a subset $I_h \subseteq I$ of the duties are made explicit. The master problem at iteration h has to solve the continuous relaxation of problem (**P**) with respect to J_h and I_h. Let us call this problem ($\mathbf{P}_h$). Consider the Lagrangean dual (**LDP**); the integrality property holds for the related subproblem. Therefore, solving ($\mathbf{P}_h$) is equivalent to solving (**LDP**) over J_h and I_h, that is maximising $\Phi(\lambda,\mu)$ with respect to J_h and I_h. Let (λ_h,μ_h) be the optimal Lagrangean multipliers returned at iteration h: they equal the optimal variables of the dual of ($\mathbf{P}_h$) associated with the relaxed constraints.

Once the master problem at iteration h has been solved, the subproblem generates runs and duties with negative reduced costs (if any) with respect to (λ_h,μ_h)

in order either to improve J_h and I_h or to assert the optimality of the current solution. This problem is modelled as the search for (constrained) shortest paths on G^C and G^V, respectively, as in all of the approaches based on column generation previously described. The only difference here is that variables of two different sets are being generated during this phase.

The sequence of multipliers $\{(\lambda_h,\mu_h)\}$ will eventually converge to the actual optimal multipliers (λ^*,μ^*) and, for complementary slackness, the duties and runs in the primal optimal continuous solution have zero reduced cost w.r.t. (λ^*,μ^*). Now we would like to devise a greedy procedure that scores variables on the basis of reduced cost. Nevertheless, because of the duality gap, we do not search for the real optimal (λ^*,μ^*) which would lead the greedy procedure towards the primal continuous solution, but rather look for some "good" multipliers close to (λ^*,μ^*).

Suppose that we have reached this stage at some iteration h, i.e. after a number of steps during which the Lagrangean function has not noticeably increased. Let (λ_h,μ_h) be the multipliers at this stage and let J_h and I_h be the set of active variables (i.e. $\bar{c}_j \geq 0 \; \forall \; j \in J_h$ and $\bar{a}_i \geq 0 \; \forall \; i \in I_h$ since the master problem is solved to optimality.) Since so far we have supposed to be able to handle the entire set of pieces P we still assume that the set of nodes in G^C and G^V at step h, P_h^C and P_h^V, are still isomorphic to P.

At this stage, a feasible crew schedule is iteratively computed in a greedy fashion by way of alternating variable fixing and reoptimization phases. This process closely relates to a family of Lagrangean-based heuristics mainly developed for Set Covering problems, although here it is combined with a column generation scheme. At each step of our procedure we solve a subproblem which is defined in a subspace of the dual space of the problem at the previous step, as well as in a subspace of the primal space. This is achieved by fixing a subset of the Lagrangean multipliers to their current value and by deleting some nodes from the node sets P^V and P^C of the graphs where columns are generated as (constrained) shortest paths. At the same time, we enrich the partial solution which will provide a feasible crew schedule.

Let us analyze these steps in more detail. At each step of the greedy procedure a set of runs is selected and added to the partial solution in the following way: the set of the currently active runs is sorted according to non-increasing reduced cost and then is scanned in that order up to a threshold value of $\bar{c}_j$. When run j is selected, the variable x_j is fixed to 1 and added to the partial solution. The search is continued until run j is found such that j intersects a run which already belongs to the partial solution, violating the set partitioning constraints. Then x_j is set to zero and the search is stopped. Note that for this partial solution complementary slackness holds with respect to the current multipliers.

Now let us see how primal variable fixing affects the graph topology. First of all, note that fixing x_j to 1 involves fixing to 1 all variables y_{uv} such that $j \in J(u,v)$, and fixing to 0 each variable y_{wv} with $w \neq u$ and each variable y_{uw} with $w \neq v$. Graph G' is modified accordingly to the updating: each node corresponding to a trip v in $V(j)$ has a single entering arc and a single outgoing arc; for each piece belonging to a selected run, there will be a unique path in G' from a node in D' to a node in D'' covering the trips which belong to that piece. As a consequence, all runs other than j covering trip v in $V(j)$ must be set to zero. In order to accomplish this restriction in the forthcoming column generations, all pieces covering any v in $V(j)$ are removed from set P^C, included the pieces belonging to run j. On the other hand, regarding vehicle duties, a feasible solution may exist in which a duty is made up by a piece of run j as well as by other pieces not in run j. In fact, although we have supposed that drivers are tied to vehicles, we expect vehicles to be exploited more heavily than crews. In order to allow the generation of duties made by pieces that belong to runs already scheduled, P^V, the set of nodes in graph G^V, must be updated as follows: all pieces intersecting with pieces in run j must be deleted from P^V, except for the ones actually belonging to run j.

This process yield a subproblem to be solved during the next iteration which is defined in a subspace of the primal space of the current problem; moreover, as a result of this fixing operation, the Lagrangean multipliers λ_{uv} relative to variables y_{uv} which have been fixed either to 0 or to 1, and the multipliers μ relative to constraints (3.4) of y_{uv} fixed to 0 are set to their current values and will not be further modified, thus defining a subproblem both in the primal as well as in the dual space.

The Lagrangean function $\Phi(\lambda,\mu)$ is again approximately optimized in the resulting subspace by way of column generation with respect to the updated sets P^V and P^C, thus providing a different dual point. The process is repeated until all trips are covered, yielding a feasible crew schedule $\boldsymbol{x}^F$ a set of dh-trips $\boldsymbol{y}^F$ and a set of multipliers (λ^F,μ^F) which is in general different from the dual point (λ_h,μ_h) w.r.t. which the variable fixing phase has first begun.

4.2 Generating a Sequence of Feasible Crew Schedules

Since we are solving an SPP by way of a column generation scheme, only a subset of the primal variables is made explicit at a time; therefore, once a primal partial solution has been set through variable fixing, the optimal continuous solution of the resulting subproblem might involve variables not presently generated, corresponding to a different point in the dual space for which their reduced cost is zero. We suggest that this situation may happen as soon as the currently selected run intersects with runs in the partial solution, which is when we update our primal and dual solution by way of solving a related subproblem, as previously mentioned. Therefore the computation of a primal solution is performed with respect not to one but to a series of varying dual multipliers, which in turn implies some drawbacks.

Complementary slackness obviously holds for $\boldsymbol{x}^F$, $\boldsymbol{y}^F$ and λ^F - runs in the current solution have all zero reduced cost with respect to λ^F - but λ^F may be dual infeasible, that is other runs in J may have negative reduced cost - casting doubts on the quality of the current solution. If on one hand the duality gap partly explains it, on the other hand this is also due to the lack of feed-back in the greedy procedure. For this reason the process is repeated a few times in order to collect several solutions among which the best is eventually selected. Nevertheless, since each step is rather time consuming, the primal solution is not re-computed from scratch, but only a subset of the current solution is put back into play according to the following criterion. Starting from (λ^F,μ^F) (the dual point at the end of the computation of the primal solution), the procedure for maximising $\Phi(\lambda,\mu)$ is applied performing column generation on the whole primal space, yielding a new sub-optimal point (λ',μ'). The new point (λ',μ') differs from (λ_h,μ_h) (the point from where the greedy procedure started), since none of them is the optimal dual point (λ^*,μ^*), as will be explained later. The runs to be put back into play are those whose reduced cost, which is zero in (λ^F,μ^F), is the most changed in (λ',μ'). More formally, let J^F be the set of runs in the current schedule $\boldsymbol{x}^F$: their reduced cost with respect to (λ',μ') - say $\bar{c}'_j$ - is evaluated to identify $J' = \{j \in J^F : \bar{c}'_j \geq \varepsilon > 0\}$, the set of runs whose reduced cost has increased over a threshold ε; the partition on $V(J') = \cup_{j\in J'} V(j)$, which is the set of trips covered by runs in J', is brought back into play, while the other runs in J^F are kept in the solution.

4.3 A Modified Column Generation Scheme

In the previous paragraphs, column generation is supposed to be performed on the whole primal space, which involves computing shortest paths on the whole graph G^V and constrained shortest paths on the whole graph G^C. Now, while the first is straightforward, the second is prohibitive, which suggests a dynamic strategy to handle set P.

At each step, a list of active pieces P', runs J' and duties I' is maintained. Periodically, pieces p not in P' resulting from the solution of the assignment are added to P' and column generation is performed to add runs to J' and duties to I' which involve piece p; each p has a label representing its contribution to the last evaluations of the Lagrangean function. This label is progressively decreased, except that it is reset to the initial value whenever p belongs either to the assignment or to duties and runs in J' or in I' with negative reduced cost. An overflow control based on reduced cost keeps J' and I' within a fixed size (over 5,000).

This strategy differentiates our procedure from “classical column generation schemes” since here the master problem and the subproblem deeply intertwine, causing the polyhedral description of the Lagrangean function to vary dynamically according to the updating of P', J' and I'. Note that the bundle algorithm used to maximize $\Phi(\lambda,\mu)$ can handle this fact. Moreover, the knowledge of the cost of a previous primal feasible solution can be exploited within bundle algorithms providing an upper bound on the value of $\Phi(\lambda,\mu)$ which improves the convergence rate of the algorithm.

This procedure is integrated with the algorithm described in the previous section, introducing diversification in the outcomes of the different iterations. This fact can be seen as a dual variant of cost perturbation, which is quite popular in heuristic approaches to SP/C problems to avoid degeneracy. In fact, in those cases in which it is almost impossible to compute the real monetary cost of each individual run, and when the main target consists of covering the service with the minimum number of crews, the objective function is a unicost function which does not discriminate among feasible runs; the resulting problem is heavily degenerate, which is where one most resorts to cost perturbation techniques.

5 A Feasible Solution for the Vehicle Scheduling Problem

Once all pieces of work induced by the crew scheduling solution are set, vehicle duties must be devised. Note that our problem is slightly different from a usual MD-VSP since blocks are already defined. Consider the network described for the multicommodity flow model, where now each node corresponds to a piece of work. Since here pieces coincide with blocks, each piece starts and ends at some specific depot; therefore, compatibility holds between two nodes only if the associated pieces share the same connecting depot.

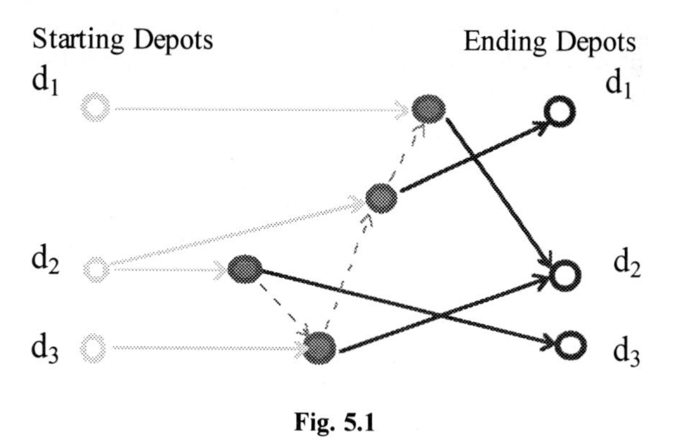

Fig. 5.1

In Fig. 5.1 depot connections are depicted, with dotted arcs representing compatibility between pieces. Note how compatibility is affected by depot connections. A feasible vehicle duty is a path in which the ending depot of the last node is the same as the starting depot of the first node. The resulting network is much smaller than in the original MD-VSP and also very sparse, potentially allowing for good performance of exponential solution algorithms; on the other hand the problem might be infeasible, i.e. a disjoint feasible path cover satisfying depot capacities might not exist - note that the crew schedule itself provides a feasible solution as long as it meets capacity constraints. In our experience, a feasible solution to the MD-VSP is straightforwardly reached by solving a set partitioning problem with side constraints with column generation performed on this network.

6 Computational Results

The integrated approach was implemented in C++ language and run on a Power PC 604, 180Mh. In order to compare it with the sequential approach, our prototype was tested against the commercial package MTRAM®, which is currently in use at several Italian municipal mass transit companies. MTRAM contains two scheduling applications, namely VS_Alg® and BSD_Alg®; the former solves MD_VSP while BDS_Alg solves CSP. VS_Alg and BSD_Alg were used in pipeline in order to implement the sequential approach.

On behalf of a fair comparison, our prototype obviously included all union constraints dealt with by the two commercial packages. Union constraints regarding feasible runs, also according to the urban case, mainly concern bounds on spread time, working time and driving time, and involve meal-breaks; peculiar to the ex-urban setting is a new type of run, the so-called *extended spread-time run* together with a global constraint allowing no more than 20% of extended runs over the set of runs assigned to each depot. We found that this new feature increases the computational burden, consistently with the widespread finding that global constraints increase the fractionality of the continuous solution of SPP.

The values of the cost coefficients that we used for this testing reflect the priorities of the transit companies in this particular setting: the main target is saving on the crews, while minimising the number of vehicles is a secondary target. Therefore, we set the cost of a run approximately equal to twice of the cost of a vehicle. There are different types of vehicles, and for each trip the *suggested* type is known, corresponding to the most suitable type for that service. The set of all types of vehicle that can feasibly operate the trip is also given for each trip. The existence of the compatibility arc between trip u and trip v is subordinated to the existence of a non-empty intersection of the feasible type sets of u and v. The cost of the arc (u,v), if present, is such that it encourages the scheduling of dh-trips between trips of the same suggested type. Moreover, the cost of a dh-trip (u,v) depends on the distance covered by the dh-trip, from the end location of u to the start location of v, and on the duration of the layover, i.e. the idle time after trip u

and before trip v. Therefore, the cost associated with the vehicles models capital cost, while the costs of the dh-trips model operational costs of the vehicle schedule.

In the computational results tables we have omitted the value of the objective function and only report the number of crews and the number of vehicles of the solutions which are the target of this study.

The two approaches were tested on samples drawn from two different data sets, S1 and S2, which are of different nature. Set S1 is a sample of typical ex-urban service, while set S2 concerns sub-urban service. The aim of this testing was to compare the performance of the two approaches in these two different settings, in order to identify under which conditions either of the two is more suitable, if any.

More specifically, set S1 refers to the ex-urban service of Bologna, Italy, where there are 28 depots at an average distance of 25 km between depots, which is quite typical for Italy. Here, MTRAM is currently used by the municipal company to schedule just the the urban service, while the ex-urban service is still scheduled manually. Eight samples of increasing size were drawn from the whole trip set.

In Tables 6.1 and 6.2 the number of trips is shown for each sample, D-Seq and R-Seq stand for the number of vehicle duties and runs obtained by MTRAM, respectively, while D-Int and R-Int refer to the number of vehicle duties and runs obtained by our approach, respectively. When MTRAM failed to provide a feasible solution, a number of trippers were produced and their number is reported. Actually, these do not necessarily correspond to the usual trippers which are runs much shorter than the minimum allowed, but rather are portions of vehicle duties that could not be included in any feasible run or real tripper.

Let us comment on results in Table 6.1. MTRAM was not able to produce a feasible crew schedule in five tests out of eight, while our prototype always was. For example, in test 1, VS_Alg yielded four duties which could not be staffed since two involved tasks too long to be contained in a piece and another two duties had tasks covering the meal break period. Moreover, note that when MTRAM is able to solve the problem, as in tests 3, 5, and 6, our approach solves it as well or even outperforms it, since it yields the same number of duties and saves a few runs.

Table 6.1: Results for samples from data set S1

Test	Trips	D-Seq	R-Seq	Tripper	D-Int	R-Int
1	60	15	17	6	15	21
2	79	21	31	4	21	33
3	117	41	27	-	41	27
4	125	18	24	3	20	23
5	139	23	30	-	23	28
6	180	27	35	-	27	33
7	200	41	58	4	44	65
8	257	34	49	2	36	45

These results encourage the use of an integrated approach in the ex-urban setting, although we have to mention that running time is still critical for the prototype we implemented. Cpu time for test 8 (the largest in size) is over 24 hours and on average it varies from two to six times that required by MTRAM, which in turn ranges from 2 to 6 hours on a Power PC 604, 180 Mhz.

Moreover, we should also mention that the running time of our approach is obviously affected by the fact that we iterate at least 20 times the computation of a feasible solution, although, in most computational tests, the best solution was found within the first 10 iterations. In fact, we lack a real lower bound to evaluate the quality of a single solution, since column generation is never computed on the whole set of pieces. Therefore, as a stopping condition for the overall procedure, we adopted the lack of improvement of the currently best solution for the next 5 iterations, which is rather time consuming.

As mentioned before, the approach was also tested on a mixed urban and sub-urban service and results are illustrated in Table 6.2. The set of data S2 refers to a case with four residence depots plus ten relief points and more than one thousand trips: six sets of data were drawn from it, considering not necessarily disjoint subsets of the whole service, with trips ranging from 120 to 135. In these tests, the characteristics of the samples are further described in terms of the number of pairs of compatible trips, which corresponds to the cardinality of set A, and the number of allowed depot connections, which corresponds to the size of set A^D. We noticed a positive correlation between the incidence of urban service in the sample and the number of pairs of compatible trips, which is plausible since in the urban setting the service is concentrated in a relatively small area and, on average, travel distance between start and end locations of trips is short.

Table 6.2: Results for samples from data set S2

Test	Trips	$\|A\|$	$\|A^d\|$	D-Seq	R-Seq	D-Int	R-Int
1	121	2188	630	23	41	23	42
2	130	608	1475	23	45	24	44
3	110	821	703	21	40	21	39
4	135	681	1320	25	47	27	45
5	127	1147	609	24	42	24	42
6	120	871	806	22	43	23	42

On this set of data our prototype shows uneven behavior. In tests 2, 3, 4, and 6 it saves a few runs at the cost of introducing additional vehicles, while in tests 1 and 5 it equals the result of the sequential approach, or is even worse than it. This may be explained by the fact that these two samples include a larger percentage of the urban service, for which the integrated approach is not suitable. In fact, the number of trips has a combinatorial effect on the number of pieces unless blocks are already defined, as happens in the sequential approach where vehicles are scheduled first, and a piece is necessarily contained within a block.

Therefore, further research should tackle the case of sub-urban service, for which neither of the two tested approaches proved fully successful.

7 Conclusions and Further Research

In this paper we have described an integrated approach for the scheduling of vehicles and crews in an ex-urban setting, and tested a prototype of our approach against a commercial package which implements the sequential approach.

We can state that in the ex-urban setting specific tools are required to solve scheduling problems. The results are rather encouraging: also according to several other papers in the literature, it now seems possible to address a problem heuristically that previously had appeared to be out of reach. Our approach proved to be rather suitable for this specific purpose, although computational time is still an issue. It is reasonable to believe that some improvement in this respect could be achieved by a professional rewriting of the code. On the other hand, it is a matter of fact that the solution of the assignment problem required by the evaluation of the Lagrangean function represents a bottleneck of our approach; moreover, the lack of implementable global optimality conditions requires additional computation in order to satisfy stopping conditions based on lack of improvement.

Concerning computational time, it is hard to compare with the two other approaches recently published in the literature which tackle the integrated solution of vehicle and crew scheduling. Computational time is a key issue also in Friberg/Haase (1997): there, the LP relaxation of their model is solved exactly in almost 6 hours on a SUN Sparc-station 10/40 for a test problem of 30 trips. Interesting results are reported in Freling (1997), where instances up to 148 trips are heuristically solved within 1 hour on a Pentium 90 pc with 32Mb, with most of the time devoted to the approximate solution of a Lagrangean Dual of his model. Nevertheless, we should remark that both approaches address the single depot case, while we deal with the general case, and this fact is exploited in their solution procedures. On the other hand, we exploit the assumption that crew dead-headings correspond to vehicle dead-headings, and therefore we actually solve a very specific problem which is particularly suited for the ex-urban case, but probably does not fit the sub-urban setting, as results in Table 6.2 suggest.

In order to overcame the drawbacks of the approach described here, we should bypass the explicit generation of pieces and generate runs directly from trips. This leads us back to the reverse approach, where crews are scheduled before vehicles, (the independent problem in Freling (1997)). We are currently testing this option and found, as expected, a considerable speed up in the computational time although the single column generation step is more time consuming, together with the risk of imposing conditions which violate the depot capacity constraints. A very simple example is shown in Fig. 7.1.

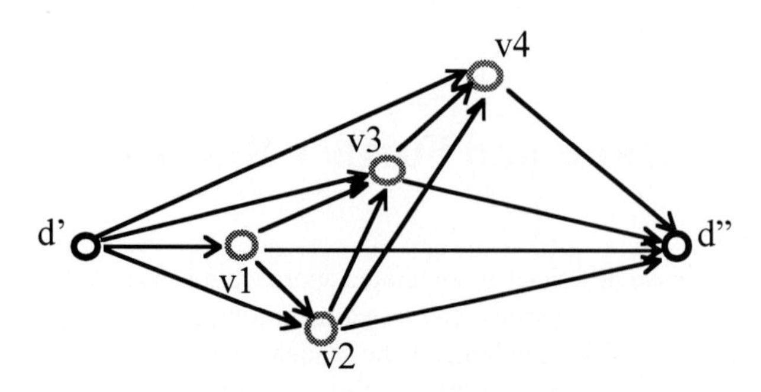

Fig. 7.1: A tricky instance for the reverse approach

Suppose that four trips are given with compatibility as depicted in Fig. 7.1; five pieces are possible, namely (d,v1,v2,d), (d,v1,v3,d), (d,v2,v3,d), (d,v2,v4,d), and (d,v3,v4,d). Moreover, suppose that it is possible to drive back to the depot after trip v2 in time to start trip v3. The solution involving the first and the last piece is the only one which requires a single vehicle, but the reverse approach does not discriminate, since vehicle duties are not considered.

Finally, we feel that integrating vehicle and crew scheduling is a challenging field of research, both for modelling and for techniques to tackle large scale linear integer problems.

References

Beasley, J.E. (1990): A Lagrangean heuristic for Set Covering problems. in: Naval Research Logistics 37, 151 - 164

Balash, E./Carrera, M.C. (1996): A dynamic subgradient based branch and bound procedure for set covering. in: Operations Research 44 (6), 875 - 890

Balash, E./Ho A. (1980): Set Covering algorithms using cutting planes, heuristics and subgradient optimization: a computational study. In: Mathematical Programming 12, 37 - 60

Ball, M./Bodin, L./Dial, R. (1983): A matching based heuristic for scheduling mass transit crews and vehicles. in: Transportation Science 17, 4 - 13

Bertossi, B./Carraresi, P./Gallo, G. (1987): On some matching problems arising in vehicle scheduling models. in: Networks 17, 271 - 281

Caprara, A./Fischetti, M./Toth, P. (1997): A heuristic procedure for the Set Covering problem. in: Operation Research 45

Carraresi, P./Frangioni A./Nonato, M. (1995): Applying bundle methods to the optimization of polyhedral functions: an application-oriented development. in: Ricerca Operativa 25 (74): 4 - 49

Carraresi, P./Gallo, G. (1984): Network Models for Vehicle and crew Scheduling. in: European Journal of Operational research 38, 121 - 150

Carraresi, P./Girardi, L./Nonato, M. (1995): Network models, Lagrangean relaxation and subgradient bundle approach to crew scheduling problems. in: Daduna, J.R./ Branco, I./Paixão, J.P. (eds.), 188 - 212

Ceria, S./Nobili, P./Sassano, A. (1995): A Lagrangean-based heuristic for large-scale Set Covering problems. Report #406, University of Roma La Sapienza

Clement, R./Wren, A. (1995): Greedy genetic algorithms, optimizing mutations and bus driver scheduling. in: Daduna, J.R./ Branco, I./Paixão, J.P. (eds.), 188 - 212

Daduna, J.R./Branco, I./Paixão, J.P. (eds.) (1995): Computer aided transit scheduling. (Springer) Berlin, Heidelberg, New York, London, Paris, Tokyo

Daduna, J.R./Paixão, J.P. (1995): Vehicle scheduling for public mass transit: an overview. in: Daduna, J.R./ Branco, I./Paixão, J.P. (eds.), 76 - 90

Daduna, J.R./Wren, A. (eds.) (1988): Computer aided transit scheduling. (Springer) Berlin, Heidelberg, New York, London, Paris, Tokyo

Desaulniers, G./Desrosiers, J./Gamache, M./Soumis, F. (1997): Crew scheduling in air transportation. in: Crainic, T./Laporte, G. (eds.): Fleet management and logistics. Montreal

Desaulniers, G./Desrosiers, J./Ioachim, I./Solomon, M.M./Soumis, F./ Villeneuve, D. (1997): A unified framework for deterministic time constrained vehicle routing and crew scheduling problems. in: Crainic, T./Laporte, G. (eds.): Fleet management and logistics. Montreal

Desaulniers, G./Desrosiers, J./Lasry, A./Solomon, M.M. (1997): Crew pairing for a regional carrier. Paper presented at the Seventh International Workshop on Computer-Aided Scheduling, Boston.

Desrochers, M./Rousseau, J.M. (eds.) (1992): Computer aided transit scheduling. (Springer) Berlin, Heidelberg, New York, London, Paris, Tokyo

Desrochers, M./Soumis, F. (1989): A column generation approach to the urban transit crew scheduling problem. in: Transportation Science 23, 1 - 13

Du Merle, O./Villeneuve, D./Desrosiers, J./Hansen, P. (1997): Stabilisation dans le cadre de la generation de colonnes. Les Cahiers du GERAD G-97-08, Montreal

Fisher, M./Kedia, P. (1990): Optimal solution of Set Covering/ Partitioning problems using dual heuristics. in: Management Science 36 (6), 674 - 688

Freling, R. (1997): Models and techniques for integrating vehicle and crew scheduling. Tinbergen Institute Research Series 157, Amsterdam

Friberg, C./Haase, K. (1997): An exact algorithm for the vehicle and crew scheduling problem. Paper presented at the Seventh International Workshop on Computer-Aided Scheduling, Boston.

Hoffman, K.L./Padberg, M. (1993): Solving airline crew scheduling problems by branch-and-cut. in: Management Science 39, 657 - 682

Kwan, A./Kwan, R./Wren, A. (1997): Driver scheduling using genetic algorithms with embedded combinatorial traits. Paper presented at the Seventh International Workshop on Computer-Aided Scheduling. Boston.

Lavoie, S./Minoux, M./Odier, E. (1988): A new approach for crew pairing problems by column generation with an application to air transportation. in: European Journal of Operational research 35, 45 - 58

Marsten, R.E./Shepardson, F. (1981): Exact solution of crew problems using Set Partitioning model: recent successful applications. in: Networks 11, 165 - 177

Patrikalakis, I./Xerocostas, D. (1992): A new decomposition scheme of the urban public transport scheduling problem. in: Desrochers, M./Rousseau, J.-M. (eds.), 407 - 425

Ribeiro, C./Soumis, F. (1991): A column generation approach to the multiple depot vehicle scheduling problem. in: Operations Research 42 (1), 41 - 52

Rousseau, J.-M. (ed.) (1985): Computer scheduling of public transport 2. North Holland, Amsterdam

Ryan, D.M./Falkner, J.C. (1988): On the integer properties of scheduling Set Partitioning models. in: European Journal of Operational research 35, 442 - 456

Scott, D. (1985): A large scale linear programming approach to the public transport scheduling and costing problem. in: J.-M. Rousseau (ed.), 473 - 491

Tosini, E./Vercellis, C. (1988): An interactive system for the extra-urban vehicle and crew scheduling problems. in: Daduna, J.R./Wren A. (eds.), 41 - 53

Producing Train Driver Schedules Under Differing Operating Strategies

Ann S.K. Kwan, Raymond S.K. Kwan, M.E. Parker and Anthony Wren
Scheduling and Constraint Management Group, School of Computer Studies, University of Leeds, Leeds, LS2 9JT, United Kingdom

Abstract: The privatization of British Rail into twenty five train operating companies and three freight companies has highlighted the need for each company to have efficient operating schedules. Manpower costs are a significant element in any transport organization, and the ability to minimize these is seen as crucial to the effective running of these companies. In addition, the need to try out different operating strategies is gaining in importance as the search for cost cutting progresses.

Following a feasibility study conducted with British Rail in 1990-91 the paper describes more recent work in which a new driver scheduling system TRACS II has been developed to take account of the conditions relating specially to the rail situation.

Taking data from several train operating companies and talking to many others, both passenger and freight, a database of train work exhibiting a wide variety of different operating strategies has been established. This has enabled the development of new software tools and indicated new areas where further research could be useful.

The driver schedules produced for the train companies have shown savings in manpower requirements over those currently in operation. This situation has been achieved consistently, and savings improved as new techniques developed through the course of the work. The ease with which the drivers' labor agreement could be changed has greatly impressed the staff in the companies with whom we have worked and further research has been independently promoted. The paper will show how these savings have been achieved and indicate the types of computer techniques employed.

1 Introduction

Automatic scheduling of train drivers using computers is rare. This may seem surprising when computer scheduling of bus drivers, which is similar in theory, is already widely practiced. Our first insight into the complexity of scheduling train drivers was gained from a project during 1990-91 in collaboration with the Operational Research Unit of British Rail (Wren/Kwan/Parker (1995)). In that project, part of the IMPACS (Smith/Wren (1988)) bus driver scheduling system was adapted for quickly estimating cost implications associated with options in restructuring driver work rules.

Between September 1994 and August 1996, we were engaged in a two-year project sponsored by the Engineering and Physical Sciences Research Council (EPSRC) in the UK to investigate further the special driver scheduling conditions applicable to train operators with the aim of ultimately producing a system suitable for private rail operators. The aim of this project was to undertake scheduling problems from a number of UK train operating companies and learn through real scheduling experience with them, thereby further enhancing the pilot system. The new system is called TRACS II. These case studies are described later in the paper.

Rather to our surprise, TRACS II was able to produce very early in the research driver schedules that were comparable in quality with, and often better than, those being operated. Many train companies appreciate the speed and accuracy of TRACS II, and several of them commissioned projects using it for investigating alternative scenarios and for assisting in production of operational driver schedules.

TRACS II continues to evolve to incorporate the experience gained from direct exposure to real train driver scheduling problems, as well as to exploit results of other on-going research at Leeds. The complex issues of train driver scheduling will be described and some of the recent developments will be reported.

2 Complexity of the Train Driver Scheduling Problem

2.1 Multi-Depot Schedules

In the U.K. bus industry drivers are usually restricted to vehicles from their home depots, so that where several depots are involved there are effectively a number of separate scheduling problems. All the train companies with whom we have worked have had several (as many as fifteen) depots, with drivers allowed in principle to drive trains from any depot provided that they return to the same starting depot at the conclusion of their shifts. Constructing a set of shifts, each of which satisfies the conditions at some depot, is non-trivial. There may be various constraints such as route and traction knowledge (see later). Some rules and parameters (e.g. time allowances for signing on) are depot-related, and some of them are adjustable in order to satisfy other rules. For the schedule as a whole, there may be shift-type specific capacity constraints to be satisfied for some depots.

2.2 Problem Size

A driver scheduling problem may be large on two different counts: it may be large because there is much work to be covered requiring many shifts, e.g. in excess of 150; it may also be large, computationally speaking, because there are very many relief opportunities close together in time. In the latter case the number of shifts required to operate the train schedule may not necessarily be large, e.g. 50 shifts. In the former case, the problem may be decomposed intelligently into smaller subproblems. In the latter case, some of the relief opportunities will be de-selected using heuristics so that the problem size is reduced. Some problems may require both decomposition and relief opportunity de-selection.

Four out of the six major case studies described in section 5 have schedules with the number of shifts in excess of 150. All four required decomposition and one also required relief opportunity de-selection

2.3 Drivers Traveling as Passengers

In order to optimize overall efficiency it is essential to transport drivers between relief points. In such circumstances the driver becomes a passenger, either on trains which are the subject of the current schedule, or on trains run by other operators, or even on other modes of transport such as taxis. A good scheduling system must be able to consider all such possibilities. Early versions of TRACS II catered for drivers travelling between relief points either by assuming a standard travel time, or by assigning more than one driver to cover the work; the excess drivers would in fact be passengers. Both methods have their shortcomings. Assigning more than one driver is achieved by having overlapping shifts; the train work assigned in these circumstances is said to be overcovered. The driver travelling as a passenger is subject to the constraints imposed upon *drivers*, which will be more stringent than those for travelling as a passenger. In addition it only allows drivers to travel on trains for which they have the appropriate route and traction knowledge. Using standard travelling times may give an inaccurate time allowance which might either be too generous or unachievable in practice.

2.4 Shifts of More than Three Spells

Initially TRACS II allowed drivers to work on at most three train units. In rail operation a shift may be more fragmented, with work on several different units. In such circumstances, the number of possible shift combinations increases exponentially with the number of train units allowed in a shift. We have extended TRACS II to allow up to four units in a shift, and we may extend it further for certain circumstances.

2.5 Route and Traction Knowledge

Route knowledge considerations are vital in preparing rail driver schedules. Rail drivers have to be trained to operate on every route on which they may drive, and have to maintain that knowledge by being assigned frequently to work every route within their operating domain. For this reason, drivers are generally restricted to certain sections of the total operating area dependent on their home depots. These sections usually overlap, so that it is not possible to subdivide the problem into areas of specific knowledge. The same argument applies to traction knowledge where certain drivers may have the operating knowledge of some, if not all, of the traction types at a particular depot or depots.

2.6 Constraints on Schedule Composition

There are often rules governing the types of shift present in a schedule. These are not so much concerned with when in the day the shifts operate, as is often the case in the bus industry, but are more concerned with the lengths of the shifts. The most common constraint on the overall shape of a schedule concerns the proportions of shifts of different lengths. For example, one constraint is that not more than 50% of the shifts should have a length (from sign on to sign off) greater than 8 hours and not more than 20% should exceed 8-1/2 hours. Examples of other constraints on schedule composition are bounds on the number of shifts for each depot and bounds on average shift length.

2.7 Meal Break Rules

There may be either one or two meal breaks in a shift. There are complicated rules governing when the meal break should occur, e.g. it must occur between the third and fifth hour relative to the start of the shift if its length is less than 8 hours; a different set of rules applies if the length is more than 8 hours; variations on these rules apply if there are two meal breaks. Some manipulation of the starting or finishing times of a shift may be necessary in order to ensure that the meal break falls in an acceptable position. This is done so that shifts which are crucial to the formation of a good schedule are included.

2.8 Other Aspects

Other features of train driver scheduling include:

- train work which could form the first part of an early shift or the last part of a night shift from the previous day, depending on how it best contributes to overall efficiency.
- the requirement that some operators might want schedules with a mixture of shifts of wide ranging lengths, e.g. 4 hours to 12 hours.

3 The TRACS II Scheduling Suite

The new scheduling system has been developed, both to combine perceived needs of bus and rail operators, and to improve the inherent heuristic and mathematical processes. This new system, TRACS II, formed the basis of the current work. In this section TRACS II, as it was at the start of the project, is outlined. This was essentially a bus driver scheduling system. It consisted of a suite of programs performing separate functions, as described below. Later sections of the paper present enhancements made during the project.

The system is driven by three main parameter sets. The first set describes the train work for which drivers are to be provided, and specifies geographic data, such as when and how drivers can travel between points of the network. The second gives the scheduling rules, for example, lengths of shifts of different types, position and duration of meal breaks, etc. The third establishes some global constraints and governs the running of the system.

The system works in two main stages. In the first stage very many possible legal shifts which cover all the train work are constructed. In the second, the minimum number of shifts which between them cover all this work are selected.

Clearly very many shifts may be constructed in the first stage and some information in the second parameter set is designed to prevent an excessive number of shifts being formed: this may describe the minimum length of shifts, the minimum driving time on one unit, the maximum length of meal break, etc. There is also a special routine within the shift construction process in which it is possible to accommodate any rules which cannot be dealt with in the data. It is sometimes necessary to have a unique version for each operating company. Provided all the train work has been covered, there may still be very many thousands of shifts from which to make a selection. Sensible heuristics have been devised to discard those shifts which are unlikely to make a contribution to the final schedule.

The second stage of the process takes all those shifts remaining after the elimination of unlikely ones and employs a set covering algorithm to select a set of shifts which together cover all the train work in a near optimal way.

3.1 Original System Components

BUILD is the major process of the first stage and generates many thousand potential shifts each satisfying all the legal requirements. It was founded on developing sensible chains of meal breaks.

COMPARE examines the shifts just built, eliminating any shift which is fully contained in another. (This means that some shifts may overlap after the later scheduling processes, but such overlaps can be eliminated by the user specifying which of the shifts concerned is to carry out the driving work: the other shift or shifts can then either be reduced in length or have its driver booked to travel as passenger.)

EVEN is a crude process used in IMPACS (the original bus driver scheduling package) to eliminate some shifts whose work is entirely covered by many other shifts. It has been replaced by SIEVE (see later).

SCHEDULE is a set covering system fully described in Smith/Wren (1988) and Wren/Smith (1988). Using a specially developed linear programming tool, it selects that subset of shifts which covers all the work in the most suitable way consistent with the problem definition.

SPRINT optionally manipulates the schedule to make marginal improvements by swapping halves of shifts or adjusting hand-over times, and prints the changes made. This may reinstate some shifts eliminated by the COMPARE or EVEN routines. It may also make improvements in overall quality where costs are identical. It continues by printing the schedule. The optional manipulation has not yet been extended to deal with the complexities of rail operation.
These inter-linked programs are normally run serially.

4 Recent TRACS II Developments

It was always recognized that a significant number of enhancements would have to be made to the system under the EPSRC sponsored research. The contacts with train operating companies added to the urgency of this. In this section we describe the principal adjustments. The suite is run corresponding to the following major phases:

4.1 Pre-Processing of Scheduling Data

This phase is responsible for working out the possibilities for drivers to travel as passengers.

The original system catered for drivers travelling between relief points either by assuming a standard travel time, or by assigning more than one driver to a unit as part of a shift. With the lower frequency of train services, and with the need to travel on trains operated by other companies, a special feature had to be added to determine possibilities of travel on any timetabled trains, or indeed on other transport modes. The use of a standard travelling time is far from ideal.

A new program called TRAVEL has been developed to output automatically a timetable for each relief opportunity. TRAVEL uses train information from the existing problem as well as that of other companies or modes, e.g. train times on other networks, underground times, taxi journeys etc. TRAVEL produces two different sets of travelling times for each relief opportunity: the first contains the times needed to travel *from* the relief opportunity to other relief points; the other is the travelling time needed to travel *to* the relief opportunity from other relief points. These times will be used by BUILD to enable the driver to travel as a passenger if required. The calculation of these travelling times follows a simplified shortest path algorithm. When this feature is required, the user may indicate if there are any points between which travelling is highly unlikely or impossible because of the distance or time involved.

A recently added feature is the ability to allow for 'leeway' whenever the driver has to change train, either on duty or travel as a passenger, as a safety margin against train delays.

4.2 Building a Large Shift Set

The fundamental approach of TRACS II is to generate a set of all the valid and potentially good shifts which must satisfy certain union agreements or work practices and together cover all the train work, and from this set select the minimum number of shifts which cover this work at least once. A number of heuristic rules are employed in the BUILD phase so that the number of shifts constructed is not excessive, but it is important to ensure that no shifts vital to an efficient schedule are omitted.

Before we present the BUILD process it is important that we define certain terms. In common parlance, a *train* is a single journey undertaken by a *set* of rolling stock. When journeys are relatively short we may wish to allocate a driver to a sequence of journeys; we may also wish to allocate a driver to part of a journey. We define a *train diagram* as a sequence of journeys undertaken by an individual set of rolling stock.

The BUILD process first looks at the work on each train diagram and marks blocks of consecutive work known as *spells*. A spell may include several train journeys, and may start or finish at an intermediate point of a journey, if reliefs are allowed there. The length of a spell is governed by the union agreement. BUILD then transforms these spells into *stretches* either by using a single spell or combining two spells linked with a gap. This gap must be no shorter than the minimum time required to travel between the relevant points. If there is enough time for a meal, this may be allowed; otherwise the gap is called a joinup. These stretches may also be used to form single stretch shifts.

BUILD will avoid forming stretches which are obviously inefficient, but will not reject stretches which are included in other stretches. Stretches which are obviously inefficient are those which have two spells, separated by a joinup, on the same train diagram. The reason for not rejecting stretches which are included in other stretches is that these smaller stretches might be used to form some vital combinations with another stretch to form a shift.

After all the possible stretches are formed, each of them, if possible, will be linked with another stretch with a gap to form a valid shift. This gap in the middle is usually a meal break but can also be a joinup. Each shift can have a maximum of four spells with at least one meal break. The general strategy is to impose progressively more restrictive heuristics in the shift formation process according to the number of spells allowed in a shift, so that for example only exceptionally efficient four-spell shifts would be formed.

Since the potential number of combinations is enormous, heuristics are used to remove shifts which are inefficient. Where there are two shifts containing work on identical train diagrams, with one of them having a meal break starting later and finishing earlier than the other, all other things being equal, the shift with a longer meal break will be rejected. Another heuristic cuts down the number of three-spell shifts by comparing them with similar two-spell shifts.

In order to produce shifts with more than four spells special heuristics will be required. This feature will be added later.

4.2.1 Single Stretch Shifts

Single stretches of work can be turned into shifts provided signing on and signing off times are feasible. These shifts can be considered as 'part' shifts. When this work started, part shifts were not allowed in the UK, but they have since become an option. 'Part' shifts exist in effect when a meal break is allocated either at the front or, more usually, at the end of the shift, with the remaining portion designated as work allocated by the depot supervisor. 'Part' shifts may not be as efficient as full ones but they can be useful in completing a schedule where there is a combination of long and short lengths of shifts.

4.2.2 Meal Break Rules

There are rules governing when a meal break should occur. These rules differ according to the spreadover ranges. Some shifts which are crucial in forming a good schedule might be excluded because of these rules. In order to accommodate them, certain intelligence has to be incorporated into BUILD. BUILD, therefore, allows a driver to sign on earlier than would have been necessary in order to fit the meal break within its allotted period. Sometimes BUILD extends the sign off time instead if it is more beneficial to force the spreadover of the shift into a different spreadover range which might have a slightly more relaxed meal break rule.

4.2.3 Spreadover Restrictions

It is standard practice in British rail operation to restrict the number of shifts over a certain length (spreadover). These restrictions apply to individual depots. BUILD classifies shifts into spreadover range categories. Restrictions on the numbers of shifts in each category may be given to the SCHEDULE component.

4.2.4 Overnight Work

In bus operation night shifts are generally created in advance of the computer process, since the occurrence of this type of work is not common. There might be only one or two instances in the schedules in which they occur. While this advance assignment is possible in train operation, since it occurs more frequently, it removes some of the flexibility afforded by use of the computer. Initially, we stipulated that vehicle work which was to be included in a night shift should be defined as such by using an extension to the 24-hour clock, e.g. 2600 representing 2 a.m. This meant that the computer would have the ability of forming late starting shifts including this work, and would have some flexibility in determining the nature of these shifts. However, we still had to specify which work was to be so treated, and it would have been preferable to allow the computer to choose the work for night shifts. BUILD now provides a facility to mark portions of work so that the computer may consider them as early or late portions, and to form shifts with them in either position.

4.2.5 Increased Shift Lengths

Most rail companies would like to assess the effect on their schedules of an increase in the range of shift lengths. This would create very many combinations of shifts and BUILD would take some time to run. To prevent such long execution times, shifts of different lengths can be built separately according to different governing parameters. These different sets of shifts can then be combined before the shift selection phase.

4.2.6 Route and Traction Knowledge

Route knowledge was required for most schedules produced during the EPSRC financed work, namely for Regional Railways North East, West Anglia Great Northern and South Wales and West. It was also needed for later contract work described below. The ability to define the nature of the link between consecutive relief opportunities is seen as a major enhancement to the TRACS II system. From information supplied to us by the train company an internal list of routes with route numbers is compiled. These routes relate directly to knowledge pertaining to individual driver depots. To every link between relief opportunities the correct route number is ascribed. For every depot a list is compiled of those routes for which the drivers have no knowledge. When the BUILD process has constructed a potential shift and the depot to which the shift will belong has been determined, every link in every portion of the shift is checked against the list of prohibited routes for that depot. If a match is found then the shift must be discarded unless it is suitable for another depot.

Traction knowledge is handled in a similar way. For every type of traction unit in the train schedule an internal list of traction types with type numbers is compiled. For every train diagram in the train schedule, a traction type number is assigned. Traction knowledge was required for Regional Railways North East.

4.3 Reduction of the Shift Set

The original TRACS II followed the shift generation process by a process of discarding certain shifts, all of whose work was covered by many other shifts. The process of doing this using the EVEN program mentioned earlier was rather crude, and there was always a danger that some essential shift might be removed if we attempted to discard a significant portion of those generated. With the advent of much larger problems in which many more shifts were being generated for all styles of railway operation, a new method of appraising and eliminating shifts was sought. This new phase, known as SIEVE, works under the principle that inefficient shifts whose work components are heavily covered by other shifts are deemed to be of less importance.

Each potential shift is ranked using a combination of three attributes: an index formulated to reflect its cost effectiveness, a least number and an average number of other shifts covering the individual pieces of work making up the shift. The lowest ranked shifts are discarded and the latter two attributes which affect the rankings are updated continuously, until a prespecified target number of shifts remain. SIEVE also allows the user to reinstate some of the discarded shifts which have favorable cost effectiveness indices.

In situations where shifts with vastly different shift lengths have been constructed and they have been built in separate sets, the sets are combined after SIEVE. Depending on the total number of shifts concerned, it might be necessary to remove more shifts in the combined set in order to reduce it to the most appropriate size.

At the time that most of the work described here was undertaken, a typical size for a consolidated shift set was about 20,000 shifts; however, the current system allows up to 100,000 (see Fores/Proll (1997)).

4.4 Mathematical Selection from Shift Set

This phase employs the set covering integer linear programming model (Wren/Smith (1988)) to select from a large (e.g. 20,000) shift set the most economical subset (the schedule) that covers all the work and satisfies any other constraints present. This component of TRACS II is called SCHEDULE.

It was originally developed in the 1970s (Ryan (1980)), but has since undergone several phases of significant revision and enhancement. Since the development of TRACS II, the program has been restructured and brought up-to-date to the FORTRAN 77 standard. Some on-going research results which improve the algorithms used for solving the ILP have been incorporated. Some adaptations are specific to rail operations, for example facilities for constraining numbers of shifts in particular ranges of lengths. This type of constraint can be met by classifying shifts into specific types and then placing limits on these shift types during the mathematical phase of TRACS II. The capability of SCHEDULE for quickly solving large problems has been demonstrated in the recent case studies.

Recent PhD research (Willers (1995), Fores (1996)) has resulted in improvements to many parts of the SCHEDULE process In particular, Fores has developed a powerful column generation strategy. This allows much larger problems to be solved as a single process, and enables better solutions to be found to other problems by forming many more potential shifts in BUILD and retaining a large portion of them through SIEVE.

4.5 Adjustments and Production of Schedule Documents

In this phase, some adjustments to the mathematically-based schedule might be performed. This might be to resolve overlapping work due to the set covering model used. A decision has to be made as to which shift should actually be covering the overlapped work in practice. There might also be some swapping of work pieces between shifts to enhance the intrinsic quality of the schedule, which could not be easily handled mathematically.

In addition, there are a number of programs for managing subproblems when it is necessary to subdivide a large problem.

5 Case Studies

The system has been used on behalf of about a dozen train operators in the last three years. The complexity of the problems has varied considerably reflecting the wide range of conditions of the train operating companies. The system has been used both to assess the efficiency of existing schedules (TRACS II has always been able to produce schedules at least as efficient as those previously in operation) and to predict the distribution of work amongst depots, as well as to construct potential new schedules under a range of different conditions. Once basic data has been established it is very easy to adjust it to cater to different scenarios, and to re-run the programs. This helps to evaluate the costs of alternative operating conditions by producing full schedules for a range of strategies, such as:

- to consider a depot closure
- to adjust the maximum shift length
- to change the positioning of meal breaks
- to add to the route knowledge of a depot
- to effect a wider range of shift lengths
- to achieve a target average shift length
- to create a new depot, or relief point.

We have worked directly with five operating companies, and through Rail Operational Research Ltd (ROR) who have run our programs for other clients. In one case we have taken on board the full scheduling work for one of the clients of ROR. Where ROR has undertaken work themselves we have acted as advisers and have modified the programs as appropriate to deal with particular customer needs. In this paper we present our work for the six companies for whom we have produced the schedules ourselves.

Five of the problems which we have investigated have been undertaken as part of the EPSRC research contract. The others have been tackled under contract to the operating companies, including one company for whom we had initially tackled a problem under the EPSRC research contract. In this paper we cannot detail the conclusions reached as a result of the exercises, but in each case we obtained schedules matching current conditions which were at least as good as those in operation, and often better.

It should be noted that in all the case studies, the train schedules have been fixed. It is the normal practice in the UK that driver schedules should be constructed after the train schedules have been agreed. This is particularly the case following the split up of British Rail. The train operating companies first agree their timetables with Railtrack, the infrastructure owner, and only after agreeing the consequent train schedules can driver shifts be produced. The exercises dealt with a single midweek day, apart from one where initially Friday was chosen.

Drivers must return to their 'home' depot at the conclusion of their shifts. It is not sufficient to make an allowance of time for them to do this from wherever they finish their allocated work. In some circumstances because of the distances involved and the relative infrequency of the train service it is necessary to allocate them to particular train journeys. In order to achieve this it is necessary for the driver to travel as a passenger on a nominated train. Similar circumstances may prevail at the start of a shift.

Great North Eastern Railway and South Wales and West have a restriction on the total mileage covered in a shift. Currently this is 450 miles. We do not at present have a mechanism for dealing with this sort of restriction (as we do not capture mileage in our data sets) and it did not cause a problem in our case studies. However in our experimental work some hard coding was required to ensure that only valid shifts were generated.

A request by most of the operators involved in our case studies was how restrictions on the numbers of drivers at particular depots might be accommodated, and we have achieved this by defining work from a critical depot as a particular shift type. Most of our current work, however, concentrates on restricting the number of shifts per schedule having a particular spreadover, as this was a basic requirement by British Rail; in this we therefore distinguish shift types by spreadover lengths. The combination of these two restrictions has yet to be implemented. Another request was how it might be possible to remove a driver depot from the schedule and this is easily handled simply by removing the depot from the depot list in the data.

5.1 Great North Eastern Railway Company (formerly InterCity East Coast)

The first problem on this network was tackled around the end of 1994. The system is a long-distance fast operation along a 400-mile track from London Kings Cross via Doncaster, York and Newcastle to Edinburgh (see Fig. 5.1). There are spurs from Doncaster to Hull and Leeds. For most of the day there are hourly services between Kings Cross and Edinburgh and between Kings Cross and the spur point Leeds, while there are slower services also hourly between Kings Cross and some of the intermediate points. The fastest trains take about 4-1/2 hours from Kings Cross to Edinburgh, and two hours from Kings Cross to Leeds. Some of the fast trains continue beyond Leeds and Edinburgh to a number of other points. The basic routes from Kings Cross to Edinburgh and Leeds are electrified, as is one of the extensions from each of Edinburgh and Leeds, but diesel train sets have to be used on services which continue to the other extensions.

Drivers can operate either diesel or electric units, but two drivers were needed while the electric trains exceeded 110 m.p.h. between Kings Cross and Newcastle. There are driver depots at Kings Cross, Doncaster, Newcastle, Edinburgh and Leeds, with train depots one or two miles from each, except Doncaster which does not have a train depot.

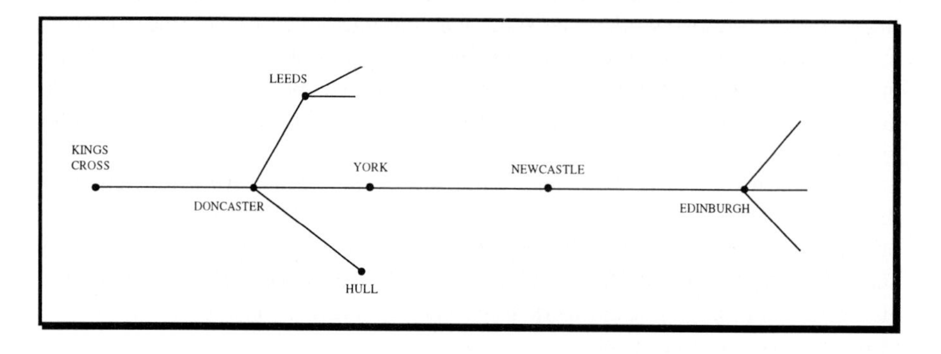

Fig. 5.1 The Great North Eastern Railway network (portion)

The driver schedule falls into two parts because of the way the double manning operation is carried out. Some drivers based at two of the depots operate exclusively between London and Newcastle, between which there is a requirement for double manning, and work as pairs throughout their entire shift. The remainder of the drivers pair up on an ad hoc basis and need not be based at the same driver depot. The driver schedules were thus constructed separately according to this criterion, with the second category being the larger. Another reason for treating these two parts separately is that there are different rules concerning the length of the break in the middle of the shift when there are two drivers compared with just one.

In order to produce a schedule where drivers team up on an arbitrary basis it is necessary to specify all the train work requiring double manning twice and to regard the shifts produced as totally independent of each other. Double manning is only required for the wheel turning part of the shift; two drivers are not required for preparing the train unit before the journey or for ensuring that the unit is left in a safe manner at the conclusion of the journey.

This problem involves a large geographic area with the opportunities to relieve drivers remotely spaced both in time and distance, a feature that is quite different to intensive urban bus operation with which we had previously dealt. This had an effect on the allocation of drivers to the five different driver depots and also in the determination of where to place the meal break in the shift. In our work with bus drivers we define two types of break, one where drivers take meals and one where they do not. In this long distance schedule because of the nature of the relief opportunities it is sometimes the case that the second type of break is long enough for a meal. Working within the rules laid down by British Rail as to where in the shift the meal break is to occur showed that our existing BUILD procedure was not always constructing the most appropriate shifts. This was to provide some incentive to establish a new BUILD process. The shift formation process was straightforward and the time required to run the whole process was short. The first schedule was produced before many enhancements had been incorporated in TRACS II. The schedules produced were comparable with those produced manually. Later, the problem was re-run using an up-to-date version of TRACS II and the schedules produced used two shifts fewer than those produced manually (see Table 5.1).

Table 5.1 Summary of results for Great North Eastern Railway

Date	Depots	Relief opportunities	Manual shifts	TRACS II
Dec 1994	5	500	91	91
Mar 1995	5	379[a]	76	74

[a] this was a sub-problem extracted from the whole problem and does not contain any double manning work.

5.2 Thameslink

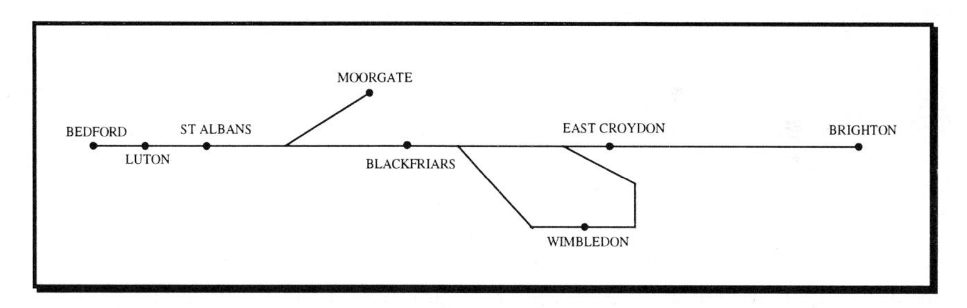

Fig. 5.2 The Thameslink network

The company operates relatively intensive commuter suburban and middle distance services over about 150 miles from Brighton to Bedford via Luton, St. Albans, Blackfriars (London) and East Croydon (see Fig. 5.2). There is a short spur leaving the main line just before Blackfriars to a terminus at Moorgate, and a loop line branching after Blackfriars through Wimbledon to return towards Bedford. The running time from end to end is about 140 minutes, and there are trains about every 15 minutes over the central section. There are driver and train depots at

Bedford and Brighton, and also at Selhurst which is close to East Croydon. To get to or from Selhurst, drivers have to travel as passengers on trains of another company unless they are driving to or from the depot. Drivers from Brighton cannot drive between St. Albans and Bedford, while drivers from Selhurst cannot drive between East Croydon and Brighton.

This problem was tackled at a very early stage before the passenger travelling feature was incorporated. Drivers' travel times between relief points were estimated and put into a table of 'standard' allowances. While these allowances might be realistic during usual commuting hours they become impracticable during late night or early morning hours when the train service is infrequent. Route knowledge was simple and did not cause any major concern. TRACS II was able to use marginally fewer shifts (see Table 5.2).

This exercise demonstrated the ability of TRACS II to produce schedules of similar quality to manual schedules, especially on the distribution of work amongst depots. The company appreciated that TRACS II could provide a very realistic forecast very quickly and they later requested us to investigate several 'what-if' scenarios. These 'what-if' exercises were completed in a short period of time and management found the results useful in evaluating different options.

Table 5.2 Summary of results for Thameslink

Date	Depots	Relief opportunities	Manual shifts	TRACS II
Mar 1995	3	615	115	112

5.3 London Underground

London Underground is a major metropolitan operator. This exercise covered the Piccadilly Line with trains every 2-1/2 minutes over the busiest period. Trains operate from Cockfosters through Oakwood, Arnos Grove and Wood Green at one end of the line, through the central area to Acton Town, where the line splits, one branch passing Northfields to Heathrow, and the other passing South Harrow and Rayners Lane to Uxbridge (see Fig. 5.3). Cockfosters to Wood Green takes 13 minutes, Wood Green to Acton Town 42 minutes, Acton Town to Heathrow and Uxbridge 25 and 35 minutes respectively. The end-to-end times are 70 or 80 minutes. Cockfosters, Oakwood and Arnos Grove are all used as turning points, while at the other end of the line most trains terminate at Heathrow or Rayners Lane. The extension to Uxbridge is used only at peak times. There are train depots near Oakwood and Northfields, while six trains are stabled late at night at South Harrow, restarting in the early morning. Driver depots, i.e. reporting points, are at Oakwood, Wood Green, Acton Town and Northfields. Acton Town and Northfields are only four minutes apart, while drivers taking trains to or from South Harrow travel by taxi to Acton Town or Northfields.

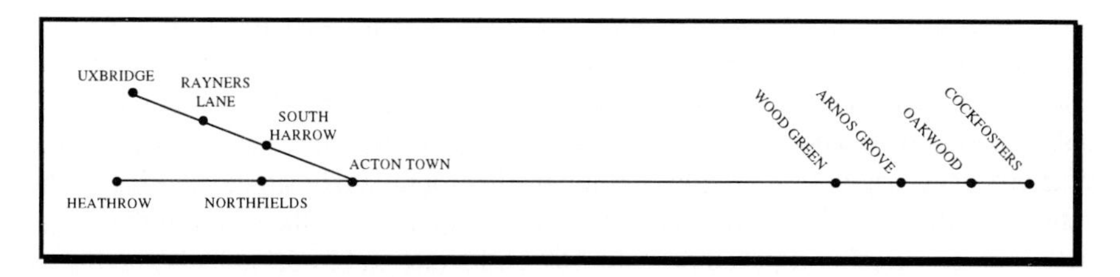

Fig. 5.3 The Piccadilly Line

At the time this was the largest problem we had ever tackled, with 2500 relief opportunities. There were a number of special conditions which we had never met before; the ability to sign on or sign off at points other than the depots at certain times of day, a limit on the earliest possible starting time at one of the depots. Travelling between places during day time could be estimated because the train service was so frequent. For late night work, extra consideration had to be given for drivers to travel by taxis. All these special circumstances were handled for this demonstration exercise by hard coding into the routine which checks the validity of shifts.

There were 72 train sets in use daily, of which 18 returned to a depot between the peaks. There were thus 90 train blocks, of which 18 operated for up to five hours over the morning peak, and 18 started before the afternoon peak. The earliest train started at around 0445, and the latest finished around 0130; trains did not necessarily finish work at the depot from which they had started. The current schedule uses 169 shifts.

Hitherto the largest problem we had tackled had had about 700 relief opportunities, so it was clear that steps would have to be taken to reduce the problem size. In our bus driver scheduling work we have developed heuristics to eliminate unlikely relief opportunities and to decompose large problems into sensible smaller units (Wren/Smith (1988)), but both had been developed from our specialized knowledge of bus driver scheduling from a single depot and were unlikely to work in the current complex situation. Manual analysis resulted in a reduction of about 300 relief opportunities that were unlikely to be useful in the early morning and late evening.

We were now faced with decomposing a problem of about 2200 relief opportunities. We formed the train blocks into two groups; group A consisted of blocks which started in the morning at Oakwood or started in the afternoon anywhere and finished at Oakwood, while group B consisted of the rest (which were associated similarly with Northfields or South Harrow). These groups were of approximately equal size, and it was observed that about half the current shifts were entirely on trains from a single group, indicating no significant correlation between driver depot and train depot. We then formed four subproblems, A1, B2, B3 and A4, of decreasing size; each consisted of similar ratios of peak, off-peak and evening trains. Following the strategy used in bus driver scheduling the first subproblem was solved and the work contained in the least efficient shifts was removed from the solution and merged with the second subproblem. This cascading process was continued automatically by the system through to the fourth subproblem, with about 30% of the work being carried forward each time (the reason for the original subproblems being made progressively smaller). Finally the retained shifts from the first three subproblems were added to the result of the fourth subproblem, yielding a schedule with 167 shifts (see Table 5.3). This saving of two shifts was accepted in principle by the company, who, however, pointed out that three shifts required the driver to travel as passenger back to the other end of the line to sign off, which would be unpopular. We therefore added a constraint forbidding this type of shift and obtained a solution with 168 shifts; the company then accepted that our original solution with 167 was preferable.

Table 5.3 Summary of results for London Underground

Date	Depots	Relief opportunities	Manual shifts	TRACS II
Sept 1995	4	2500	169	167

5.4 Regional Railways North East

A wide range of services is operated over a complex network (see Fig. 5.4). The longest routes operate from Newcastle, Middlesborough, Scarborough and Hull through York, Leeds and Huddersfield to Manchester, Manchester Airport and Liverpool taking about four hours, Cleethorpes through Doncaster and Sheffield to Manchester and Manchester Airport also taking about four hours, Middlesborough and York via Leeds and Bradford to Blackpool taking about four hours, Newcastle to Carlisle (1–1/2 hours), Leeds to Carlisle (three hours), in addition to many local inter-urban and rural services. There are train depots at Newcastle and Leeds and driver depots at Newcastle, Darlington, Scarborough, York, Hull, Doncaster, Hull, Cleethorpes, Sheffield, Leeds, Harrogate, Huddersfield, Blackpool, Manchester, Skipton and Carlisle.

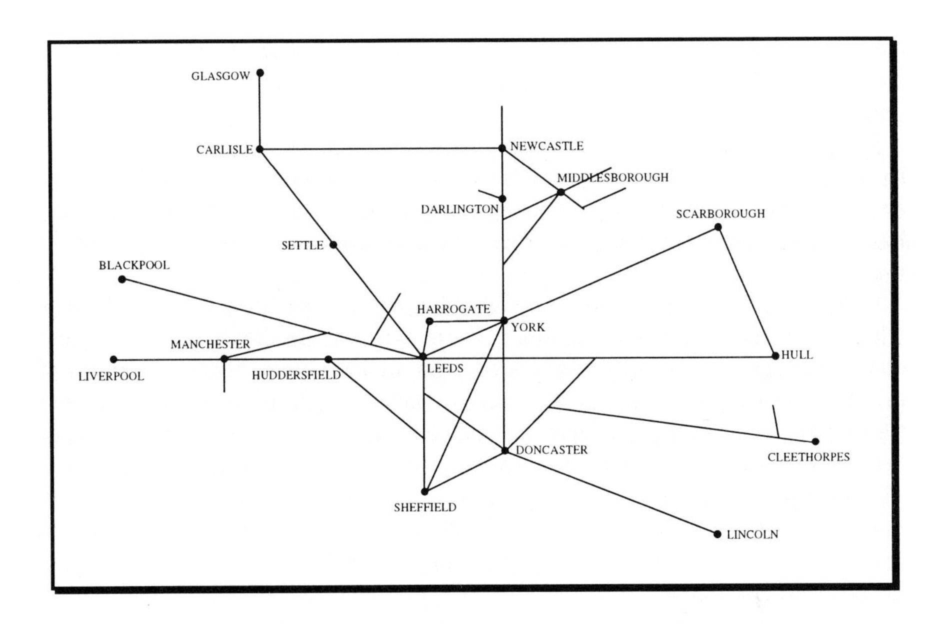

Fig. 5.4 The Regional Railway North East Network

Regional Railways North East (RRNE) first assisted our research in April 1995 by providing us with driver shifts for Monday to Friday based on Newcastle, Darlington and Carlisle depots, together with the rules on which the shifts were based. The initial study combined an urban commuting area with a rural area and there were more opportunities to relieve drivers compared to those for GNER but not as many as for London Underground. Relief points are close together both in time and distance. However an extra constraint placed upon the scheduling process is that not all the drivers can drive on all parts of the network. Certain assumptions had to be made about times allowed to travel as passenger between points on the system, as TRACS II could not at that stage deal with drivers travelling other than as drivers on trains which were being scheduled. Also we were not yet able to create shifts of more than three spells. Despite these limitations, the schedules produced were satisfactory after minor adjustments, and actually used four fewer shifts than the existing operation. An interesting aspect of this operation was that it was very similar to an exercise which we carried out as part of the 1990-91 project. Since that time the driver depot at Middlesborough had been closed. All early morning trains starting from Middlesborough have thus to be covered by drivers from Darlington travelling as passengers on earlier starting trains.

RRNE approached us in October 1995, asking us to repeat the exercise we had already carried out, but using a range of different operating scenarios. This work was undertaken in the period up to March 1996, and assisted the company to narrow the choice of scenario to a single set of rules affecting such matters as signing on and off, preparation and disposal allowances and meal breaks, with three options for other rules.

By the spring of 1996, TRACS II had been extended to allow the formation of shifts involving more than three spells, and to incorporate provision for passenger travel on trains of any appropriate company or indeed by taxis, provided that these had all been coded into the data. The following work was undertaken based on this new version.

In April we were commissioned by RRNE to apply alternative scenarios to their whole Monday to Friday operation. This was by far the largest and most complicated problem we had ever tackled, with around twenty depots or groups of drivers, and over 400 daily driver shifts. There were restrictions on route and traction knowledge for each depot, so that any individual portion of train work between relief opportunities could only be driven from a small number of depots. The route knowledge of neighboring depots did however overlap considerably, so that the whole operation was a single interacting process rather than a set of individual depot-based problems. The numbers of shifts to be assigned to certain depots were also limited.

Ideally we should have liked to have had time to develop a solution strategy based on the TRACS II capability of subdividing a large problem, solving the first subdivision and carrying forward badly scheduled work to the next subproblem. This is done automatically in our earlier bus driver scheduling system. However, in bus driver scheduling, the subdivision is carried out by consideration of how fragments of work from a single depot best hang together, and a different set of rules would have had to be developed for train operation. We could of course have followed the same strategy, but carried out the subdividing processes manually as we did with London Underground. We might first have considered the northernmost part of the operation. After creating shifts based on this, we could have removed the poorer shifts, automatically adding their work to later subproblems and continued the process over the whole operation.

However, RRNE required us to obtain schedules as quickly as possible, and we knew that in this first exercise over the whole network we would meet a wide range of new situations in coding data and in deciding how some local difficulties should be tackled. We would not therefore have had time to carry out the work in sequence as above. Instead, we first coded data and developed shifts for a set of relatively self-contained work. This enabled us to assess the magnitude of the whole problem and to make some decisions as to how to proceed.

We then subdivided the whole operation into five subproblems based on suitable combinations of depots, route and traction knowledge. These subproblems were then solved in parallel. For each subproblem many tens of thousands of possible shifts were generated according to the most restrictive of the possible scenarios. The shift generation program provides information on any train work which cannot be covered according to the rules being followed. This work was examined carefully, and dealt with by a number of actions.

Sometimes work was uncovered due to errors in coding or to inconsistencies in allocation of work to subproblems. In other cases, work which was in theory suitable for one subproblem could not be fitted together with other work in that subproblem to make an efficient shift according to the parameters being followed. Our next task was therefore to correct the errors and to exchange problematic work between subproblems.

However, in some cases the problems had arisen because of real difficulties at certain points. For example, in order to prevent the generation of very large numbers of unlikely shifts, TRACS II imposes maximum lengths on the amount of idle time between components of a shift. At certain places it was not possible to find matches for some incoming trains according to these rules. We therefore had to allow the rules to be bent in certain circumstances. In other cases, examination of the problem situation revealed that there was no way of covering certain work according to the given operating scenario, and some of the parameters originally stipulated had to be altered by agreement with RRNE.

The first sets of shifts covering two scenarios were sent to RRNE about six weeks after the start of the exercise. Following examination by RRNE some errors in the data were corrected. There was also a new requirement for the system to allow for a specified leeway whenever a driver has to change train either to carry on their duty or to travel as passenger. This leeway feature was used as a safety measure for possible delays and was later implemented as a standard feature of TRACS II. The revised schedules for the whole network were dispatched to the company three weeks later (see Table 5.4).

The company operates an intensive service over a system which is broadly H-shaped, with several additional spurs and loops (see Fig. 5.5). The legs London Kings Cross, Hitchin and Peterborough and London Liverpool Street, Cambridge and Kings Lynn are joined between Hitchin and Cambridge, and are linked by services from Kings Cross through Hitchin and Cambridge to Kings Lynn. Our remit was to concentrate on the Kings Cross/Moorgate to Kings Lynn/Peterborough sections, though some drivers on this section also work into Liverpool Street. This intensive service gives rise to very many relief opportunities which are closely packed together. The total number of weekday shifts is about 145 from four driver depots at Kings Cross, Hitchin, Peterborough and Kings Lynn. The journey times between Kings Cross and Peterborough,

Table 5.4 Summary of results for Regional Railways North East

Date	Depots	Relief opportunities	Manual shifts	TRACS II
May 1995	5	389	68	64
June 1996	15	2041	425	375 to 400[a]

[a] no direct comparison was made with the manual schedule; the new schedules were obtained using a range of new conditions.

Cambridge, and Kings Lynn are about 90 minutes, 70 minutes and 100 minutes respectively. The journey times between London Moorgate and Welwyn Garden City and Hertford are about 45 minutes each. There is a combination of nine express and stopping trains each hour resulting in three trains for Cambridge, one train for Peterborough, one train for Huntingdon, and one train for Kings Lynn on the Kings Cross to Kings Lynn and Peterborough sections, and from Moorgate, two trains to Welwyn Garden City and one to Hertford. Peterborough drivers do not have route knowledge from Hitchin to Cambridge and Kings Lynn, and Cambridge and Kings Lynn drivers do not have route knowledge from Hitchin to Peterborough.

5.5 West Anglia and Great Northern

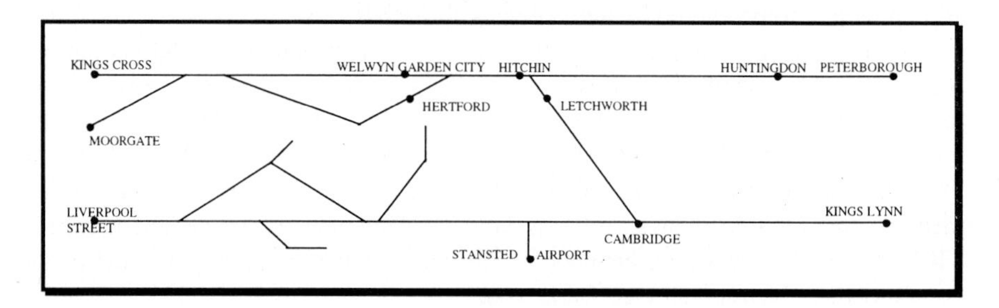

Fig. 5.5 The WAGN network

The train work was presented in two parts: that covering work from Kings Cross and Moorgate to Welwyn Garden City, Letchworth and Hertford and that covering Kings Cross to Peterborough and Kings Lynn. The traction types were different for each part. It was observed that there were few drivers who drove trains from both parts and this seemed a reasonable way to subdivide the problem. Although dividing the train work in this way would lose some of the flexibility afforded by a computerized approach, it was felt to be worth pursuing. Two traction types are used but there are no traction type restrictions. The first subproblem was solved and the work of the least efficient shifts was carried forward to the second subproblem. This resulted in a schedule with 141 drivers, four fewer than actually used (see Table 5.5).

One main feature of the manual schedule is the extensive use of passenger travelling. A few of the passenger journeys involved travelling on three different trains together with a walk. Currently, TRACS II only searches for passenger links involving a maximum of two different vehicles plus further walking to or from points. Hence, some of the passenger journey times have to be adjusted manually.

Another feature was that it was preferable for shifts to sign on and off outside the 'unsocial hours' period of 0000 - 0459 inclusive, otherwise penalty payments would be incurred.

Table 5.5 Summary of results for West Anglia Great Northern

Date	Depots	Relief opportunities	Manual shifts	TRACS II
Oct 1996	4	975	145	141

5.6 South Wales and West

The exercise was undertaken in collaboration with ROR, although the majority of work was done in Leeds. The company operates over a wide geographic area with thirteen driver depots (see Fig. 5.6). The rail network has a mixture of long distance, commuter, rural and branch line traffic. A particular feature here is that there is a very significant amount of travel as passenger within shifts, and that trains of other companies are sometimes used for this purpose. (There are many other companies' rail links not shown here between points on this network.) We therefore had to extend our arrangement for passenger travel to allow the system to read details of trains which it was not going to schedule but could be used for the movement of drivers.

We first scheduled a part of the company's operation covering the West Country which is a large, but relatively isolated, geographic area consisting of a main line from Bristol to Penzance with many branches. This part of the operation was covered principally from seven depots, but we included the work of six shifts which operated into the area in question from three other depots. One further depot in the area was expected to cause difficulties in meeting the relevant conditions, and its work was initially excluded from the schedule. After also removing from consideration the work of two night shifts which consisted of many very short shunting movements which could be well defined in advance, there were 49 shifts in the current schedule; TRACS II produced a satisfactory schedule also using 49 shifts (see Table 5.6). Experiments were then carried out with a number of alternative scheduling rules.

We were then commissioned to extend the work to the whole of the company, and initially this was to be divided into several areas as subproblems, one of which was the West Country area which had already been investigated and was relatively self-contained. The remaining areas have a total of around 150 shifts. Although the size of the remaining subset is big in terms of number of shifts, the huge geographic area it involved and the route restrictions helped in restricting the number of possible combinations. It was then decided to divide the problem into two subproblems only; one contained the West Country area and the other contained the remaining area. The schedule of the whole network produced by TRACS II has three fewer shifts than that of the manual process. We then tested a number of scenarios on the whole operation. The whole exercise was completed within about one month.

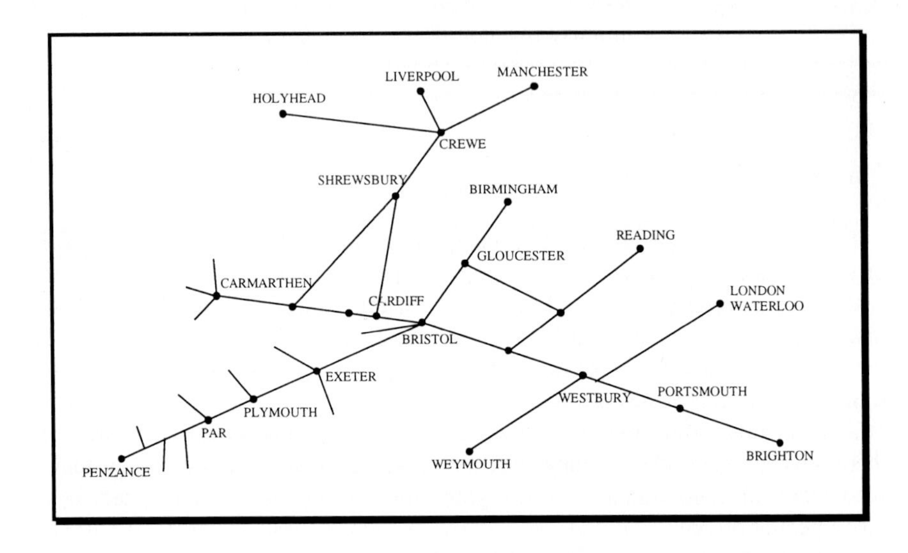

Fig. 5.6 The South Wales and West network

Table 5.6 Summary of results for South Wales and West

Date	Depots	Relief opportunities	Manual shifts	TRACS II
Dec 1995	7	399	49	49
Jan 1996	13	778	148	145

5.7 Other Exercises Using TRACS II

It had become important with the division of British Rail into many separate companies that they should each be able to determine the set of driver scheduling conditions that best met their geographical and operational patterns. We have therefore undertaken in partnership with ROR six further scheduling exercises. These are:

- InterCity West Coast (on catering staff)
- Mersey Rail

- North London Railways
- GNER
- Great Western
- Cardiff Valleys

In these we had been constrained by having to make available the program suite before it was fully developed in order to meet the urgency with which operating companies were having to change their scheduling practices. In this joint work with ROR the normal pattern has been for ROR to liaise with the client and to undertake the computer scheduling runs, referring to us when, as is natural in a developing system, difficulties arose. We also extended TRACS II at the request of ROR so that it could handle certain complex problems. In the case of the largest company we undertook most of the scheduling work at the University because we had the detailed software system knowledge necessary to expand TRACS II to meet their needs, and to apply it successfully to their very difficult problems.

6 Conclusions

It should be stressed that the EPSRC sponsored research project was for a period of two years starting in September 1994, and was intended to produce a prototype system which would require further development before it could be used on real problems. The fact that our earliest experiments resulted in successful schedules meant that we could respond rapidly to company requests for scheduling help. Thus we have been feeding the results of our research into the system while solving urgent problems in advance of our planned system release time. This has inevitably caused difficulties from time to time, but the evolving system has responded well to the demands placed upon it.

We have demonstrated that it is possible to produce efficient schedules for train drivers according to widely differing and complex circumstances using TRACS II. In all cases these have been acceptable to management, and usually cheaper than existing schedules, often saving more than one shift. Some very large problems, with twenty different driver depots and overlapping work, have been satisfactorily solved. The system can be used both to compile production schedules and to investigate alternative operating scenarios, such as using a different set of depots to cover the operations, or adopting different working rules. It has proved a very valuable tool in helping train companies to determine a set of operating rules appropriate to their special circumstances. This has been particularly valuable in the UK where British Rail, with a standard nationwide operating agreement, has been divided into more than 25 different companies with very different types of operation. As rail operation in the UK is competitive, we are unable to reveal the conclusions reached by the various companies, but we can say that TRACS II generally indicated that significant savings could be made if drivers would work mixtures of long and short days.

We believe that the RRNE exercise is the first time anywhere in the world where train driver schedules for non-urban situations have been produced on such a large scale. We have demonstrated that the system is also suitable for intensive urban operations. It is a very powerful tool for producing schedules under alternative operating scenarios. Unlike a manual scheduler, the system has no preconceived ideas of how the portions of train work should be put together and this might lead to better shift formulation.

Since the completion of the EPSRC project we have continued to enhance the system, and we have concluded an agreement with Comreco Rail Ltd. to integrate and market what is now TRACS III with their train planning software.

Acknowledgements: We are grateful to the staff of all the train operating companies with which we worked at the start of this project for having the foresight initially to provide us with information and help which might not have been expected to yield them practical results. We would also like to thank those who have commissioned us to undertake practical scheduling tasks which have helped us to enhance the system. Particular thanks are due to ROR, and in particular to Andrew Sparkes who has persuaded many companies to use the system, and whose team has taken part in many of the practical projects.

Thanks are also due to the EPSRC for providing funding to allow us to pursue this project in the first instance.

References

Ryan, D.M. (1980): ZIP - a zero-one integer programming package for scheduling. Computer Science and Systems Division, Atomic Energy Research Establishment, Harwell, Oxfordshire.

Willers, W.P. (1995): Improved algorithms for bus crew scheduling. University of Leeds, PhD thesis.

Fores, S. (1996): Column generation approaches to bus driver scheduling. University of Leeds, PhD thesis.

Wren, A./Kwan, R.S.K./Parker, M.E. (1995): Modelling the scheduling of train drivers. in: Daduna, J.R./Branco, I./Paixao, J.M.P. (eds.): Computer-aided transit scheduling. (Springer-Verlag) 359 - 370.

Smith, B.M./Wren, A. (1988): A bus crew scheduling system using a set covering formulation. in: Transportation Research 22A, 97 - 108.

Wren, A./Smith, B.M. (1988): Experiences with a crew scheduling system based on set covering. in: Daduna, J.R./Wren, A. (eds.): Computer-aided transit scheduling. (Springer-Verlag) 104 - 118.

Fores, S./Proll, L./Wren, A. (1997): An improved ILP system for driver scheduling. Paper presented at the Seventh International Workshop on Computer-Aided Scheduling of Public Transport, Cambridge MA.

Integrated Scheduling of Buses and Drivers

Anthony Wren[1] and Nicolau D. Fares Gualda[2]

[1] Scheduling and Constraint Management Group, School of Computer Studies, University of Leeds, Leeds LS2 9JT, United Kingdom

[2] Departamento de Engenharia de Transportes, Escola Politécnica, Universidade de São Paulo, Brasil

Abstract: In Brazil, bus services, which have to meet extraordinary levels of demand and traffic congestion, are generally planned by a city transport authority, and franchised out to several private companies for operation. The nature of the scheduling problem is different from most of those described at previous workshops, because vehicle and driver schedules are developed simultaneously, but on a line by line basis. The present paper describes work carried out on representative Brazilian city schedules using enhancements of the Leeds scheduling systems reported at previous workshops. Although the precise nature of the problem varies among the cities studied, the approaches adopted in this research are similar. The Leeds scheduling software, adapted for the Brazilian situation, is used in a novel way to produce efficient combined bus and driver schedules, with buses being switched between lines to improve efficiency. Already savings have been indicated. The paper includes a detailed exposition of the processes used in one of the cities, serving to emphasize the practical problems which must be addressed in any implementation.

1 Introduction

In Brazil, bus services play the most important role in providing commuter and other urban transport. Dependency on public transport is high compared to other advanced countries, and there is little rail rapid transit. Roads in major cities are very congested, while some cities are extremely large, so that commuter journey times by bus are high. Services are generally planned by a city transport authority, and franchised out to several private companies for operation. Bus and driver schedules are generally determined by the city authority, although the operating companies may have some flexibility, provided that they adhere to the specified journeys.

The nature of the scheduling problem is different from most of those described at previous workshops, because vehicle and driver schedules are developed simultaneously, but on a line by line basis. The law in Brazil allows shifts with meal breaks of up to one hour, and shifts without meal breaks are allowed provided that they are no longer than six hours. Shifts with longer breaks are not allowed. Advantage is taken of shifts without meal breaks where appropriate, usually to cover work between the end of an early shift and the beginning of a late shift.

The general rules are subject to changes by agreement with the unions, but since the unions are weak in most cities, there is considerable flexibility. Agreed minimum meal break durations range from ten to thirty minutes according to the organization. The usual scheduling strategy is first to develop bus schedules with built-in periods when the buses are parked so that meal breaks can be taken, the driver remaining with the vehicle. The work of each bus is then divided into drivers' shifts, but in at least some cases the formation of good shifts is inhibited by the prior allocation of breaks to the bus schedule.

The city authorities should have an interest in reducing costs by more efficient scheduling, possibly switching vehicles and drivers between lines or introducing empty balancing journeys against the peak flow, practices seldom followed at present. The operating companies also have an interest in efficiency, as they may enhance their profits by operating the required journeys with fewer resources than anticipated by the authorities.

This paper describes investigations of schedules of several Brazilian cities, using enhancements of the Leeds scheduling systems reported at previous workshops. Although the precise nature of the problem varies among the cities studied, the approaches adopted or considered in this research are similar.

As in many other countries, departure times of bus journeys are initially computed from statistics on passenger demand. However, in only one of the present studies could this be done, as the statistical information was not forthcoming in the other cases.

In all three studies, a bus schedule was first developed to fit the journey times using state-of-the-art software; the system indicates how vehicle numbers may be reduced by slight revisions to journeys, and by interworking between lines. This bus schedule differs from that normally produced by the authorities in that it has no provision for meal breaks, which might otherwise have been placed in inefficient positions by traditional methods.

Driver scheduling software, adapted for the Brazilian situation, was next used to determine good shifts; meal breaks were accommodated at this stage by switching drivers between vehicles and, where allowed, between lines.

Finally, the bus schedule was revised so that the buses followed the driver shifts created, allowing most drivers to remain with the vehicle throughout the shift, as is normal in Brazil. However, in one case by agreement with the authority, an experiment was undertaken to show the effect of allowing some crews to change vehicle at a meal break.

This work has yielded in most cases combined bus and crew schedules which are more efficient than those achieved manually, although lack of feedback on apparent inconsistencies in the manual schedule has made strict comparisons difficult.

Three cities, Fortaleza, São Paulo and Sorocaba, have been studied to different extents. The work with Fortaleza, completed in 1994, used earlier versions of the Leeds scheduling systems. The other two projects used experimental versions current in early 1997; these have since been improved and installed as Openbus™ for Reading Buses in the United Kingdom.

2 Fortaleza

Fortaleza is the capital of the state of Ceará in northeastern Brazil. It has 1.9 million inhabitants, and is the center of a metropolitan region with a population of 2.5 million. There were about 1250 buses operated by 23 franchised companies at the time of the study (1994). There are about 150 bus lines, each scheduled separately.

The work reported here was carried out by a student who had previous knowledge of the transport system of Fortaleza. We give a summary here; for a full description see Azevedo Filho/Kwan/Wren (1994).

As is normal in Brazil, bus and driver schedules are drawn up simultaneously, with breaks (in this case of 15 minutes) inserted in the vehicle schedule to allow for crews' meals. These breaks are between 0900 and 1100, and between 1500 and 1700. A single cross-city line about 34 km long and currently using 16 buses was studied, where breaks could be taken at either terminus; the existing bus schedule placed these breaks rather arbitrarily at one or other terminus, at the first opportunity after the morning peak, and again in the afternoon. Crews consist of two staff. In practice, some of the breaks were as short as 12 minutes, apparently violating the rules.

The transport authority did not in fact create the crew schedule; this was left to the operating company, but the authority assumed that each bus would be covered by two crews. This provided a target of 32 crews, and since the minimum daily payment to a crew was 7 hours and 20 minutes, a target cost might be 32 times this, i.e., 234:40. In practice this might not be achievable, and drivers are allowed to work up to two additional hours, paid at 50% extra.

In order to tackle this Brazilian process of building part of the crew scheduling activity into the bus schedule while using UK software, the following steps were designed for the computer process (they were also subsequently used in the other studies described later):

1 Create a set of one-way bus trips according to observed passenger demand and indicated journey times (this had been done in Brazil before the start of the project);
2 Apply vehicle scheduling software to cover the above trips efficiently without allowing for meals, but provisionally leave idle vehicles at a terminus rather than return them to the depot, except at the end of the day;
3 Apply crew scheduling software to create shifts based on the above, providing for meals of appropriate duration by changing buses. Provide sufficient signing-on and -off time to allow for vehicles being transferred to and from the depot if necessary;
4 Revise the bus schedule so that the vehicles follow the crews, being laid up at termini during meal breaks, and either returning to the depot at the end of a shift or being handed over to start another crew.

Before starting his research in Leeds, Azevedo had worked with the transport authority, and had developed a bus scheduling system, ALOCA, which included gaps in the schedule for meals, and applied it to the line in question; the ALOCA schedule had been approved by the authority. He had also manually prepared a crew schedule based on this. These schedules deviated from the strict rules in ways deemed acceptable by the authorities.

The work in Leeds was carried out in three exercises:

- The BUSPLAN and CREWPLAN components of the BUSMAN system (see Wren/Chamberlain (1988)) were used for steps (2) and (3) of the above process. As we no longer had access to the proprietary code of CREWPLAN we could not amend it to take account precisely of all the rules, but made some assumptions which may have been over-generous;
- The IMPACS suite (see Smith/Wren (1988)) was used for step (3), applied to the BUSPLAN vehicle schedule. Although CREWPLAN had been derived from IMPACS around 1985, the systems had since diverged, and IMPACS had been upgraded to take account of more recent research (see Fores/Proll/Wren (1997)). IMPACS was also more adaptable, and was adjusted to take account of the precise rules from Fortaleza;
- IMPACS was applied to the ALOCA bus schedule, with straight-through shifts being allowed when they contained bus work in which a period suitable for a meal had been set aside.

In all the above cases step (4) of the process was carried out manually, following a previously devised algorithm.

Table 2.1 summarizes the schedules obtained both in Fortaleza and in Leeds, together with the probably unattainable targets (as previously quoted) displayed in the first row. All the bus schedules used 16 vehicles.

The bus and crew schedules produced by the various computer processes have been discussed with the former head of the Fortaleza authority's scheduling group, who judged them to be operable, although there might be some difficulty in getting the drivers to accept the 30-shift solution because it contained a greater number of split shifts than the current schedule. The BUSPLAN/IMPACS schedule was believed to be best, as it contained fewest violations of the given rules, and because it had an acceptable number of split shifts.

Table 2.1 Comparison of manual and computer schedules for Fortaleza

Bus schedule	Crew schedule	No. of shifts	Total cost	Mean shift cost
	Targets:-	32	234:40	7:20
Manual	Manual	32	274:30	8:35
ALOCA	Manual	32	263:19	8:14
BUSPLAN	CREWPLAN	31	235:48	7:36
BUSPLAN	IMPACS	31	241:30	7:47
ALOCA	IMPACS	30	236:33	7:53

The important result established was that the use of a good computer system to produce combined bus and crew schedules could result in very significant savings in crew costs, of the order of 12%. It should be recognized, however, that crew costs in Brazil are substantially lower than in many countries, and that a saving of one vehicle (3.1%), if achievable by concentrating on buses, could be more important than apparently large savings on crews.

3 São Paulo

São Paulo is the largest city in the Southern hemisphere, with a metropolitan region population of 12 million. It is the focus of the most prosperous state in Brazil. However, the vast majority of the population has no access to private transport.

There is a limited modern metro system of three lines (one very short), and some suburban railways. The highway system includes a network of multi-lane roads which can provide fast transport, but are subject to severe and often unpredictable congestion. By far the largest number of work and leisure journeys are by bus, and there are about 12,000 buses in total. It is common to find streets entirely clogged by buses. At the mayoral election in November 1996 the issue of public transport featured prominently in the winning candidate's manifesto, with ambitious plans for novel transportation systems.

Bus lines, vehicle and crew schedules are drawn up by the city authority, SPTrans, while the operation of individual lines is franchised to private companies, of which there are about fifty. Virtually all the lines operate as round trips from a suburb to the city center and back. There is no space for layover to be taken in the city center, so although there are notional departure times for outward journeys, these often cannot be adhered to, and buses set off again as soon as they arrive in the center; the unpredictable nature of journey durations forces the use of built-in slack, but this must be entirely at the outer termini. The scheduled round trip journey time on individual lines is often in excess of four hours.

Even during the afternoon peak, extra service is introduced from the outer terminal, so that there is an apparent wastage of an inbound trip (and a corresponding wastage of an outbound trip after the morning peak). However, the franchises normally operate from depots near the outer termini, so that the cost of servicing these trips is not as high as might be expected. The diversification of employment throughout the city region allows some workers making relatively short journeys to take advantage of these trips.

Vehicle schedules for individual lines are drawn up by SPTrans using graphical methods to link the round trips. The graph is used to assist the placing of meal breaks of thirty minutes at the outer terminals after certain journeys, usually immediately after the peaks. Crews remain with, or close to, their vehicles while meals are taken, and once a chain of meal breaks has been provided, surplus buses are returned to the depot. A meal is given before the depot return if one is not already present in the bus schedule.

Each bus is normally covered by two shifts in any day. These shifts may be consecutive, with a handover of crews at the outer terminus, or the first may return to the depot with the bus after its morning work, while the second covers the afternoon peak and the evening.

The cost of the planned schedules produced by SPTrans forms the basis of the payments to the individual companies. However, if the companies can find cheaper ways of providing the agreed journeys, they are free to do so.

Although gains in efficiency might in theory be obtained through switching vehicles and crews between lines, few if any lines meet at outer terminals, and the uncertain nature of journey times would make this impracticable at the city center. Further, franchises are awarded on a line-by-line basis.

A short study was carried out on one of the smaller lines, line 6032, with 94 round trips whose running time varied from 80 to 122 minutes, while the headway varied from 20 down to 8 minutes. There are 17 buses in the morning peak, 8 between the peaks, and 13 in the afternoon peak. These are operated by thirty shifts, of which 19 include meal breaks between shift portions, while the other 11 finish with a meal break. The eight buses operating all day are each completely covered by two shifts, with the handover sometime between 1145 and 1520. Since no split shifts are allowed, it is easy to see that thirty is the minimum number of shifts required to cover the peaks.

Taking the times of each of the round trips as fixed, the BOOST program (see Kwan/Rahin (1997)) was used to devise a new bus schedule based on the existing minimum layover of four minutes at the outer terminus. In this very simple situation where the only opportunity for linking trips was at this one terminus it is not surprising that the same number of buses were used during each of the main periods in both the manual and computer schedules.

The next stage was to create crew schedules. This was done using the TRACS II program (see Kwan/Kwan/Parker/Wren (1997)). Although this was specifically developed for rail operation, it incorporates knowledge of bus driver scheduling built up over thirty years in Leeds, and is also suitable for this mode. As the actual movements of buses to and from the depot would depend on the details of the crew schedule, these movements were removed before running TRACS II, allowance being made for them in the signing-on and signing-off times specified at the outer terminus.

Several crew schedules were formed, using different parameters, as described below. In each case, the bus schedule was then revised so that the buses followed the crews, either returning to the depot after completing a shift, or providing a handover at the outer terminus to another shift.

In practice, some experimentation had to be done before deciding on the best way to use TRACS II. This was mainly because of violations of some of the rules as defined. Examination of the current schedule yielded the following:

- Maximum unbroken work; defined 5:30, actual 5:52 (5:58 in half shifts without meal breaks);
- Minimum meal break; defined 30 minutes, actual 29;
- Maximum spreadover; defined 8:30, actual 8:45.

Table 3.1 Manual and alternative computer crew schedules for São Paulo

	SP	A	B	C	D	E	F	G	H	I	J
Min meal	29	29	30	30	30	30	30	35	29	30	30
Max stretch	5:52	5:52	5:30	5:30	5:40	5:35	5:35	5:35	5:52	5:35	5:35
Max spread	8:45	8:45	8:45	8:40	8:40	8:40	8:35	8:40	8:45	8:40	8:40
Shifts	30	30	30	31	30	30	31	30	30	30	30
Full shifts	19	18	23	22	22	23	21	22	18	23	22
Half shifts	11	12	7	9	8	7	10	8	12	7	8
Paid hours	208	207	211	209	208	211	209	212	207	211	212
minutes	13	46	19	24	57	27	34	59	43	27	04
Min meal	29	29	30	30	32	30	32	38	29	30	30
Max stretch	5:52	5:48	5:13	5:12	5:39	5:32	5:32	5:32	5:50	5:32	5:32
Max spread	8:45	8:45	8:45	8:39	8:39	8:39	8:35	8:39	8:45	8:39	8:39

Table 3.1 shows the results of a number of experiments labeled A through J in which we first allowed the above rule violations, and gradually moved towards the ideal. The first three rows of the table show the parameter values used; the last three rows show the actual limits in the resulting schedules. Experiments C and F indicated that it was not possible to obtain a schedule with 30 shifts while adhering to all the rules. Some variables not shown affected the balance between full and half shifts (which explains the difference between experiments I and J). The column headed SP refers to the existing manual schedule.

The prime consideration in optimization was to minimize the number of shifts and it should be noted that the two solutions using 31 shifts had relatively low costs. In fact, use of an additional shift adds the cost of a meal break and extra signing-on or -off time, a total of about 50 minutes; it may enable other features to be improved. Half shifts are in general more efficient than full ones, since they all have meal breaks of 30 minutes, while longer breaks are often necessary in full shifts. However, a significant number of full shifts are required to ensure that all work is covered. In experiments A to G, the maximum stretch of work in a half shift was set to the existing maximum of 5:58.

This exercise on data from São Paulo serves to illustrate principally how TRACS II may be used to investigate the effect of changing parameters governing the lengths of shifts and their components. On average, each experiment on a 110MHz Pentium PC took about five minutes, about half of which was occupied by manually editing the parameters and taking care of the rather non-standard way in which the system had to be applied.

In many of the experiments some of the maximum lengths actually achieved were identical. This was because a single piece of critical bus work was covered in the same way in each. Thus the maximum stretch length of 5:32 in five of the experiments refers to the same spell of bus work in each case.

4 Sorocaba

Sorocaba is a city of about half a million inhabitants lying just within the Tropic of Capricorn, in the state of São Paulo. The inner city is surrounded by a ring road, close to which lie two diagonally opposed bus terminals. Bus services operate from the suburbs to the nearest of these terminals, and the terminals provide efficient interchanges to a few other services which link the terminals and serve the central area.

The bus services are regulated by the authority URBES, and franchised to two bus companies, each of which operates predominantly, but not exclusively, to one of the terminals. The city has been using computerized tools for transit management since the early 1980's (see Pietrantonio (1988)), but usually decides operational changes on a more empirical basis (maintaining the practice of scheduling line by line), leaving considerable room for operating companies to discuss the impact of changes and to make final adjustments to vehicle and crew schedules.

Two of the planners in URBES are associates of one of the present authors and have provided significant assistance with this project. With their help a suitable subset of the entire weekday city bus operation was identified for study. This consisted of all bus lines (a total of 31 lines) operated by just one of the companies using conventional buses to and from one of the terminals, Terminal Santo Antonio (TSA). (Certain other of the company's lines used articulated vehicles.) The lines in question used 111 buses serviced by 263 daily shifts, the busiest line using 14 buses while ten of the lines used just one bus each.

As in the other cities considered, every driver shift was entirely on a single bus. However, there were up to three shifts on a bus in the course of a day, the longest stretch of bus work being just over 20 hours.

Every line was allocated a target round trip time, including an allowance of slack time. This time was split theoretically between inward and outward journeys, but as there was no room at TSA for layovers, all layover time in the bus schedule was allocated to the outer terminus. All but a handful of journeys were, as in São Paulo, considered as round trips starting and finishing at the outer termini. The standard given round trip time varied between peak and off-peak, and was always a multiple of the headway, so that there was apparently no slack. In practice it was observed that many of the trips, even at busy times, were scheduled in shorter times than the standard, so that there appeared to be considerable slack time in other trips, which probably resulted in early arrivals at the outer termini.

Buses that remained in operation for a significant time after either of the peaks had allowances for meals built into the schedule at the relevant outer terminus by delaying subsequent departures. Officially these breaks were at least ten minutes, but several appeared to be shorter, perhaps because it was possible surreptitiously to reduce the outbound journey time. Where three shifts had been allocated to a bus in the course of a day, the middle shift normally had no meal break (but might have hidden slack at the outer terminus).

The current schedules restricted each bus to its own line, apart from two or three odd journeys, and shifts were also restricted to individual lines. However, URBES was interested in ascertaining the effect of allowing buses and drivers to switch between lines. Such switching would have to take place at TSA, which was the only common terminus, and this would be unacceptable if it resulted in buses being parked at that point, as there was no free space during busy times. This meant that if drivers had to change lines, the bus would have to be handed over immediately to a different driver. It would be possible following such a handover for a driver to take a meal off the bus at TSA.

4.1 Procedure

We would have liked to have followed the four-step process outlined in Section 2 above for Fortaleza. However, no data is available on passenger loadings and minimum journey times, and we therefore took the existing journeys as given, starting our work from Step 2.

We have not taken time to code Step 4, the revision of the bus schedule to follow the crew schedule. This is currently done manually, following an algorithm that could be coded, as in Fortaleza. We have however, checked that Step 4 can be carried out, and have demonstrated to URBES how the schedule for some selected buses would operate. We therefore describe below the bus and crew scheduling work we have carried out as Steps 2 and 3, and briefly describe the demonstration of Step 4.

4.2 Step 2 - Bus Scheduling

The bus schedule was compiled using the BOOST system (see Kwan/Rahin (1997)). This incorporates some ideas which were borrowed by our BUSMAN system (Wren/Chamberlain (1988)) from earlier work in Leeds (see Wren (1972) and Smith/Wren (1981)), but implements them in an improved way. BOOST assigns vehicles to previously defined journeys, optionally showing what adjustments to journeys would have to be made in order to save vehicles.

As a first exercise, the current departure times for all trips were taken as fixed. As turn-round times at TSA were already close to instantaneous, and no outer terminus was used by more than one line, it was not expected that any savings would result. However, the exercise would demonstrate that the computer could produce a schedule of a similar standard to the present one, and would lay the foundation for future work.

For each line it was necessary to study carefully the existing bus schedule. Many of the individual journeys violated the given trip times, and some of the meal breaks were shorter than the required ten minutes. It was tempting to specify to BOOST trip times based on the minimum times actually used (taking account of peak and off-peak observations separately), but this would have been dangerous as many of the scheduled times may have been too short for consistent use. BOOST allows trip times to change from one period of the day to another. We chose to use the given trip times as standard for peak and off-peak respectively, but had to specify considerably reduced times for some short periods so that BOOST could take advantage of them in the same way as the manual schedule.

Because of the above variations, data preparation was both tedious and error-prone. However, BOOST itself was a considerable help in ensuring that the data was entered correctly. We entered each line separately, and first used BOOST to schedule each line after its data had been entered.

BOOST has a feature whereby it first estimates the number of buses required for a schedule. This estimate is always a strong lower bound. The user can specify this as the target number of buses to be used. If no feasible schedule is possible with this number, a schedule is nonetheless constructed which is as close to feasibility as possible. This will require buses to make dead journeys and/or layover in less time than specified; in extreme cases these times may have to be negative. The graphical displays produced by BOOST on the screen highlight such infeasibilities in red, while normal links between journeys are shown in blue. It is possible to zoom in on the graph so that any time of day may be examined in detail.

Each line was scheduled by BOOST with the existing number of buses specified as the target. It will be recalled that BOOST would not adjust the schedule for meal breaks, as this would be done implicitly in step 3. However, the off-peak service interval on most lines was such that a ten-minute meal break would be created automatically because the manually produced journeys had previously been adjusted to provide for this.

Part of the screen display of a BOOST graph for the first three scheduled lines, using the existing trip times and a target of eight buses (one fewer than at present) is given in Fig. 4.1. Infeasibilities caused by the reduction in buses are shown (actually in red) at Nova Esperança (NESPER) between 0638 and 0805, when arrivals are linked to slightly earlier departures. The authorities would be advised to examine these to determine whether the line from Nova Esperança could be serviced by four buses instead of five.

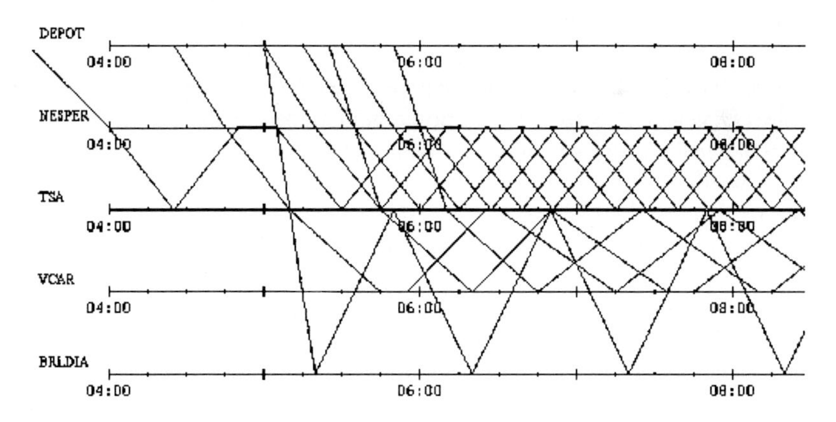

Fig. 4.1 Sorocaba lines 2, 3 and 5 using eight buses instead of nine

Inspection of BOOST's graph for any line (or groups of related lines) allowed us to see whether sensible links were being made. Critical links of three types were then investigated, each (if present) highlighting an error in data preparation:

- Infeasibilities normally implied that trip times specified in the data were longer than were used in practice, but might alternatively be due to an extra trip having been specified by mistake;
- Dead runs are shown in a different color and should only have appeared on two or three lines where some one-way journeys existed in the current schedule. Their presence in other lines usually implied that a trip in the relevant direction had been omitted;
- Layovers should have been relatively short (only a few minutes), except where meals were taken in the current schedule and the service interval did not allow a tighter linking of arrivals and departures. Therefore long layovers were investigated; where they were not due to a meal they were usually caused by some error in a departure time or journey duration.

Once these anomalies had been corrected, BOOST could produce for each line a schedule using the same number of buses as in the manual schedule. It may seem disappointing that fewer buses could not be used in some circumstances, but it should be recalled that the specified round trip times had been chosen to provide the desired headways with the minimum number of buses. In each case, BOOST was re-run with the target number of buses set one lower. This produced infeasibilities highlighted in red on the graph, but in every case these were too large to justify a recommendation that the data should be altered to allow fewer buses to be used.

Once each line had been scheduled on its own and any anomalies had been resolved, the line was added to a master file and rescheduled together with all lines already on the file. The number of buses scheduled was compared with the total number actually used on the lines in question. As already noted, we did not expect any savings in vehicle numbers, but on one occasion, the computer generated schedule used one vehicle fewer than the existing schedule. Inspection of the graph showed that one peak journey was arriving earlier at TSA than expected, and was being linked to an earlier departure (on a different line) than in the current schedule. This was due to a data error which had not been noticed when one of the individual lines was scheduled.

Results are displayed in Table 4.1. The BOOST schedule, like the manual one, used 111 buses. Parameters had been specified to BOOST so that it minimized a cost function in which dead running time was given a double weight compared to layover, and the resulting schedule used 1749 hours and 24 minutes in total. The manual schedule used 1745 hours and 22 minutes, but almost certainly allowed shorter journey times between some termini and the depot than had been specified to BOOST, so that better results might have been achieved by BOOST had full information been received from Sorocaba. As an exercise, BOOST was rerun with equal weights given to dead running and layover, producing a schedule with 1747 hours and 21 minutes.

BOOST was next run with a target of 110 buses. As expected, the schedule contained several infeasibilities (averaging four minutes, but ranging up to 10 minutes). Any decision on whether these changes can be implemented must be left to the authorities in Sorocaba, but it may be noted that even if the reduction can be achieved in only one peak period, there would be an incidental saving of one crew.

Table 4.1 Alternative bus schedules for Sorocaba

Case	Trips	Buses		Total time	Dead time	Infeasibilities	
		a.m.	p.m.			no.	extent
1	2731	111	111	1745:22	96:57	0	0
2	2731	111	111	1749:24	97:08	0	0
3	2731	111	111	1748:01	100:46	0	0
4	2731	110	110	1747:24	96:22	18	76 mins
5	2612	105	107	1698:25	122:22	0	0

1. Existing manual schedule
2. Corresponding BOOST schedule, dead running weight 2
3. Corresponding BOOST schedule, dead running weight 1
4. Corresponding BOOST schedule, one bus less
5. BOOST schedule with revised journeys (Section 4.2.1)

This exercise has served to demonstrate to URBES that it is possible to produce a good schedule using BOOST. The schedules produced by BOOST did switch some buses between lines at TSA, but apart from a few cases in the schedule with 110 buses, the switches have been made only because the version of BOOST being used did not distinguish between lines; a newer version allows maintenance of line integrity which does not cost more than a specified amount.

4.2.1 Revised Bus Schedule

With the agreement of URBES staff we have produced a revised bus schedule as described below. Ideally, we would have tailored journey times to demand and traffic speeds in a Step 1 process as mentioned in Section 2. However, we did not have the necessary information, and instead we decided to create a new set of bus trips based on the current timetables, but amended as described below.

Our aim was to separate the round trips into their two components, and transfer the slack from the outer termini to TSA so that opportunities could be taken to exchange buses there between lines. We also decided to try to provide an even interval service throughout the day, except during the peak periods when we maintained the existing trips in the direction of peak flow only.

The schedule for each line was examined closely, and realistic journey times were assessed, often slightly shorter than the current times because these allowed for slack at the outer termini. We then built up lists of departure times for the line as follows:

- Starting with inbound journeys we used current departure times until the end of the morning peak;
- We then defined a *standard* headway for the line as the current normal off-peak headway, and used this to provide a constant service interval in the inbound direction through to the early evening when the current service level fell. The departure times over the standard period were chosen to match the current departure times as closely as possible, subject to the provision of a constant headway. Thus, if current departures were at 15, 35 and 55 minutes past the hour for most of the day, we provided service at these times for the whole period, ignoring any irregularity in the current schedule produced by providing gaps for meal breaks. Generally this provided slightly more journeys than at present immediately after the morning peak;
- Moving to the outbound direction, we provided trips at times approximately equal to the current departure times until such time as the service became as frequent as the standard interval. We adjusted these times slightly in order to minimize slack at the outer terminus unless this would have resulted in very irregular service in the outbound direction. (In some lines this stage was unnecessary, as the standard service level was achieved from the beginning of the day.)
- We next scheduled subsequent outbound trips at the standard interval at such times as would allow arrivals and departures at the outer terminus to match without slack, and continued using this interval until the current schedule increased in frequency in the afternoon;
- We scheduled outbound trips for the rest of the day at the times used in the current schedule;
- We then moved to the inbound direction, scheduling evening departures from the outer terminus to match arrivals there, with some adjustment at the end of the day so that the final journey was at approximately the same time as in the existing schedule;
- We then adjusted the above in order to ensure smooth transitions between different headways without incurring any excessive gaps (i.e., ensuring that no service interval exceeded the norm for the time of day);
- BOOST was then run on the individual line, with a target number of buses equal to the number currently used on the line. Inspection of the bus graph sometimes showed infeasibilities at the outer terminus. These might be due to errors in the data, which were then corrected. Alternatively, they might be due to mismatches between trips at the standard headway in one direction and trips at the peak headway in the other; these were removed by revising one or both of the affected journeys. Large infeasibilities at TSA were treated in the same way, but small infeasibilities, except where due to data errors, were allowed to stand, as the final schedule would link arrivals at TSA with departures on different lines if appropriate.

On some lines we varied the above practice slightly because of special circumstances, but on every line we provided a service at least as frequent as that provided at present, except in the direction against the peak flow, when the off-peak daytime headway was maintained.

Once satisfactory schedules had been obtained for each line separately, BOOST was run on the whole data set, obtaining a feasible schedule with 105 buses in the morning peak and 107 in the afternoon, savings of six and four buses respectively. Statistics for this schedule are displayed in row 5 of Table 4.1. The reduction in the number of live trips is entirely due to the rescheduling of trips in the direction against the peak flow (as are the other savings). The numbers of live trips in the peak directions and during all off-peak periods are at least as great on each line as at present; on most lines this number has increased, as we have not left gaps in the schedule for meals. Thus the public should perceive a better service, and with standard service intervals should be able to remember the regular departure times, which they cannot do at present.

The increase in dead running is due to BOOST's inserting dead journeys against the peak flow to compensate for the imbalance in service levels. Some of these dead journeys could in fact be replaced by additional service journeys where time permits, particularly before the morning peak and after the evening peak, but we have not examined this possibility in detail.

Although there might at first appear to be a danger that buses would need layovers at TSA and therefore add to the congestion there, in practice with about 1350 journeys arriving at TSA in the course of a day, linking arrivals and departures on a first-in, first-out basis and returning surplus vehicles immediately to other points ensures that there is no excessive layover. At present congestion at TSA is reduced by delaying some journeys from the outer termini, and this could also be done by slight adjustment of the proposed schedule if necessary.

The BOOST schedule still has some large slack periods at outer termini due to mismatches between trip ends during the peak. If the principles behind this revised schedule are accepted by URBES, it would be appropriate to revise carefully the allowances for dead running (which have been set at conservative levels in the present exercise) and to check any slack to determine whether further savings could be made by adjusting trip times to close the gaps.

Thus BOOST has been applied to Sorocaba schedules;

- first, to show that it can produce schedules to existing conditions which are comparable to current schedules (and incidentally to demonstrate that current schedules are efficient unless existing departure times and journey durations are relaxed);
- second, to show what alterations to journeys would be necessary to achieve a saving of one bus out of 111. This would also be dependent on a few buses switching between lines;

- third, to investigate the effects of revising the schedules to remove slack from the outer termini and to remove surplus contra-flow journeys, producing savings of six and four buses in the respective peaks.

The use of a predecessor of BOOST to effect savings by similarly rationalizing the bus service in contra-flow directions was first described by Wren (1982) in the Scottish city of Dundee and accepted by the bus authority.

4.3 Step 3 - Crew Scheduling

It was agreed with URBES first to demonstrate the bus crew scheduling capabilities of TRACS II (see Kwan/Kwan/Parker/Wren (1997)) based on the existing manually produced bus schedule. Although this had built-in meal breaks at the outer termini, we would also allow meals at TSA, and would allow crews to change there to other lines after having a meal. It was hoped that the additional flexibility gained would allow efficiency improvements.

There were some peculiarities of the situation in Sorocaba which would ideally have been treated by making some minor adjustments to TRACS II, but since the precise nature of these peculiarities would have taken some time to discover, it was decided to approximate them by certain parameter settings as discussed below. The parameters were chosen so that all the constraints of Sorocaba could be met; some relaxations which occurred in practice were denied to TRACS II, which might otherwise have produced even better results.

Crew shifts in Sorocaba may contain up to 8 hours and 40 minutes' work. Every shift is paid for at least 6 hours and 40 minutes, and time in excess of this is charged at 50% extra. In the existing schedule, three shifts exceed 8:40, with one shift of 9:20. As we had no means of knowing the circumstances under which such violations were allowed, we restricted TRACS II to a maximum duration for straight shifts of 8:40.

Strictly, split shifts are not allowed, but the existing schedule contains 26 of these, with work content up to 9:50 and some spreadovers above 16 hours. We restricted the spreadover to 16 hours, and in the absence of information as to how such long spreadovers were paid, we gave each split shift a cost equal to its spreadover. All buses which re-enter service for the evening peak return to the depot immediately after that peak (up to about 2100). In practice, there is no accepted mechanism for transferring drivers between the depot and other points, so each of the 26 buses which return to the depot between the peaks forms a split shift, the spreadover being determined by the operational considerations of the line in question.

In theory every shift requires a meal break of at least 10 minutes, but in practice there were many shifts during the midday period, some of nearly nine hours' duration, which had no allocated meal break; presumably there is enough slack in the bus schedule for short breaks to be taken in such shifts.

Most buses leave the depot before 0600, and the last one leaves at 0653, so the peak period is relatively early by many standards. The normal pattern of work on a bus which operates from early morning to late evening is for it to be serviced by three crews, one having a break after the morning peak (perhaps as early as 0800), one working from mid-morning to late afternoon or early evening without a break, and one taking over from that crew and working until the end of the day, with a break after the evening peak. The normal pattern on a bus which operates through the day and returns to the depot after the evening peak is for an early shift including a break as above to be followed by a shift which continues until the end of the operation without a break.

There are some relatively short shifts without a break (from two and a half hours upwards); they are all costed at 6:40.

As this was a rather quick demonstration project, we could not take account of all the vagaries of the actual schedules, and we did not adjust the code of TRACS II to allow work in excess of 6:40 to be paid 50% extra. The effect of this has been to prohibit TRACS II from forming some shifts used in practice, and to reduce its ability to minimize extra payments. However, since TRACS II does give each shift a cost of at least 6:40, it will prefer to form two shifts of 6:40 (costing 13:20) rather than one of 5:40 and one of 7:40 (costing TRACS II 14:20, and actually costing 14:50).

TRACS II operates by generating a large set of potential shifts, reducing this heuristically to a manageable size, and then using integer linear programming (ILP) to select a subset of the shifts which together form a legal schedule. In order to cope with the widely varying types of shift we took advantage of the TRACS II facility to generate separate sets of shifts and to merge them before entering the ILP stage.

First, TRACS II generated 33300 standard shifts with meal breaks. This number was then reduced to 28900 by the refining processes which eliminate selected shifts whose work is covered by many other shifts. It was possible to cover most of the work by the generated shifts, but the peak only buses cannot be covered by standard shifts because there is no provision for moving crews to and from the depot to re-enter service after a meal break. TRACS II has a facility whereby a new set of parameters can be used to generate shifts covering work previously left out, and this was used to generate 810 further possible split shifts.

Next, we generated continuous shifts on the same bus without meal breaks, starting at various times from 0700 onwards, as in the existing schedule; there were 1343 of these. Finally, we generated shorter continuous shifts (durations of 2:30 upwards); there were 961 of these.

The merged set of 32014 shifts was then presented to the ILP, which produced a schedule with 252 shifts, compared to the manual schedule of 263 shifts. It will be recalled that TRACS II has been prevented from forming some existing shifts which violate the conditions; had it been allowed to form such shifts, an even better schedule might have been formed. The results are compared in Table 4.2.

It is clear that the reduction in shifts has been achieved at the expense of additional hours in excess of 6:40, but with an overall saving in hours paid. The reduction is due to the ability of crews to switch routes and to have meals at TSA, since in the existing schedule every bus is constrained to have precisely one, two or three shifts, and this can lead to very inefficient situations. It is possible that some lesser saving might have been achieved on the busier lines if crews had been allowed to change buses and have meals at TSA while still being restricted to their own lines, but this was not attempted.

Table 4.2 Alternative crew schedules for Sorocaba

Case	Shifts	Paid time	Time over 6:40
1	263	1918:47	110:18
2	252	1903:18	148:52
3	258	1916:20	130:53
4	245	1879:14	163:56
5	251	1901:20	152:00

1. Existing manual schedule
2. TRACS II on existing bus schedule
3. TRACS II on existing bus schedule, constrained to 258 shifts
4. TRACS II on BOOST bus schedule
5. TRACS II on BOOST bus schedule, constrained to 251 shifts

In order to try to reduce the number of shifts in excess of 6:40, the ILP component of TRACS II was rerun with the number of shifts constrained to be at least 258. As Table 4.2 shows, the total paid time increased to nearly the current level. However, the costs do not take account of savings in overheads due to reductions in crews, and this schedule may still be a considerable improvement against the current schedule which uses five more crews.

The BOOST bus schedule was next coded for TRACS II, allowing relief opportunities every time a bus entered the main terminal at TSA. Additionally, relief opportunities were provided wherever there was sufficient slack for a meal break at an outer terminus. This gave much more flexibility in the schedule, and the first set of shifts generated by TRACS II contained about 80000 possibilities; these were refined by the heuristics to a sufficiently low number to provide a total of about 30000 shifts entering the ILP after merging with the other three generation runs as described above.

The resultant schedule had 245 shifts. The BOOST bus schedule had contained some buses which re-entered service before the evening peak and stayed on the road until late evening, unlike the existing schedule in which such buses returned to the depot immediately after the peak. As we had restricted split shifts, as in the current schedule, to start and finish both portions of work at the depot, these buses had to be covered by late straight shifts. Were it not for this, a further reduction in crews might have been achieved. However, we did not know whether split shifts which did not finish at the depot would be allowable.

Again, in order to try to reduce the number of extra hours, we reran the ILP with an increased lower bound (of 251) on the number of shifts in the schedule.

As shown in Table 4.2, all the TRACS II schedules yielded savings in numbers of shifts and in costs. The paid time includes the time over 6:40 multiplied by 1.5. The first TRACS II driver schedule based on the BOOST bus schedule yielded savings of 6.8% in number of shifts and 2.1% in paid time.

4.4 Step 4 - Combined Schedule

As previously indicated, the current crew schedule maintains each shift on a single bus. However, URBES had agreed for the purpose of the above experiments that meals could be taken at TSA. As there was no room to park buses there, crews would hand their vehicles over when taking a meal, and would resume work on another vehicle.

Had all the TRACS II shifts apart from the split ones been on two buses, separated by a meal at TSA or at an outer terminus, there would have been no need for further refinement of the schedule. However, some shifts were in three portions (with one very short change of vehicle at TSA), while others on only two buses had no meal, but a change of vehicle at TSA. The schedule using 245 shifts required some relatively small alterations to the bus schedule in order to adjust 24 shifts so that the buses followed the crews. There was no effect on the total numbers of buses or crews, or on costs.

5 Integration

An important feature of this work has been integration, both of bus and driver schedules and of manual and computer processes. The paper serves to emphasize the importance of undertaking a proper study of the detailed schedules and scheduling requirements when faced with non-standard situations. It also shows what can be achieved by the combination of scheduling experience in the consulting team with powerful computer methods. In many localities where scheduling practice is less well developed, it is essential to view computer scheduling tools as aids within an overall process rather than as models which can by themselves produce efficient solutions.

In Fortaleza, the fact that the exercises were carried out by someone with direct experience of the local scheduling situation under the direction of a team with wide experience of both computer and manual scheduling enabled a sound overall approach to be developed, while allowing sensible decisions to be made regarding some of the softer constraints.

The São Paulo work started from a graphical layout of the journeys to be scheduled, from which the current bus and crew schedules could be deduced. Again, the fact that the work was done by someone with wide scheduling and computer experience enabled sensible decisions to be made in preparation of data and choice of parameters. The manual work here took about two days in total, while all the computer runs were accomplished in under an hour in total.

Sorocaba was by far the most time-consuming exercise, taking about four person-weeks of work spread over two months. The manual processes have been described in some detail in Section 4, and were carried out by one of the present authors (Wren). Once more, general experience of scheduling problems led to a productive integration of manual and computer processes without which it is doubtful whether any sensible results would have been obtained.

In all the cases, decisions as to how best to simulate the present integration of the bus and crew scheduling processes were informed by practical experience of using both computer and manual methods to construct real schedules in a vast range of situations over thirty years. While many transport authorities apply computer methods with little or no manual intervention, there are others where suitable combinations of manual and automatic processes will provide the ideal solution in the medium term. In the longer term, lessons learned from this will enable better integrated systems to be developed for the types of situation described in this paper.

Even in advanced countries there are organizations which are reluctant to use computer methods because the systems they have investigated do not entirely meet their needs. This situation can be rectified by a true synergy between the developer and users of systems. This is best achieved where the systems supplier has taken the trouble to obtain substantial experience of manual scheduling situations. The systems described here are perhaps extreme examples of the need for this.

6 Conclusions

In three Brazilian cities, bus services and their drivers have been scheduled using software developed in the University of Leeds linked with manual analysis. The current Brazilian practice of scheduling buses and crews simultaneously has been simulated by:

- Scheduling buses without reference to crews using the BOOST system;
- Scheduling crews using IMPACS or TRACS II;
- Modifying the bus schedule so that buses follow the crews where appropriate.

In the largest exercise, new potential ways of operating were identified and agreed with the authority.

In one of the cities percentage savings of 3.6% in vehicle numbers were shown; in the other two cases the exercises involved single relatively simple bus lines, and no saving in bus numbers was anticipated or achieved.

In two cities, savings of 3.1% and 6.8% in crew numbers (14.1% and 2.1% in paid time) were identified; in the third very simple case, crew savings were insignificant, but effects of changing some of the parameters governing driver scheduling were demonstrated.

In the largest of the problems, effects of changing both bus and driver scheduling rules were demonstrated.

Acknowledgements: The authors are grateful to the British Council and CAPES for financing the academic exchange which made this work possible, and to staff of the three transport authorities concerned for making data available, and for discussing scheduling problems with the authors. We particularly thank Renato Gianolla and Roberto Battaglini of URBES for providing the data and giving helpful feedback during the early stages of the project in Sorocaba. We would like to thank Dr Hugo Pietrantonio for providing background information and for commenting on an early partial draft of the paper. We also thank Mario Azevedo Filho for carrying out the study on Fortaleza under the direction of one of us, and Ann Kwan, Margaret Parker and Dr Raymond Kwan for providing assistance with TRACS II and BOOST.

References

Azevedo Filho, M.A.N./Kwan, R.S.K./Wren, A (1994): A Alocacão de onibus e motoristas no Brasil: alguma experiencia pratica. (Scheduling buses and their drivers in Brazil; some practical experience. English version available.) in: Anais do VIII Congresso de Pesquisa e Ensino em Transportes, ANPET, Recife 231--242.

Fores, S./Proll, L.G./Wren, A. (1997): An improved ILP system for driver scheduling. Paper presented at the Seventh International Workshop on Computer-Aided Scheduling of Public Transport, Cambridge, MA.

Kwan, A./Kwan, R.S.K/Parker, M.E./Wren, A. (1997): A new system for scheduling train drivers. Paper presented at the Seventh International Workshop on Computer-Aided Scheduling of Public Transport, Cambridge, MA.

Kwan, R.S.K./Rahin, M.A. (1997): Object oriented bus vehicle scheduling - The BOOST system. Paper presented at the Seventh International Workshop on Computer-Aided Scheduling of Public Transport, Cambridge, MA.

Pietrantonio, H. (1988): SITCO - Sistema de Informações para o Transporte Coletivo por Ônibus. in: Anais do II Congresso de Pesquisa e Ensino em Transportes, ANPET, Brasilia.

Smith, B.M./Wren, A. (1981): VAMPIRES and TASC: Two successfully applied bus scheduling programs. in: Wren, A. (ed.): Computer scheduling of public transport, North-Holland. 97--124.

Smith, B.M./Wren, A. (1988): A bus crew scheduling system using a set covering formulation. in: Transportation Research, 22A, 97--108.

Wren, A. (1972): Bus scheduling: an interactive computer method. in: Transportation Planning and Technology, 1, 115--122.

Wren, A. (1982): VAMPIRES in Dundee - a new look at a proven bus scheduling system. in: Proc. Fourteenth Annual Seminar on Public Transport Operations Research, University of Leeds.

Wren, A./Chamberlain M. (1988): The development of Micro-BUSMAN: Scheduling on micro-computers. in: Daduna, J.R./Wren, A. (eds.): Computer Aided Transit Scheduling, (Springer-Verlag), 160--174.

Object Oriented Bus Vehicle Scheduling - the BOOST System

Raymond S.K.Kwan and Mohammad A.Rahin[1]
Scheduling and Constraint Management Group
School of Computer Studies, University of Leeds, Leeds LS2 9JT, UK

Abstract: BOOST (<u>B</u>asis for <u>O</u>bject <u>O</u>riented <u>S</u>cheduling of <u>T</u>ransport) embraces the object-oriented paradigm, which is much acclaimed for excellent conceptualisation, extensibility and reusability. The VAMPIRES algorithm, originated in the 1960s for scheduling train locomotives and later formed the basis of the BUSPLAN system within the BUSMAN package, has been updated using the object-oriented approach and used as the core scheduling algorithm in BOOST. In this paper the advantages of the object-oriented approach, as compared with the traditional procedural approach, are illustrated through the re-modelling of the VAMPIRES algorithm. Concepts and domain knowledge are abstracted at different levels dependent on the contexts. Thus the object-oriented scheduling processes are clear to understand and easy to extend. BOOST features a Windows- based graphical user interface (GUI). The styles of data management and interactive schedule manipulation utilities are described. Results of testing BOOST against the conventional BUSPLAN and recent practical applications of BOOST are reported.

1 Introduction

Bus scheduling is a complex combinatorial optimisation problem, which is generally reviewed in Daduna/Paixao (1995) and Wren (1981). Research into transport vehicle scheduling at the University of Leeds started in the 1960s. Initially train locomotives were scheduled (see Wolfenden/Wren (1966)). The heuristics used were adapted in the 1970s into what is known as the VAMPIRES

[1] Now with the Division of Computer Science of the School of Computer Studies.

algorithm for scheduling buses (see Smith/Wren (1981)). VAMPIRES is particularly good at making savings through efficient usage of vehicles on a network wide basis. In the mid-to-late 1970s, a more user-friendly system called TASC was developed (see Smith/Wren (1981)). TASC was designed for quick compilation of vehicle schedules and production of timetables/schedule documents for relatively simple and regular bus services. In the 1980s, the versatility of TASC and the optimising capability of VAMPIRES were blended together into the BUSPLAN system. BUSPLAN is a key component of the BUSMAN package (see Wren/Chamberlain (1988)) being used by many operators mainly in the UK. There was little success in the late 1980s in several attempts to bring the commercial version of the vehicle scheduling system up-to-date with new technologies and user requirements.

BOOST is a recent development embracing the object-oriented paradigm, which is much acclaimed for excellent conceptualisation, extensibility and reusability. The VAMPIRES algorithm, the object-oriented model of which forms the core of BOOST, has been overhauled. In this paper, the VAMPIRES algorithm will be outlined first in conventional terms, which typically involves embedding domain knowledge in complex logic. This will be contrasted by a description of the new object-oriented model, which abstracts much of the domain knowledge out as data (object class hierarchy). Thus the functions making up the heuristic algorithms become easier to understand and to extend.

BOOST features a Windows-based graphical user interface, the object-oriented model for which has facilitated its design and implementation. The styles of data management and interactive schedule manipulation utilities will be described.

BOOST has been tested against the conventional BUSPLAN system, and has achieved improved results. It also has been used in case studies for some bus companies.

2 Object Oriented Versus Procedural Approach

An object-oriented software system is an organisation of data entities, called *objects*, capable of carrying out some tasks or actions when suitably triggered. A procedure oriented software system is a collection of rather independent procedures and data structures, which are bound together by program control constructs. We shall briefly compare the two different paradigms.

The procedural approach is the traditional approach for software development. A procedure oriented software system is largely hierarchical in structure, which is usually developed either *top-down* or *bottom-up*. The system development cycle is *task-centric*. Data structures are developed as the need arises to support the steps of the algorithms used and the tasks to be performed. Data structures are used as information carriers and are passed among the procedures to facilitate various types of operations. Global variables together represent the state of the system. One of the main drawbacks of procedure oriented systems is that the domain knowledge captured within the system is usually not clearly and explicitly represented, often requiring a lot of untangling of program control constructs to understand. Procedure oriented systems are therefore generally error prone, difficult to maintain and difficult to enhance.

The object-oriented paradigm on the other hand is more *data-centric*. The major data entities in the problem domain are identified and modelled as objects to be used as starting points. Other abstract concepts will also be modelled as objects. The behavior and capability of individual objects rather than that of the system as a whole is the focus of modelling. The system functions through communication and interaction between objects, which is sometimes referred to as *message passing*. Each object has its own internal data structures as well as a set of tasks, called *methods,* that it can perform; these together define the object's state, behavior and capability. Upon receiving a message, an object will react to it and execute one of its own methods, which in turn would send messages either to itself or to other objects to trigger further activities in the system. Object-oriented systems are usually developed in a *distributed* manner, i.e. one would model an object independently, but from time to time branch out to model other objects in interaction. Fig. 2.1 illustrates the comparison at a high level.

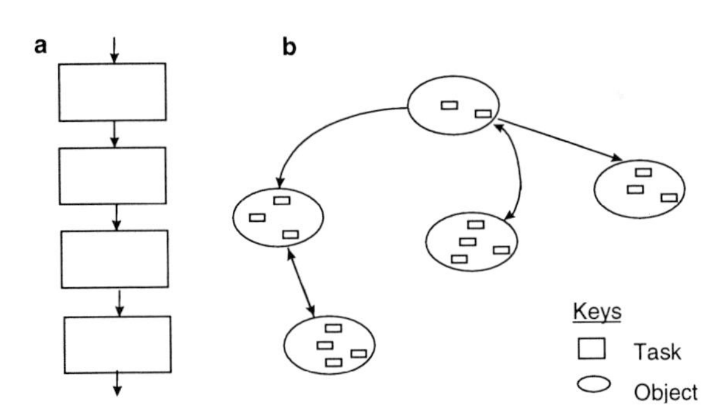

Fig. 2.1 High level comparison, **a**: Procedure; **b**: Object oriented

By Wagner's definition (see Wagner (1987)), a programming language must support the following three concepts to be considered object-oriented:

- Objects
- Classes
- Inheritance

We may further state that an object-oriented software system is one that is designed and implemented using the above concepts. *Objects* as explained earlier are the basic system components. Computations are performed via the creation of new objects and the communication between them. A *class* is a set of objects that are conceptually alike. The definition of a class may be regarded as a template for the creation of objects belonging to that class. Each object is thus an *instance* of a class. *Inheritance* is a technique used to produce specialized new *derived classes* from already defined ones. A derived class has all the characteristics of its parent class plus some additional or overriding definitions. This facilitates the reuse of and extensions to existing object definitions, and is particularly powerful when a rich and well tested set of objects is available to build upon.

3 The VAMPIRES Algorithm

The VAMPIRES algorithm is an iterative improvement heuristic, which has been described in detail in Smith/Wren (1981) and is outlined in pseudo-code below:

```
PROCEDURE VAMPIRES

BEGIN_PROCEDURE
  Form list of trips D sorted by departure times
  numberOfBuses = Estimate( D )
  FormInitialSchedule( numberOfBuses, D )
  REPEAT
      FOR ( each trip_i in D )
         WHILE( New Neighbour for trip_i found )
            trip_j = Neighbour( trip_i )

            trip_r = LinkedWith( trip_i )   // Existing links
            trip_s = LinkedWith( trip_j )

            oldCost = CalculateCost( trip_i, trip_r )
                                + CalculateCost( trip_j, trip_s )
            newCost = CalculateCost( trip_i, trip_s )
                                + CalculateCost( trip_j, trip_r )

            IF( ( newCost < oldCost )
                 OR( ( newCost = oldCost ) AND
                        ( special swap condition holds ) ))
                RemoveLink( trip_i, trip_r ) // Swap links
                RemoveLink( trip_j, trip_s )
                MakeLink( trip_i, trip_s )
                MakeLink( trip_j, trip_r )
            ENDIF
         ENDWHILE
      ENDFOR ;
    UNTIL No further improvement
ENDPROCEDURE
```

The number of buses to be used can be estimated quite simply from the profile of simultaneous trips in operation during the day. However, the scheduler may request VAMPIRES to use fewer buses than estimated in order to identify any critical trips. The initial schedule is formed by linking trip arrivals to trip departures in a fairly arbitrary manner. The schedule is strictly restricted to using the number of buses specified, and therefore the initial schedule may contain links that are time infeasible, i.e. a bus arriving too late for its next scheduled departure. The links are costed by a function of any incurred time infeasibility, idling time and empty running time. For example, the cost function may be the weighted sum of any time infeasibility, idle time, and empty running time associated with a link in minutes. Typically the weight for time infeasibility is a very large constant, and the values 1 and 2 are used as weights for idle time and empty running time respectively. The iterative improvement procedure then examines alternative linkages and performs swaps that will reduce the total cost. The algorithm terminates with the least cost schedule that it can find. It is possible that the final schedule still contains some infeasibilities because the target (number of buses) is too tight. This is an advantageous feature because sometimes infeasibilities involving only a few minutes each could be removed by adjusting the trips, saving buses that otherwise would be needed.

The procedural implementation of VAMPIRES has a long history and has undergone several major revamps since the 1970s. However, the task of maintaining and enhancing the algorithm and the system built around it has proved to be increasingly difficult. The major re-development work entailed in the BUSMAN II project (see Chamberlain/Wren (1992)) during the late 1980s was a failure, and there has not been any significant new development since then.

Using the procedural approach, data structures are secondary considerations during system design. Often the design of procedural steps dictates the addition of new data structures. It is therefore easy to have settled for some seemingly, but often not, adequate data structures. For example, to satisfy the above pseudo-code procedure, it seems reasonable to use a data structure of nodes each holding information about a trip and its linkages to the trips before and after it in forming a schedule. The data structure is simple and amenable to efficient sorting of the arrival and departure events. However, a practical implementation of the VAMPIRES algorithm would require many more details than in the code shown above, e.g. temporarily return a bus to depot when it is feasible and desirable, handling of vehicle-start and vehicle-end links in swapping trials, heuristics for achieving high success rate of link swapping trials, etc. It would be necessary to go up and down the list of trip nodes to compute and test for different types and sub-types of link, e.g. whether the link is time feasible, time infeasible, has a long time gap, involves empty running, etc. Different procedural steps would be taken dependent on the type and sub-type of link being processed. Thus the procedural code would be full of conditional branches. In later sections, it will be shown how an object-oriented approach has led to a more satisfactory object design in place of the list of trip nodes achieving a higher degree of abstraction by hiding computations unimportant to the current context.

VAMPIRES, like all heuristics, cannot guarantee to yield the optimal solutions, although it has never been bettered in practical situations when properly used. However, it is possible to construct some artificial data such that simple link swapping strategies may fail. One criticism of BUSPLAN is that its implementation of the VAMPIRES algorithm is too rigid for making changes to improve the link swapping strategies. Also, BUSPLAN may sometimes take a very long time to run. Again, the computational efficiency of the VAMPIRES algorithm hinges on how good the link swapping strategies are.

4 BOOST - An Object Oriented Bus Scheduling System

BOOST (**B**asis for **O**bject **O**riented **S**cheduling of **T**ransport) uses an object-oriented framework incorporating the VAMPIRES algorithm. It has a collection of object classes, such as those for describing the route network, the bus services and the vehicle activities being scheduled. The object classes can be grouped into the following categories:

- Utilities
- Bus network
- Vehicle activities
- Link swapping rules
- System processes
- User interface

The utility classes include entities that are basic but specialized for the problem domain of bus scheduling. Examples are the classes *Time*, *TimeRange*, *Point*, *Event*. A *Time* object can represent either a clock time or a length of time. It is capable of time arithmetic so that the usual addition and subtraction operators can be applied directly. A *TimeRange* object owns two *Time* objects which delimit a period of clock time. It is capable of calculating the length of the range, testing if another given clock time is within its range and testing if it overlaps with another given range. A *Point* object owns attributes such as a short code and a full description for a location in the bus network. The *Event* class is modelled as deriving (inheriting) from the *Time* class. It has an attribute which is a pointer to a *Point* object representing the location where the event occurs. Fig.4.1 illustrates the *Event* class. Arrows are used to indicate inheritance, and in this case an *Event* object will inherit all the properties owned by a *Time* object, e.g. possessing the properties of *Time* objects, *Event* objects can be directly involved in arithmetic expressions such as *timeGap = departureY - arrivalX*, where *departureY* and *arrivalX* are both *Event* objects and the result *timeGap* is a *Time* object.

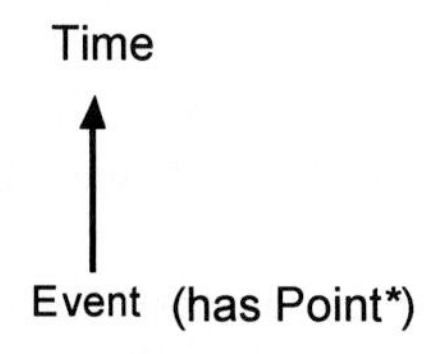

Fig. 4.1 The *Event* object class

Utility classes, as illustrated in the above example, provide a level of abstraction so that we can express the overall system model in a more succinct language relevant to the problem domain. Also, all the utility classes in BOOST have been defined with standard input and output methods, e.g. the code

```
outFile << departureY ;
```

will suitably format the time and place where the event *departureY* occurs and write them to the file *outFile*. In the case of inputting a *Time* object, the corresponding method will also automatically validate it.

Bus network classes are used for information akin to public timetables. There are bus routes (*BusRoute* objects) defined as sequences of timing points or stops (*Point* derived objects) with starting and ending termini (*Point* derived objects). Run times, which may vary according to the period of the working day, are assigned to route sections (*RouteArc* objects). The *BusRoute* class has attributes for specifying regular headways, which are common in the UK. Bus network classes have three major roles within BOOST. First, BOOST can automatically generate from them the trip data required by the core scheduling algorithm. This is far less tedious and error-prone than typing in data for each trip. Second, the bus network data provides a basis for computing a matrix of empty running times, called a *deadrun matrix*, between any pair of termini/depots. The deadrun matrix is a requirement for VAMPIRES to work. BOOST applies a shortest path algorithm using the in-service running times multiplied by a pre-specified discount factor. It will also accept deadrun times between certain locations explicitly specified by the user. Thirdly, the bus network classes are used to support the interactive scheduling facilities in BOOST (see later).

The vehicle activity classes form the backbone of the object-oriented model of VAMPIRES. They are discussed at greater length in the next section.

Attempts to swap bus links to improve the current schedule is a key feature of VAMPIRES. However, choosing bus links randomly for swapping trials would be very inefficient. We also have to guard against performing swaps in endless circles. Hence, heuristic strategies which employ a number of different swapping criteria are used. One simple criterion is to allow a swap if it reduces the total time infeasibility of bus links. Another criterion is to allow a swap if it results in creating a larger time gap in the bus links, which is a form of the first-in-last-out rule. Instead of hard coding the link swapping rules into the iterative refinement process, they are abstracted as rule objects. The heuristic processes in BOOST are designed to have generalized interfaces with the link swapping rule objects, thus making it flexible for developing new strategies and resulting in less complex processes.

The system process classes represent the major functional components in BOOST: data entry/management, VAMPIRES automatic scheduling, interactive scheduling, graphical scheduling and schedule document production. Within the system process classes are methods that can either be triggered internally or be triggered by the user through the user interface objects.

The user interface objects support two-way communication between the user and the BOOST system. They include the standard Windows GUI features such as dialog boxes, menus and icon buttons.

5 Vehicle Activity Object Classes

Bus scheduling is essentially the planning and sequencing of activities of a fleet of buses. At the start, the timetabled trips are the only known vehicle activities. The scheduling process creates auxiliary activities, called *bus links*, for chaining together trips to form the work of each bus. It will try minimising the total number of chains to cover all the trips being scheduled. Thus the intermediate steps of the scheduling process would also evaluate bus links, and destroy them when there are better alternatives. The operations on bus links are not all the same, but are dependent on their types. For example, in a *Layover* the bus remains at the same terminus between its arrival and departure. In a *Relocation*, the bus is driven without passengers to another terminus for its next departure. The evaluation operation on *Layover*'s and *Relocation*'s would generally apply different costing formulae.

The original VAMPIRES algorithm did not formally classify bus links. BOOST does so as shown in the inheritance tree in Fig. 5.1.

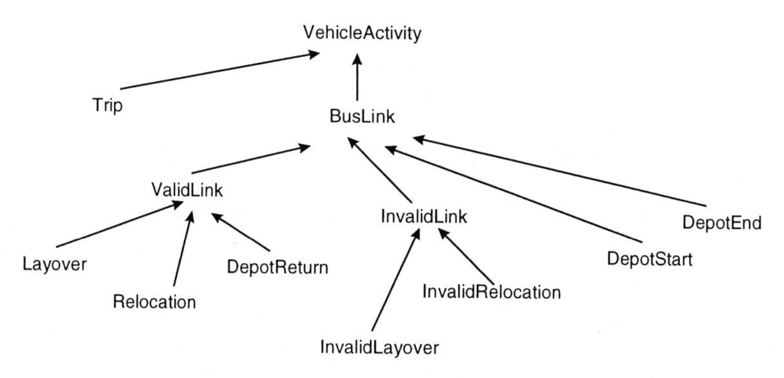

Fig. 5.1 Hierarchy of vehicle activity object classes

DepotReturn is the class of bus links which have time gaps long enough to justify temporarily returning the bus to the depot before its next departure. The names used for other classes are self-explanatory. With this classification scheme, instance variables of bus links can only be created from the "leaf node" classes on the inheritance tree (i.e. classes *Trip*, *Layover*, *Relocation*, *InvalidLayover*, *InvalidRelocation*, *DepotReturn*, *DepotStart*, *DepotEnd*). The other classes (*VehicleActivity*, *BusLink*, *ValidLink*, *InvalidLink*) provide three higher levels of abstraction. For example, a *Layover* object can also be referenced as either a *ValidLink*, a *BusLink*, or a *VehicleActivity* depending on the level of abstraction required.

Use of the abstraction levels is clearly advantageous in developing the link swapping strategies for the iterative refinement process. For example, the system may have one container holding all the *InvalidLink*s and one container holding all the *ValidLinks*. An initial phase might be to conduct swapping trials choosing one bus link from each of the containers. The implementation code does not have to deal with any of the lower level classes at all.

Another example is the object class *Bus*. *Bus* has an internal data structure declared as a list of *VehicleActivitys*, representing its sequence of work for the day (Fig. 5.2).

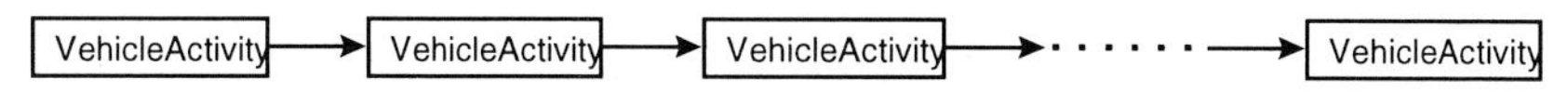

Fig. 5.2 *sequenceOfWork* declared in class *Bus*

When the data structure *sequenceOfWork* is built up, it will accept any lower level objects as its list nodes. However, a *Bus* object may work on the *sequenceOfWork* data structure using lower levels of abstraction when appropriate, e.g. given one of the bus links, a *Bus* object may have methods for retrieving the trips at positions relative to where the bus link is stored in *sequenceOfWork* and these methods would treat *sequenceOfWork* as an alternation of *BusLink* and *Trip* objects (Fig. 5.3), i.e. using a lower abstraction level.

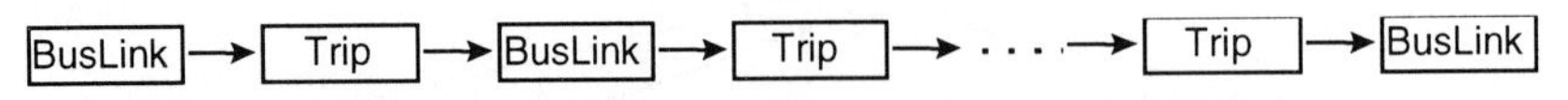

Fig. 5.3 *sequenceOfWork* used at a lower abstraction level

The above classification scheme for vehicle activities (Fig. 5.1) is very useful for managing variations of commonly possessed methods. To illustrate, consider two example methods (brackets are appended to method names as a convention to distinguish them from object names, c.f. C++ function calls without parameters): *costing( )* and *grouping( )*, which must be possessed by all the vehicle activity objects. The method *costing( )* computes and assigns a cost to the activity. The method *grouping( )* puts the activity object into the appropriate container(s) for the scheduling algorithm to work on. The scheduling algorithm needs to be flexible in fine tuning the costing formulae according to the types of activity, and the link swapping strategies may demand the grouping of activities under many different criteria. On the other hand, it would be very cumbersome if every different situation for a different costing formula or grouping criterion had to be tested and invoked explicitly in the code, as is the case in a procedural approach. Under the object-oriented approach, we can identify where in the inheritance tree a specialized method is needed and define it within the corresponding derived class. In coding the scheduling processes, there will be no need to test for which version of the specialized methods to use; the object-oriented software environment will take care of it automatically by reference to the inheritance tree, for example:

```
activityX -> costing() ;
activityX -> grouping() ;
```

invokes the two example methods and there is no need to care about which particular type *activityX* is. Fig. 5.4 shows which derived classes in the inheritance tree have been implemented with specialized versions of the *costing( )* and *grouping()* methods.

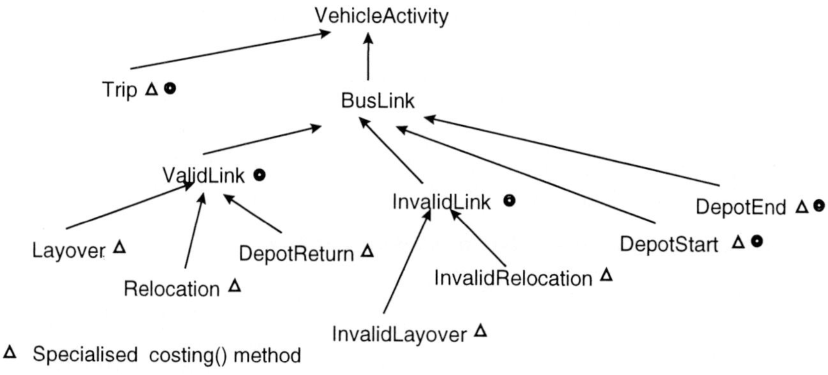

Fig. 5.4 Specialization of class methods

6 Implementation and Results

BOOST features data entry, schedule compilation and report generation in an integrated environment implemented using the Borland C++ Development Suite (Version 5) for PCs. In particular, Borland's Object Windows Library (OWL) and the Standard Template Library (STL) have been used.

Although the core scheduling processes only require trip data and a matrix of empty running times between termini (the *deadrun matrix*), the data management module accepts data in more basic forms such as bus route definitions and regular service intervals, which are converted to trip data automatically. Alternatively, trips may be entered into BOOST directly. BOOST features automatic estimation of the number of vehicles required, but that can be overridden when exploration of a different target is desired. It can automatically compute the deadrun matrix from the route definitions using a shortest path algorithm. However, the user may elect to specify the deadrun times to be used between particular pairs of termini. Usually this is done when empty running is allowed to take short cuts not defined in the route network being scheduled.

Besides using the automatic VAMPIRES algorithm, vehicle schedules may be compiled interactively. Two interactive methods are provided. The first one uses a Windows dialog box (e.g. see Fig. 6.1), where a list of trips are presented in order of their departure times.

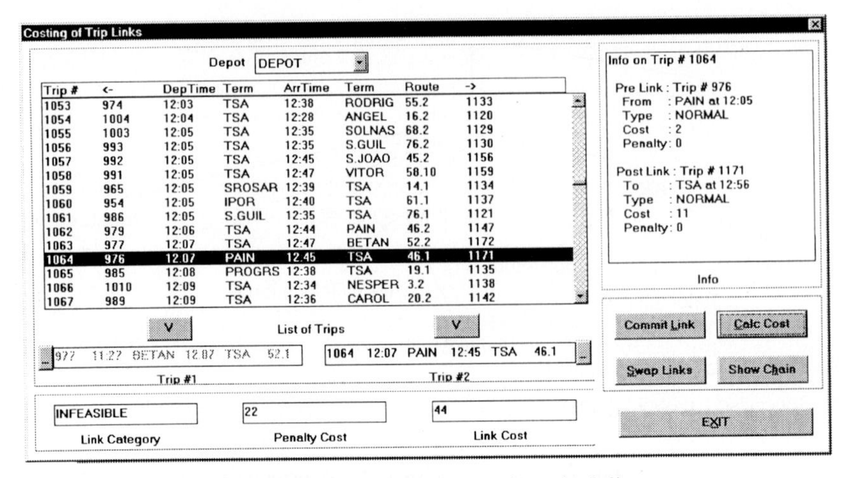

Fig. 6.1 Dialog box for interactive scheduling

The user may select any trip from the list and enquire about the status of links around it. A tentative link may be formed by selecting a pair of trips and pressing some control buttons. A new bus may be created to start from a depot and linked to a selected trip, and similarly a bus may be returned to a depot at the end of a trip. A complete schedule can thus be created interactively. The other method is graphical (illustrated in Fig. 6.2), by which bus links are formed by connecting trips drawn graphically using a mouse. The visual interactive scheduling component in BOOST supports full editing and various queries about the schedule.

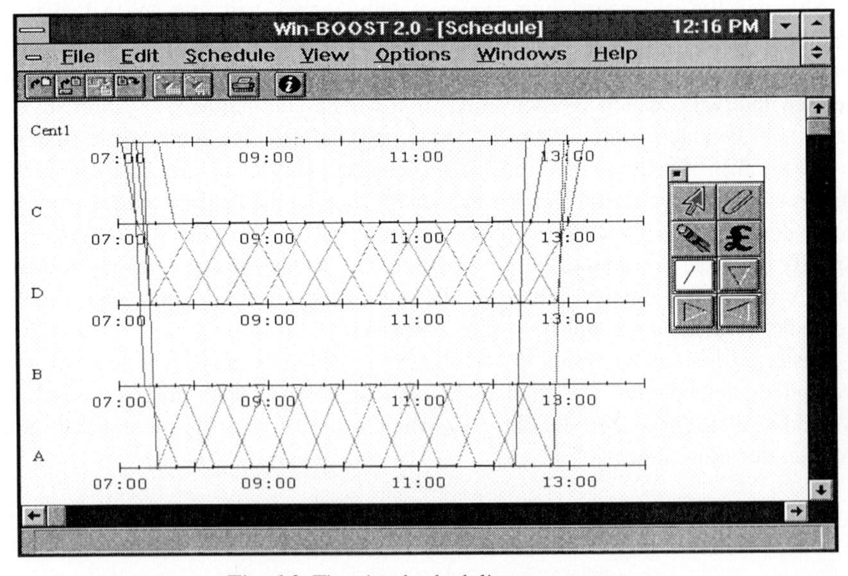

Fig. 6.2 The visual scheduling component

Each terminus and depot occupies a horizontal line, corresponding to a timescale of the working day, for highlighting events at the terminus/depot. The activities of a bus are color-coded and are traced across or on top of the horizontal time-location lines. The time-infeasible bus links are in particular highlighted in red. A "floating toolbox" window contains icon buttons for zooming, editing and querying functions. For example, the link tool (represented by the paper clip icon) is used for inserting a bus link between two selected trips. A pictorial representation helps the user easily to recognize regular patterns and anything critical in the schedule.

Although the interactive scheduling facilities can be used for producing complete schedules (rather tediously), they are more useful for visualizing and fine-tuning the schedules compiled using the VAMPIRES algorithm. What-if trials could be performed and the user can see the effects immediately.

BOOST has been tested on a number of data sets obtained from the bus industry in the past. Some results as compared with runs using BUSPLAN (the version as was in 1993) are shown in Table 6.1, which indicates that BOOST is consistently producing good schedules and in some cases performs better than BUSPLAN. The improvements are attributable to the flexibility of the object-oriented model for developing better heuristic strategies in the VAMPIRES algorithm. It is worth noting that the estimator in BOOST produces tighter lower bounds on the number of buses needed than its counterpart in BUSPLAN. Although accurate timing of runs on the two systems are not available, BOOST is noticeably faster.

Table 6.1 Comparison of results using BOOST and BUSPLAN

Data set	Total trips (Bus hrs.)	System	Buses estimated	Buses used	Total deadrun	Total layover	Depot returns	Total cost*
A	325	BOOST	18	18	16:23	44:12	10	4918
	(133:43)	BUSPLAN	16	18	16:25	58:42	10	5742
B	1702	BOOST	40	40	13:55	18:20	6	3136
	(494:16)	BUSPLAN	39	40	16:06	23:32	4	3464
C	367	BOOST	14	15	11:05	07:42	8	1894
	(135:43)	BUSPLAN	14	15	12:25	07:22	13	2322
D	1543	BOOST	39	39	13:58	23:32	3	3133
	(444:10)	BUSPLAN	39	39	15:49	25:45	2	3503
E	522	BOOST	12	12	05:42	12:41	1	1460
	(143:52)	BUSPLAN	12	12	05:42	12:41	1	1460
F	345	BOOST	9	9	03:40	04:33	1	728
	(102:33)	BUSPLAN	9	9	03:40	04:33	1	728

*Total Cost = Total deadrun x 2 + Total layover + Depot returns x 30

BOOST has also been used on data from Strathclyde Buses in Glasgow and in recent work for bus operators in Brazil. BOOST has been demonstrated to be a very useful bus scheduling tool in the Brazilian exercises. The largest run has 2612 trips using 107 buses. The Brazilian case studies are reported in detail in (Wren/Gualda (1997)). Since it would involve too much data conversion work, these data sets have not been tested using BUSPLAN and therefore no comparative results are available.

7 Conclusions

An object-oriented model of the VAMPIRES algorithm has been developed and built upon into the BOOST system, which has incorporated many useful features implemented using Windows GUI. Test results indicate that the object-oriented VAMPIRES algorithm is producing good schedules, and in some cases performed better than BUSPLAN. The object-oriented model will be a solid foundation for future extensions to cater for more complex scheduling requirements and constraints, such as multi-vehicle types and depots, and automatic retimings to achieve a well co-ordinated network of services.

Acknowledgements: Thanks are due to the Engineering and Physical Sciences Research Council for providing funding support for the work described in this paper. We also thank Professor Anthony Wren for valuable advice and comments during the project.

References

Chamberlain, M.P./Wren, A. (1992): Developments and recent experience with the BUSMAN and BUSMAN II system, in: Desrochers, M./Rousseau, J.-M. (eds.): Computer-aided transit scheduling. (Springer-Verlag) 1 - 16.

Daduna, J.R./Paixao, J.M.P. (1995): Vehicle scheduling for public mass transit - An overview, in: Daduna, J.R./Branco, I./Paixao, J.M.P (eds.): Computer-aided transit scheduling. (Springer-Verlag) 76 - 90.

Kwan, R.S.K./Rahin, R.A. (1995): Bus Scheduling with Trip Co-ordination and Complex Constraints, in: Daduna, J.R./Branco, I./Paixao, J.M.P (eds.): Computer-aided transit scheduling. (Springer-Verlag) 91 - 101.

Smith, B.M./Wren, A. (1981): VAMPIRES and TASC: two successfully applied bus scheduling programs, in: Wren, A. (ed.): Computer scheduling of public transport. (North-Holland) 97 - 124.

Wagner, P. (1987): Dimensions of object-based language design, in: OOPSLA '87 Proceedings, ACM. New York.

Wolfenden, K./Wren, A. (1966): Locomotive scheduling by computer. Proceedings of the British Joint Computer Conference, IEE Conference Publication no. 19, p.31.

Wren, A. (1972): Bus scheduling: an interactive computer method. Transportation Planning and Technology, vol. 1, 115 - 122

Wren, A. (1981): General review of the use of computers in scheduling buses and their crews, in: Wren, A. (ed.): Computer scheduling of public transport. (North-Holland) 3 - 16.

Wren, A./Chamberlain, M.P. (1988): The development of MICRO-BUSMAN: scheduling on micro-computers, in: Daduna, J.R./Wren, A. (eds.): Computer-aided transit scheduling. (Springer-Verlag) 160 - 174.

Wren, A./Gualda, N.D.A. (1997): Integrated scheduling of buses and drivers, in: Preprints of the 7th International workshop on computer-aided scheduling of public transport, Boston, US.

Solving Large-Scale Multiple-Depot Vehicle Scheduling Problems

Andreas Löbel

Konrad-Zuse-Zentrum für Informationstechnik Berlin (ZIB), Takustraße 7, D-14195 Berlin, Germany, E-mail: `loebel@zib.de`, URL: `www.zib.de`

Abstract: This paper presents an integer linear programming approach with column generation for the $\mathcal{NP}$-hard Multiple-Depot Vehicle Scheduling Problem (MDVSP) in public mass transit. We describe in detail the basic ingredients of our approach that seem indispensable to solve truly large-scale problems to optimality, and we report on computational investigations that are based on real-world instances of three large German public transportation companies. These instances have up to 25,000 timetabled trips and 70 million integer decision variables. Compared to the results obtained with one of the best planning system currently available in practice, our test runs indicate savings of several vehicles and a cost reduction of about 10 %.

1 Introduction

Solving transportation problems was and still is one of the driving forces behind the development of mathematical disciplines such as optimization and operations research (see Borndörfer/Grötschel/Löbel (1995)). Truly large transportation problems have to be solved, for instance, in airlines (aircraft and crew scheduling) and public mass transit (vehicle and duty scheduling). In the past, the corresponding transportation markets have often been protected monopolies. However, deregulation of such monopolistic markets has led to world-wide competition. It is therefore obvious that competitive participants in these markets must use computer-aided tools for their operational planning process to employ their resources as efficiently as possible. Modern sophisticated mathematical optimization techniques can help to solve such planning problems.

For instance, public transportation in the European Community is subject to such market deregulation. Monopolistic markets have become more liberal or will soon be broken up. In order to prevent their complete extinction, monopolistic transportation organizations will therefore have to change from deficit-oriented monopolies to competitive market players. One important factor in facing the challenges of a competitive market is, of course, cost reduction, which can be obtained by making intelligent use of the latest mathematical knowhow.

Vehicle scheduling is one important step in the hierarchical planning process in public transportation. The *Multiple-Depot Vehicle Scheduling Problem* (MDVSP) is to assign a fleet of vehicles, possibly stationed at several garages, to a given set of passenger trips such that operational, company-specific, technical, and further side constraints are satisfied and the available resources are employed as efficiently as possible. In the last three decades, considerable research has gone into the development of academic as well as practice-oriented solution techniques for the $\mathcal{NP}$-hard MDVSP and special, often polynomially solvable cases of it. Review articles on this topic are, for instance, Desrosiers/Dumas/Solomon/Soumis (1995), Daduna/Paixão (1995), and Bussieck/Winter/Zimmermann (1997).

The most successful solution approaches for the MDVSP are based on *network flow models* and their integer programming analogues. In the literature, there are two basic mathematical models of this type: First, a direct *arc-oriented model* leading to a *multicommodity flow problem* and, second, a *path-oriented model* leading to a *set partitioning problem*. The latter can also be derived from Dantzig-Wolfe decomposition applied to the first. Both approaches lead to large-scale integer programs, and *column generation techniques* are required to solve their LP relaxations. We shall explicitly discuss the differences between these two models in Section 3.

We investigate in this paper the solution of the *multicommodity flow formulation*. Solution techniques for models of this flavour have been discussed in various articles: Carpaneto/Dell'Amico/Fischetti/Toth (1989) describe an integer LP (ILP) formulation based on an arc-oriented assignment problem with additional path-oriented flow conservation constraints. They apply a so-called "additive lower bounding" procedure to obtain a lower bound for their ILP formulation. Ribeiro/Soumis (1994) show that this additive lower bounding is a special case of Lagrangean relaxation and its corresponding subgradient method. Forbes/Holt/Watts (1994) solve the integer linear programming formulation of the multicommodity flow model by branch-and-bound. The sizes of the problems that have been solved to optimality in these publications are relatively small involving up to 600 timetabled trips and three depots.

The *contribution of this paper* is the efficient solution of the ILP (derived from the multicommodity flow formulation) by means of LP column generation techniques. We use a new technique, called *Lagrangean pricing*, that is based on Lagrangean relaxations of the multicommodity flow model. Embedded within a branch-and-cut frame, this method makes it possible to solve problems from practice to *proven* optimality. Lagrangean pricing has been developed independently at the same time by Fischetti/Vigo (1996) and Fischetti/Toth (1996) for solving the Asymmetric Travelling Salesman Problem and the Resource-Constrained Arborescence Problem.

Our computational investigations are performed on large-scale data from three German public transportation companies: the Berliner Verkehrsbetriebe (BVG), the Hamburger Hochbahn AG (HHA), and the Verkehrsbetriebe Hamburg-Holstein AG (VHH). These instances involve problems with up to 49 depots, about 25,000 timetabled trips, and about 70 million integer decision variables. The runs on these test sets show that our method is able to solve problems of this size optimally. These problems are orders of magnitude larger than the instances successfully solved with other approaches, as far as we know.

In the following, we present our branch-and-cut approach with column generation. We start in the next section by describing our multicommodity flow version of the problem and present an ILP formulation and relaxations thereof. Section 3 discusses the model and compares it with problem relaxations often used in practice. Our algorithm is described in Sect. 4 presenting various tools to solve MDVSP problems and subproblems such as primal heuristics, a network simplex algorithm with column generation (Löbel (1996)), Lagrangean relaxations (Kokott/Löbel (1996)), linear programming relaxations (Löbel (1997d)), and the branch-and-cut approach composing all these ingredients. The tested real-world data are presented in Sect. 5, and the computational results are discussed in Sect. 6. A comprehensive report about this project can be found in Löbel (1997c).

2 The MDVSP

The following section refers to some basic terminology for MDVSPs that we quickly review here. For more details see Löbel (1997c).

The fleet of a transportation company is divided among a set of **depots** denoted by $\mathcal{D}$. With each depot $d \in \mathcal{D}$, we associate a start point d^+ and an end point d^- where its vehicles start and terminate their daily duties. Let $\mathcal{D}^+ := \{d^+ | \; d \in \mathcal{D}\}$ and $\mathcal{D}^- := \{d^- | \; d \in \mathcal{D}\}$. The number of available vehicles, the **depot capacity**, of each depot d is denoted by κ_d. A given timetable defines a set of **timetabled trips**, denoted by $\mathcal{T}$, that are used to carry passengers. We associate with each $t \in \mathcal{T}$ a first stop t^-, a last stop t^+, a departure time s_t, an arrival time e_t, and a **depot-group** $G(t) \subseteq \mathcal{D}$. Each $G(t)$ includes those depots whose vehicles are allowed and able to service trip t. Let $\mathcal{T}_d := \{t \in \mathcal{T} | \; d \in G(t)\}$, $\mathcal{T}^- := \{t^- | \; t \in \mathcal{T}\}$, and $\mathcal{T}^+ := \{t^+ | \; t \in \mathcal{T}\}$.

There are further types of trips, which do not carry passengers: A **pull-out trip** connects a start point d^+ with a first stop t^-, a **pull-in trip** connects a last stop t^+ with an end point d^-, and a **dead-head trip** connects a last stop t^+ with a succeeding first stop t'^-. For simplicity, these trips are all called **unloaded trips**.

For two trips $p, q \in \mathcal{T}$, let $\Delta_{p,q} \geqslant 0$ be given. In the literature, $\Delta_{p,q}$ is often used as the time (travel plus layover time) from the last stop p^+ to the first stop q^- (for example, see Dell'Amico/Fischetti/Toth (1993), Ribeiro/Soumis (1994), and Daduna/Paixão (1995)). However, our operating partners use such a definition of $\Delta_{p,q}$ only for those dead-head trips for which the idle time or the difference $s_q - e_p$ does not exceed a predefined maximum ranging from 40 to 120 minutes. Otherwise, $\Delta_{p,q}$ is set to infinity. We will show that such a restriction in the degree of freedom can lead to a higher vehicle demand and, therefore, to suboptimal solutions. To make it possible to use such links in spite of this, we set $\Delta_{p,q} := s_q - e_p$ whenever it is possible to park a vehicle between p and q at the depot. We call these special dead-head trips **pull-in-pull-out trips**. They were first described in Bokinge/Hasselström (1980). Whenever $e_p + \Delta_{p,q} \leqslant s_q$ is satisfied, the corresponding dead-head trip is called **compatible**.

A **vehicle schedule** or (duty) is a chain of trips such that the first trip is a pull-out trip, the last trip is a pull-in trip, and the timetabled and unloaded trips occur alternately. A vehicle schedule is called **feasible** if all its trips belong to the same depot. A circulation is also called a **block** (or **rotation**) if it includes no pull-in-pull-out trip.

For each depot $d \in \mathcal{D}$, we introduce the following sets of trips: $A_d^{\text{t-trip}} := \{(t^-, t^+) | \, t \in \mathcal{T}_d\}$ (timetabled trips) and $A_d^{\text{u-trip}} := \{(d^+, t^-), (t^+, d^-) | \, t \in \mathcal{T}_d\} \cup \{(p^+, q^-) | \, p, q \in \mathcal{T}_d \wedge e_p + \Delta_{p,q} \leqslant s_q\}$ (unloaded trips). With each unloaded trip $a \in A_d^{\text{u-trip}}$, we associate a weight $c_a^d \in \mathbb{Q}$ representing its operational costs. In addition, we add to the weight of each pull-out trip a sufficiently large $\mathbf{M}$, representing the capital costs, that it is larger than the operational costs of any feasible solution. The minimization of this "two-stage" objective function first minimizes the fleet size and, then, the operational costs among all minimal fleet solutions. With this terminology, the MDVSP is to find a weight minimal set of feasible vehicle schedules such that each timetabled trip is covered by exactly one vehicle schedule.

The MDVSP can be stated as an integer multicommodity flow problem as follows. For each depot $d \in \mathcal{D}$, let (d^-, d^+) denote an additional **backward arc** (on which depot capacities can be controlled) and let $A_d := A_d^{\text{t-trip}} \cup A_d^{\text{u-trip}} \cup \{(d^-, d^+)\}$. Let $D = (V, A)$ be a digraph with node set $V := \mathcal{D}^+ \cup \mathcal{D}^- \cup \mathcal{T}^- \cup \mathcal{T}^+$ and arc set $A := \dot{\bigcup}_{d \in \mathcal{D}} A_d$. Figure 1 illustrates a small example with $\mathcal{D} = \{\text{r,g}\}$, $\mathcal{T} = \{\text{a,b,c,d,e}\}$, $\mathcal{T}_{\text{r}} = \{\text{a,c,d}\}$, and $\mathcal{T}_{\text{g}} = \{\text{a,b,c,e}\}$.

2.1 An Integer Linear Program

We introduce an integer variable x_a^d for each $a \in A_d$ and each $d \in \mathcal{D}$. x_a^d denotes a decision variable indicating whether a vehicle of depot d runs trip a, unless a denotes the backward arc. In this case, x_a^d counts all vehicles used at depot d. The variables x_a^d are combined into vectors $x^d := (x_a^d)_{a \in A_d} \in \mathbb{R}^{A_d}$, $d \in \mathcal{D}$, and these into $x := (x^d)_{d \in \mathcal{D}} \in \mathbb{R}^A$.

Given a node $v \in V$, let $\delta^+(v)$ denote all arcs of A with tail in v and, accordingly, $\delta^-(v)$ denote all arcs with head in v. For a given set $\tilde{A} \subset A$, we

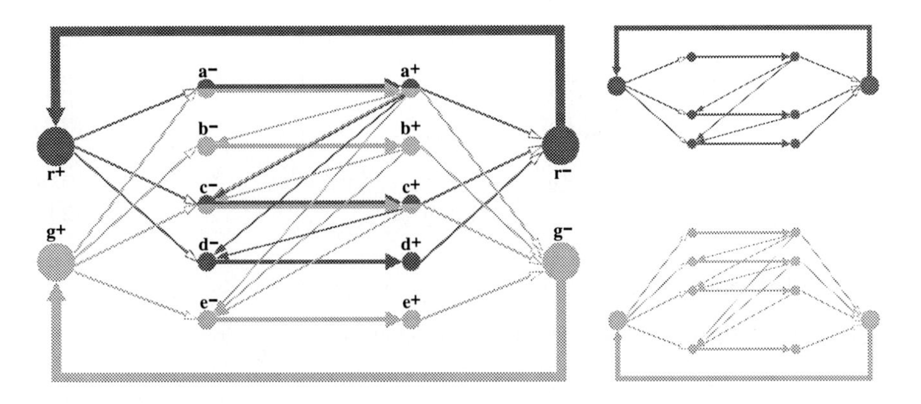

Figure 1: Digraph (V, A) and its single-depot graphs

define $x^d(\tilde{A}) := \sum_{a \in \tilde{A} \cap A_d} x_a^d$ and $x(\tilde{A}) := \sum_{d \in \mathcal{D}} x^d(\tilde{A})$. The ILP formulation of the MDVSP is ILP-1:

$$\min \sum_{d \in \mathcal{D}} \sum_{a \in A_d^{\text{u-trip}}} c_a^d x_a^d \tag{1a}$$

subject to

$$\begin{aligned}
x\big(\delta^+(t^-) \cap \delta^-(t^+)\big) &= 1, \quad \forall\, t \in \mathcal{T}, & (1\,\text{b})\\
x^d\big(\delta^+(v)\big) - x^d\big(\delta^-(v)\big) &= 0, \quad \forall\, v \in \mathcal{T}_d \cup \{d^+, d^-\}\ \forall\, d \in \mathcal{D}, & (1\,\text{c})\\
x^d_{(d^-, d^+)} &\leqslant \kappa_d, \quad \forall\, d \in \mathcal{D}, & (1\,\text{d})\\
x_a^d &\geqslant 0, \quad \forall\, a \in A_d\ \forall\, d \in \mathcal{D}, & (1\,\text{e})\\
& x \text{ integral}. & (1\,\text{f})
\end{aligned}$$

Note that $x\big(\delta^+(t^-) \cap \delta^-(t^+)\big) = \sum_{d \in G(t)} x^d_{(t^-, t^+)}$. Constraints (1 b), the **flow conditions**, ensure that each timetabled trip is serviced exactly once. Constraints (1 c), **flow conservation**, guarantee that the total flow value of each depot d entering some node $v \in V$ must also leave v.

ILP-1 includes many redundant constraints that can be eliminated by performing some preprocessing steps as shown in Löbel (1997c). The main idea is to shrink t^- and t^+ to one node t, for all $t \in \mathcal{T}$, and to shrink d^- and d^+ to one node d, for all $d \in \mathcal{D}$. This corresponds to eliminating each arc not belonging to some unloaded trip and leads to the following equivalent ILP, ILP-2:

$$\min \sum_{d \in \mathcal{D}} \sum_{a \in A_d^{\text{u-trip}}} c_a^d x_a^d \tag{2a}$$

subject to

$$
\begin{aligned}
x(\delta^+(t)) &= 1, && \forall\, t \in \mathcal{T}, && \text{(2b)}\\
x^d(\delta^+(t)) - x^d(\delta^-(t)) &= 0, && \forall\, t \in \mathcal{T}_d\ \forall\, d \in \mathcal{D}, && \text{(2c)}\\
x^d(\delta^+(d)) &\leqslant \kappa_d, && \forall\, d \in \mathcal{D}, && \text{(2d)}\\
x^d_a &\geqslant 0, && \forall\, a \in A^{\text{u-trip}}_d\ \forall\, d \in \mathcal{D}, && \text{(2e)}\\
& x \text{ integral}. && && \text{(2f)}
\end{aligned}
$$

2.2 Relaxations

The natural relaxation of ILP-2 is of course its LP relaxation LP-3:

$$
\min \quad \sum_{d\in\mathcal{D}} \sum_{a\in A^{\text{u-trip}}_d} c^d_a\, x^d_a. \tag{3}
$$

subject to (2b)-(2f).

Let $\nu \in \mathbb{R}^{\mathcal{T}}$, $\pi := (\pi^d \in \mathbb{R}^{\mathcal{T}_d})_{d\in\mathcal{D}}$, and $0 \leqslant \gamma \in \mathbb{R}^{\mathcal{D}}$ denote the dual multipliers for (2b), (2c), and (2d). Consider a subset $\tilde{A} \subseteq A^{\text{u-trip}}$. The LP containing just the columns corresponding to $\tilde{A}$ is called the **restricted** LP of LP-3 and is denoted by **RLP**.

We briefly give two possible Lagrangean relaxations for the MDVSP. For notational simplicity, we use the same symbols for the dual variables of LP-3 and for the Lagrangean multipliers of the two following Lagrangean relaxations.

Let $\pi := (\pi^d \in \mathbb{R}^{\mathcal{T}_d})_{d\in\mathcal{D}}$ and $0 \leqslant \gamma := (\gamma^d)_{d\in\mathcal{D}} \in \mathbb{R}^{\mathcal{D}}$ denote the Lagrangean multipliers according to the flow conservations (2c) and the depot capacities (2d). Relaxing (2c) and (2d), we obtain a Lagrangean dual LR_{fcs} reading $\max_{\pi;\gamma\geqslant 0} L_{\text{fcs}}(\pi,\gamma)$ with inner minimization problem

$$
\begin{aligned}
L_{\text{fcs}}(\pi,\gamma) := \min \sum_{d\in\mathcal{D}} \Bigg\{ \sum_{a\in A^{\text{u-trip}}_d} c^d_a\, x^d_a \;-\; \sum_{t\in\mathcal{T}_d} \pi^d_t \Big(x^d(\delta^+(t)) - x^d(\delta^-(t)) \Big) \\
- \gamma^d \Big(\kappa_d - x^d(\delta^+(d)) \Big) \Bigg\} \qquad \text{(4a)}
\end{aligned}
$$

subject to

$$
x \text{ satisfies (2b), (2e), (2f), and } - x(\delta^-(t)) = -1,\ \forall\, t \in \mathcal{T}. \tag{4b}
$$

The subscript "fcs" of L_{fcs} and LR_{fcs} stands for **F**low-**C**on**S**ervation. Note, for fixed arguments, L_{fcs} is a minimum-cost flow problem.

The second Lagrangean relaxation is based on the ILP-1 (1). Let $\nu := (\nu_t)_{t\in\mathcal{T}} \in \mathbb{R}^{\mathcal{T}}$ denote the Lagrangean multipliers for to the flow conditions

(1 b). Relaxing (1 b), we obtain a Lagrangean dual $\mathrm{LR_{fcd}}$ reading $\max_\nu L_{\mathrm{fcd}}(\nu)$ with inner minimization problem

$$L_{\mathrm{fcd}}(\nu) := \nu^{\mathrm{T}}\mathbb{1} \; + \; \sum_{d\in\mathcal{D}} \min\Big\{ \sum_{a\in A_d^{\text{u-trip}}} c_a^d\, x_a^d \; - \; \sum_{t\in\mathcal{T}_d} \nu_t\, x_{(t^-,t^+)}^d \Big\} \tag{5a}$$

subject to

$$x \text{ satisfies (1c)–(1f) and } x_{(t^-,t^+)}^d \leqslant 1, \;\; \forall\, t \in \mathcal{T}_d \;\; \forall\, d \in \mathcal{D}. \tag{5b}$$

The subscript "fcd" of L_{fcd} and $\mathrm{LR_{fcd}}$ stands for **F**low-**C**on**D**ition. Note that L_{fcd} decomposes a constant part $\nu^{\mathrm{T}}\mathbb{1}$ into $|\mathcal{D}|$ independently solvable minimum-cost flow circulation problems.

It is easy to see that the additional constraints in (4 b) and (5 b) are redundant in ILP-1 and ILP-2, respectively, but necessary to obtain convenient inner minimization problems that are efficiently solvable minimum-cost flow problems.

3 Discussion of the Model

In this section, we discuss the relation and differences between the arc-oriented multicommodity flow and the path-oriented set partitioning formulations. We will also distinguish our multicommodity flow formulation from some other (arc-oriented) models that have been presented in the literature.

3.1 Multicommodity Flow and Set Partitioning Models

Arc-oriented multicommodity flow and path-oriented Dantzig-Wolfe (DW) set partitioning formulations are usually used to model the MDVSP. Applied to practical vehicle scheduling problems, their corresponding ILP formulations can involve several million integer variables. Solving such large ILPs requires column generation techniques.

For the arc-oriented model, column generation can be seen as an implicit pricing technique (see Schrijver (1989)): one works on restricted subsets of active arcs that are generated and eliminated in a dynamic process. For the DW decomposition, column generation usually leads to pricing problems in the form of constrained shortest path problems. Many researchers automatically associate the term "column generation" with the solution process used in a DW decomposition (see Soumis (1997)). To distinguish this use of the term "column generation" from those as a general LP pricing technique in the sense of Schrijver, DW column generation is also called *delayed column generation* as proposed in Chvátal (1980). To avoid misunderstandings, we will use in this paper the term "column generation" as a general LP pricing technique in the sense of Schrijver.

Direct approaches to the multicommodity flow formulation can be used if all side constraints can be formulated solely in terms of the arcs of the network. This is the case for the MDVSP considered here. *DW decomposition* is needed particularly for problems that involve path constraints. It applies not only to vehicle scheduling problems, but also to similar applications such as crew and airline scheduling. For a survey on DW set partitioning approaches to such problems, we refer the reader to (for example) Desrosiers/Dumas/Solomon/Soumis (1995) and Soumis (1997).

3.2 Differences from Other Arc Oriented Models

Practice-oriented methods for the MDVSP are in most cases based on a single-commodity minimum-cost flow relaxation within a **schedule first – cluster second** approach (see Daduna/Paixão (1995) for a detailed description of this approach). The multiple-depot formulation is reduced to a single-depot relaxation. Unlike multicommodity flow formulations, however, those single-depot relaxations have two significant drawbacks:

Depot-groups and flow conservation: It is only possible to consider a single (depot independent) dead-head trip – we will call it a *link* – between two timetabled trips. Such a link (t, t') is considered to be feasible with respect to the depot-groups if $G(t) \cap G(t') \neq \emptyset$. But if depot-groups must only be satisfied locally between two trips, the intersection of the depot-groups of a generated block may be empty. In other words, the solution would be infeasible, see Fig. 2. Splitting such an infeasible block into its feasible parts can lead to suboptimal solutions.

Figure 2: Invalid block.

To avoid falling into such traps, the MDVSP should be modelled as a multicommodity flow problem. Many researchers have considered the MDVSP as a multicommodity flow problem long before we started our investigations. The need for many real-world applications to consider different depot-groups, however, was realized only recently (see Forbes/Holt/Watts (1994)). It is obvious that multicommodity flow formulations are natural for this kind of scheduling problems.

Limited duration for dead-head trips: It is often the case that single-depot relaxations consider dead-head trips with a limited duration (see Daduna/Mojsilovic/Schütze (1993)). It is therefore only possible to generate blocks that must be linked to vehicle schedules in a succeeding step. Based on heuristic ideas, the main objective is to use as many dead-head trips as possible and, secondarily, to minimize operational costs. Obviously, this objective function does indeed minimize the number of blocks if depot-groups are handled correctly. At the same time, it is assumed that a block minimal solution provides also a fleet minimal solution. It can be shown, however, that this is not true in general, see Fig. 3. The blocks, which have been determined by this strategy are assigned to the depots and then linked to vehicle schedules. These links correspond to pull-in-pull-out trips. It is clear that such a problem decomposition into two successive steps can lead to suboptimal solutions.

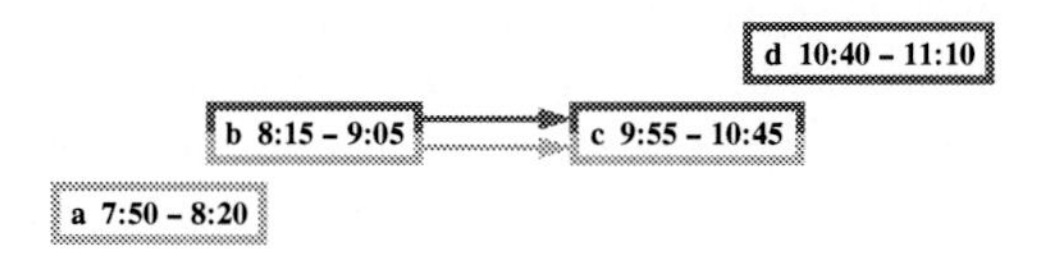

Figure 3: Minimal block solution is not fleet minimal.

Figure 3 displays a multiple-depot instance with two depots for which the fleet minimal solution cannot be obtained with a block minimal solution: The first depot can service trips "b, "c", and "d", and the second depot can service "a", "b", and "c". Two timetabled trips may be linked by a pull-in-pull-out trip if the depot-groups are satisfied and if the two timetabled trips do not overlap. The maximum allowed duration of a dead-head trip is set to 60 minutes such that for both depots just the dead-head trip between "b" and "c" is possible. The block minimal number is three ("d" is assigned to the first depot, "a" is assigned to the second depot, and "b→c" is assigned to the first or the second depot) and requires three vehicles, but two vehicles are optimal ({a,c} and {b,d}) if each timetabled trip defines its own block.

Since we cannot generate a fleet minimal solution in such a two step approach, linking blocks optimally and selecting user-defined unloaded trips must be done simultaneously. Pull-in-pull-out trips translate the decision of linking blocks into the terminology of dead-head trips. Therefore, using pull-in-pull-out trips makes it possible to generate a fleet minimal solution with minimum operational costs in one step.

Each pull-in-pull-out trip stands for a pull-in trip followed by a pull-out trip. The set of all pull-in-pull-out trips represents all feasible possibilities to link blocks to vehicle schedules. If we enlarge the user-defined unloaded trips by the pull-in-pull-out trips, the number of necessary vehicles is nothing but the number of used pull-out (or pull-in) trips. Vice versa, if we replace each pull-in-pull-out trip of a vehicle schedule by the corresponding pull-in and pull-out trip, it is always possible to assign all resulting blocks of this vehicle schedule to a single vehicle.

4 Solving MDVSPs

This section sketches the branch-and-cut method to solve the MDVSP. We give here only a brief summary of the basic ingredients that were required to solve our test instances: because of limited space, it is not possible to explain the details. We refer the reader to Löbel (1997c) for a detailed description.

Real-world problems of large cities such as Berlin have up to 25,000 daily timetabled trips and 70 million unloaded trips. At first glance, it seems impossible to solve such large ILPs and their LP relaxations exactly using commercial or publicly available software, even on the newest and fastest workstations or supercomputers. Nonetheless, with an intelligent combination of available LP and minimum-cost flow software together with implementation of many concepts of combinatorial optimization and integer programming, it is now possible to solve such problems to optimality by column generation and branch-and-cut on fast workstations. The basic components and concepts are:

- Lagrangean relaxations to obtain fast and tight lower bounds for the minimum fleet size and the minimum operational costs as close as possible to the integer optimum value.

- Primal opening heuristics to obtain a first integer feasible solution and a good starting point for the LP relaxation.

- The LP relaxation approach with a column generation scheme including Lagrangean pricing.

- *LP-plunging* to exploit the information compiled in each (R)LP and its optimal solution.

- Branch-and-cut to solve a problem to proven optimality.

- The workhorses: MCF combined with a column generation and the LP solver CPLEX.

Our basic method to solve the MDVSP is to solve the integer linear programming formulation by primal and dual heuristics, column generation, and branch-and-cut (see Fig. 4).

First, we determine a fast and tight lower bound c_L by Lagrangean relaxation and compute an upper bound c_U using opening heuristics. Second, the LP relaxation is solved to optimality using a column generation and column elimination scheme. The column generation procedure is based on new Lagrangean pricing and on standard reduced cost pricing, the column elimination procedure uses only the reduced cost criterion. Within this iterative process, we optionally call an LP-plunging heuristic to find a better integer solution. If the upper bound has been improved by the LP-plunging, we check

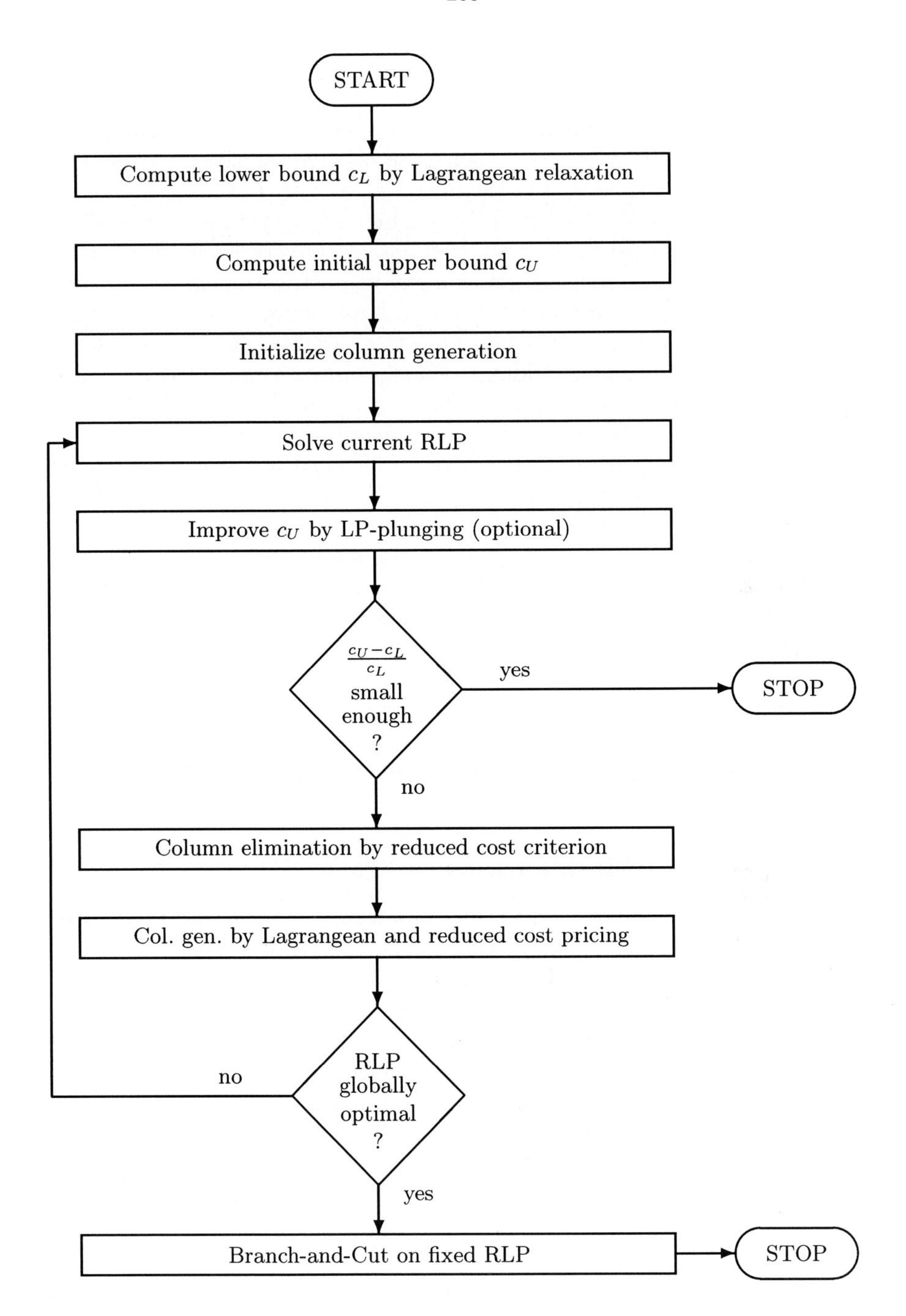

Figure 4: Solving MDVSPs: Flow chart.

whether $\frac{c_U}{L_{\mathrm{fcs}}(0,0)}$ is "small enough" from a practical point of view and stop in this case.

Up to the point where the LP relaxation has been solved to optimality, our method generates for most test instances an optimal solution or a minimal fleet solution with a small gap in the operational costs. For many instances, the current solution obtained by LP-plunging is (almost) optimal, and we have already terminated the optimization process. Otherwise, let c_{LP} denote the optimal LP value. We generate as many non-active columns as possible (respecting the primary memory limits) that have reduced costs smaller than $c_U - c_{LP}$. Note that none of the other inactive variables can have a positive value in an integer solution yielding a smaller objective value than c_U. The resulting RLP is then fixed and solved by branch-and-cut. Of course, branch-and-cut solves the complete problem to proven optimality only if all necessary variables have been generated that may be included in an optimal integer solution. Otherwise, branch-and-cut is only a heuristic that solves the integer version of the last RLP to optimality. Our branch-and-cut approach turned out to be not a very important part of our method. Therefore, it is not described in this paper, but we refer the interested reader to Löbel (1997c).

We have also investigated a Dantzig-Wolfe decomposition. It turned out that such a decomposition approach is unsuitable for the MDVSP, at least for the test set that we have investigated. The major obstacle here is that the continuous master problem relaxations become too hard to solve efficiently. Especially for problems with more than one thousand timetabled trips, the LU factorization in solving a restricted master problem takes far too much time.

4.1 Lagrangean Relaxations

The first important use of the Lagrangean relaxations $\mathrm{LR}_{\mathrm{fcs}}$ and $\mathrm{LR}_{\mathrm{fcd}}$ is to compute quickly lower bounds as close as possible to the integer optimum value. The trivial problem relaxation $L_{\mathrm{fcs}}(0,0)$, which is equivalent to neglecting the flow conservation constraints and the depot capacities, already gives very good lower bounds. These lower bounds can be improved using a computationally expensive subgradient method.

Let c_U denote the value of an integer feasible solution and c^* denote the optimal integer solution value. Since $0 \leqslant L_{\mathrm{fcs}}(0,0) \leqslant c^* \leqslant c_U$, the percentage deviation between c_U and c^* can be approximated and estimated by

$$0 \leqslant \frac{c_U - c^*}{c^*} \leqslant \frac{c_U - L_{\mathrm{fcs}}(0,0)}{L_{\mathrm{fcs}}(0,0)}$$

Thus, as long as the LP relaxation is not solved to optimality, the somewhat weaker lower bound $L_{\mathrm{fcs}}(0,0)$ can be used to estimate the quality of integer feasible solutions that have been generated by the opening or LP-plunging heuristics.

4.2 Opening Heuristics

We have implemented two opening heuristics considering depot capacities:

- ND: a cluster first – schedule second method based on a nearest depot heuristic. It simply assigns each trip $t \in \mathcal{T}$ to the depot $G(t)$ that provides the cheapest pull-out and pull-in trips.
- SCR: a schedule – cluster – reschedule method. This heuristic is not based on an assignment approach, but rather on minimum-cost flow that retains all degrees of freedom. The individual depot nodes in $\mathcal{D}^+$ and $\mathcal{D}^-$ are contracted to single depot nodes $\mathcal{D}$, and the individual depot flow conservations are aggregated to one flow conservation. This relaxed system is solved to optimality. The resulting vehicle schedules may violate some individual flow conservation constraints, and these violations can be repaired heuristically: the vehicle schedules are pieced together into feasible blocks defining a cluster that considers depot capacities. This cluster can than be (re-)scheduled optimally. This procedure is embedded in a tabu search that forbids the use of certain dead-head trips that create the infeasibilities of a current (relaxed) solution.

Complexity theory tells us that it is $\mathcal{NP}$-hard just to find a feasible solution if depot capacities are considered (see Löbel (1997c)). However, it turned out that depot capacities are often soft constraints and can thus sometimes be violated somewhat since vehicles can often be shifted from one garage (or depot) to another.

4.3 Solving the LP Relaxation

Our computational investigations on real-world test data have shown that *the hard part* in solving the MDVSP to proven optimality *is to solve the LP relaxation.* Standard software alone (for example pure CPLEX) as well as standard approaches from integer linear programming (e. g., column generation and column elimination schemes based on the reduced cost criterion) are unable to solve larger instances with several thousand timetabled trips. We have developed a column generation method including **new** techniques based on the two Lagrangean relaxations LR_{fcs} and LR_{fcd}. Lagrangean pricing is a new idea that makes it possible to solve the LP-3 relaxation of even large-scale instances.

The basic idea of Lagrangean pricing is to approximate the LP relaxation with all active and inactive variables. It is important that dual information compiled in the last RLP are used as in the reduced cost pricing method: Let $\tilde{\nu}$, $\tilde{\pi}$, and $\tilde{\gamma}$ denote the values of the dual LP multipliers according to the flow conditions (2b), the flow conservations (2c), and the depot capacities (2d) of the last basis of the current RLP. We evaluate L_{fcs} and L_{fcd} at $(\tilde{\pi}, \tilde{\gamma})$

and $\tilde{\nu}$, respectively, using the complete variable set. The solutions L_{fcs} and L_{fcd} can be interpreted as a set of vehicle schedules and/or unloaded trips that seem to be advantageous for the given shadow prices of the current RLP relaxation. Each still non-active variable corresponding to such an unloaded trip is therefore generated and added to the next RLP. The new idea of Lagrangean pricing is to generate, in addition to columns with negative reduced costs, also those that have non-negative reduced costs, but are necessary to complete (almost) optimal solutions.

Computational tests have shown that the LP relaxations of our problems can only be solved if we use advanced starting solutions that yield a value as close as possible to the LP optimum. Therefore, the set of all unloaded trips, used hitherto by the solutions of $L_{\text{fcs}}(0,0)$ and the opening heuristics, define the first restricted arc set $\tilde{A}$ of the initial RLP.

4.4 LP-Plunging

Our real-world MDVSP instances in practice exhibit a nice "almost-integral property": solutions x of the LP-3 relaxation or an RLP include few fractional variables. It is often the case that x is integral or there exists some integral solution yielding (almost) the same objective value. Moreover, the gap between the optimal LP or RLP value and its optimal integer value is often small or zero. LP-plunging makes use of this property by iteratively rounding up and fixing components of the LP solution and reoptimizing the enlarged LP.

Given an LP-3 or an RLP and a nonintegral feasible vector x. Let $\Delta \in (0.5, 1.0)$ denote some threshold value for which all fractional variables having a value within $(\Delta, 1)$ are rounded up and fixed to one, and let $\alpha \in (0.5, 1.0)$ denote some shrink factor for Δ. The standard values for Δ and α are 0.95 and 0.9. As long as the current x is nonintegral and the current (R)LP is primal feasible, the following steps are performed:

1. All variables $x_a^d \in (\Delta, 1)$ are rounded up and fixed to 1.
2. If no variable was fixed to 1 and if $\alpha\Delta > 0.5$, we reset $\Delta := \alpha\Delta$ and go to 1. Otherwise, each fractional variable yields $x_a^d \leqslant 0.5$, and we fix the first variable x_a^d to 1 yielding the largest fractional value.
3. Logical implications are followed, i. e., for each variable x_{ij}^d being fixed to one, we fix the variables of all arcs $(\delta^+(i) \cup \delta^-(j)) \cap A_d^{\text{u-trip}}$ and $(\delta(i) \cup \delta(j)) \setminus A_d^{\text{u-trip}}$ to zero.
4. The LP enlarged by the variable fixings is reoptimized with the dual simplex algorithm.

If the LP-plunging succeeds, the clustering defined by the last (integral) x is depot-wise rescheduled to optimality using all possible unloaded trips.

Since the restricted column set of an RLP generally includes only a small part of $A^{\text{u-trip}}$, the LP-plunging generates in many cases only poor or infeasible integer solutions. If this is the case, we enlarge the current RLP parameter controlled by inactive columns (such that the probability to find a better integer solution is presumably increased, but the dual feasibility of the optimal basis of the RLP is not destroyed and the main memory limit of the workstation is not exceeded) and apply the LP-plunging a second time.

4.5 The Workhorses: Minimum-Cost Flow and LP

Solving the MDVSP with our algorithm requires the efficient solution of minimum-cost flow problems and LPs at several steps: the minimum-cost flow problems stem from single-depot subproblems and the RLPs from LP-3.

Standard tools in vehicle scheduling are, of course, network flow models and algorithms, which have been thoroughly investigated and are well understood. We have implemented a network simplex algorithm, called MCF, and *combined it with column generation.*[1] This implementation allows solving the single-depot problems and subproblems to optimality in a few seconds. The Lagrangean functions can also be evaluated in a few seconds up to a few minutes, depending on the problem size. For instance, the Lagrangean function L_{fcs} of the problem with 70 million unloaded trips can be exactly evaluated in about 15 minutes. MCF (without column generation) is available free of charge for academic use via WWW at www.zib.de/Optimization (see Löbel (1997b)).

We solve the linear programs with the primal as well as the dual simplex solver of CPLEX, currently version 4.0.9 CPLEX (1997). CPLEX turned out to be a reliable and robust method for our degenerate (R)LP problems.

---

[1]Due to the very special ("almost transportation") structure of these minimum cost flow problems and the importance of fast solutions, an anonymous referee proposed trying more specialized algorithms like augmenting path methods. We have doubts that such methods could improve the performance of our branch-and-cut method: First, the portion of the total run time spend in the minimum cost flow subroutines can be neglected. Second, the considered minimum cost flow problems are *not* assignment or transportation problem although they include such substructures. Therefore, very specialized assignment or transportation algorithms *cannot* be used. Third, the cost coefficients of the inner minimizations problems L_{fcs} and L_{fcd} can also be negative. Augmenting path methods require a nonnegative objective function, otherwise, nontrivial network transformations are necessary. A general purpose network simplex code, however, can handle an arbitrary objective function easily. Last, we believe that augmenting path methods *cannot* handle up to 70 million variables efficiently. We have also compared MCF with other efficient network flow solver such as RELAX IV and the cost scaling code CS 2 (see Löbel (1996)). For our special minimum-cost flow problems, MCF turned out to be, on the average, the fastest code.

5 Test Data

Our computational investigations are based on real-world data from the city of Berlin (BVG), the city of Hamburg (HHA), and the region around Hamburg (VHH). Different parameter settings and optimization aspects yielded in the test instances, which are illustrated in Table 1 ($\varnothing G := \sum_{t \in \mathcal{T}} G(t)/|\mathcal{T}|$ denotes the average depot-group size). Note that the number of equations of LP-3 is equal to the number of flow conditions and flow conservations.

Test Sets	$\vert\mathcal{D}\vert$	$\vert\mathcal{T}\vert$	$\vert A^{\text{u-trip}}\vert/1,000$		$\varnothing G$	no. of equations
			User[a]	All		
Berlin 1	44	24,906	846	69,700	4.03	125,255
Berlin 2	49	24,906	304	13,200	1.56	63,641
Berlin 3	3	1,313	77	2,300	2.33	4,370
Berlin-Spandau 1	9	2,424	164	3,700	4.94	14,418
Berlin-Spandau 2	9	3,308	327	8,800	5.49	21,470
Berlin-Spandau 3	13	2,424	39	590	1.92	7,103
Berlin-Spandau 4	13	3,308	72	1,530	2.25	10,753
Berlin-Spandau 5	13	3,331	75	1,550	2.25	10,834
Berlin-Spandau 6	13	1,998	28	380	1.90	5,798
Berlin-Spandau 7	7	2,424	145	3,300	4.16	12,506
Berlin-Spandau 8	7	3,308	283	7,800	5.02	18,376
Hamburg 1	12	8,563	1,322	10,900	2.23	27,696
Hamburg 2	9	1,834	99	1,000	2.02	5,549
Hamburg 3	2	791	30	200	1.32	1,835
Hamburg 4	2	238	2	23	1.04	487
Hamburg 5	2	1,461	85	580	1.31	3,379
Hamburg 6	2	2,283	176	1,600	1.33	5,323
Hamburg 7	2	341	6	34	1.32	795
Hamburg-Holstein 1	4	3,413	230	4,000	1.68	9,167
Hamburg-Holstein 2	19	5,447	1,054	9,400	3.65	25,334

[a]The unloaded trips without pull-in-pull-out trips.

Table 1: Real-world test sets.

Currently, BVG maintains 9 garages and runs 10 different vehicle types resulting in 44 depots. For a normal weekday, about 28,000 timetabled trips have to be serviced. Since BVG outsources some trips to third-party companies, this number reduces to 24,906. Using all degrees of freedom, these 25,000 trips can be linked with about 70 million unloaded trips.

Berlin 1: This is the complete BVG problem with all possible degrees of freedom.

Berlin 2: This problem is based on the timetabled trip set of Berlin 1, but the depots and the dead-head trips are generated with different rules resulting in fewer degrees of freedom.

Berlin 3: This is a small test instance including 9 lines from the south of Berlin and 3 depots from one single garage.
Berlin-Spandau 1 – 8: All the test sets denoted by Berlin-Spandau are defined on the data of the district of Spandau for different weekdays and different depot generation rules.

HHA together with some other transportation companies maintain 14 garages with 9 different vehicle types resulting in 40 depots. More than 16,000 daily timetabled trips must be scheduled with about 15.1 million potential unloaded trips. This problem decomposes into a 12-depot problem, a 9-depot problem, five smaller 2-depot problems, and nine small 1-depot problems.
Hamburg 1 – 7: Here we consider the multiple-depot subproblems of HHA.

VHH currently plans 10 garages with 9 different vehicle types. The garage-vehicle combinations define 19 depots. The 5,447 timetabled trips of VHH can be linked with about 10 million unloaded trips.
Hamburg-Holstein 1: This is a subset of VHH containing not all its depots and trips.
Hamburg-Holstein 2: This test set is based on the complete data of VHH.

6 Computational Results

In the following, we want to show the effectiveness of our developed and implemented method to solve large MDVSP instances from practice. All the computational tests used a SUN Model 170 UltraSPARC with 512 MByte main memory and 1.7 MByte virtual memory. We were the only user on this machine during our test runs. All linear programs have been solved with CPLEX, version 4.0.7 and 4.0.9, all minimum-cost flow problems and single-depot subproblems have been solved with our network simplex code MCF combined with a column generation.

The following objective values (fleet sizes and operational weights) are given in Table 2: (i) the lower bounds obtained with $L_{\text{fcs}}(0,0)$ and the LP relaxations; (ii) the integer optimum or, if the optimum is still unknown, the best integer solution values; (iii) the upper bounds obtained by our opening heuristics as well as by our branch-and-cut method starting with SCR or ND and terminating after a maximum run time limit of 10 hours (and 16 hours for Berlin 1 starting with ND). The largest problem, Berlin 1, has not been solved to optimality. Berlin 2 and Berlin-Spandau 2 and 8 have been solved fleet minimally, but not to proven cost minimality.

The run times that have been required to solve the function $L_{\text{fcs}}(0,0)$, the LP relaxation pure without LP-plunging, the opening heuristics SCR and ND, and our exact method (with and without using the optional LP-plunging within the column generation) are given in Table 3.

Table 2: Fleet sizes and operational weights (optimal int. values are in bold).

Test Sets[b]	Lower bounds				Optimum or best solution		Upper bounds						
	$L_{fcs}(0,0)$		LP relaxation				SCR heuristic[a]				ND heuristic		
							pure		+ LP method		pure[c]	+ LP method	
	Fleet	Weight	Fleet	Weight	Fleet	Weight	Fleet	Weight	Fleet	Weight	Fleet	Fleet	Weight
B 1	1323	715714	1323	759162[d]	1329	850680	1347	1317379	1335	1118287	1575	1356	982914
B 2	1350	715623	1353.7	797919	**1354**	777823	1366	1318085	**1354**	809611	1655	**1354**	788958
B 3	**69**	14043	**69**	**14119**	**69**	**14119**	**69**	14122	**69**	**14119**	70	**69**	**14119**
BS 1	**125**	65585	**125**	65611	**125**	**65611**	**125**	125786	**125**	65835	139	**125**	65901
BS 2	184	78947	184.5	79110	**185**	79052	**185**	289262	**185**	80430	207	**185**	92249
BS 3	**127**	90514	**127**	**93745**	**127**	**93745**	**127**	152109	**127**	**93745**	135	**127**	**93745**
BS 4	**191**	195844	**191**	**230846**	**191**	**230846**	192	395891	**191**	**230846**	222	**191**	**230846**
BS 5	**191**	191141	**191**	**227580**	**191**	**227580**	194	393922	**191**	**227580**	220	**191**	**227580**
BS 6	**98**	91109	**98**	**101075**	**98**	**101075**	**98**	132650	**98**	**101075**	109	**98**	**101075**
BS 7	**125**	65585	**125**	65611	**125**	**65611**	**125**	105853	**125**	**65611**	139	**125**	65724
BS 8	184	78947	184.5	79110	**185**	79093	**185**	259406	**185**	79273	207	**185**	79959
H 1	**432**	66874	**432**	71068	**432**	**71069**	446	70291	434	73066	489	**432**	71270
H 2	**103**	15356	**103**	**16070**	**103**	**16070**	104	16792	**103**	**16070**	114	**103**	**16070**
H 3	**39**	5557	**39**	**5860**	**39**	**5860**	**39**	6298	**39**	**5860**	41	**39**	**5860**
H 4	**6**	**1358**	**6**	**1358**	**6**	**1358**	**6**	**1358**	**6**	**1358**	**6**	**6**	**1358**
H 5	**62**	12092	**62**	**12502**	**62**	**12502**	**62**	13535	**62**	**12502**	65	**62**	**12502**
H 6	**111**	15705	**111**	**15791**	**111**	**15791**	**111**	16588	**111**	**15791**	**111**	**111**	**15791**
H 7	**15**	2832	**15**	**2961**	**15**	**2961**	16	2836	**15**	**2961**	16	**15**	**2961**
HH 1	**201**	28697	**201**	**29027**	**201**	**29027**	**201**	30497	**201**	**29027**	213	**201**	**29027**
HH 2	360	51084	**362**	52788	**362**	**52788**	363	72700	**362**	52986	393	**362**	53090

[a] Results obtained with SCR and ND: using only the heuristics ("pure"), and using each as the opening method within our exact LP method ("+ LP method"). In addition, we used a run time limit of ten hours and 16 hours for Berlin 1 starting with ND.

[b] B = Berlin, BS = Berlin-Spandau, H = Hamburg, HH = Hamburg-Holstein

[c] The results of the operational weights are not satisfying and, because of a lack of space, omitted. They are given in Löbel (1997c).

[d] Best known feasible LP value.

Test Sets	Lower bounds			Upper bound (or optimum)					
	Lagrangean relaxation: $L_{\text{fcs}}(0,0)$	Pure LP times starting with		SCR heuristic			Nearest depot heuristic		
				pure	+ Exact method LP-plunging[a]		pure	+ Exact method LP-plunging[b]	
		SCR	ND		always	b&c		always	b&c
Berlin 1	916	—[b]	—[b]	12386	—[b]	—[b]	171	—[b]	—[b]
Berlin 2	229	34795	32767	3810	—[b]	35202[c]	79	30985[c]	33248[c]
Berlin 3	17	431	311	25	389	435	7	249	330
B-Spandau 1	27	43777	66501	343	59337	44053	15	68487	134386
B-Spandau 2	93	112337	165048	1939	—[b]	138212[c]	39	—[b]	240852[c]
B-Spandau 3	9	975	739	42	1626	990	7	953	758
B-Spandau 4	25	6014	4384	334	6228	6077	14	6773	4433
B-Spandau 5	31	5264	5618	354	6896	5304	15	13821	5666
B-Spandau 6	17	162	227	7	211	173	6	188	244
B-Spandau 7	23	24717	46597	259	26606	24793	14	45810	52031
B-Spandau 8	67	82284	62041	1238	146284[c]	84725[c]	36	—[b]	63215[c]
Hamburg 1	185	—[b]	50246	2868	—[b]	—[b]	29	88767	53971
Hamburg 2	12	875	685	103	732	902	7	926	708
Hamburg 3	4	35	31	16	37	41	2	34	38
Hamburg 4	2	3	3	1	3	3	1	3	3
Hamburg 5	10	258	155	86	288	279	5	119	174
Hamburg 6	18	148	84	30	158	181	11	124	86
Hamburg 7	2	8	9	2	11	10	1	8	11
H-Holstein 1	40	2619	2087	199	2166	2695	18	2445	2158
H-Holstein 2	101	46673	64489	1696	55915	54604	40	68534	71562

[a] LP-plunging is only used in the branch-and-cut part on a fixed RLP or always whenever an RLP has been solved.
[b] Not solved to optimality since, for instance, the objective progress was too small or stalling occurred.
[c] The problem has been solved fleet minimally, but not to proven cost minimality.

Table 3: Run times in seconds.

6.1 Lower Bounds

To obtain lower bounds by Lagrangean relaxations, we have only considered $L_{\mathrm{fcs}}(0,0)$. For $L_{\mathrm{fcs}}(0,0)$, let ν^+ and ν^- denote optimal dual variables associated with the flow conditions $x(\delta^+(t)) = 1$ and $-x(\delta^-(t)) = -1$, respectively. We have shown in Löbel (1997c) that $L_{\mathrm{fcd}}(\nu^+ - \nu^-)$ and $L_{\mathrm{fcs}}(0,0)$ yield the same optimal value. The values obtained by $L_{\mathrm{fcs}}(0,0)$ give excellent approximations. The average gap between the approximation and the minimum integral fleet size is a mere 0.06%. It is remarkable that this trivial problem relaxation – simply neglecting flow conservation – gives such tight approximations. For 15 out of our 20 instances, the fleet sizes can be approximated exactly. Ignoring for those problems the values for the fleet size, the gap between the operational costs of $L_{\mathrm{fcs}}(0,0)$ and the optimum is at most 16 % and 5 % on the average. The computing times are quite fast: For Berlin 1, $L_{\mathrm{fcs}}(0,0)$ with 70 million variables can be evaluated in about 15 minutes; all the other instances together were computed in 15 minutes.

All LP relaxations, except for Berlin 1, have been solved to optimality. To find a fleet minimal LP value for Berlin 1, our column generation requires about 200 hours cpu time. The values obtained by the LP relaxation give lower bounds quite close to the integer optimal values. For 12 out of the 20 considered instances, the LP relaxation already provides the integer optimal value, and for 3 instances, it can be obtained by rounding up the LP value to the next integer value. For Berlin 1, we do not know the minimal number of vehicles, but expect that the fleet size lower bound provided by the LP relaxation is also tight. Whenever the LP relaxation provides an exact fleet size, it also provides the minimal operational weights.

We have seen that the LP values are quite tight. A similar phenomenon is observed by Forbes/Holt/Watts (1994): 22 of their 30 test instances with up to 600 trips have integral LP solutions, and the largest gap between the LP value and the integral optimum is at most 0.003 % for the remaining problems. So, this observation does not seem to be a *small scale phenomenon*.

The value of the operational weights in the objective value of the lower bounds do not necessarily define lower bounds for the integer optimal weights among all minimal fleet solutions. To estimate the quality of the operational weights requires that the lower bound of the fleet size is tight. For all problems that do not satisfy this condition, however, we believe that they nevertheless give good estimated values for the minimal operational weights.

Comparing the run times of the Lagrangean and LP relaxation, it is obvious that Lagrangean relaxations $L_{\mathrm{fcs}}(0,0)$ are the faster method to obtain good lower bounds. The better lower bounds provided by the LP relaxation require long run times that are only justified by a succeeding branch-and-cut method. The solution produced by SCR are always significantly better than those of ND. On the average, however, SCR used as the opening heuristic for the branch-and-cut algorithm does not provide better starting points. It is worth mentioning that starting without any heuristically generated solution, our LP method is unable to solve any of our larger problem instances at all.

6.2 Upper Bounds

We will now consider the upper bounds obtained by the two opening heuristics (SCR and ND) and obtained by the exact branch-and-bound method starting with SCR and ND, using LP-plunging between two RLPs, and terminating after a given run time limit of 10 hours (16 hours for Berlin 1 starting with ND).

The trivial opening heuristic ND already delivers good results: The fleet size gap is, on the average, about 10 % with a standard deviation of 6 %. From a practical point of view, however, the operational costs of these solutions are not acceptable. The better results are obtained from the SCR heuristic: The average fleet size gap is 0.8 % with a standard deviation of 1.6 %. The operational costs of these solutions are comparable to the results obtained by the best codes currently used in practice.

We almost always obtain optimal results if we apply our exact branch-and-cut method with a time limit of 10 hours. The objective gaps are, on the average, less than 0.12 %. It does not make any difference which opening heuristic we use for the exact method since the run times are comparable for both. The run times of our exact method may be decreased if we use both opening heuristics together to determine the first RLP. This may be the basis for further computational tests.

Figures 5–8 display the development of the upper bound values (fleet sizes and operational weights) obtained by the LP-plunging heuristic in proportion to the integer optimal (or lower bound) values. Starting our method with the solution obtained with ND, the fleet sizes can be approximated in two hours with a gap less than 3 %, in 4 four hours with gap of about 1 %, and in 6 hours with a gap less than 1 % for all problems except Berlin 1, see Fig. 5. Starting with the solution obtained with SCR, the fleet sizes can be approximated in one hour with a gap less than 2 % and in 10 hours with a gap less than 1 %, see Fig. 7. There is also a positive development of the operational costs: compared with the optimal integer costs of fleet minimal solutions, the operational costs can be approximated with a small gap, see Figs. 6 and 8. If the run time limit is 10 hours or more, the four figures show that it is irrelevant which opening heuristic is used, the results are in any case comparable. However, if there is a stronger time limit of two or three hours, starting with SCR provides better results.

6.3 Optimal Solutions

Without any run time limit, each instance of our test set, with the exception of the problem Berlin 1, can be solved to proven fleet minimality. With the exceptions of the problems Berlin 2, Berlin-Spandau 2, and Berlin-Spandau 8, each instance can be solved to proven fleet and cost optimality.

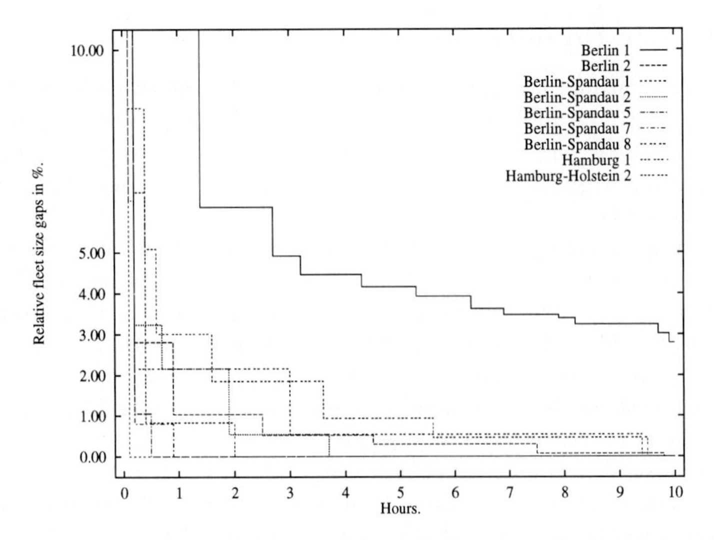

Figure 5: Development of fleet size upper bounds of problems requiring more than 2 hours run time to obtain a fleet minimal solution; starting with the ND heuristic.

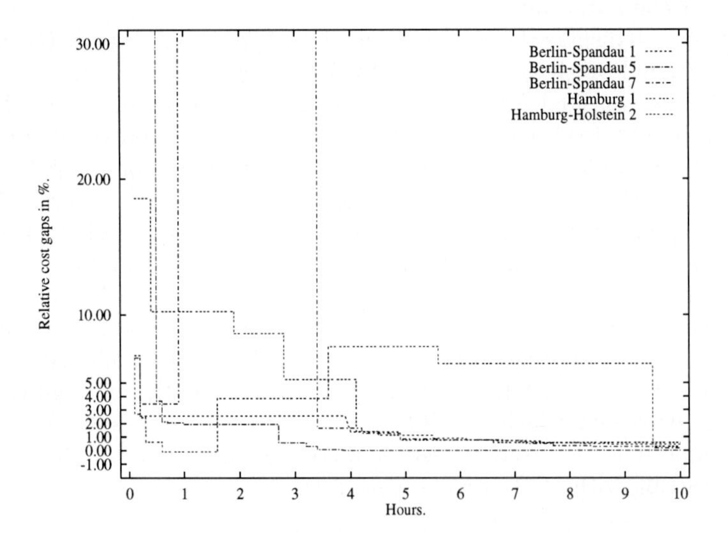

Figure 6: Development of operational weight upper bounds of problems requiring more than 2 hours run time to obtain a fleet minimal solution and knowing the minimum weight among all minimal fleet solutions; starting with the ND heuristic.

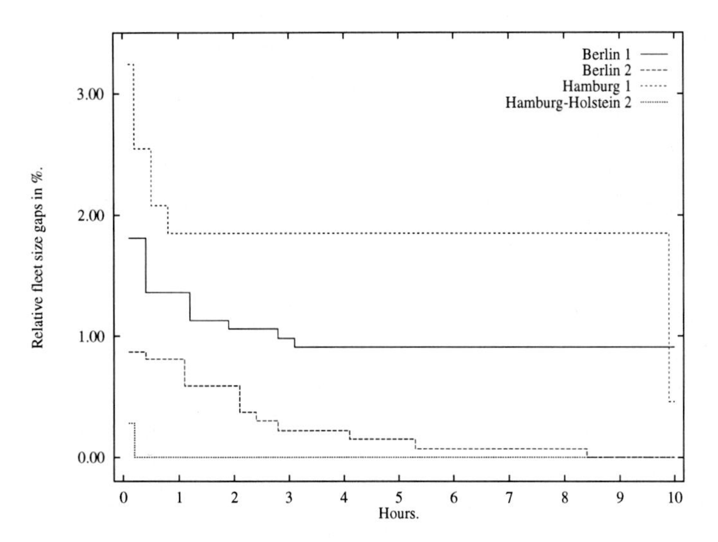

Figure 7: Development of fleet size upper bounds of problems requiring more than 2 hours run time to obtain the optimum; starting with the SCR heuristic.

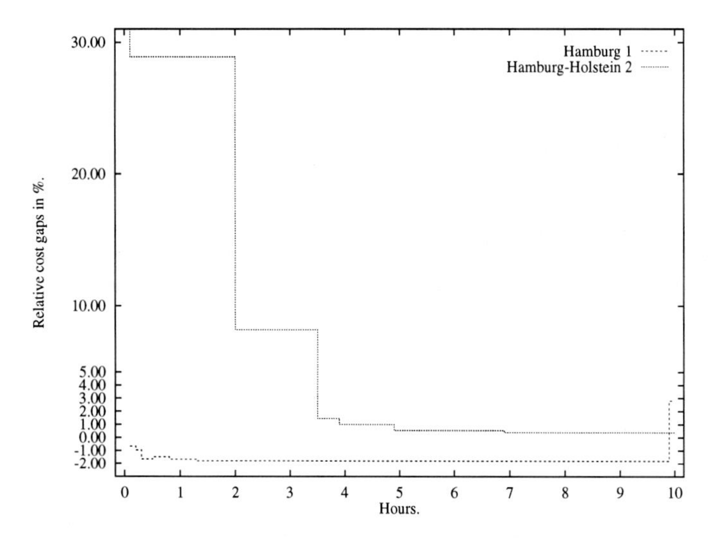

Figure 8: Development of operational weight upper bounds of problems requiring more than 2 hours run time to obtain the optimum and knowing the minimum weight among all minimal fleet solutions; starting with the SCR heuristic.

With the current version of our branch-and-cut method, solving really large-scale problems to proven optimality leads to impractical run times. In particular, solving Berlin 1 with 70 million variables to optimality is still a challenge to us. Nevertheless, the results obtained with our methods are currently the best obtainable. Solutions providing possibly a gap of a few vehicles, but with reasonable operational weights can be computed in acceptable run times.

7 Conclusions

This paper is devoted to the Multiple-Depot Vehicle Scheduling Problem (MDVSP). We have presented a branch-and-cut method for its solution. A well-chosen combination of these methods turned out to be able to solve (almost) all problems of practical interest in accepatable running times. The success of the implementations, of course, gains from the (in the recent years drastically increased) computing power of modern workstations and sophisticated commercial optimization software (such as the LP solver CPLEX). We summarize some of our findings:

Upper bounds that have been generated with the schedule – cluster – reschedule heuristic (SCR) can be computed quickly and are of high quality. Compared with the optimal integer solutions, SCR provides solutions with a fleet size and operational weight gap of less than 1.25 % and 5.2 %, respectively.

Lagrangean relaxations allow to compute tight lower bounds even for large multiple-depot instances. Lagrangean relaxations can be used to quickly simulate fleet and cost effects of different parameter settings and, thus, to easily find out a useful scenario.

Branch-and-cut is capable of solving even very large multiple-depot instances to optimality, see Table 2.

Lagrangean pricing is a good idea to solve the large degenerate LPs that come up in solving multiple-depot instances with branch-and-cut. Our initial code used the well known standard reduced cost pricing techniques. However, this did not work at all because of stalling. To cure stalling, we introduced (what we call) Lagrangean pricing. We propose it as one of the basic ingredients of an effective method to solve multiple-depot vehicle scheduling problems. Similar positive results have been observed by Fischetti/Toth (1996) and Fischetti/Vigo (1996) also dealing with large degenerate LPs. We believe that variable pricing based on Lagrangean relaxation is a useful tool that can help to solve many combinatorial optimization problems.

Computational breakthrough: to our knowledge, at present no other implementation is able to solve MDVSPs with more than 1,000 timetabled trips to optimality. Our code has successfully produced optimal solutions of various real-world problem instances with up to 25,000 timetabled trips. The integer multicommodity flow problems arising this way are orders of magnitude larger than what other codes are able to handle. The largest real instance we encountered gave rise to an integral multicommodity flow problem with about 125,000 equations and 70 million integer variables. We could not produce an optimal solution, but found a solution with a fleet size gap of less than 0.5 %.

Possible savings indicated by our test runs are immense. Compared with a manual planning process, the SCR heuristic indicates savings of about 19 % of the vehicles and about 14 % of the operational costs. Compared with an assignment heuristic, our branch-and-cut method indicates savings of several vehicles and about 10 % cost reduction. However, the final evaluations of the SCR generated solutions have not been finished by BVG, HHA, and VHH yet. It still needs to be seen whether our vehicle schedules provide a useful input for duty scheduling, the next step in the hierarchical planning process. It is therefore not clear how much of these indicated savings can be obtained in practice. Nonetheless, our methods can solve large problems optimally. The Berliner Verkehrsbetriebe, for instance, expect to save about DM 100 million per year with our SCR heuristic (see Schmidt (1997)).

There is a high demand within industry for efficient methods for the MDVSP. Parts of our system have been purchased by BVG for their planning system BERTA, by IVU for MICROBUS II, and by the research department of the SIEMENS AG in Munich.

Acknowledgements: We are grateful to Manfred Völker and Anna Neufeld of HanseCom GmbH for their support to model the MDVSP and providing us with real-world problems from the Hamburger Hochbahn AG and the Verkehrsbetriebe Hamburg-Holstein AG. We are grateful to Uwe Strubbe of IVU GmbH for providing us with real-world problems from the Berliner Verkehrsbetriebe. We are grateful to the Berliner Verkehrsbetriebe, the Hamburger Hochbahn AG, and the Verkehrsbetriebe Hamburg-Holstein AG for their kind permission to use and publish their data. We are indebted to Martin Grötschel, Ralf Borndörfer, and Alexander Martin (all at ZIB), and Bob Bixby for their helpful discussions. We are also grateful to Bob Bixby and ILOG CPLEX Division for regularly providing us access to the newest versions of CPLEX and the CPLEX Callable Library. This work has been supported by the German Federal Ministry of Education, Science, Research, and Technology grant no. 03-GR7ZIB -7.

References

Ball, M.O. / Magnanti, T.L. / Monma, C.L. / Nemhauser, G.L. (editors) **(1995):** *Network Routing*, volume 8 of *Handbooks in Operations Research and Management Science.* Elsevier Science B.V.

Bokinge, U. / Hasselström, D. (1980): Improved vehicle scheduling in public transport through systematic changes in the time-table. *European Journal of Operational Research*, 5:388–395.

Borndörfer, R. / Grötschel, M. / Löbel, A. (1995): Alcuin's transportation problems and integer programming. Preprint SC 95-27, Konrad-Zuse-Zentrum für Informationstechnik Berlin. Available at www.zib.de. To appear in Butzer, P.L. / Jongen, H.T. / Oberschelp, W. (editors), *Charlemagne and his Heritage: 1200 Years of Civilization and Science in Europe, Volume II: The Mathematical Arts*, Brepols Publishers.

Bussieck, M. / Winter, T. / Zimmermann, U.T. (1997): Discrete optimization in public rail transport. In Liebling, T.M. / de Werra, D. (editors), *Mathematical Programming: A Publication of the Mathematical Programming Society*, pages 415–444. Elsevier Science B.V.

Carpaneto, G. / Dell'Amico, M. / Fischetti, M. / Toth, P. (1989): A branch and bound algorithm for the multiple depot vehicle scheduling problem. *Networks*, 19:531–548.

Chvátal, V. (1980): *Linear programming.* W. H. Freeman and Company, New York.

CPLEX (1997): *Using the CPLEX Callable Library.* ILOG CPLEX Division, 889 Alder Avenue, Suite 200, Incline Village, NV 89451, USA. Information about CPLEX available at www.cplex.com.

Daduna, J.R. / Branco, I. / Paixão, J.M.P. (editors) **(1995):** *Computer-Aided Transit Scheduling*, Lecture Notes in Economics and Mathematical Systems. Springer Verlag.

Daduna, J.R. / Mojsilovic, M. / Schütze, P. (1993): Practical experiences using an interactive optimization procedure for vehicle scheduling. In Du, D.-Z. / Pardalos, P.M. (editors), *Network Optimization Problems: Algorithms, Applications and Complexity*, volume 2 of *Series on Applied Mathematics*, pages 37–52. World Scientific Publishing Co. Pte. Ltd.

Daduna, J.R. / Paixão, J.M.P. (1995): Vehicle scheduling for public mass transit – an overview. In Daduna/Branco/Paixão (1995).

Dell'Amico, M. / Fischetti, M. / Toth, P. (1993): Heuristic algorithms for the multiple depot vehicle scheduling problem. *Management Science*, 39(1):115–125.

Desrosiers, J. / Dumas, Y. / Solomon, M.M. / Soumis, F. (1995): *Time Constrained Routing and Scheduling.* In Ball/Magnanti/Monma/Nemhauser (1995), chapter 2, pages 35–139.

Fischetti, M. / Toth, P. (1996): A polyhedral approach to the asymmetric traveling salesman problem. Technical report, University of Bologna. To appear in *Management Science.*

Fischetti, M. / Vigo, D. (1996): A branch-and-cut algorithm for the resource-constrained arborescence problem. *Networks*, 29:55–67.

Forbes, M.A. / Holt, J.N. / Watts, A.M. (1994): An exact algorithm for multiple depot bus scheduling. *European Journal of Operational Research*, 72(1):115–124.

Freling, R. / Paixão, J.M.P. (1995): Vehicle scheduling with time constraint. In Daduna/Branco/Paixão (1995).

Kokott, A. / Löbel, A. (1996): Lagrangean relaxations and subgradient methods for multiple-depot vehicle scheduling problems. Preprint SC 96-22, Konrad-Zuse-Zentrum für Informationstechnik Berlin. Available at www.zib.de.

Löbel, A. (1996): Solving large-scale real-world minimum-cost flow problems by a network simplex method. Preprint SC 96-7, Konrad-Zuse-Zentrum für Informationstechnik Berlin. Available at www.zib.de.

Löbel, A. (1997a): Experiments with a Dantzig-Wolfe decomposition for multiple-depot vehicle scheduling problems. Preprint SC 97-16, Konrad-Zuse-Zentrum für Informationstechnik Berlin. Available at www.zib.de.

Löbel, A. (1997b): *MCF Version 1.0 – A network simplex implementation.* Available for academic use free of charge at www.zib.de.

Löbel, A. (1997c): *Optimal Vehicle Scheduling in Public Transit.* PhD thesis, Technische Universität Berlin.

Löbel, A. (1997d): Vehicle scheduling in public transit and Lagrangean pricing. Revised Preprint SC 96-26, Konrad-Zuse-Zentrum für Informationstechnik Berlin. Available at www.zib.de.

Ribeiro, C.C. / Soumis, F. (1994): A column generation approach to the multiple-depot vehicle scheduling problem. *Operations Research*, 42(1):41–52.

Schmidt, V. A. (1997): *Auf Sparkurs zum Ziel.* Rheinischer Merkur, number 39, page 37, 26th September 1997. In German.

Schrijver, A. (1989): *Theory of Linear and Integer Programming.* John Wiley & Sons Ltd., Chichester.

Soumis, F. (1997): *Decomposition and Column Generation.* Chapter 8 in Dell'Amico, M. / Maffioli, F. / Martello, S. (editors), *Annotated Bibliographies in Combinatorial Optimization*, pages 115–126. John Wiley & Sons Ltd, Chichester.

Exact Algorithms for the Multi-Depot Vehicle Scheduling Problem Based on Multicommodity Network Flow Type Formulations

Marta Mesquita[1] and José Paixão[2]

1 Instituto Superior de Agronomia, Dep. de Matemática, Tapada da Ajuda, 1300 Lisboa, Portugal

2 DEIO, Faculdade de Ciências de Lisboa, Bloco C2-Campo Grande, 1700 Lisboa, Portugal

Abstract: We compare the linear programming relaxation of different mathematical formulations for the multi-depot vehicle scheduling problem. As a result of this theoretical analysis, we select for development a tree search procedure based on a multicommodity network flow formulation that involves two different types of decision variables: one type is used to describe the connections between trips, in order to obtain the vehicle blocks, while the other type is related to the assignment of trips to depots. We also develop a branch and bound algorithm based on the linear relaxation of a more compact multicommodity network flow formulation, in the sense that it contains just one type of variable and fewer constraints than the previous model. Computational experience is presented to compare the two algorithms.

1 Introduction

This paper is concerned with the multi-depot vehicle scheduling problem (MDVSP) which consists of grouping a set of timetabled trips into vehicle blocks and, at the same time, assigning these vehicle blocks to depots. Each vehicle block represents a set of trips that can be carried out by one vehicle.

A set of n timetabled trips $T_1,...,T_n$ has to be operated by vehicles housed at k different depots, $D_1,...,D_k$, at the ℓth of which d_ℓ vehicles are stationed ($\ell = 1,...,k$). The vehicles are treated as identical. Each trip T_i is characterized by its starting and ending times and locations. An ordered pair of trips (T_i, T_j) is said to be compatible if the bus released after the completion of trip T_i can be assigned to trip

T_j. The objective is to group compatible trips into vehicle blocks and at the same time assign these vehicle blocks to depots, in order to minimize the cost associated with the schedule.

The MDVSP has been shown to be NP-Hard (Bertossi/Carraresi/Gallo (1987)). Several mathematical formulations and different solution approaches, both heuristics (see Dell'Amico/Fischetti/Toth (1993), Mesquita/Paixão (1992), Lamatsch (1992), Bertossi/Carraresi/Gallo (1987), El-Azm (1985), Smith/Wren (1981), Ceder/Stern (1981)) and exact methods (see Forbes/Holt/Watts (1994), Ribeiro/Soumis (1994), Carpaneto/Dell'Amico/Fischetti/Toth (1989)) have been proposed for the MDVSP. The last survey on the subject can be found in Ball/Magnanti/Monma/Nemhauser (1995).

In this paper we focus on exact methods for solving the MDVSP. In particular, we discuss different branching strategies for a tree-search scheme based on a multicommodity network flow formulation that involves two different types of decision variables: one is associated with the connections between trips while the other is related to the assignment of trips to depots. We also consider a more compact multicommodity network flow formulation, in the sense that it contains just one type of variable and less constraints than the previous model. Both formulations are used to compare different branch and bound procedures.

The paper is organized as follows. First, in Sect. 2 we review the mathematical formulations proposed for the MDVSP. In Sect. 3 we compare the linear programming relaxation for the different mathematical formulations mentioned above. As a result of this theoretical analysis, in Sect. 4, two branch and bound procedures based on different multicommodity network flow type formulations are presented. Computational experience with real data from a bus operator, Carris-Lisbon, is shown in Sect. 5 and is analyzed in Sect. 6. Finally some conclusions are drawn in Sect. 7.

2 Integer Programming Models of the MDVSP

The integer programming models presented in this section can be grouped into three basic approaches: 1) multicommodity flow formulations; 2) single-commodity flow formulations, namely, an assignment and a quasi-assignment formulation; 3) set partitioning formulation.

2.1 Multicommodity Flow Formulations

Bertossi/Carraresi/Gallo (1987) formulated the MDVSP as a multicommodity network flow problem in a complete bipartite graph. They consider the complete bipartite graph (S,T,A) where nodes $i \in S, i = 1,...,n$ correspond to the ending locations of trips, and nodes $j \in T, j = 1,...,n$ correspond to the starting locations of trips. The arc set, A is partitioned into two subsets A_1 and A_2. An arc $(i,j) \in A_1$ represents the deadhead trip from the end of trip T_i to the start of trip T_j, while an arc $(i,j) \in A_2$ denotes a sequence of two deadhead trips, from the end of trip T_i to the depot and from the depot to the start of trip T_j. If costs c_{ij}^{ℓ} are given, the MDVSP can be defined as the problem of finding a partition of $\{1,...,n\}$, $I_1,...,I_k$ and k arc sets $M_1,...,M_k$ such that:

(i) M_ℓ, $\ell = 1,...,k$ is a perfect matching for (S, T, A);

(ii) $\sum_{\ell \in K} \sum_{(i,j) \in M_\ell} c_{ij}^{\ell}$ is minimum.

Each set $I_\ell, \ell = 1,...,k$ corresponds to the set of trips assigned to each depot ℓ whereas M_ℓ defines the schedule of such trips.

Next, we define the mathematical formulation presented by Bertossi/ Carraresi/Gallo (1987), upon a sparse graph, such as the graph used by Ribeiro/Soumis (1994).

Let $N = \{1,...,n\}$ represent the set of trips, $I \subseteq N \times N$ denote the set of compatible pairs of trips and $K = \{n+1,...,n+k\}$ represent the set of depots. With no loss of generality, we assume that trips are ordered by increasing value of their starting times.

A graph $G^\ell = (V^\ell, A^\ell)$ is associated with depot ℓ, where $V^\ell = N \cup \{n+\ell\}$ and $A^\ell = I \cup (\{n+\ell\} \times N) \cup (N \times \{n+\ell\})$, $\ell = 1,...,k$.

The following example will be used to illustrate the different network flow approaches. Consider a problem with four trips and two depots, each one having a single vehicle. The starting and ending time of each trip is:

Trip	1	2	3	4
start time	9:00	9:30	11:00	12:30
end time	10:30	12:00	13:30	14:00

Figure 2.1 shows graph G^1, associated with depot D_1, represented by full lines and graph G^2, associated with depot D_2, represented by dotted lines.

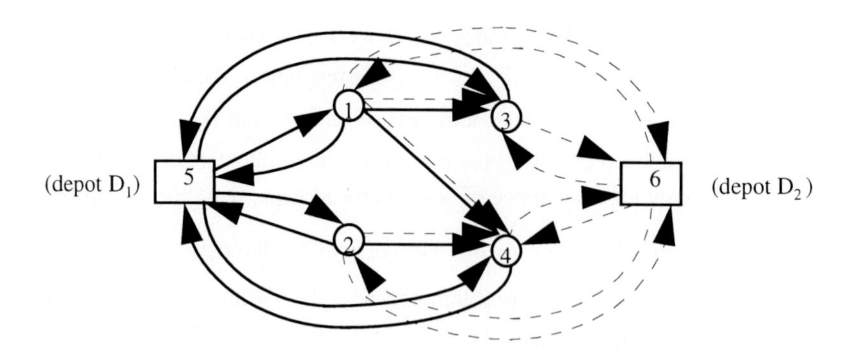

Fig. 2.1 Example of a graph for the multicommodity flow models.

Costs c_{ij}, $(i,j) \in A^{\ell}$ are known and are independent of depot ℓ if $(i,j) \in I$, while $c_{n+\ell,i}$ and $c_{i,n+\ell}$, $i \in N$, are usually dependent on depot ℓ. Note that, in the multicommodity network flow model presented by Bertossi/Carraresi/Gallo (1987) costs c_{ij}^{ℓ}, $(i,j) \in I$, always depend on depot ℓ that supplies the vehicle to perform the deadhead trip. Further, consider the binary variables: x_{ij}^{ℓ}, $(i,j) \in I$, $\ell \in K$, which indicate whether trips i and j are run in order by the same vehicle housed at depot ℓ, $x_{ij}^{\ell} = 1$, or not, $x_{ij}^{\ell} = 0$; $x_{i,n+\ell}$ ($x_{n+\ell,i}$), $i \in N, \ell \in K$, which indicate whether the bus immediately returns to depot ℓ after trip i, $x_{i,n+\ell} = 1$, or not, $x_{i,n+\ell} = 0$(whether depot ℓ directly supplies a bus for trip i), and y_i^{ℓ}, which indicate whether trip i is assigned to depot ℓ, $y_i^{\ell} = 1$, or not, $y_i^{\ell} = 0$. Then the mathematical formulation presented in Bertossi/Carraresi/Gallo (1987), will take the form:

$$(MF_{xy}) \quad \min \sum_{\ell \in K} \sum_{(i,j) \in I} c_{ij} x_{ij}^{\ell} + \sum_{\ell \in K} \sum_{i \in N} \left(c_{i,n+\ell} x_{i,n+\ell} + c_{n+\ell,i} x_{n+\ell,i} \right)$$

$$s.t. \quad \sum_{j:(i,j) \in I} x_{ij}^{\ell} + x_{i,n+\ell} = y_i^{\ell} \qquad \forall \ell \in K, \forall i \in N \tag{2.1}$$

$$\sum_{i:(i,j) \in I} x_{ij}^{\ell} + x_{n+\ell,j} = y_j^{\ell} \qquad \forall \ell \in K, \forall j \in N \tag{2.2}$$

$$\sum_{j \in N} x_{n+\ell,j} \le d_{\ell} \qquad \forall \ell \in K \tag{2.3}$$

$$\sum_{\ell \in K} y_i^\ell = 1 \qquad \forall i \in N \qquad (2.4)$$

$$x_{ij}^\ell \in \{0, 1\} \qquad \forall \ell \in K, \forall (i,j) \in I \qquad (2.5)$$

$$x_{i,n+\ell}, x_{n+\ell,i}, y_i^\ell \in \{0, 1\} \qquad \forall \ell \in K, \forall i \in N \qquad (2.6)$$

The objective function for (MF_{xy}) consists of two major components. The first is related to the linking of trips and may account for both the corresponding travel costs and eventual penalties imposed by the user, for instance on idle times. The second refers to the costs of linking trips and depots which, besides the travel costs, may include a penalty associated with the use of a vehicle.

The constraint set of (MF_{xy}) includes two types of decision variables: the assignment variables, y_i^ℓ, which determine the assignment of trips to the depots and the scheduling variables, x_{ij}^ℓ, $x_{i,n+\ell}$, $x_{n+\ell,i}$ which establish the assignment between trips. Constraints (2.1) and (2.2) relate for each trip the corresponding two types of assignment variables. Constraints (2.3) refer to the number of vehicles available at each depot. Constraints (2.4) guarantee that each trip will be assigned to only one depot.

Since trips are ordered by increasing value of their starting times, set I contains only arcs (i,j) with $i < j$. Therefore, no circuit containing only vertices $i \in N$ exists.

Ribeiro/Soumis (1994) also presented an integer multicommodity network formulation which is more compact, in the sense that it uses only one of the sets of variables involved in (MF_{xy}) and contains less constraints than the integer program described above.

$$(MF_x) \quad \min \sum_{\ell \in K} \sum_{(i,j) \in I} c_{ij} x_{ij}^\ell + \sum_{\ell \in K} \sum_{i \in N} \left(c_{i,n+\ell} x_{i,n+\ell} + c_{n+\ell,i} x_{n+\ell,i} \right)$$

$$s.t. \quad \sum_{\ell \in K} \sum_{\{j:(i,j) \in I\}} x_{ij}^\ell + \sum_{\ell \in K} x_{i,n+\ell} = 1 \qquad \forall i \in N \qquad (2.7)$$

$$\sum_{i:(i,j) \in I} x_{ij}^\ell + x_{n+\ell,j} - \sum_{i:(j,i) \in I} x_{ji}^\ell - x_{j,n+\ell} = 0 \qquad \forall \ell \in K, \forall j \in N \qquad (2.8)$$

$$\sum_{j \in N} x_{n+\ell,j} \le d_\ell \qquad \forall \ell \in K \qquad (2.9)$$

$$x_{ij}^{\ell} \in \{0, 1\} \qquad \forall \ell \in K, \forall (i, j) \in I \quad (2.10)$$

$$x_{i,n+\ell}, x_{n+\ell,i} \in \{0, 1\} \qquad \forall \ell \in K, \forall i \in N \quad (2.11)$$

Constraints (2.7) ensure that each vertex $i \in N$ is visited exactly once, while constraints (2.8) are flow conservation constraints, for each vertex $j \in N$. For example, a feasible solution for the problem stated above is shown in Fig. 2.2.

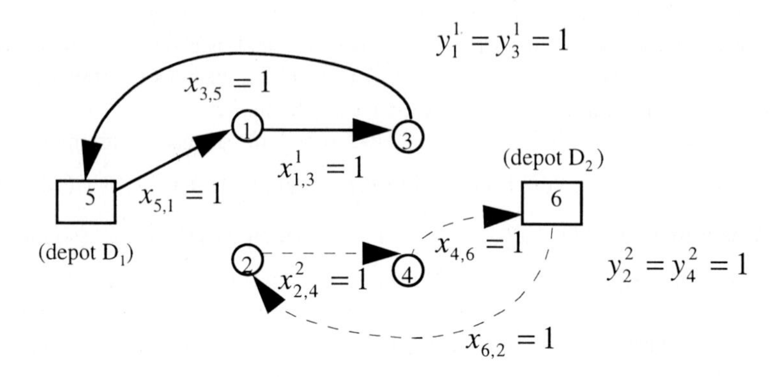

Fig. 2.2 A feasible solution.

The remaining variables, not shown in Fig. 2.2, are equal to zero. Variables $x_{ij}^{\ell}, x_{i,n+\ell}, x_{n+\ell,i}$ are shared by formulations (MF_{xy}) and (MF_x) while variables y_i^{ℓ} only exist in (MF_{xy}).

2.2 Single-Commodity Flow Formulations

Next, we are going to describe two alternative single-commodity flow formulations. The first is based on the assignment problem while the second is based on the quasi-assignment problem.

Consider a graph $G = (V, A)$, in which the vertex set $V = \{1, ..., n+k\}$, is partitioned into $N = \{1, ..., n\}$ corresponding to trips and $K = \{n+1, ..., n+k\}$ referring to depots, and the arc set $A = I \cup (K \times N) \cup (N \times K)$. A cost c_{ij} is associated with each arc $(i, j) \in A$. Fig. 2.3, below, shows graph $G = (V, A)$ for the example.

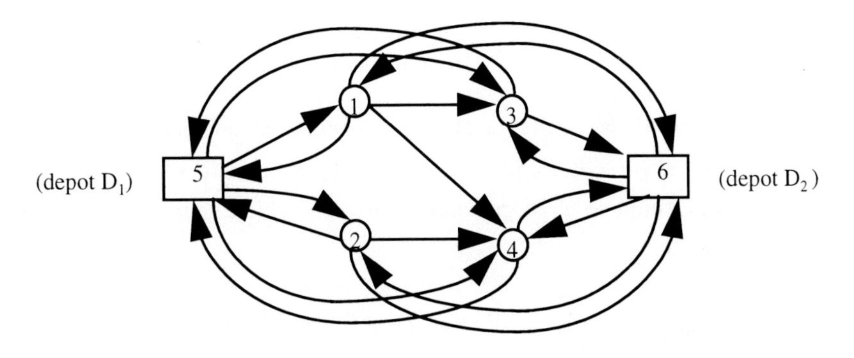

Fig. 2.3 Example of a graph for the single-commodity flow models.

Carpaneto/Dell'Amico/Fischetti/Toth (1989) split each depot vertex, $n+\ell$, into d_ℓ vertices and present a mathematical formulation based on the assignment model with subtour breaking constraints. In order to facilitate comparisons with other formulations, we present this formulation in a slightly different form. Consider the binary variables $x_{ij}=1$, $(i,j)\in I$, if trip i is directly connected to trip j, $x_{ij}=0$ otherwise, and $x_{i,n+\ell}, x_{n+\ell,i}$, y_i^ℓ, $i\in N$, $\ell\in K$, as defined for models (MF_{xy}) and (MF_x). Let Π denote the family of all elementary paths P joining different vertices in K.

$$(MA)\quad \min \sum_{(i,j)\in I} c_{ij}x_{ij} + \sum_{\ell\in K}\sum_{i\in N}\left(c_{i,n+\ell}x_{i,n+\ell} + c_{n+\ell,i}x_{n+\ell,i}\right)$$

$$s.t. \quad \sum_{j:(i,j)\in I} x_{ij} + \sum_{\ell\in K} x_{i,n+\ell} = 1 \qquad \forall i\in N \qquad (2.12)$$

$$\sum_{i:(i,j)\in I} x_{ij} + \sum_{\ell\in K} x_{n+\ell,j} = 1 \qquad \forall j\in N \qquad (2.13)$$

$$\sum_{j\in N} x_{n+\ell,j} \le d_\ell \qquad \forall \ell\in K \qquad (2.14)$$

$$\sum_{(i,j)\in P} x_{ij} \le |P|-1 \qquad P\in\Pi \qquad (2.15)$$

$$x_{ij}\in\{0,1\} \qquad \forall (i,j)\in A \qquad (2.16)$$

Then, the subtour breaking constraints, (2.15), ensure that no path P joining different vertices in K is allowed, or equivalently, that each vehicle block is assigned to exactly one depot.

On the other hand, Mesquita/Paixão (1992) proposed an integer formulation where variables y_i^ℓ, $\ell \in K, i \in N$, and extra constraints are added to a quasi-assignment formulation so as to guarantee that each vehicle returns to the source depot.

$$(QA) \quad \min \sum_{(i,j)\in I} c_{ij} x_{ij} + \sum_{\ell \in K} \sum_{j \in N} \left(c_{i,n+\ell} x_{i,n+\ell} + c_{n+\ell,i} x_{n+\ell,i} \right)$$

$$s.t. \quad \sum_{j:(i,j)\in I} x_{ij} + \sum_{\ell \in K} x_{i,n+\ell} = 1 \qquad \forall i \in N \qquad (2.17)$$

$$\sum_{i:(i,j)\in I} x_{ij} + \sum_{\ell \in K} x_{n+\ell,j} = 1 \qquad \forall j \in N \qquad (2.18)$$

$$\sum_{j \in N} x_{n+\ell,j} \le d_\ell \qquad \forall \ell \in K \qquad (2.19)$$

$$x_{n+\ell,j} - y_j^\ell \le 0 \qquad \forall \ell \in K, \forall j \in N \qquad (2.20)$$

$$x_{i,n+\ell} - y_i^\ell \le 0 \qquad \forall \ell \in K, \forall i \in N \qquad (2.21)$$

$$y_i^\ell + x_{ij} - y_j^\ell \le 1 \qquad \forall \ell \in K, \forall (i,j) \in I \qquad (2.22)$$

$$y_j^\ell + x_{ij} - y_i^\ell \le 1 \qquad \forall \ell \in K, \forall (i,j) \in I \qquad (2.22a)$$

$$\sum_{\ell \in K} y_i^\ell = 1 \qquad \forall i \in N \qquad (2.23)$$

$$x_{ij} \in \{0,1\} \qquad \forall (i,j) \in A \qquad (2.24)$$

$$y_i^\ell \in \{0,1\} \qquad \forall \ell \in K, \forall i \in N \qquad (2.25)$$

Typical assignment constraints are valid for the vertices corresponding to trips, while multiple assignments are allowed for the vertices representing the depots.

The constraint set {(2.20), (2.21), (2.22), (2.22a), (2.23)} is related to the assignment of trips to depots. That is, constraints (2.20) and (2.21) ensure that if a trip is directly connected to a depot then it has to be assigned to that particular depot. Constraint set (2.22) requires that if trip i, which is assigned to depot ℓ, is directly connected to trip j then trip j must also be assigned to that particular depot. Constraint set (2.22a) is redundant in the integer model (QA) but, is relevant in the linear relaxation of (*QA*), since the resulting lower bounds are better than those produced without this extra constraint. Constraint set (2.23) ensures that each trip is assigned to exactly one depot. As an example, a feasible solution for the MDVSP is shown in Fig. 2.4.

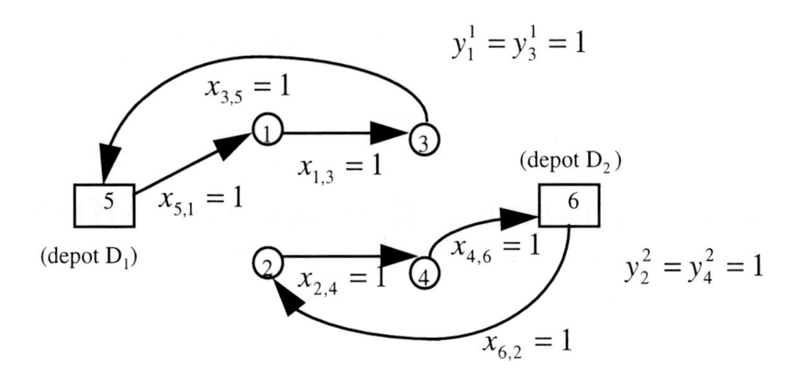

Fig. 2.4 A feasible solution.

Variables $x_{ij}, x_{i,n+\ell}, x_{n+\ell,i}$ are shared by formulations (*MA*) and (*QA*) while variables y_i^ℓ only exist in (*QA*).

2.3 Set Partitioning Formulation

Recently, Ribeiro/Soumis (1994) give a new mathematical formulation of the MDVSP as a set partitioning problem with side constraints. Taking graphs $G^\ell = (V^\ell, A^\ell)$, $\ell = 1,\ldots,k$ as defined for (MF_x), they denote by Ω_ℓ the set of paths leaving depot ℓ, visiting some nodes of *N* and returning to the same depot. For every path $p \in \Omega = \bigcup_{\ell=1}^{k} \Omega_\ell$ they represent c_p as the sum of the costs of its arcs. Let $q_{jp} = 1$ if path p contains node $j \in N$, otherwise $q_{jp} = 0$. If binary variables y_p are associated with every path a new mathematical formulation is obtained:

$$(MP) \quad \min \sum_{\ell \in K} \sum_{p \in \Omega_\ell} c_p y_p$$

$$s.t. \quad \sum_{\ell \in K} \sum_{p \in \Omega_\ell} a_{jp} y_p = 1 \qquad \forall j \in N \qquad (2.26)$$

$$\sum_{p\in\Omega_\ell} y_p \le d_\ell \qquad \forall \ell \in K \qquad (2.27)$$

$$y_p \in \{0,1\} \qquad p \in \Omega \qquad (2.28)$$

The set partitioning constraints (2.26) guarantee that each trip $j \in N$ belongs to exactly one vehicle block.

3 Linear Programming Relaxations of MDVSP Formulations

Since all the integer mathematical formulations presented in the previous section describe the MDVSP we have:

$$v_{IP}\left(MF_{xy}\right) = v_{IP}\left(MF_x\right) = v_{IP}\left(MA\right) = v_{IP}\left(QA\right) = v_{IP}\left(MP\right)$$

where $v_{IP}(.)$ denotes the optimal value for the integer programming model. In this section, we are going to relate the bound produced by the linear relaxation of *(QA)* with the bounds produced by the linear relaxations of (MF_{xy}), (MF_x) and (MA).

Denoting by $v_{LP}(.)$ the optimal value of the linear programming relaxation, we can prove that the equivalence between (MF_{xy}) and (MF_x) does not depend on the integrality of the decision variables.

Proposition 3.1: $v_{LP}(MF_{xy}) = v_{LP}(MF_x)$.

proof. Summing up (2.1) for $\ell = 1,\ldots,k$, we obtain (2.7). Combining the left side of (2.1) and the left side of (2.2) we get (2.8). Conversely, if we denote by y_i^ℓ the inflow (or outflow) of type ℓ in vertex *i*, (2.8) can be rewritten as,

$$\sum_{i:(i,j)\in I} x_{ij}^\ell + x_{n+\ell,j} = \sum_{i:(j,i)\in I} x_{ji}^\ell - x_{j,n+\ell} = y_i^\ell, \ \forall \ell \in K, \ \forall j \in N,$$

and we can derive (2.1) and (2.2). Now adding (2.1) for $\ell = 1,\ldots,k$ and using (2.7), we obtain (2.4) and the proof is complete.

Our second result relates $v_{LP}(MF_{xy})$ and $v_{LP}(QA)$.

Proposition 3.2: $v_{LP}\,(MF_{xy'}) \geq v_{LP}\,(QA)$.

proof. For the sake of simplicity, let us denote by z_{ij} variables $x_{ij}, (i,j) \in A$, of model (*QA*). The decision variables of model (*QA*), z_{ij}, $(i,j) \in I$, represent the binary flow in arc $(i,j) \in I$ while the decision variables x_{ij}^{ℓ} of model $(MF_{xy'})$ represent the flow of type ℓ in arc $(i,j) \in A^{\ell}$. Thus, concerning the linear relaxation of $(MF_{xy'})$, the total flow that moves from i to j, $(i,j) \in I$, is given by $\sum_{\ell \in K} x_{ij}^{\ell}, (i,j) \in I$, and we have the following relation between the decision variables involved in the linear relaxation of the above models:

$$z_{ij} = \sum_{\ell \in K} x_{ij}^{\ell}, \forall (i,j) \in I\,;\; z_{i,n+\ell} = x_{i,n+\ell},\quad z_{n+\ell,i} = x_{n+\ell,i} \quad \forall i \in N, \forall\, \ell \in K$$

Remember that the decision variables y_i^{ℓ} are the same in both models.

We prove the result by deriving the constraint set of (*QA*) from the constraint set of $(MF_{xy'})$. Summing up (2.1) and (2.2) for $\ell = 1,...,k$, we obtain (2.17) and (2.18). On the other hand (2.1) and (2.2) imply (2.20) and (2.21), respectively. It remains to be shown that (2.22) can also be deduced from the constraint set of $(MF_{xy'})$. We can rewrite (2.1) as

$$\sum_{j:(i,j)\in I} x_{ij}^{\ell} + x_{i,n+\ell} = y_i^{\ell} \Leftrightarrow \sum_{j:(i,j)\in I} x_{ij}^{\ell} = y_i^{\ell} - x_{i,n+\ell} \Leftrightarrow$$

$$x_{ij}^{\ell} = y_i^{\ell} - \sum_{\{j':(i,j')\in I, j'\neq j\}} x_{ij'}^{\ell} - x_{i,n+\ell}$$

in order to get $z_{ij} = \sum_{\ell \in K} x_{ij}^{\ell} = 1 - \sum_{\ell \in K} \sum_{\{j':(i,j')\in I, j'\neq j\}} x_{ij'}^{\ell} - \sum_{\ell \in K} x_{i,n+\ell}$.

Replacing variables z_{ij} in (2.22a) we obtain

$$y_i^{\ell'} + z_{ij} - y_j^{\ell'} = y_i^{\ell'} + 1 - \sum_{\ell \in K} \sum_{\{j':(i,j')\in I, j'\neq j\}} x_{ij'}^{\ell} - \sum_{\ell \in K} x_{i,n+\ell} - y_j^{\ell'}$$

Using (2.1), substitution of $y_i^{\ell'}$ in (2.22) we get

$$\sum_{j':(i,j')\in I} x_{ij'}^{\ell'} + x_{i,n+\ell'} + 1 - \sum_{\ell\in K} \sum_{\{j':(i,j')\in I, j'\neq j\}} x_{ij'}^{\ell} - \sum_{\ell\in K} x_{i,n+\ell} - y_j^{\ell'} \leq$$

$$\sum_{j':(i,j')\in I} x_{ij'}^{\ell'} + 1 - \sum_{\ell\in K} \sum_{\{j':(i,j')\in I, j'\neq j\}} x_{ij'}^{\ell} - y_j^{\ell'} \leq 1$$

The first inequality is derived from the fact that the decision variables are non-negative and $x_{i,n+\ell'} \leq \sum_{\ell\in K} x_{i,n+\ell}$. On the other hand, the second inequality is due

to $\sum_{j':(i,j')\in I} x_{ij'}^{\ell'} - \sum_{\ell\in K} \sum_{\{j':(i,j')\in I, j'\neq j\}} x_{ij'}^{\ell} \leq x_{ij}^{\ell'} \leq y_j^{\ell'}$, and we finally prove that (2.22) can be deduced from the constraint set of (MF_{xy}). For constraints (2.22a) the proof is similar.

If we look at the computational results presented in Sect. 5, we see that there are instances for which $v_{LP}(MF_{xy}) > v_{LP}(QA)$.

Proposition 3.3: $\upsilon_{LP}(QA) \geq \upsilon_{LP}(MA)$.

proof. We prove the result by showing that $F_{LP}(QA) \subseteq F_{LP}(MA)$, where $F_{LP}(P)$ denotes the set of feasible solutions for the linear relaxation of P. But, $F_{LP}(QA) \subseteq F_{LP}(MA)$ is equivalent to $\overline{F_{LP}(MA)} \subseteq \overline{F_{LP}(QA)}$, where $\overline{F_{LP}(P)}$ is the complementary set of $F_{LP}(P)$. So, we prove the result by showing that if solution (x_{ij}) is not feasible in the linear relaxation of (*MA*) no feasible solution (x_{ij}, y_i^{ℓ}) exists for the linear relaxation of (*QA*).

Since constraints (2.12) and (2.13) in (*MA*) are the same as (2.17) and (2.18) in (*QA*), if (x_{ij}) violates (2.12) or (2.13) then it violates (2.17) or (2.18). Hence, it is enough to consider the case where (x_{ij}) satisfies constraints (2.12) and (2.13) of (*MA*) but violates constraints (2.15). That is, (x_{ij}) contains a path P that starts and ends at different depots and $\sum_{(i,j)\in P} x_{ij} > |P| - 1$. Without loss of generality, let us represent the above path P by

$$n+1 \to 1 \to 2 \to \ldots \to r-1 \to r \to n+2$$

where $n+1$ and $n+2$ are vertices corresponding to two different depots, $n+1 \neq n+2$, and $|P| = r+1$.

Since variables $x_{ij}, (i, j) \in A$, are the same in both models, we want to construct (x_{ij}, y_i^{ℓ}) feasible for constraints (2.20), (2.21), (2.22), (2.22a) and (2.23).

The ending depot of the above path P is represented by vertex n+2. For a trip i, included in path P, the minimum value y_i^2 can take is given by:

$$y_r^2 = x_{r,n+2}, \qquad y_{r-1}^2 = y_r^2 + x_{r-1,r} - 1, \qquad \cdot\cdot \qquad y_2^2 = y_3^2 + x_{2,3} - 1,$$
$$y_1^2 = y_2^2 + x_{1,2} - 1$$

On the other hand, referring to the assignment of trip i, included in path P, to the starting depot, the minimum value y_i^1 can assume is the following:

$$y_1^1 = x_{n+1,1}, \qquad y_2^1 = y_1^1 + x_{1,2} - 1, \qquad \cdot\cdot \qquad y_{r-1}^1 = y_{r-2}^1 + x_{r-2,r-1} - 1,$$
$$y_r^1 = y_{r-1}^1 + x_{r-1,r} - 1$$

Specifically for vertex 1, we have

$$y_1^1 + y_1^2 = \sum_{(i,j)\in P} x_{ij} - (r-1) > |P| - 1 - |P| + 2 = 1.$$

From $y_i^{\ell} \geq 0$, we have $\sum_{\ell \in K} y_i^{\ell} > 1$.

We proved that the set of feasible solutions defined by the linear relaxation of (MA) contains the projection into the $x_{ij}, (i, j) \in A$, subspace of the set of feasible solutions defined by the linear relaxation of (QA). Hence, if solution (x_{ij}, y_i^{ℓ}) is feasible in the linear relaxation of (QA) then solution (x_{ij}) is feasible in the linear relaxation of (MA) and, since the two formulations have the same objective function value, we conclude, $\upsilon_{LP}(QA) \geq \upsilon_{LP}(MA)$.

Proposition 3.4: There are instances with $\upsilon_{LP}(QA) > \upsilon_{LP}(MA)$.

proof. Our aim is to present a solution (x_{ij}) feasible in the linear relaxation of (*MA*) for which there is no (y_i^ℓ) such that (x_{ij}, y_i^ℓ) is a feasible solution for the linear relaxation of (*QA*).

Consider the example below, depicted in Fig. 3.1, where the value of each variable x_{ij} is near the corresponding arc.

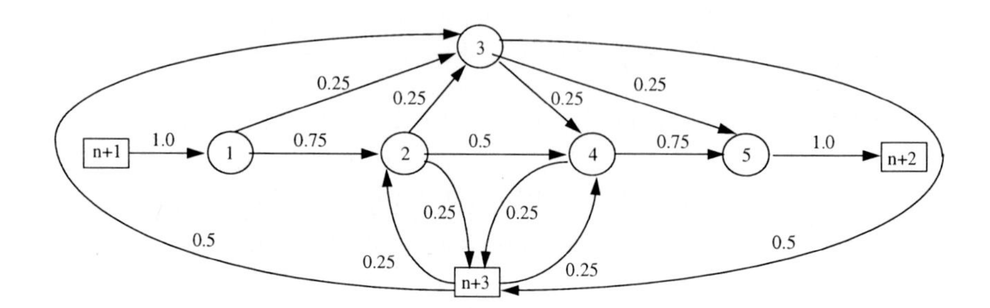

Fig. 3.1 Example of a feasible solution for the linear relaxation of (*MA*) but not feasible for the linear relaxation of (*QA*)

Note that the solution shown in Fig. 3.1 can be easily completed so that flow conservation constraints are valid for depots *n*+1 and *n*+2.

Consider path *P*, drawn from the above solution

Using (2.20), (2.21) and (2.23) we establish:

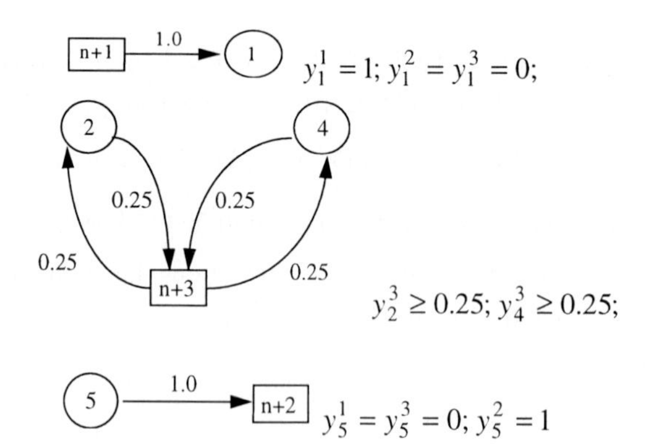

From (2.22) and (2.22a) we have

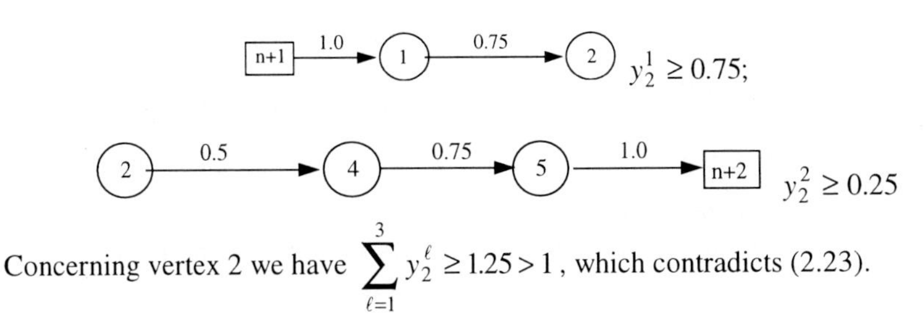

Concerning vertex 2 we have $\sum_{\ell=1}^{3} y_2^\ell \geq 1.25 > 1$, which contradicts (2.23).

Note that the solution presented in Fig. 3.1, which is not feasible in the linear relaxation of (*QA*), is feasible in the linear relaxation of (*QA*) without the constraint set (2.22a).

4 Branch and Bound

In the previous section we saw that the bound obtained by the linear relaxation of (MF_{xy}) is equal to the bound obtained by the linear relaxation of (MF_x). Now, we describe two branch and bound algorithms based on the linear relaxation of these formulations.

If the optimal solution of the linear relaxation has fractional valued integer variables then a fractional integer variable is chosen for branching and two new subproblems are generated. One node with this reference variable value fixed to 1 and the other with this variable value fixed to 0. At each node of the tree the linear relaxation of the corresponding subproblem is solved. A node is fathomed if the corresponding subproblem is infeasible, or if the corresponding lower bound is greater than or equal to the best known upper bound, or the optimal solution for its linear programming relaxation is integer. The process is repeated until all nodes of the branching tree are fathomed.

If the values of the assignment variables y_i^ℓ are known then the values of the scheduling variables x_{ij}^ℓ, $x_{i,n+\ell}$, $x_{n+\ell,i}$ are easily computed by solving k vehicle scheduling problems with one depot. On the other hand, if the values of variables x_{ij}^ℓ, $x_{i,n+\ell}$, $x_{n+\ell,i}$ are known, then the values of variables y_i^ℓ are determined as follows: $y_i^\ell = 1$, if trip i belongs to a vehicle block that starts its duty at depot ℓ, $y_i^{\ell'} = 0$, $\ell' \neq \ell$, $\ell' \in K$.

Referring to (MF_{xy}), we can consider different branching strategies, depending on what variable type is chosen to branch on. The first strategy consists of branching only on the subset of variables y_i^ℓ, the second allows branching only on the subset of variables x_{ij}^ℓ, $x_{i,n+\ell}$, $x_{n+\ell,i}$ while the third option permits branching on all the decision variables of the model, x_{ij}^ℓ, $x_{i,n+\ell}$, $x_{n+\ell,i}$, y_i^ℓ. Since there are less variables y_i^ℓ than variables x_{ij}^ℓ, $x_{i,n+\ell}$, $x_{n+\ell,i}$, we expect the first option to give the optimal solution in the shorter time.

Concerning (MF_x), any of the decision variables involved in the model can be chosen to branch on.

5 Computational Results

The branch and bound methods have been computationally evaluated using real data from Carris, the public transport company of Lisbon (Portugal). These test problems correspond to trips covering peak hours and include two classes of objectives:

Class A: The objective is to minimize the total number of vehicles + deadhead costs;

Class B: The objective is to minimize the total number of vehicles + deadhead costs + idle time costs.

The codes were written in the C language and we used the Mixed Integer Programming Package CPLEX. All the tests were run on a SUN SPARCstation 10-52 with 48 Mbytes of RAM-50MHz.

All instances have four depots and in order to ensure the use of the minimum number of vehicles, a penalty of 5000 is added to costs associated with links to and from a depot. Table 5.1 shows the characteristics of the test problems.

In Tables 5.2 - 5.5, we report on test problems where no limit is imposed on the number of vehicles available at a depot, which corresponds to the company's real situation. Tables 5.6 - 5.9 refer to computational experience carried out with the same instances but imposing depot capacity constraints. These capacities were established as follows: 1) we solve a multi-depot vehicle scheduling problem without requiring that each vehicle returns to the source depot; 2) the capacity of depot ℓ, $\ell = 1,...,k$ is equal to the number of vehicles starting, or finishing, their duty at depot ℓ, in the optimal solution of 1).

Tables 5.2, 5.3, 5.6, and 5.7 give the results for the linear relaxation. Whenever the optimal solution is not reached a gap is defined by:

$$gap = (v_{IP}(.) - v_{LP}(.)) \,/\, (v_{IP}(.))$$

Tables 5.4, 5.5, 5.8, 5.9 compare the performance of the branch and bound method for the different mathematical models and for the different branching strategies. The notation $(MF_{xy})/y$, $(MF_{xy})/x$, $(MF_{xy})/xy$, stands for the mathematical model (MF_{xy}) with branching allowed only on variables y_i^ℓ, only on variables x_{ij}^ℓ, $x_{i,n+\ell}$, $x_{n+\ell,i}$ and on all the decision variables, respectively. Referring to the mathematical formulation (MF_x), branching is allowed on all variables, x_{ij}^ℓ, $x_{i,n+\ell}$, $x_{n+\ell,i}$.

The notation *sec* refers to computing times in seconds, excluding input and output. For the branch and bound procedure *sec* includes the time spent at the root node.

Table 5.1 Characteristics of the test problems

Problem reference	n	k	depot capacities for constrained problems
P1 / P13	192	4	12-7-10-20 / 12-7-11-19
P2 / P14	192	4	11-7-10-21 / 12-8-10-19
P3 / P15	192	4	11-8-10-20 / 12-8-10-19
P4 / P16	192	4	11-8-10-20 / 12-8-10-19
P5 / P17	255	4	14-14-14-24 / 14-14-13-24
P6 / P18	255	4	14-13-13-24 / 14-13-13-24
P7 / P19	255	4	14-13-13-24 / 14-13-13-24
P8 / P20	255	4	14-13-13-24 / 14-13-13-24
P9 / P21	264	4	10-8-11-21 / 10-8-11-21
P10 / P22	264	4	10-8-11-21 / 10-8-11-21
P11	352	4	11-13-15-29
P12 / P23	352	4	11-13-13-28 / 11-13-13-28

Table 5.2 Linear relaxation. Problem class A (no constraints on depot capacities)

	$v_{LP}(QA)$	(QA) *sec*	gap $\times 10^{-4}$	$v_{LP}(MF_{xy})$ $v_{LP}(MF_x)$	(MF_{xy}) *sec*	(MF_x) *sec*	gap $\times 10^{-4}$
P1	491640.5	891	1.05	491692.0*	38	188	0.00
P2	491499.3	2675	1.01	491549.0*	84	179	0.00
P3	491470.9	2926	1.08	491524.0*	105	316	0.00
P4	491465.1	6335	1.04	491516.0*	105	360	0.00
P5	652337.0	5741	1.18	652405.8	264	825	0.13
P6	642291.0	9119	1.03	642357.0*	238	907	0.00
P7	642278.1	9549	1.12	642350.0*	371	831	0.00
P8	642276.5	12577	1.14	642350.0*	359	1144	0.00
P9	501077.4	197	0.25	501090.0*	68	54	0.00
P10	491050.0	341	0.37	491068.0*	46	126	0.00
P11	681594.3	87	0.16	681605.0*	8	10	0.00
P12	651483.4	1120	1.31	651537.0	414	325	0.11

Table 5.3 Linear relaxation. Problem class B (no constraints on depot capacities)

	$v_{LP}(QA)$	(QA) *sec*	gap $\times 10^{-4}$	$v_{LP}(MF_{xy})$ $v_{LP}(MF_x)$	(MF_{xy}) *sec*	(MF_x) *sec*	gap $\times 10^{-4}$
P13	491894.5	740	1.13	491950.0*	86	137	0.00
P14	491832.0	836	1.48	491904.3	51	197	0.01
P15	491829.7	1560	1.47	491902.0*	52	398	0.00
P16	491829.7	2271	1.47	491902.0*	63	296	0.00
P17	652745.2	4727	1.08	652815.7	166	923	0.01
P18	642741.8	10188	1.09	642812.0*	146	925	0.00
P19	642741.8	17389	1.09	642812.0*	140	1247	0.00
P20	642741.8	8301	1.09	642812.0*	200	1107	0.00
P21	501085.2	222	0.12	501091.0*	27	34	0.00
P22	491076.6	332	0.17	491085.0*	18	34	0.00
P23	651528.9	889	0.62	651569.0	156	130	0.00

Table 5.4 Branch and Bound. Problem class A (no constraints on depot capacities)

	opt. sol. value	$(MF_{xy})/y$ *sec* / nodes	$(MF_{xy})/x$ *sec* / nodes	$(MF_{xy})/xy$ *sec* / nodes	(MF_x) *sec* / nodes	opt. sol. vehi. by dep
P5	652414	1706 / 127	723 / 53	581 / 21	1051 / 52	12-13-14-26
P12	651544	436 / 11	1262 / 17	633 / 22	1085 / 18	11-13-15-26

Table 5.5 Branch and Bound. Problem class B (no constraints on depot capacities)

	opt. sol. value	$(MF_{xy})/y$ *sec* / nodes	$(MF_{xy})/x$ *sec* / nodes	$(MF_{xy})/xy$ *sec* / nodes	(MF_x) *sec* / nodes	opt. sol. vehi. by dep
P14	491905	127 / 10	183 / 8	131 / 8	321 / 7	11-9-11-18
P17	652816	303/ 4	402 / 3	398 / 7	890 / 6	12-13-15-25
P23	651569	159 / 1	481 / 1	237 / 1	958 / 1	12-13-15-25

Table 5.6 Linear relaxation. Problem class A (constraints on depot capacities)

	$v_{LP}(QA)$	(QA) *sec*	gap $\times 10^{-4}$	$v_{LP}(MF_{xy})$ $v_{LP}(MF_x)$	(MF_{xy}) *sec*	(MF_x) *sec*	gap $\times 10^{-4}$
P1	491509.0	2874	3.78	491695.0*	46	204	0.00
P2	491509.6	2882	1.00	491559.0*	87	174	0.00
P3	491474.7	2748	1.04	491526.0*	106	337	0.00
P4	491469.5	7318	0.99	491518.0*	88	376	0.00
P5	652339.8	5718	1.35	652420.1	258	947	0.12
P6	642292.9	9294	1.25	642371.5	364	935	0.02
P7	642280.1	9884	1.23	642359.0*	420	839	0.00
P8	642278.4	14800	1.26	642359.0*	402	1143	0.00
P9	501087.7	235	0.78	501123.0	46	64	0.08
P10	491058.3	401	0.42	491079.0*	46	127	0.00
P11	681610.5	94	0.30	681631.0*	9	11	0.00
P12	651489.2	1190	1.12	651556.5	275	354	0.08

Table 5.7 Linear relaxation. Problem class B (constraints on depot capacities)

	$v_{LP}(QA)$	(QA) *sec*	gap $\times 10^{-4}$	$v_{LP}(MF_{xy})$ $v_{LP}(MF_x)$	(MF_{xy}) *sec*	(MF_x) *sec*	gap $\times 10^{-4}$
P13	491900.9	640	1.22	491959.00	43	143	0.04
P14	491836.4	1685	1.62	491910.50	55	198	0.11
P15	491835.7	1836	1.57	491908.00	64	421	0.10
P16	491835.7	2098	1.57	491908.00	65	314	0.10
P17	652748.0	5762	1.13	652822.00*	222	946	0.00
P18	642743.8	10690	1.26	642824.00	392	956	0.02
P19	642743.3	11667	1.27	642824.00	142	1281	0.02
P20	642743.3	9688	1.27	642824.00	185	1141	0.02
P21	501097.6	323	0.71	501133.00*	23	61	0.00
P22	491085.4	281	0.26	491098.00*	21	49	0.00
P23	651535.3	981	0.85	651589.00	148	142	0.03

Table 5.8 Branch and Bound. Problem class A (constraints on depot capacities)

	opt. sol. value	$(MF_{xy})/y$ *sec* / nodes	$(MF_{xy})/x$ *sec* / nodes	$(MF_{xy})/xy$ *sec* / nodes	(MF_x) *sec* / nodes
P5	652428	2009 / 128	604 / 24	778 / 31	1081 / 52
P6	642373	1184 / 27	1520 / 33	914 / 11	1235 / 4
P9	501127	105 / 5	143 / 4	160 / 7	80 / 2
P12	651562	397 / 6	1259 / 8	511 / 2	362 / 6

Table 5.9 Branch and Bound. Problem class B (constraints on depot capacities)

	opt. sol. value	$(MF_{xy})/y$ *sec* / nodes	$(MF_{xy})/x$ *sec* / nodes	$(MF_{xy})/xy$ *sec* / nodes	(MF_x) *sec* / nodes
P13	491961	88 / 2	120 / 3	92 / 5	222 / 2
P14	491916	162 / 12	196 / 9	156 / 16	364 / 16
P15	491913	195 / 8	264 / 7	179 / 5	359 / 9
P16	491913	200 / 8	259 / 7	152 / 5	350 / 9
P18	642825	410 / 4	531 / 4	484 / 6	1195 / 1
P19	642825	430 / 2	495 / 3	364 / 3	1161 / 3
P20	642825	574 / 18	990 / 4	534 / 5	991 / 4
P23	651591	167 / 2	229 / 2	263 / 2	989 / 2

6 Analysis of the Results

In all test problems, the bound provided by the linear relaxation of (QA) is below the bound given by the linear relaxation of (MF_{xy}) or (MF_x). Furthermore, excessively high computing time is needed for solving the linear relaxation of (QA), possibly due to the high number of constraints.

In terms of the computational experience, note the excellent quality of the lower bounds provided by the linear relaxations of (MF_{xy}) and (MF_x). For 23 test problems, without constraints on depot capacities, 18 were solved to optimality without going to branch and bound. For the same problem set, but imposing constraints on depot capacities, eight problems were solved to optimality without going to branch and bound. The largest gap between the bound provided by the linear relaxation of (MF_{xy}), or (MF_x), was 0.000013. Ribeiro/Soumis (1994) and Forbes/Holt/Watts (1994) presented an argument that partly explains the quality of such lower bounds. They say that the structure of the set partitioning constraints of (MP) eliminates many fractional solutions from its optimal solutions. That is, if p_1, p_2, p_3 are three columns, each associated with a circuit, and suppose p_1 includes trips T_i before T_j, p_2 includes trips T_i before T_m and p_3 includes trips T_j before T_m. Then there is a circuit p_4 that includes T_i, T_j and T_m and the column corresponding to p_4, with negative reduced cost, is a suitable candidate to enter in the linear programming problem.

It is very interesting to compare the time spent for the different linear relaxations (see Figs. 6.1 - 6.4).

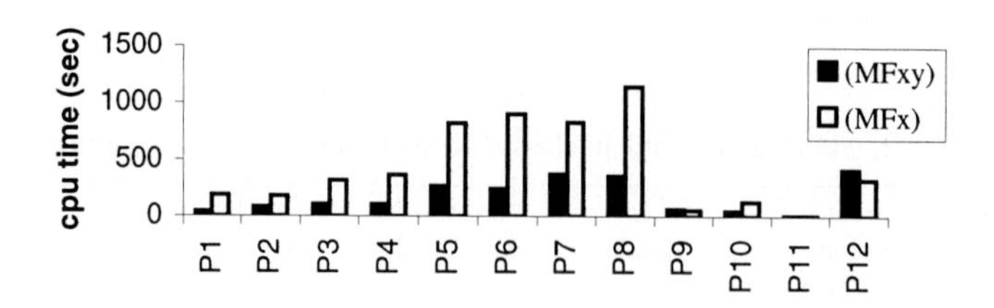

Fig. 6.1 Linear relaxation. Results from table 5.2.

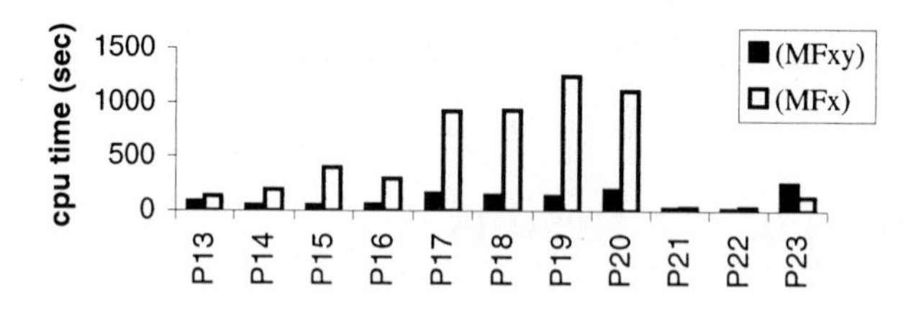

Fig. 6.2 Linear relaxation. Results from table 5.3.

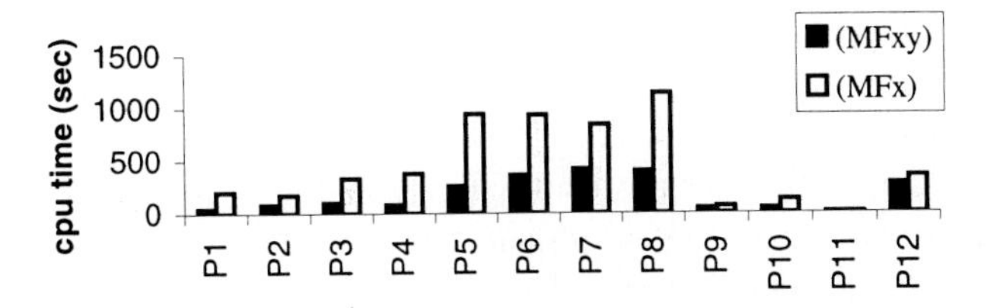

Fig. 6.3 Linear relaxation. Results from table 5.6.

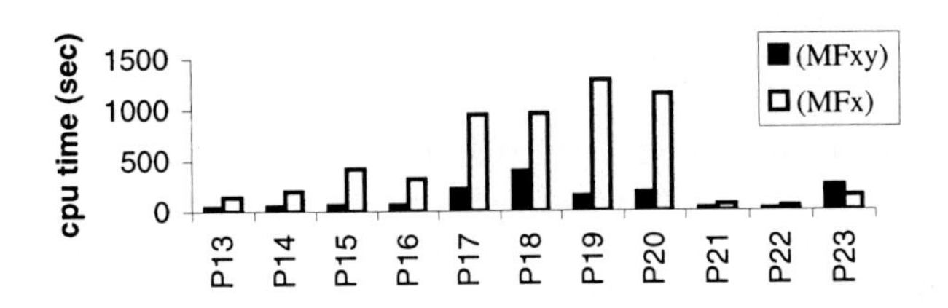

Fig. 6.4 Linear relaxation. Results from table 5.7.

The linear relaxation of (MF_{xy}) takes less time than the linear relaxation of (MF_x), which is a more compact model since it contains fewer variables and constraints than (MF_{xy}). A heuristic explanation for these results follows from the fact that the decision variables y_i^ℓ, which are not essential in a multicommodity flow type formulation, help in the computation of the x_{ij}^ℓ variable values, which are fundamental in such mathematical approaches. In the integer formulation (MF_{xy}) the number of variables x_{ij}^ℓ is greater than the number of variables y_i^ℓ, but whenever the y_i^ℓ variable values are known then the x_{ij}^ℓ variable values are easily determined by solving k vehicle scheduling problems with one depot.

If we compute the average time for the different branch and bound algorithms, we can see that the best time was attained for model (MF_{xy}) with branching on variables x_{ij}^ℓ, y_i^ℓ. If instead of the average time we compute the median, which is less sensitive to outliers, we conclude that the best strategy is to use model (MF_{xy}) and to branch on variables y_i^ℓ.

7 Conclusions

The first exact method for the MDVSP was presented by Carpaneto/Dell'Amico/Fischetti/Toth (1989). The authors formulated the MDVSP as an assignment problem with side constraints and proposed a branch and bound algorithm based on the computation of the lower bounds through an additive lower bound procedure. Computational experience is presented for problems with up to 70 trips and three depots. Alternatively, Ribeiro/Soumis (1994) proposed an exact algorithm based on the continuous relaxation of (*MP*). The authors report the solution of test problems with up to 300 trips and six depots. At the same time, Forbes/Holt/Watts (1994) presented an exact algorithm based on the linear relaxation of the multicommodity network flow type formulation (MF_x). They consider test problems with three depots and varying in size from 100 to 600 trips.

Ribeiro/Soumis (1994) proved that the linear relaxation of (MF_x) is equivalent to the linear relaxation of (*MP*). Now, we have compared the bounds given by linear programming relaxations of the different MDVSP formulations presented in Sect. 2 together with the bound given by the linear relaxation of the (*QA*) model. The main results obtained in this paper can be compiled in the following scheme.

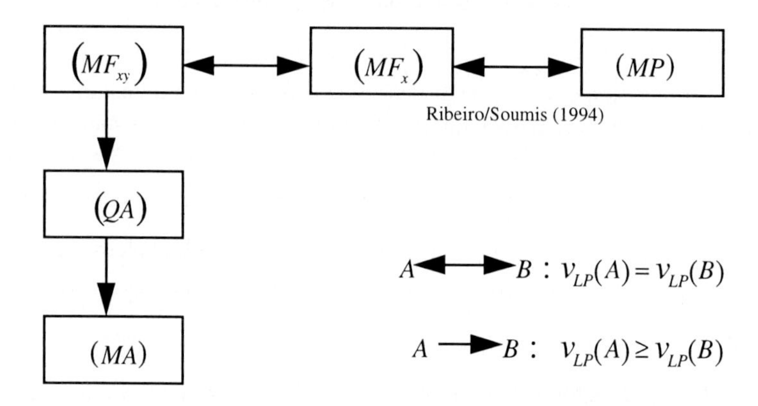

Fig. 7.1 Classification of the MDVSP linear relaxations

As a result of this theoretical analysis we selected the multicommodity network flow formulation (MF_{xy}) for use in a tree search procedure. Model (MF_{xy}) allows different branching strategies according to the subset of variables on which branching is possible. We also develop a branch and bound algorithm based on the linear relaxation of a more compact multicommodity network flow formulation, (MF_x).

Computational results show that, concerning the computational time required either for solving the linear relaxation or for performing a tree search procedure based on this linear relaxation, model (MF_{xy}) outperforms model (MF_x).

Acknowledgments: The authors would like to thank the referees for their helpful suggestions.

References

Ball, M.O./Magnanti, T.L./Monma, C.L./Nemhauser, G.L. (1995): Network routing. (North Holland) Handbooks in Operations Research and Management Science, 8.

Bertossi, A. A./Carraresi, P./Gallo, G. (1987): On some Matching Problems arising in vehicle scheduling models. in: Networks 17, 271 - 281.

Carpaneto, G./Dell'Amico, M./Fischetti, M./Toth, P. (1989): A branch and bound algorithm for the Multiple Vehicle Scheduling Problem. in: Networks 19, 531 - 548.

Ceder, A./Stern, H. I. (1981): Deficit function bus scheduling with deadheading trip insertions for fleet size reduction. in: Transportation Science 15, 338 - 363.

Dell'Amico, M./Fischetti, M./Toth, P. (1993): Heuristic algorithms for the Multiple Depot Vehicle Scheduling Problem. in: Management Science, Vol. 39, No 1, 115 - 125.

El-Azm (1985): The minimum fleet size problem and its applications to bus scheduling. in: Rousseau, J. M. (ed.): Computer Scheduling of Public Transport 2. (North-Holland) Amsterdam, 493 - 512.

Forbes, M. A./Holt, J. N./Watts, A. M. (1994): An exact algorithm for Multiple Depot Bus Scheduling. in: European Journal of Operational Research 72, 115 - 124.

Lamatsch, A. (1992): An approach to vehicle scheduling with depot capacity constraints. in: Desrochers, M./Rousseau, J.M. (eds.): Computer-Aided Transit Scheduling, Lecture Notes in Economics and Mathematical Systems 386. (Springer Verlag) Berlin, Heidelberg, 181 - 195.

Mesquita , M./Paixão, J. (1992): Multiple Depot Vehicle Scheduling Problem: A new heuristic based on quasi-assignment algorithms. in: Desrochers, M./Rousseau, J.M. (eds): Computer-Aided Transit Scheduling, Lecture Notes in Economics and Mathematical Systems 386. (Springer Verlag) Berlin, Heidelberg, 167 - 180.

Ribeiro, C./Soumis, F. (1994): A column ceneration cpproach to the Multiple Depot Vehicle Scheduling Problem. in: Operations Research 42, 41 - 52.

Smith, B/Wren, A. (1981): VAMPIRES and TASC: two successfully applied bus scheduling programs. in: Wren, A. (ed.): Computer Scheduling of Public Transport: Urban Passenger Vehicle and Crew Scheduling. (North-Holland) Amsterdam, 97 - 124.

Timetable Synchronization for Buses

Avishai Ceder and Ofer Tal
Transportation Research Institute, Faculty of Civil Engineering,
Technion - Israel Institute of Technology, Haifa 32000, Israel

Abstract: The objective of this work is to create a timetable for a given network of buses, in order to maximize the synchronization among the buses, namely, to maximize the number of simultaneous arrivals of buses at the connection nodes (transfer nodes) of the network. The data for the problem are as follows: a bus route network, including connection nodes of the bus routes; the number of bus departures for every bus route in the network, during a given time period [O,T]; the traveling time of each bus route to the nodes of the network; maximal and minimal headways between two adjacent departures for each bus route. The decision variables are the departure times of all buses for each route. Creating a timetable with maximal synchronization enables the transfer of passengers from one route to another, with minimum waiting time at the transfer nodes. This problem is a major concern to the public transportation designers, taking into account the satisfaction and convenience of the system's users. In this work, the problem is formulated as a mixed integer programming problem, and a heuristic algorithm is developed to solve the problem in polynomial time.

1 Introduction and the Scope of the Problem

One of the major roles of transit schedulers (see Ceder/Wilson (1986)) is to create timetables for the bus routes of a given bus network. According to Ceder (1986), there are three levels of decision problems that have to be addressed prior to the actual scheduling when constructing such timetables: 1) selecting the type of headway (even or uneven headways); 2) selecting a method for setting the frequencies (maximum load or load profile); 3) selecting one or more objective functions for the scheduling process (for example, minimizing operator cost while providing adequate service, minimizing operator and user cost through weighing factors).

This paper concentrates on one objective function: maximal synchronization. This is a rather important objective from both the operators and the users perspectives, involving as it does creating timetables that will maximize the number of simultaneous arrivals of buses at the connection (transfer) nodes. Any route design undertaking at the network level attempts to eliminate a large number of transfer points, due to their adverse effect on the user (see Ceder/Wilson (1986)). No doubt, there is a trade-off between this elimination and the efficiency of the bus route network from the operating cost perspective. In order to allow for an adequate bus level-of-service, the schedulers face the synchronization task to ensure a maximal number of situations in which passengers can switch from one route to another without waiting time. This task can be extended, also minimizing the waiting time for those passengers who require connections. By doing so, the scheduler creates a more attractive transit system that generates the opportunity for increasing the number of riders. This is presented schematically in Fig. 1.1.

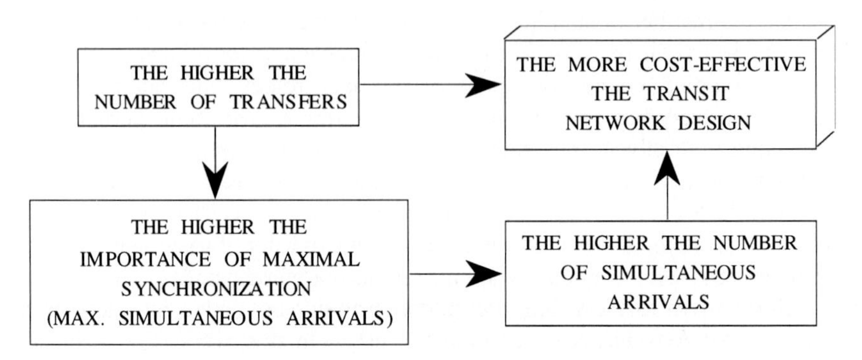

Fig. 1.1 The relative importance of synchronization

Actually, synchronization is the most difficult task of transit schedulers, and is currently addressed intuitively. The scheduler, in fact, attempts to create the departure times in the timetable while complying with: 1) the required frequency; 2) efficient assignment of trips to a single bus chain, and 3) synchronization of certain arrivals. This paper presents an effective mathematical procedure for maximal synchronization to be employed as a useful tool for the scheduler in the process of creating timetables.

This paper presents a model which enables transit schedulers to set restrictions on the headways for each route, to introduce different frequencies for every route, and to apply other constraints. The objective function is to maximize the number of simultaneous bus arrivals in the network. The departure times are set in such a way as to achieve this goal. It is worth mentioning that the model presented is also capable, with a small extension, to assign different weights on each different simultaneous arrival (two different lines or time periods). That is, if the scheduler wishes to provide different importance levels for each synchronization situation, he may then introduce the weights, and the objective function will change to maximize the sum of all weights. The entire study including many examples, more algorithms than appear here and comparison between the heuristic and optimal solutions is presented in the final report by Ceder/Golani/Tal (1991).

Sect. 2 of this paper provides two mathematical formulations of the problem – a nonlinear programming and a mixed integer linear programming. Sect. 3 presents a heuristic algorithm that solves the problem, and a variation of this algorithm. Examples in which the heuristic algorithm are compared to optimal solutions are presented in Sect. 4. Summary and conclusions are given in Sect. 5.

2 Formulating the Model

The given bus network is represented by a directed graph, $G = \{A,N\}$, where:

A = A set of arcs representing the bus routes;

N = A set of transfer nodes in the network.

The problem data are the following:

T = The planning horizon (the departure times of the buses can be set in the interval [0,T] which is a discrete interval);

M = The number of bus routes in the network;

N = The number of transfer nodes in the network;

$Hmin_k$ = The minimum headway (operator's requirements) between two adjacent bus departures on route k ($1 \le k \le M$);

$Hmax_k$ = The maximal headway (policy headway) permitted between two adjacent bus departures on route k ($1 \le k \le M$);

F_k = The number of departures to be scheduled for route k during the interval [0,T] ($1 \le k \le M$);

T_{kj} = The travel time from the starting point of route k to node j ($1 \le k \le M$, $1 \le j \le N$) (travel times are considered deterministic, and can be set to the mean travel times).

The following should be noted: (a) this paper assumes throughout that the first departure on each route k must take place in the interval $[0,Hmax_k]$; (b) the case where a route k does not pass through a node j is represented by $T_{kj} = -1$, and (c) the problem is infeasible unless the following constraints hold, for each k:

$$Hmax_k \ge Hmin_k \tag{2.1}$$

$T \geq (F_k - 1) \cdot Hmin_k$ (2.2)

$T \leq (F_k - 1) \cdot Hmax_k$ (2.3)

The decision variables are the following:

(a) X_{ik} represents the departure time of the i-th bus on route k ($1 \leq i \leq F_k$);

(b) Z_{ikjqn} is a binary variable with value 1 if the i-th bus on route k meets the j-th bus on route q at node n; otherwise, it has the value 0;

Let $A_{kq} = \{n : 1 \leq n \leq N, T_{kn} \geq 0, T_{qn} \geq 0\}$.

The initial formulation of the problem is as follows:

$$\text{Max} \sum_{k=1}^{M-1} \sum_{i=1}^{F_k} \sum_{q=k+1}^{M} \sum_{j=1}^{F_q} \sum_{n \varepsilon A_{kq}} Z_{ikjqn}$$

S.T.

$$X_{1k} < Hmax_k, \; 1 \leq k \leq M \quad (2.4)$$

$$X_{F_k k} \leq T, \; 1 \leq k \leq M \quad (2.5)$$

$$Hmin_k \leq X_{(i+1)k} - X_{ik} \leq Hmax_k, \; 1 \leq k \leq M, \; 1 \leq i \leq F_k - 1 \quad (2.6)$$

$$Z_{ikjqn} = \text{Max}\,[1 - |\,(X_{ik} + T_{kn}) - (X_{jq} + T_{qn})\,|\,, 0] \quad (2.7)$$

Constraint (2.4) ensures that the first departure time will not be beyond the maximal headway from the start of the time horizon, while constraint (2.5) treats the last departure similarly. Constraint (2.6) indicates the headway limits, and constraint (2.7) defines the binary variable of the objective function.

This model can be simplified by defining a variable, Y_{kq}, representing the overall number of simultaneous arrivals of buses on route k with buses on route q. The model is changed as follows:

$$Max \sum_{k=1}^{M-1} \sum_{q=k+1}^{M} Y_{kq}$$

S.T.

$$Y_{kq} = \sum_{n \varepsilon A_{kq}} \sum_{i=1}^{F_k} \sum_{j=1}^{F_q} Max[1 - |(X_{ik} + T_{kn}) - (X_{jq} + T_{qn})|, 0] \quad (2.8)$$

Constraints (2.4), (2.5), (2.6) and (2.7) remain unchanged.

The last formulation represents a nonlinear programming problem. It can be reformulated as a mixed integer linear programming problem, which can be solved (up to certain sizes) by several software packages. The nonlinear constraint is (2.8). Let D_{nijkq} denote a binary variable (defined over the same domain as Z_{ikjqn}), and B a large number $(B = T + \max_{i,j} T_{ij})$.

Constraint (2.8) is replaced with the following constraints:

$$B \cdot D_{nijkq} \geq X_{ik} + T_{kn} - (X_{jq} + T_{qn}) \tag{2.9}$$

$$B \cdot D_{nijkq} \geq X_{jq} + T_{qn} - (X_{ik} + T_{kn}) \tag{2.10}$$

$$Y_{kq} < \sum_{n \varepsilon A_{kq}} \sum_{i=1}^{F_k} \sum_{j=1}^{F_q} (1 - D_{nijkq}) \tag{2.11}$$

If $X_{ik} + T_{kn} = X_{jq} + T_{qn}$, there is a simultaneous arrival of the i-th bus in route k with the j-th bus on route q at node n. The variable D_{nijkq} can yield the value 0, and Y_{kq} is increased by one, according to (2.11).

If $X_{ik} + T_{kn} \neq X_{jq} + T_{qn}$, the arrivals do not coincide, and D_{nijkq} must yield the value to satisfy constraints (2.9) and (2.10). The number of simultaneous arrivals between buses of routes k and q (Y_{kq}) is not increased in (2.11).

An upper bound on the number of possible simultaneous arrivals in a given bus network is:

$$Z^* = \sum_{k=1}^{M-1} \sum_{q=k+1}^{M} \sum_{n \varepsilon A_{kq}} \mathrm{Min}(F_k, F_q)$$

The number of integer variables in an MIP problem is generally a good index of its complexity. The variable D_{nijkq} represents the simultaneous arrival of the i-th bus on route k and the j-th bus on route q at node n. This means an integer variable for every combination of two buses on different routes that intersect at node n. Let F be $\mathrm{Max}(F_k)$; the number of integer variables in the worst case is $0(NM^2F^2)$, which is a very large number. However, in a more realistic setting, this number is $0(M^2F^2)$ where N can be replaced by the average number of nodes shared in common by any two routes (normally 1 or 2).

3 Heuristic Algorithm

In the previous section, the problem was formulated as a large MIP problem. Running small network examples (5 routes, 5 nodes), using GAMS software (see Brooke/Kendrick/Meeraus (1988)) on an IBM-PC (ps2 model 80) required hours and even days of running time! This motivated the development of heuristic algorithms that would solve such problems in a reasonable time. This section will present a heuristic algorithm and a variation of this algorithm. The algorithm was implemented in Turbo-Pascal (see Borland (1989)), and many examples were checked and compared to the optimal solutions obtained by using GAMS. The algorithm is based on the selection of nodes within the network. In each step, the next node is selected, providing that at this node, not all the departure times have yet been determined. Once the departure time is resolved, all its corresponding arrival times are also set.

Definition 1: A node is defined as "possible" if the following holds:

(a) There is at least one bus route that passes through the node, and not all the departure times for that route are set;

(b) It is possible to create more simultaneous arrivals at the node.

Definition 2: A node is defined as "new" if no arrival times have been set for it. The basic algorithm uses several procedures as described in flow chart form in Fig. 3.1.

Step 1: Initialization: check whether the problem is feasible, and create the data structures. Mark all nodes as "possible";

Step 2: Select the next node, NO, from the possible nodes;

Step 3: If NO is "new," perform procedure FIRST, otherwise perform procedure MIDDLE;

Step 4: If there is any possible node, go to Step 2; if there are any more routes, perform procedure CHOOSE and go to Step 2, otherwise stop.

Step 1 contains a check of whether the problem is feasible (constraints (2.1), (2.2), and (2.3)), and two data structures are built:

(a) A structure called Route for each bus route i, which includes $Hmin_i$, $Hmax_i$, F_i, the number of nodes the route passes through, and the departure times that have already been set;

(b) A structure called Node for each node n, which includes the number of routes passing through the node; the route with the maximum travel time to the node, and the number of simultaneous arrivals at the node at each time point in interval

$$[0, T + \max_{i,j} T_{ij}]$$

All the nodes are marked possible.

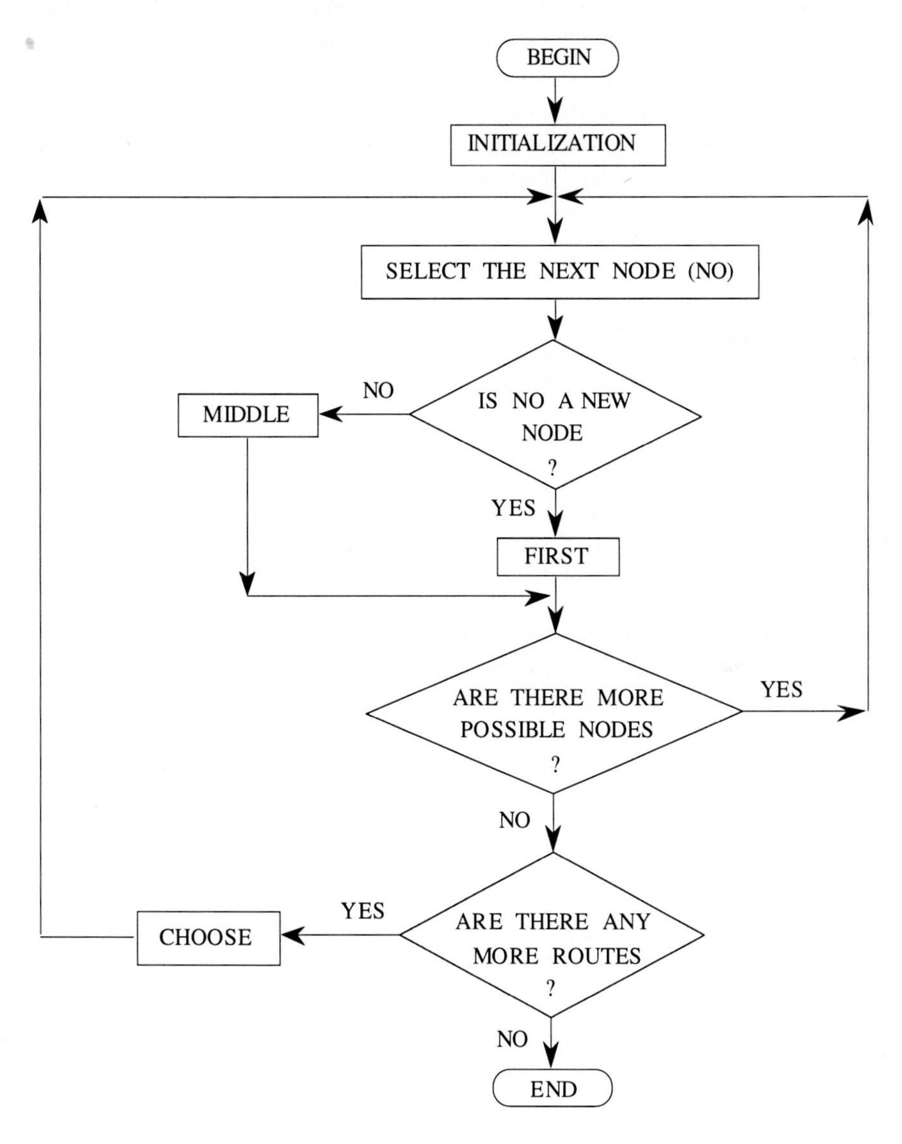

Fig. 3.1 A flow chart of the basic algorithm

In Step 2, the next node, NO, is selected from among the possible nodes. Node NO must satisfy the following:

(a) The number of different bus arrival times at the node is maximal; in such a node, there is a greater probability that another bus departure can be set so that it will arrive at NO by any one of the (already set) arrival times;

(b) Among the nodes satisfying (a), NO is that through which a maximal number of routes pass; in such a node, there is a greater potential for simultaneous arrivals;

(c) Among the nodes that satisfy (b), NO minimizes the maximum travel time of all routes from their origin to the node (after the departure times of buses are set in order to meet at NO, there is still a potential for simultaneous arrivals at distant nodes).

A distinction is made in Step 3 between a "new" and another node. For a "new" node, procedure FIRST attempts to set the departure times of buses that pass through it, such that the buses will arrive at the node at the earliest time possible, and simultaneous arrivals will continue to be created at the node according to the Hmin, Hmax of the routes. For example, buses on three routes that arrive simultaneously at a certain node at time t_0: if the next departure time for each of these routes can be set at a fixed difference, d, from the last departure time of the route $\left(\min_{i=1,2,3} \mathrm{H}_{\min i} \leq d \leq \min_{i=1,2,3} \mathrm{H}_{\max i} \right)$, then there will be additional simultaneous arrivals at the node at time $t_0 + d$. Procedure FIRST finds the minimal possible d. If parameter d cannot be set (a situation where $Hmin_i > Hmax_j$), the next departure times of buses on these routes will not be resolved at this step.

For a node that is not "new," there is an attempt to set the departure times of buses on routes passing through it, such that the buses will arrive at the node at the earliest time among all the arrival times already set in that node. If no more simultaneous arrivals are available at the node, the node is marked "not possible." This procedure is called MIDDLE.

Step 4 contains a test of whether there are any more "possible" nodes. If not, there may be routes on which not all the departure times of the buses were set. In such cases, the route which passes through the maximum number of nodes is chosen, and its next departure time is set by using the difference $Hmin_i$ from the last departure. In such a manner, the algorithm sets additional bus arrivals for the maximum number of nodes possible. All the nodes through which route i passes are marked "possible," and the algorithm returns to Step 2. This procedure is called CHOOSE. The complexity of the basic algorithm is, in the worst case, $0(NTFM^2)$.

One variation of this algorithm is to run the algorithm N times, each time choosing a different node to be handled first (the first time the node is chosen in the basic algorithm, it is selected according to the principles explained in Step 2; there are examples where this selection is unsuccessful). The complexity of this variation is, in the worst case, $0(TFN^2M^2)$.

4 Examples

This section presents three bus network examples, with one presented in detail for the sake of clarity.

Fig. 4.1 presents a simple network example combining four transfer points covering two routes. The numbers on the arcs are travel times (say, in minutes).

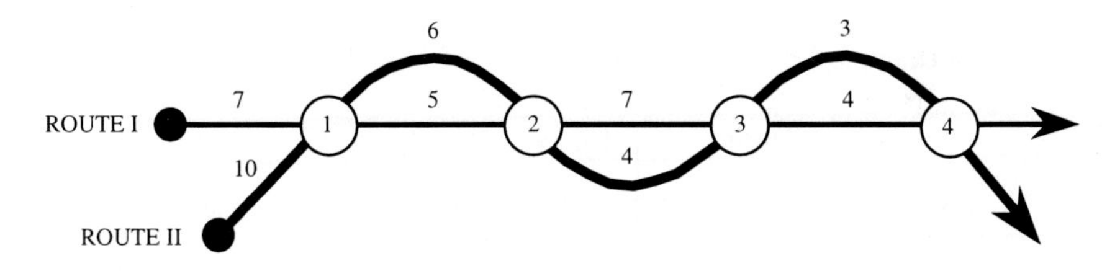

Fig. 4.1 The basic network for example 1

The data for example 1 are as follows:

i	$Hmin_i$	$Hmax_i$	F_i	T
Route I	6	10	4	30
Route II	3	5	6	

The algorithm ends with the final results:

Dep. time		Meeting time at node 1	Meeting time at node 4	Total No. meetings
Route I	Route II			
3	0	10		
9	3	16		
15	6	22		6
21	9		26	
	12		32	
	15		38	

In this simple example, the optimal and heuristic procedures coincide.

Fig. 4.2 presents a more complex example, combining 4 nodes with 4 routes.

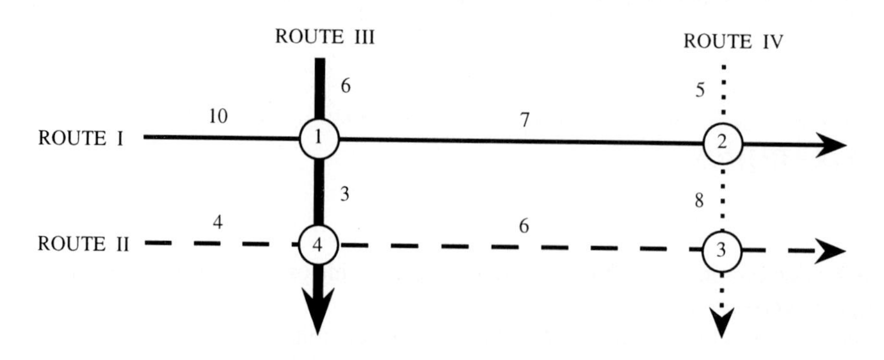

Fig. 4.2 The network for example 2

The data for example 2 are as follows:

i	$Hmin_i$	$Hmax_i$	F_i	T
Route I	8	15	2	30
Route II	10	15	3	
Route III	10	15	3	
Route IV	14	20	2	

Step 1: In structure (a), the number of nodes each route passes through is two. Structure (b):

Node n	No. Routes	Route with max. T_{in}
1	2	I
2	2	I
3	2	IV
4	2	III

Step 2: Select NO:

Number of dep. times (=0) for all nodes.
Number of crossing routes (=2) for all nodes.
Max. travel times for the nodes are 10, 17, 13, 9, respectively.
The minimum is 9, so NO = 4.

Step 3: Procedure FIRST. The first possible meeting time at node 4 is nine. The parameter d is 10 (max $(10,10) \leq d \leq$ min. $(15, 15)$).

Procedure FIRST results:

Dep. time, Route II	Dep. time, Route III	Meeting time
5	0	9
15	10	19
25	20	29

Step 4: There are more possible nodes – go to Step 2.

Step 2: Select NO:

In the previous steps, the algorithm set three arrival times for nodes 1 and 3. The number of routes passing through these nodes is the same; therefore NO=1, by the min-max criteria.

Step 3: Procedure MIDDLE is performed.

The bus arrival time (route III) at node 1, as set by procedure FIRST, are 6, 16, 26. Procedure MIDDLE sets the departure time of route I buses to arrive at node 1, and meet route III buses as early as possible.

Procedure MIDDLE results in the following:

Dep. time, Route I	Meeting time, node 1
6	16
16	26

Step 4: There are more possible nodes – go to Step 2.

Step 2: Select NO.
The number of bus arrivals at node 2 is two (route I buses).
The number of bus arrivals at node 3 is three (route II buses).
Hence, NO = 3.

Step 3: Procedure MIDDLE is performed.
This procedure sets the bus departure time of route IV buses to meet the route II buses at node 3.
The arrival times of node 3 are 15, 25, 35. The procedure yields two more meetings by the following :

Dep. time, Route IV	Meeting time, node 3
2	15
22	35

Step 4: No more nodes or routes. Stop!
The algorithm ends with this timetable:

Dep. time			
Route I	II	III	IV
6	5	0	2
16	15	10	22
	25	20	

The total number of meetings is seven (which is the optimal solution for this network), as follows:

Node	Time of Meeting
4	9
4	19
4	29
1	16
1	26
3	15
3	35

The third example is a simple network shown in Fig. 4.3.

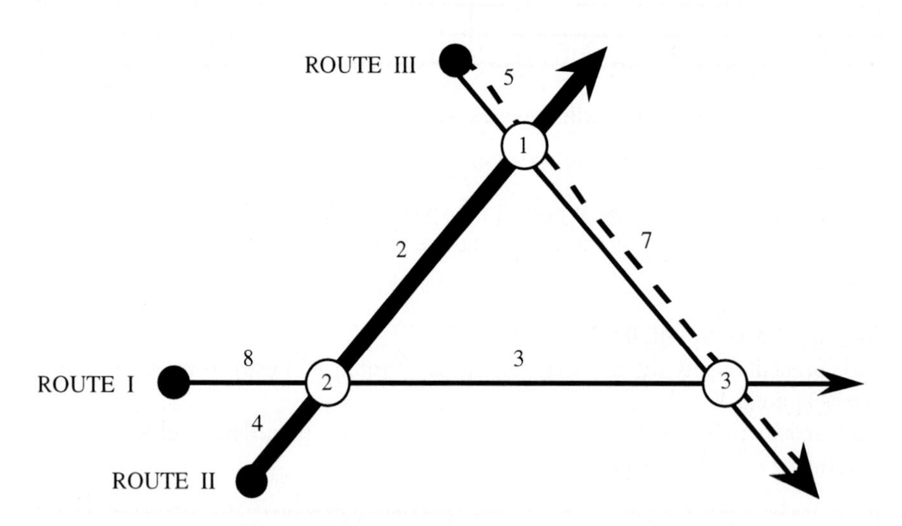

Fig. 4.3 The basic network for example 3

The network comprises three routes and three nodes. Each pair of routes intersect at a single node. The numbers on the arcs are the travel time (in minutes) to the nodes. The data for example 3 are as follows:

i	$Hmin_i$	$Hmax_i$	F_i	T
Route I	3	8	3	20
Route II	4	7	4	
Route III	5	12	4	

The optimal solution obtained by GAMS for this example is 10 simultaneous arrivals. The basic algorithm yielded only 7 simultaneous arrivals. An optimal timetable (with 10 simultaneous arrivals) is shown below.

A Timetable for Fig. 4.3

Route	Departure Times			
I	2	8	14	
II	0	6	12	18
III	1	7	13	19

5 Summary and Conclusions

This paper has described the problem of maximal synchronization in creating bus timetables; that is maximizing the number of simultaneous bus arrivals at transfer nodes. The mathematical model described, with deterministic travel times, was formulated as a mixed integer linear programming problem, which is known to be an NP problem. For large networks, it is not practical to solve the problem by using software for MIP problems, and that was the motivation for developing heuristic algorithms.

The heuristic algorithm developed was implemented with Turbo-Pascal programs, and it achieved very good results compared to the optimal solution (see Ceder/Golani/Tal (1991)).

The model described can be extended in various ways; for example, defining weights for each node or for a simultaneous arrival between each pair of bus routes (describing situations in which there are major nodes, or differences in the importance of bus routes). Another option is to define a "simultaneous arrival" in different ways. For example, a simultaneous arrival can also be defined as an arrival of two buses within an interval of t minutes or at a time difference of at least t minutes (in situations where the goal is to prevent simultaneous arrivals).

The paper presented a heuristic approach to the synchronization problem and a mathematical tool for solving the issue. It has applications for transit schedulers who face this problem frequently while creating the timetables.

References

Borland International (1989): Turbo Pascal version 5.0.

Brooke, A./Kendrick, D./Meeraus, A. (1988): GAMS, A User's Guide.

Ceder, A. (1986): Methods for creating bus timetables. in: Trans. Research, Vol. 21A, No. 1, 59 - 83.

Ceder, A./Wilson, N.H.M. (1986): Bus network design. in: Trans. Research, Vol. 20B, No. 4, 331 - 344.

Ceder, A./Golani, B./Tal, O. (1991): Maximal Synchronization in Designing Bus Timetables. Transportation Research Institute, Technion, Project 115-119, Final Research Report No. 911-169, 95.

Transportation Service Network Design: Models and Algorithms

Daeki Kim
i2 Technologies, White Plains, New York 10606

Cynthia Barnhart
Massachusetts Institute of Technology, Cambridge, MA 02139

Abstract: We focus on the problem of large scale service network design for transportation carriers requiring the determination of the cost minimizing or profit maximizing set of transportation services and their schedules, given resource constraints. Examples include determining the set of flights and their schedules for an airline; determining the routing and scheduling of tractors and trailers in a trucking operation; and jointly determining the aircraft flights, ground vehicle and package routes and schedules for a provider of express shipment service. In this paper, we survey network design literature, and present a general modeling framework and solution approach for large-scale transportation service network design.

1 Introduction

The focus of our research is to develop models and solution algorithms for service network design problems faced by transportation carriers. Service network design problems arising at railroads, airlines, trucking firms, intermodal partnerships, transit agencies, etc. require the determination of the cost minimizing or profit maximizing set of services and their schedules, given limited resources. Examples of service network design problems include determining the set of flights and their schedules for an airline; determining the routing and scheduling of tractors and trailers in a trucking operation; and jointly determining the aircraft flights, ground vehicle and package routes and schedules for an express freight delivery provider. Our objective is to develop general models and solution procedures for these transportation service network design problems. We want to use our procedure to make *design decisions* answering questions such as:

- Which transportation services should be offered (and at what times) such that available resources are best utilized and profits are maximized?
- What is the best location and size of terminals such that overall costs are minimized?
- What is the best fleet size and composition such that service requirements are met and profits are maximized?

There are two major shortcomings of current service network design models when applied to the problems arising in the transportation industry. First, the interactions between design decisions in transportation scheduling problems are more complicated than in many other application areas. For example, selecting service between two points implies that a vehicle of some type departs some location and arrives at another. This, in addition, implies that other services must be selected to ensure conservation of flow, or flow balance, for that vehicle. Another complexity resulting from these interactions involves fixed costs. Since multiple services may be performed by a single vehicle, fixed costs are associated with *sets* of design decisions and not a single design decision. The second major issue associated with transportation scheduling problems is size. In transportation related applications, since time and space both must be modeled, network size can be massive. State-of-the-art service network design methods simply are not designed for problems of the immense size encountered in transportation. Our focus, then, is to develop a general modeling framework and solution approach for large scale transportation service network design problems.

Outline of The Paper

In Section 2, we provide a review of the network design literature, focusing on specific applications to transportation service network design and related transportation problems. In Section 3, we describe various representations of the networks underlying transportation service network design problems. Using these networks, we present and compare different models for the service network design problem in Section 4. Finally, exact solution algorithms and heuristics for service network design are described in Section 5.

2 Problem Description and Literature Review

A variety of network design problems in transportation, distribution, communication, and several other problem domains require trade-offs between variable operating costs and fixed design costs. Before reviewing the literature, we present a generic network design model.

A Network Design Model

Let $G = (N, A)$ be a directed graph, where N is the node set and A is the arc set. A commodity is defined for each origin-destination pair with nonzero flow. We let k denote the set of commodities K and b^k denote the quantity of commodity k to be shipped from its origin, denoted $O(k)$, to its destination, $D(k)$, for each $k \in K$. The general service network design model contains two types of variables - one modeling integer design decisions and the other continuous flow decisions. Let u^f denote the capacity of service type $f \in F$. Let y_{ij}^f be an integer variable that indicates the number of times an arc $(i,j) \in A$ of service type $f \in F$ is included in the solution. Let x_{ij}^k denote the fraction of b^k on arc (i,j). Let h_{ij}^f be the fixed cost of including arc (i,j) with service type f in the solution once and c_{ij}^k be the cost per unit flow of k along $(i,j) \in A$. Then, the node-arc representation of the network design problem (NDP) is:

(NDP)

$$\min \quad \sum_{f \in F} \sum_{(i,j) \in A} h_{ij}^f y_{ij}^f + \sum_{k \in K} \sum_{(i,j) \in A} c_{ij}^k b^k x_{ij}^k \tag{1}$$

$$\sum_{k \in K} b^k x_{ij}^k \leq \sum_{f \in F} u^f y_{ij}^f \qquad \text{for all } (i,j) \in A \tag{2}$$

$$\sum_{j \in N} x_{ij}^k - \sum_{j \in N} x_{ji}^k = \begin{cases} 1 & \text{if } i = O(k) \\ 0 & \text{if } i \neq O(k), D(k) \\ -1 & \text{if } i = D(k) \end{cases} \qquad \text{for all } i \in N,\ k \in K \tag{3}$$

$$x_{ij}^k \geq 0 \qquad \text{for all } k \in K,\ (i,j) \in A \tag{4}$$

$$y_{ij}^f \geq 0 \text{ and integer} \qquad \text{for all } (i,j) \in A,\ f \in F \tag{5}$$

The objective function is to find the cost minimizing set of services and routing of flows over these services. Constraints (2), the *forcing constraints*, limit the amount of flow on an arc to its capacity, and arc capacity is determined by the value of the design variables. Constraints (3), *flow conservation constraints*, ensure that commodity is transported from its origin to its destination. Finally, constraints (4) and (5) ensure nonnegativity of commodity flows and integrality and nonnegativity of design variables.

Network design problems have been studied extensively, with comprehensive surveys presented in Minoux (1989) and in Magnanti/Wong (1984). Even though network design problems have been studied extensively, limited capability exists to solve large-scale transportation applications.

Transportation Service Network Design Literature

Billheimer/Gray (1973) use link elimination/insertion heuristics to solve uncapacitated network design problems with an application to transit network design. Crainic/Rousseau (1986) use decomposition heuristics and column generation principles for uncapacitated freight transportation service network design problems. Their decomposition-based algorithm heuristically solves the following two subproblems iteratively - service design and traffic routing. Lamar/ Sheffi/ Powell (1990) study uncapacitated, multicommodity network design problems in the less-than-truckload motor carrier industry and present a new lower bound for the problem and an efficient implementation scheme using shortest path and linearized knapsack programs. Farvolden/Powell (1994) present local-improvement heuristics for a Service Network Design problem encountered in Less-Than-Truckload (LTL) common carrier applications. The add/drop heuristics are based upon subgradients derived from the optimal dual variables of the shipment routing subproblem. Powell (1986), Powell/Sheffi (1989) also apply add/drop heuristics in LTL motor carrier applications. Barnhart/Schneur (1996) develop models and algorithms to solve an express shipment service network design problem. Leung/Magnanti/Singhal (1990) study a point-to-point route planning problem that arises in many large scale delivery systems. Their approach exploits the structure of the problem in order to decompose the problem into two smaller subproblems that are each amenable to solution by a combination of optimization and heuristic techniques based on Lagrangian relaxation. Recently, Newton (1996) studies network design problems under budget constraints with an application to railroad blocking problems.

Facility Location Problems

The *facility location problem (FLP)* is a specialization of the network design problem in which the design decisions concern the location and sizing of facilities at nodes of a distribution or transportation network. Van Roy (1986) provides a comprehensive survey. Early work of Erlenkotter (1978) on uncapacitated *FLP's* provides lower bounds from the dual problem that allow large combinatorial *FLP* problems to be solved. Guignard/Spielberg (1979) show that inclusion of variable upper bounds give very tight lower bounds and sparse search trees. Van Roy (1986) presents an implementation of the Cross Decomposition method to solve the capacitated facility location problem, which combines Benders decomposition and Lagrangian relaxation into a single framework that involves successive solutions to a transportation problem and an uncapacitated plant location problem. Guignard (1988) uses Benders inequalities generated in a Lagrangian dual ascent procedure to solve uncapacitated plant location problems.

Vehicle Routing Problems

Consider a network design problem with a single source node e, and with a fixed capacity $u_{ij} = U$, on all arcs. In addition, suppose that the following assignment constraints on the design variables are imposed:

$$\begin{aligned} \sum_{j \in N} y_{ij} &= 1 \qquad \text{for all } i \in N \backslash \{e\} \\ \sum_{i \in N} y_{ij} &= 1 \qquad \text{for all } j \in N \backslash \{e\} \end{aligned}$$

as well as the constraint

$$\sum_{j \in N} y_{ej} \leq n.$$

Then, the linear cost *NDP* becomes a *vehicle routing problem* (*VRP*) for a homogeneous fleet of n vehicles, each domiciled at the source node (or depot) e and each having capacity UComprehensive surveys by Magnanti (1981), Magnanti/Wong (1984) and the recent vehicle routing and scheduling survey by Desrosiers/Dumas/ Solomon/Soumis (1992) and Desaulniers/Desrosiers/Ioachim /Solomon/Soumis (1994) summarize much of this field. A harder class of *VRP's* involve *multiple depot, heterogeneous vehicle routing problems* (*MDHTVRP*). Let f be the set of vehicles and E denote a set of depots ($e \in E$), then *MDHTVRP* can be formulated as:

(MDHTVRP)

$$\min \quad \sum_{f \in F} \sum_{(i,j) \in A} h_{ij}^f y_{ij}^f \tag{6}$$

$$\sum_{k \in K} b^k x_{ij}^k \leq \sum_{f \in F} u^f y_{ij}^f \qquad \text{for all } (i,j) \in A \tag{7}$$

$$\sum_{j \in N} x_{ij}^k - \sum_{j \in N} x_{ji}^k = \begin{cases} 1 & \text{if } i = O(k) \\ 0 & \text{if } i \neq O(k), D(k) \\ -1 & \text{if } i = D(k) \end{cases} \qquad \text{for all } i \in N,\ k \in K \tag{8}$$

$$\sum_{j \in N} y_{ij}^f - \sum_{j \in N} y_{ji}^f = 0 \qquad \text{for all } i \in N,\ f \in F \tag{9}$$

$$\sum_{f \in F} \sum_{j \in N} y_{ij}^{f} = 1 \qquad \text{for all } i \in N \backslash E \tag{10}$$

$$\sum_{f \in F} \sum_{i \in N} y_{ij}^{f} = 1 \qquad \text{for all } j \in N \backslash E \tag{11}$$

$$\sum_{j \in N} y_{ej}^{f} = \sum_{j \in N} y_{je}^{f} \leq n^{f} \qquad \text{for all } e \in E \tag{12}$$

$$x_{ij}^{k} \geq 0 \qquad \text{for all } k \in K,\ (i,j) \in A \tag{13}$$

$$y_{ij} \geq 0 \text{ and binary} \qquad \text{for all } (i,j) \in A \tag{14}$$

The objective function (6) is to find the cost minimizing set of vehicle routings, subject to constraints (7) - (14). Constraints (7), the *forcing constraints*, limit the amount of flow on an arc to its capacity. Constraints (8), *flow conservation constraints*, ensure that each commodity, from its origin to its destination, is fully serviced. Constraints (9) are the *design balance constraints* that require the number of each vehicle type entering a node to equal the number leaving, for each node in the network. Constraints (10) and (11) assure that each demand is serviced exactly once given the assumption of $b^k \leq \max_f (u^f)$ for all $k \in K$. Constraints (12) require that each vehicle starting at a depot must return to that same depot. Finally, constraints (13) and (14) ensure nonnegativity of commodity flows and integrality and nonnegativity of design variables. Desrosiers/Dumas/Solomon/Soumis (1992) and Desaulniers/Desrosiers/ Ioachim/ Solomon/Soumis (1994) provide comprehensive surveys of time constrained routing and scheduling problems, including fixed schedule problems, traveling salesman problems, shortest path problems, vehicle routing problems, pickup and delivery problems, multicommodity network flow problems, bus driver scheduling problems and airline crew scheduling problems.

Ziarati/Soumis/Desrosiers/Gelinas/Saintonge (1995) study the problem of assigning a sufficient number of locomotives of different types (heterogeneous consists) to trains to operate a pre-planned schedule. The original large-scale scheduling problem is decomposed and each smaller problem is then solved using a Dantzig-Wolfe decomposition method.

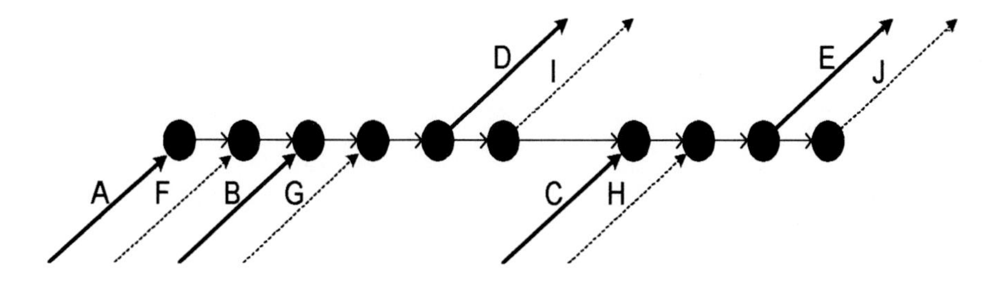

Figure 1: Time-Line Network

3 Service Network Representation

In many transportation related scheduling problems, time and space are essential elements. In order to evaluate the system properly, the conventional modeling practice is to build dynamic models. At their core, these models have a time-space structure, with nodes representing time and space and arcs representing movement in time and possibly, space. Two different representations of time-space networks for service design are the *connection network* and the *time-line network.*

In both the time-line and connection networks, a node i represents a location, denoted l_i, and a time, denoted t_i, and there are two types of arcs. In the time-line and the connection network, a *"design"* arc is placed between every pair of nodes i and j if a design variable from location l_i to location l_j with elapsed time $t_j - t_i$ is possible. The second type of arc in the time-line network, the *"time-line"* arc, exists between every pair of adjacent nodes at the same locations, where node j is adjacent to node i if $t_j - t_i \geq 0$, $j \neq i$, and $t_{j'} - t_i \geq t_j - t_i$, $\forall$ $j' \neq i$. The arcs represent the movement in time at a location (*see* Figure 1). The second type of arc in the connection network, the *"connection"* arcs, potentially exists between every pair of nodes i and j at the same location(*see* Figure 2).

A comparison of the relative strengths and weaknesses of the two network types can be summarized as follows:

- The time-line network usually requires fewer arcs than the connection network; and
- The connection network is better able to model complicating constraints, such as restrictions on sets of design variables selected.

The best network to use, then, is problem specific.

We can reduce dramatically the size of time-line networks using a network reduction method, called *node consolidation.* Without destroying feasibility of

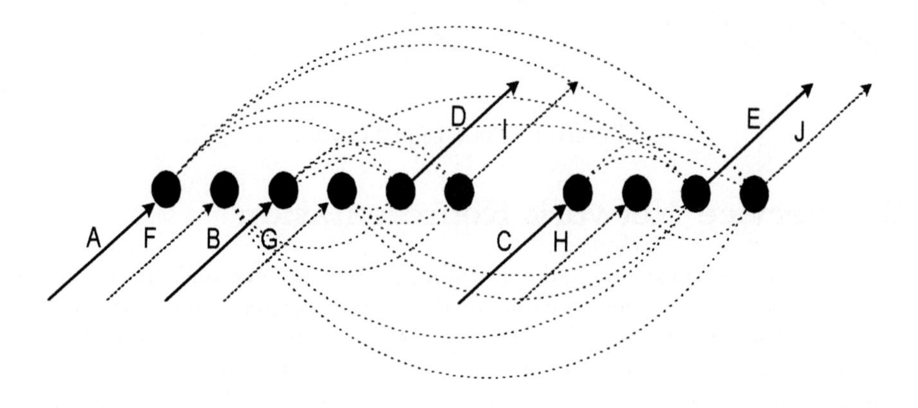

Figure 2: Connection Network

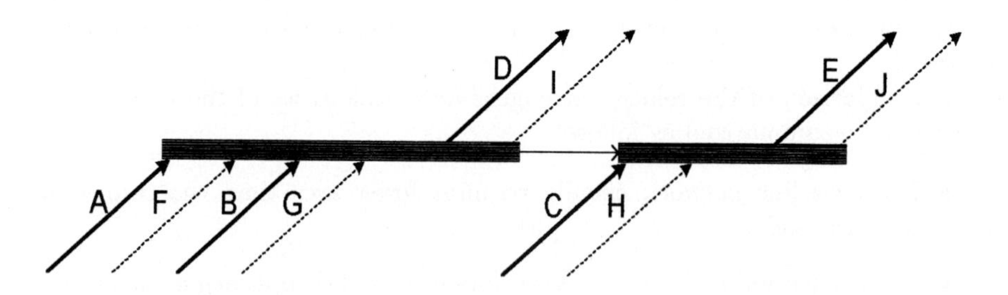

Figure 3: Node Consolidation of Time-Line Network

any path in the network, we consider all nodes representing some location i and arrange the nodes in increasing order of their associated time. We aggregate each sequence of adjacent arrival nodes (departure nodes) that are uninterrupted by a departure node (an arrival node), into a single super arrival (departure) node. Then, we merge a super arrival node followed by a super departure node into a single node, called a *super node.* Figure 3 shows how the network of Figure 1 is represented after node consolidation. The node consolidation method has been previously applied by Hane/Barnhart/Johnson/Marsten/ Nemhauser/Sigismondi (1994), Nemhauser (1994) and Shenoi (1996) in applications for the airline industry. Another network reduction method, referred to as *link consolidation* is detailed in Kim (1997).

4 Service Network Design Models

The objective of the transportation service network design problem (*SNDP*) can be viewed as finding *a set of vehicle routes* (they determine the capacity in the network) minimizing total system costs (i.e., the sum of fixed design costs and variable flow costs) while serving all demands. In general, the service network design problem (*SNDP*) is similar to the capacitated network design problem (*NDP*) except *SNDP* has added complexity. Compared to *NDP*, *SNDP* requires that *transportation* assets be assigned to network arcs and the assignments must achieve balance of these assets. *Balance* requires that the supply of vehicles plus the number of vehicles entering a location equals the number departing plus the demand for vehicles at that location. For example, Figure 4 shows a network that consists of three nodes and two available service (or vehicle) types. The capacity and fixed design cost of service type 1 is (1,10) and is (2, 15) for service type 2. Assume that there is a unit demand from node 1 to $2, 1$ to $3, 2$ to 1, 2 to 3, and 3 to 1 and that there are no variable flow costs involved. The optimal *NDP* solution, with value of 70, includes service type 2 links $(1,2)$, $(2,3)$, and $(3,1)$ and service type 1 link $(2,3)$. The optimal *SNDP* solution, however, includes the additional service type 1 links $(1,2)$ and $(3,1)$ to achieve balance by service type. The resulting *SNDP* solution value is 90.

In this section, we present three different but equivalent models for service network design problems. All of the models produce the same optimal solution values; however, they may differ in the numbers of constraints and the numbers of decision variables they contain. The result is a potentially significant difference in computational hardware requirements and solution run times.

4.1 Node-Arc Formulation

The node-arc representation of the service network design problem is:

(SNDP_Node-Arc)

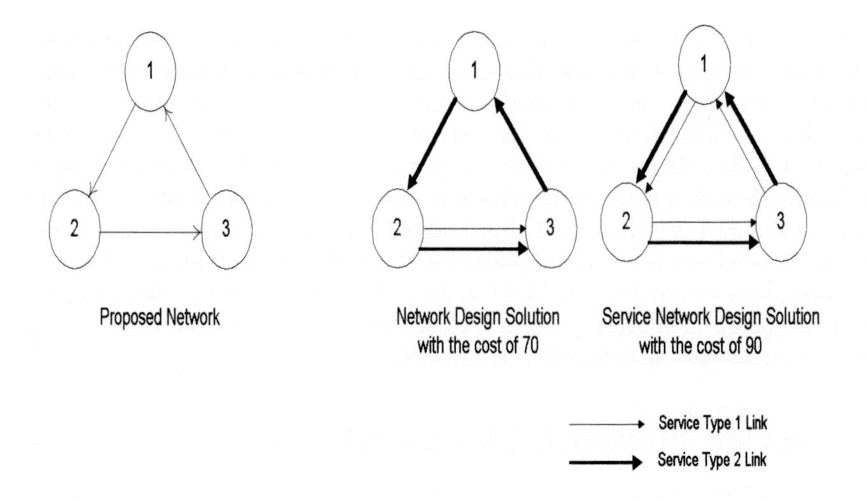

Figure 4: Service Network Design vs. Network Design Solution

$$\min \qquad \sum_{f\in F}\sum_{(i,j)\in A} h_{ij}^{f} y_{ij}^{f} + \sum_{k\in K}\sum_{(i,j)\in A} c_{ij}^{k} b^{k} x_{ij}^{k} \tag{15}$$

$$\sum_{k\in K} b^{k} x_{ij}^{k} \leq \sum_{f\in F} u^{f} y_{ij}^{f} \qquad \text{for all } (i,j)\in A \tag{16}$$

$$\sum_{j\in N} x_{ij}^{k} - \sum_{j\in N} x_{ji}^{k} = \begin{cases} 1 & \text{if } i = O(k) \\ 0 & \text{if } i \neq O(k), D(k) \\ -1 & \text{if } i = D(k) \end{cases} \qquad \text{for all } i\in N,\ k\in K \tag{17}$$

$$\sum_{j\in N} y_{ij}^{f} - \sum_{j\in N} y_{ji}^{f} = 0 \qquad \text{for all } i\in N,\ f\in F \tag{18}$$

$$x_{ij}^{k} \geq 0 \qquad \text{for all } k\in K,\ (i,j)\in A \tag{19}$$

$$y_{ij}^{f} \geq 0 \text{ and integer} \qquad \text{for all } (i,j)\in A,\ f\in F \tag{20}$$

The objective function (15) is to find the cost minimizing set of services and routing of flows over these services, subject to constraints (16) - (20). Constraints (16), the *forcing constraints*, limit the amount of flow on an arc to its capacity, and arc capacity is determined by the value of the design variables. Constraints (17), *flow conservation constraints*, ensure that each commodity is transported from its origin to its destination. Constraints (18) are the *design balance constraints* that distinguish service network design problems from conventional network design problems. The balance constraints require the number of services of each type entering a node to equal the number leaving, for each node in the network. Finally, constraints (19) and (20) ensure nonnegativity of commodity flows and integrality and nonnegativity of design variables.

4.2 Path Formulation

A design route r is a set of design variables of some type f that form a sequence that is balanced everywhere except possibly at the start and end of the sequence. We let R^f represent the set of all design routes for service type f. $h_r^f = \sum_{(i,j)\in A} h_{ij}\alpha_{ij}^r$ denotes the cost of design route r of type f, $c_p^k = \sum_{(i,j)\in A} c_{ij}^k \delta_{ij}^p$ is the cost per unit flow of k along path p, α_{ij}^r equals 1 if design variable (i,j) is included in design route r and equals 0 otherwise, δ_{ij}^p equals 1 if path p includes (i,j) and equals 0 otherwise, β_i^r equals 1 if node $i \in N$ is the start node of design route r, equals -1 if i is the end node of r and equals 0 otherwise.

By substitution in *SNDP_Node-Arc*, we obtain the following equivalent *SNDP_Path* formulation (for details *see* Ahuja/Magnanti/Orlin 1993):

(SNDP_Path)

$$\min \quad \sum_{f\in F}\sum_{r\in R^f} h_r^f y_r^f + \sum_{k\in K}\sum_{p\in P^k} \left(c_p^k b^k\right) x_p^k \tag{21}$$

$$\sum_{k\in K}\sum_{p\in P^k} \left(\delta_{ij}^p b^k\right) x_p^k \leq \sum_{f\in F}\sum_{r\in R^f} u^f y_r^f \alpha_{ij}^r \qquad \text{for all } (i,j)\in A \tag{22}$$

$$\sum_{p\in P^k} x_p^k = 1 \qquad \text{for all } k \in K \tag{23}$$

$$\sum_{r\in R^f} \beta_i^r y_r^f = 0 \qquad \text{for all } i \in N,\ f \in F \tag{24}$$

$$x_p^k \geq 0 \qquad \text{for all } k \in K,\ p \in P^k \tag{25}$$

$$y_r^f \geq 0 \text{ and integer} \qquad \text{for all } f \in F,\ r \in R^f \tag{26}$$

Constraints (22) are the forcing constraints for each design variable $(i,j) \in A$. Constraints (23) are the equivalent flow conservation constraints that ensure each commodity has its flow assigned completely from origin to destination. Constraints (24) enforce balance at the start and end of the design route (balance at intermediate locations is guaranteed by definition). Finally, constraints (25) and (26) ensure nonnegativity of flows and integrality and nonnegativity of design routes.

Compared to *SNDP_Node-Arc,* the number of *flow conservation constraints* in *SNDP_Path* is reduced from $|N| \times |K|$ to $|K|$ - a reduction that can have a big impact, especially for large-scale applications containing an extensive network and/or nearly complete demand pattern.

4.3 Tree Formulation

If arc costs do not vary by commodity, we can reduce the number of flow conservation constraints in the model using an alternative formulation called the *tree formulation,* presented in Jones/Lustig/Farvolden/Powell (1993). The idea is to aggregate commodities with the same origin into a single *super commodity*[1]. Each super commodity $s_{O(k)}$ in the set S of super commodities corresponds to the set of commodities $k \in K$ originating at $O(k)$. Since each commodity $k \in s_{O(k)}$ may flow along several paths between its origin and its destination, each super commodity $s_{O(k)}$ can flow along several *trees,* denoted by the set of trees $Q^{s_{O(k)}}$. We let Γ_{ij}^q equal 1 if tree q contains arc (i,j) and equal 0 otherwise. Each tree $q \in Q^{s_{O(k)}}$ is rooted at $O(k)$ and contains one $O(k)$ to $D(k)$ path, denoted p_q^k, for only those $k \in s_{O(k)}$. The flow on each path in a tree is a constant w, with $0 \leq w \leq 1$, of b^k for every commodity k in the tree. $c_q^{s_{O(k)}}$ for tree $q \in Q^{s_{O(k)}}$ and super commodity $s_{O(k)} \in S$ is the total cost of sending b^k units of k along the path p_q^k, for all $k \in s_{O(k)}$.

Given costs that are not commodity specific, Jones/Lustig/Farvolden/Powell (1993) show equivalence of the path and tree solutions for multicommodity flow problems and Kim (1997) shows equivalence for transportation service network design problems.

(SNDP_Tree)

$$\min \quad \sum_{f \in F} \sum_{r \in R^f} h_r^f y_r^f + \sum_{s_{O(k)} \in S} \sum_{q \in Q^{s_{O(k)}}} c_q^{s_{O(k)}} w_q^{s_{O(k)}} \tag{27}$$

[1]Alternatively, super commodities can be aggregated by destination.

$$\sum_{s_{O(k)} \in S} \sum_{q \in Q^{s_{O(k)}}} \left(\sum_{k \in s_{O(k)}} \Gamma_{ij}^{q} b^{k} \right) w_q^{s_{O(k)}} \leq \sum_{f \in F} \sum_{r \in R^f} \alpha_{ij}^{r} u^f y_r^f \qquad \text{for all } (i,j) \in A \tag{28}$$

$$\sum_{q \in Q^{s_{O(k)}}} w_q^{s_{O(k)}} = 1 \qquad \text{for all } s_{O(k)} \in S \tag{29}$$

$$\sum_{r \in R^f} \beta_i^r y_r^f = 0 \qquad \text{for all } i \in N, \ f \in F \tag{30}$$

$$w_q^{s_{O(k)}} \geq 0 \qquad \text{for all } q \in Q^{s_{O(k)}}, \text{ for all } s_{O(k)} \in S \tag{31}$$

$$y_r^f \geq 0 \text{ and integer} \qquad \text{for all } f \in F, \ r \in R^f \tag{32}$$

Using the tree formulation, we further reduce the number of flow conservation constraints (23) from $|K|$ in the path formulation to $|S|$, where $|S|$ corresponds to the number of super commodities, or equivalently, the number of origins.

Model Comparisons

In general, path or tree formulations (*SNDP_Path* and *SNDP_Tree*) have several advantages over link (or equivalently node-arc) formulations. First, we can easily incorporate non-linear and flow-dependent link or route costs in path or tree models but not in link models. Second, certain classes of constraints that are difficult, if not impossible, to write for link models are easily defined in path or tree formulations. For example, it is hard (although not impossible) to control the number of arcs a design route contains with link formulations, but it is easy with path or tree formulations since only design routes satisfying this condition are included in the model. Third, the number of constraints in path and tree models is greatly reduced compared to the number in link models. For example, the number of flow balance constraints in link formulations is $|N| \times |K|$, in path formulations is $|K|$, and in tree formulations is $|S|$. This is important since LP solution time slows with increases in the number of constraints. The disadvantage, however, is that the decrease in the number of constraints comes at a cost of an exponential increase in the number of decision variables. In the next section, we describe some of the solution methods for handling large numbers of decision variables.

4.4 An Approximate Service Network Design Model

When flow variable costs are negligible compared to fixed design variable costs, Magnanti/Mirchandani (1993) show that under certain conditions, the network design problem can be modeled without considering flows explicitly. This result does not apply to problems containing multiple commodites, however, we can use their modeling concept to construct *approximate SNDP* models. An approximate model allows potentially near-feasible solutions to be determined considering only design set variables and ignoring the huge number of flow variables and the large number of constraints associated with the flow variables. Run time and memory requirements for this approximate model are reduced considerably, and the solution may be used to provide a lower bound on the optimal solution value of the original problem, or as an advanced start solution for the exact models.

In our approximate models, we use the concept of *O/D cutsets.* We let S (T) denote a subset of nodes, some of which correspond to origin (destination) nodes for some commodities and we define the O/D cutset $\{S,T\}$ as a partitioning of the node set N into two nonempty disjoint sets N and $N \backslash S$, with $S \subset N$ and $T = N \backslash S$. An arc (i,j) belongs to cutset $\{S,T\}$ if nodes i and j belong to different sets S and T. We let $Y^f_{S,T}$ equal the total number of type f design variables loaded on the cutset arcs, i.e., $Y^f_{S,T}$ equals $\sum_{(i,j)\in\{S,T\}} y^f_{ij}$ for the SNDP_Node-Arc formulation and equals $\sum_{r\in R^f} \sum_{(i,j)\in\{S,T\}} y^f_r \alpha^r_{ij}$ for both the SNDP_Path and SNDP_Tree formulations. The aggregate demand, $D_{S,T}$, denotes the demand of all commodities with $O(k) \in S$ and $D(k) \in T$ or $O(k) \in T$ and $D(k) \in S$.

Given values for the design variable vectors, y, which specify capacities on the arcs, a feasible flow of b^k units from $O(k)$ to $D(k)$ is possible only if the capacity of every $O(k)/D(k)$ cutset is at least b^k. In general, for any feasible solution to the problem, the aggregate capacity across the cutset must be no less than the demand across the cutset. These *aggregate capacity demand inequalities* (see Magnanti/Mirchandani/Vachani (1995) for details) are expressed as:

$$\sum_{f\in F} \left(u^f Y^f_{S,T}\right) \geq D_{S,T} \qquad \text{for all } O/D \text{ cutsets } \{S,T\}, \qquad (33)$$

Our approximate service network design models include these inequalities and are formulated as:

(ASNDP-Cut_Node-Arc)

$$\min \quad \sum_{f \in F} \sum_{(i,j) \in A} h_{ij}^f y_{ij}^f \tag{34}$$

$$\sum_{f \in F} u^f Y_{S,T}^f \geq D_{S,T} \qquad \text{for all } O/D \text{ cutsets } \{S,T\} \tag{35}$$

$$\sum_{j \in N} y_{ij}^f - \sum_{j \in N} y_{ji}^f = 0 \qquad \text{for all } i \in N,\ f \in F \tag{36}$$

$$y_{ij}^f \geq 0 \text{ and integer} \qquad \text{for all } (i,j) \in A,\ f \in F \tag{37}$$

(ASNDP-Cut_Route)

$$\min \quad \sum_{f \in F} \sum_{r \in R^f} h_r^f y_r^f \tag{38}$$

$$\sum_{f \in F} u^f Y_{S,T}^f \geq D_{S,T} \qquad \text{for all } O/D \text{ cutsets } \{S,T\} \tag{39}$$

$$\sum_{r \in R^f} \beta_i^r y_r^f = 0 \qquad \text{for all } i \in N,\ ,f \in F \tag{40}$$

$$y_r^f \geq 0 \text{ and integer} \qquad \text{for all } f \in F,\ r \in R^f. \tag{41}$$

With only fleet design variables included, the objective functions (34), (38) are to find the cost-minimizing set of service routes satisfying the cutset inequalities (35), (39) and satisfying constraints (36), (37) and (40), (41), as defined for *SNDP_Node-Arc* and *SNDP_Path.*

Inequalities (33) are knapsack inequalities that can be strengthened in several ways. One way is to use a simple integer rounding procedure that produces *Chvátal-Gomory (C-G) cuts.* Alternatively, we can strengthen the inequalities by generating *cutset inequalities*, detailed in Magnanti/Mirchandani/Vachani (1993,1995) and Magnanti/Mirchandani (1993). In some cases, the C-G cuts dominate the cutset inequalities and in other cases the result is reversed.

5 Service Network Design Problem Solutions

Typically, the size of service network design applications is prohibitive, and direct solution is usually unachievable even with state-of-the-art LP/IP solvers (such as, CPLEX (1995), OSL (1992), MINTO (Nemhauser/Salvelsbergh /Sigismondi 1994), etc.). The number of constraints in the path and tree SNDP models is greatly reduced compared to the number in the link model, however, the number of decision variables is increased exponentially. Further, in order to achieve tight LP relaxations, we add valid inequalities and increase the number of constraints exponentially.

In this section, we review *column and cut generation* and *branch-and-bound* methods (see, Bradley/Hax /Magnanti (1977) and Nemhauser/Wolsey 1988) to solve huge linear and integer programs. Our aim is to overview, rather than detail, these solution methods. More detailed descriptions are provided in Bradley/Hax/Magnanti (1977), Nemhauser/Wolsey (1988), Barnhart/Johnson/Nemhauser/Savelsbergh/Vance (1998) and Desrosiers/ Dumas/ Solomon/Soumis (1992).

5.1 LP Solutions

Column Generation

Column generation methods achieve optimal solutions to LP's containing a huge number of decision variables, without explicitly considering all decision variables. The *master problem (MP)* refers to the original problem containing all decision variables and the *restricted master problem (RMP)* is the master problem with many of the variables eliminated. To check if an optimal solution to the RMP is also optimal for the MP, a subproblem, called the *pricing problem*, is solved. The outcome is that either optimality is proved or previously ignored variables (columns) that may improve the solution are identified. If such columns are found, we add the new columns to the RMP and the RMP is re-optimized. The entire process is repeated until no additional columns are identified by the subproblem solver and optimality of the MP is proved. Figure 5 shows the general steps of column generation. In the worst case, all variables may be added and the original MP is solved at the last step, however, optimality is typically proven by considering only a relatively small number of variables.

To start column generation, artificial variables with arbitrarily large costs might need to be introduced to ensure feasiblility of the RMP at each iteration. If the MP is feasible, artificial variables will not be in an optimal solution. If some artificial variables are in the optimal solution, the MP is infeasible.

The difficulty of solving the pricing subproblem determines the practicality of applying column generation approaches, since the subproblem is solved hundreds or thousands of times. *Implicit column generation* refers to solving the

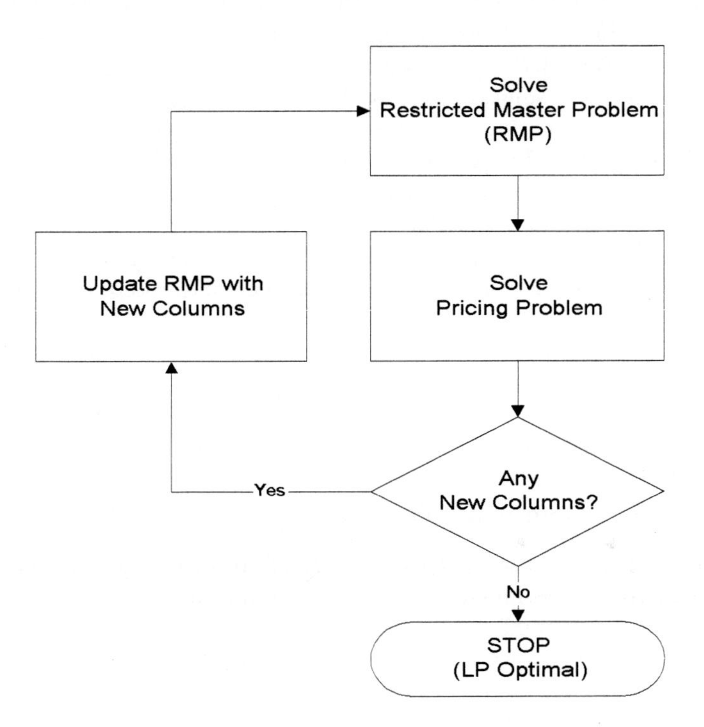

Figure 5: Column Generation Procedure

pricing subproblem *without* evaluating all columns (e.g., by solving a shortest path problem), while *explicit column generation* refers to solving the pricing subproblem by computing the reduced cost of *all* variables.

Cut Generation

Cut generation methods achieve optimal solutions to LP's with huge numbers of constraints or achieve better LP relaxations (i.e., an improved bound on the optimal objective function value) by selectively adding valid inequalities to the LP relaxation. The methods start with a small number of constraints, called the relaxed problem (*RP*), by eliminating a large number of constraints from the master problem *MP*. By ignoring some constraints and/or valid inequalities, the RP optimal solution may be LP infeasible or may be a weak bound on the optimal IP solution. To check the optimality of the current solution, a subproblem, called the *separation problem*, is solved to identify cuts (i.e., violated constraints or valid inequalities) to add to the current RP. If one or more cuts are found, they are added to the RP and the RP is re-optimized. The entire process is repeated until the separation problem finds no cuts and/or a sufficiently tight bound is achieved. Figure 6 depicts the steps of cut generation algorithms.

Explicit cut generation is necessary when the separation problem cannot be solved efficiently. In this case, cuts are identified through explicit evaluation, for any RP optimal solution. *Implicit cut generation* refers to the case when the separation problem solution is achieved without explicit evaluation.

Synchronized Column and Cut Generation

When an LP contains both a huge number of constraints and a huge number of variables, it may be necessary to apply column and cut generation. These methods start with a small number of columns and constraints from the original master problem MP, to form the restriced master problem *RMP*. Applying column generation first,[2] the process repeats until all columns have non-negative reduced costs or some stopping criterion is satisfied. Then, the separation problem is solved repeatedly until no cuts are identified or until some stopping criterion is satisfied. After completing cut generation, we repeat the entire procedure. The algorithm terminates with an LP optimal solution when there are no columns or violated constraints to add.

Figure 7 shows the steps of a synchronized column and cut generation algorithm and Figure 8 shows how our *RMP* changes as the algorithm of Figure 7 progresses. Implementing synchronized column and cut generation procedures is a nontrivial task. The pricing (separation) problem can be much harder after additional cuts (columns) are added, because the new rows (columns) can

[2] Alternatively, cut generation can be performed first.

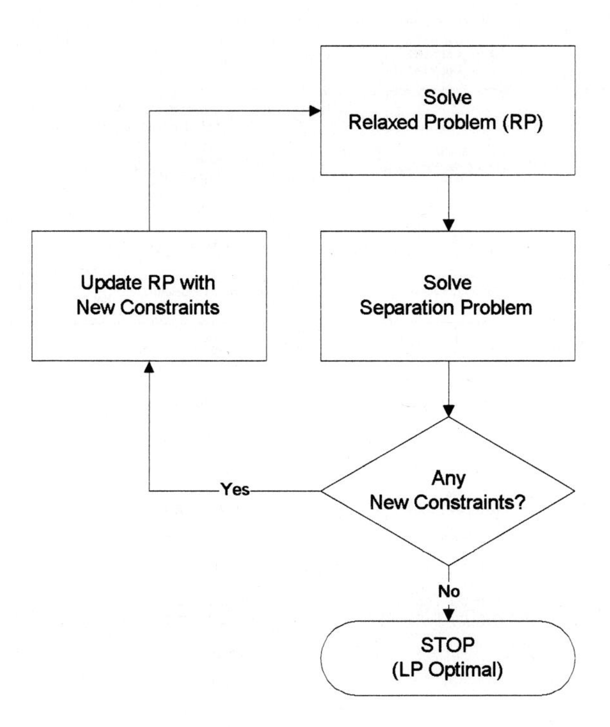

Figure 6: Cut Generation Procedure

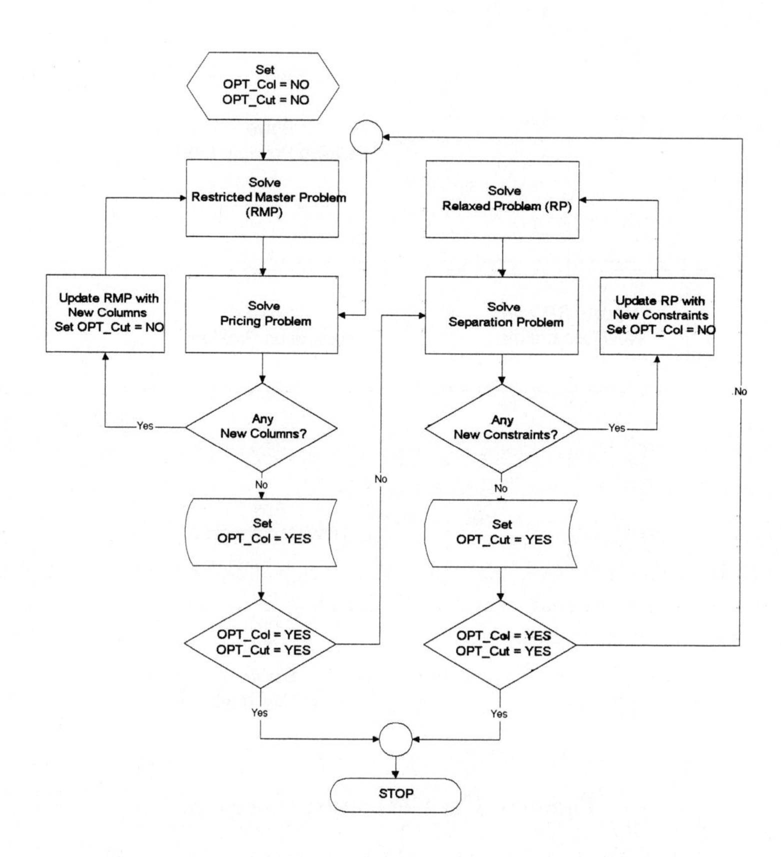

Figure 7: Synchronized Column and Cut Generation

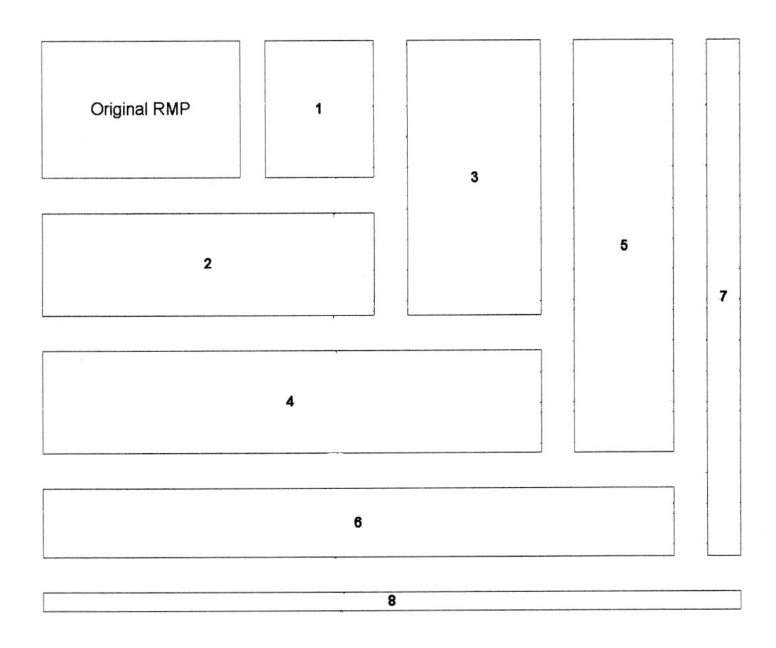

Figure 8: Restricted Master Problem Changes

destroy the structure of the pricing (separation) problem. This *compatibility* issue may be avoided when *explicit* column generation and *explicit* cut generation are used, provided that explicit column and cut generation procedures are computationally inexpensive.

5.2 IP Solutions

Branch-and-bound (B&B), a strategy of *divide and conquer,* obtains integer solutions by solving a series of LP subproblems (Bradley/Hax/Magnanti 1977). When the problem contains a huge number of decision variables, a modified form of B&B, called *branch-and-price* (B&P), should be employed. B&P uses column generation in solving each LP at each node of the B&B tree. Branching occurs when no columns price out to enter the basis and the LP solution does not satisfy the integrality conditions. The major challenge of B&P is to maintain *compatibility* between the rules used for branching and the pricing problem (Barnhart/Johnson/Nemhauser/Savelsbergh/Vance 1998). This means that a branching rule should be enforceable without changing the structure of the pricing problem since LP's at each of the new branches are solved using column generation. For example, when the binary decision variables x are path- based, enforcing $x_j = 0$ requires a special type of network representation in order not to destroy tractability of the subproblem. Moreover, a branching strategy based on variable dichotomy might be *unbalanced* since setting $x_j = 1$ forces $x_k = 0$, for any $k \neq j$; whereas setting $x_j = 0$ has much less impact since there are a huge number of decision variables. Parker/Ryan (1994), Desrochers/Soumis (1989), Desrosiers/Dumas/ Solomon/Soumis (1992), Barnhart/Johnson/Nemhauser/Savelsbergh /Vance (1998) and Vance/Barnhart/ Johnson/ Nemhauser (1994) suggest alternative branching strategies.

When the problem contains a huge number of constraints (or many valid inequalities), another variant of B&B, called *branch-and-cut* (B&C), can be used. B&C uses cut generation in solving each LP at each node of the B&B tree. Branching occurs when no violated inequalities are found or some stopping criteria is satisfied. The difficulties of B&C are similar to those of B&P.

When the problem contains both a huge number of columns and a huge number of valid inequalities, branch-and-price and branch-and-cut can be combined to form a *branch-and-price-and-cut* (B&P&C) solution procedure. There are fundamental challenges in obtaining optimal (or even reasonably good feasible) solutions of huge IP's with B&P&C. In addition to the difficulties arising from B&P and B&C individually, the principal difficulty of B&P&C is *incompatibility* between column generation and cut generation. As a result, some heuristic solution approaches (Barnhart/Johnson/ Nemhauser/Savelsbergh/Vance 1998) solve the root node LP relaxation by synchronized column and cut generation and solve the remaining LP's with B&B using only the columns and constraints generated at the root node. When there are a sufficient number of columns and cuts generated at the root node, the heuristic may be able to generate good feasible solutions. However, finding good, or even feasible solutions, to the problem is not guaranteed.

6 Summary of Contributions

In this paper, we summarize our contributions as:

- We provide a review of network design literature, focusing on specific applications to transportation service network design and related transportation problems such as the vehicle routing problem and the facility location problem.
- We present general service network design models that apply to service network design problems arising at railroads, airlines, trucking firms, intermodal partnerships, transit agencies, etc. All of these problems require the determination of a cost minimizing or profit maximizing set of services and their schedules, given resource constraints. Depending on the application, additional side constraints, including fleet balance constraints, fleet size constraints, capacity constraints, etc,. may be added. We present

three different but equivalent service network design models: node-arc (or link) formulation, path formulation and tree formulation. We use *route based decision variables* in order to capture complex cost structures (i.e., non-linear and flow-dependent link or route costs) and to satisfy complicated operating rules. Further, route based decision variables reduce the number of constraints in the models.

- We present general methods that exploit network structure and achieve dramatic decreases in model size. These reduction techniques may transform the problem from unsolvable to solvable, without losing optimality of the model.

- We detail a general branch-and-bound algorithm for huge service network design problems (containing millions, or even billions of variables). The algorithm synthesizes *column generation* and *row generation* techniques. Column generation overcomes the difficulties of a huge number of decision variables and row generation overcomes the difficulties associated with a weak LP relaxation, namely, the need to generate potentially many violated inequalities. By generating columns and rows on an "as-needed" basis, service network design problems can be solved using only a small portion of the constraint matrix.

References

Ahuja, R.K./Magnanti, T.L./Orlin, J.B. Network Flows - Theory, Algorithms, and Applications, Prentice Hall, 1993

Barnhart, C. /Johnson, E.L. /Nemhauser, G.L. /Savelsbergh, M.W.F. /Vance, P.H., Branch-and-Price: Column Generation for Solving Huge Integer Programs. Operations Research, 46:3, 1998

Barnhart, C./Schneur, R.R., Air Network Design for Express Shipment Service. Operations Research, 44:6, 1996

Bradley, S.P./Hax, A.C./Magnanti, T.L., Applied Mathematical Programming. Addision-Wesley Publishing Company, 1977

CPLEX Optimization, Inc., Using the CPLEX Callable Library, version 4.0. CPLEX Optimization, Inc. Suite 279, 930 Tahoe Blvd., Bldg. 802, Incline Village, NV 89451-9436, 1995

Crainic, T.G./Rousseau, J.M., Multicommodity, Multimode Freight Transportation: A General Modeling and Algorithmic Framework for the Service Network Design Problem. Transportation Research-B, 20B:3, 1986

Desaulniers, G./Desrosiers, J./Ioachim, I./Solomon, M./Soumis, F.,

Unified Framework for Deterministic Time Constrained Vehicle Routing and Crew Scheduling Problems. Working Paper, GERAD, 1994

Desrochers, M./Soumis, F., A Column Generation Approach to the Urban Transit Crew Scheduling Problem. Transportation Science 23, 1989

Desrosiers, J./Dumas, Y./Solomon, M./Soumis, F., Time Constrained Routing and Scheduling. In Handbooks in Operations Research and Management Science, Volume on Networks, North Holland, 1992

Erlenkotter, D., A Dual-Based Procedure for Uncapacitated Facility Location. Operations Research, 26:6, 1978

Farvolden, J.M./Powell, W.B., Subgradient Methods for the Service Network Design Problem. Transportation Science, 28:3, 1994

Guignard, M./Spielberg, K., A Direct Dual Method for the Mixed Plant Location Problem with Some Side Constraints. Mathematical Programming, 17, 1979

Guignard, M., A Lagrangean dual ascent algorithm for simple plant location problems. European Journal of Operations Research, 35, 1988

Hane, C.A. /Barnhart, C. /Johnson, E.L. /Marsten, R.E. /Nemhauser, G.L. /Sigismondi, G., The Fleet Assignment Problem: Solving a Large-Scale Integer Program. Tech. Rep. COC-92-04, School of Industrial and Systems Engineering, Georgia Institute of Technology, June 1994

Jones, K.L./Lustig, I.J./Farvolden, J.M./Powell, W.B., Multicommodity Network Flows: The Impact of Formulation on Decomposition. Mathematical Programming, 62, 1993

Kim, D., Large Scale Transportation Service Network Design: Models, Algorithms and Applications. Ph.D. Thesis, Massachusetts Institute of Technology, 1997

Lamar, B.W./Sheffi, Y./Powell, W.B., A Capacity Improvement Lower Bound for Fixed Charge Network Design Problems. Operations Research, 38:4, 1990

Leung, J.M.Y./Magnanti, T.L./Singhal, V., Routing Point-to-Point Delivery Systems: Formulation and Heurisitcs. Transportation Science, 24, 1990

Magnanti, T.L., Combinatorial Optimization and Vehicle Fleet Planning: Perspectives and Prospects. Networks, 11, 1981

Magnanti, T.L./Wong, R.T., Network Design and Transportation Planning: Models and Algorithms. Transportation Science, 18:1, February 1984

Magnanti, T.L./Mirchandani, P., Shortest Paths, Single Origin- Desigination Network Design, and Associated Polyhedra. Networks, 23, 1993

Magnanti, T.L./Mirchandani, P./Vachani, R., The Convex Hull of Two Core Capacitated Network Design Problems. Mathematical Programming, 60, 1993

Magnanti, T.L./Mirchandani, P./Vachani, R., Modeling and Solving the Two- Facility Capacitated Network Loading Problem. Operations Research, 43:1, 1995

Minoux, M., Network Systhesis and Optimum Network Design Problems: Models, Solution Methods and Applications. Network, 19, 1989

Nemhauser, G.L./Wolsey, L.A., Integer and Combinatorial Optimization. A Wiley-Interscience Publication, John Wiley & Sons, Inc., 1988

Nemhauser, G.L./Salvelsbergh, M.W.P./Sigismondi, G.C., MINTO, a Mixed INTeger Optimizer. Operations Research Letters, 15, 1994

Nemhauser, G.L., The Age of Optimization: Solving Large-Scale Real-World Problems. OR Forum, Operations Research, 42:1, 1994

Newton, H.N., Network Design Under Budget Constraints with Application to the Railroad Blocking Problem. Ph.D. Thesis, Industrial and Systems Engineering, Auburn University, Auburn, Alabama, 1996

OSL, IBM Optimization Subroutine Library Guide and Reference. IBM Systems Journal, SC23-0519, Issue 1, 31, 1992

Parker, M./Ryan, J., A Column Generation Algorithm for Bandwidth Packing. Telecommunications Systems, 1994

Powell, W.B., A Local Improvement Heuristic for the Design of Less-than-Truckload Motor Carrier Networks. Transportation Science, 20:4, 1986

Powell, W.B./Sheffi, Y., Design and Implementation of an Interactive Optimizatin System for Network Design in the Motor Carrier Industry. Operations Research, 37, 1989

Shenoi, R. G., Integrated Airline Schedule Optimization: Models and Solution Methods. Ph.D. Thesis, Massachusetts Institute of Technology, 1996

Van Roy, T.J., A Cross Decomposition Algorithm for Capacitated Facility Location. Operations Research, 34:1, 1986

Vance, P.H./Barnhart, C./Johnson, E.L./Nemhauser, G.L., Solving Binary Cutting Stock Problems by Column Generation and Branch-and-Bound, Computational Optimization and Applications, Computational Optimization and Applications, 3, 111-130, 1994

Ziarati, K./Soumis, F./Desrosiers, J./Gelinas, S./Saintonge, A., Locomotive Assignment with Heterogeneous Consists at CN North America. Working Paper, GERAD and École Polytechnique de Montreal, 1995

Locomotive Assignment Using Train Delays

Koorush Ziarati[1], François Soumis[1]
and Jacques Desrosiers[2]

[1] École Polytechnique and GERAD, 3000 Ch. de la Côte-Ste-Catherine, Montréal, Canada, H3T 2A7
[2] École des Hautes Études Commerciales and GERAD, 3000 Ch. de la Côte-Ste-Catherine, Montréal, Canada, H3T 2A7

Abstract: The locomotive assignment problem is to provide at minimum cost sufficient motive power to pull all the trains of a timetabled schedule. Optimization methods for large scale problems are based on multi-commodity flow models with additional restrictions, solved on a rolling horizon. Branch-and-bound strategies are used to find integer solutions. When there is an insufficient number of locomotives, some companies rent additional units while others prefer to postpone train departures. In this paper, we propose a method that finds a feasible solution by delaying the departure of some trains, according to their types. The numerical results using data from the company CN North America show that with a total of about 38 hours of delay time, all the train requirements could be satisfied for a one-week planning problem involving 1988 train segments. When the train requirements are satisfied, the cost penalties associated with undercovering are significantly reduced, and consequently, the integrality gap is decreased. In our numerical experiments, this gap decreases by 2% on average.

1 Introduction

The locomotive assignment problem consists of providing each train segment of a timetabled schedule with sufficient motive power. Train segments are small portions of a train route that take place between two power change points. The heterogeneous version of the problem which may require several locomotives of various types on each train segment can be formulated as a multi-commodity flow model with additional constraints. Florian/Bushell/Ferland/Guérin/Nastansky (1976) were among the first to make use of such a formulation and to propose a solution method based on a Benders decomposition method (see Benders (1963)). Unfortunately their implementation does not converge rapidly, even for small problems. Recently, an interesting heuristic decomposition scheme has been considered by Ziarati (1997). First of all, the problem is split into strategic and tactical levels. The strategic level determines a power dispatching pattern by day and by locomotive type at power change points while the tactical level involves the specific dispatching of each locomotive over a 24-hour horizon while considering more detailed constraints. In their numerical experiments, Ziarati/Soumis/Desrosiers/Gélinas/Saintonge (1997) split a very large scale strategic problem into smaller overlapping problems solved over a rolling horizon. Each small problem is in turn solved using a Dantzig-Wolfe decomposition method (see Dantzig/Wolfe (1960)), where subproblems are formulated as constrained or unconstrained shortest path problems depending on the locomotive type. They obtain a 6% improvement in terms of the number of locomotives used as compared to a solution provided by the company CN North America.

Simplified versions of the locomotive assignment problem where all locomotives are of the same type can be solved as minimum cost flow problems (see Ahuja/Magnanti/Orlin (1993)). Booler (1980), Wright (1989) and Forbes/Holt/Watts (1991) consider a version involving several types of locomotives but where each train segment must be pulled by a single locomotive. This problem is equivalent to the Multiple Depot Vehicle Scheduling Problem for which efficient heuristics and exact methods exist (see Fischetti/Toth (1989), Ribeiro/Soumis (1994)). Finally, the interested reader is referred to the recent survey of Cordeau/Toth/Vigo (1998) for a comprehensive exposition of recent train routing and scheduling applications.

Locomotives of each type are generally available in limited numbers and when these are not sufficient to supply all the train power demands, some companies rent additional units while others prefer to delay the departure of some train segments. In this paper, we propose an optimization based method to determine the train delays.

The paper is organized as follows. Sect. 1 presents and discusses the previous results obtained by Ziarati/Soumis/Desrosiers/Gélinas/Saintonge (1997) for the strategic module. In Sect. 2, we propose a method to find the train delays according to two train types. We next examine in Sect. 3 a variant of the model which permits us to compute a lower bound on the solution cost: this allows us to evaluate the quality of any feasible solution. Our conclusions complete this paper.

2 The Strategic Module

A consist is the mix of locomotives assigned to a train segment. The locomotive assignment problem with heterogeneous consists is modeled as a multi-commodity flow problem with supplementary variables and constraints in Ziarati/Soumis/Desrosiers/Gélinas/Saintonge (1997). This formulation is defined on an acyclic time-space network (typically a one-week horizon) where each commodity is seen as a locomotive type. The model is solved by using a Dantzig-Wolfe decomposition (or column generation) procedure embedded in a branch-and-bound search tree.

Without going into the mathematical details of the formulation, let us look at the important features of the model. Train coverage requirements are expressed in two ways: minimum number of locomotives, and either horsepower or tonnage. A limited number of each locomotive type is also provided as well as outpost (a branch of the rail network where locomotives are dispatched for local pick-up and delivery duties) and maintenance demands. The above constraints constitute the linking constraints used in the Dantzig-Wolfe decomposition procedure. The subproblem structure depends on the type of locomotives. On the one hand, critical locomotive types for which maintenance is due shortly use a time window constrained shortest path problem involving time variables. On the other hand, for all the remaining locomotive types, the subproblem is a classical shortest path problem.

The objective function minimizes the *fixed* cost of locomotive utilization as well as the *routing* costs. Most of the time, it is not possible to find a feasible consist which exactly matches the requested horsepower or tonnage demand of a given train segment. For example, a 6500 *hp* demand to be fulfilled with locomotives of 3000 *hp* or 4000 *hp* requires at least two locomotives and the best possible value, 7000 *hp*, results in over coverage of 7.7%. This phenomenon leads to relatively large integrality gaps that make the solution of this type of problem more complicated than that for usual vehicle routing problems such as aircraft scheduling for which a single unit is required to cover a flight leg.

The solution to the above model must result in the routing and scheduling of the locomotives so that the company's schedule is satisfied. However to ensure a feasible solution even if there are not enough locomotives, non-negative slack variables are added to the locomotive number and the horsepower or tonnage covering constraints. Therefore a train demand can be partially satisfied, at the price of a certain penalty on the corresponding slack variable. In practical situations, a train segment is considered as covered if at least 90% of its demand (units, horsepower or tonnage) is satisfied. A 10% shortage is usually not enough to prevent a train from running; the trip time becomes longer but the lateness remains acceptable. The penalty for uncovered demand is modeled as a convex piecewise linear function and is described in Ziarati (1997).

Computational Experiments: Table 2.1 presents some results obtained with the strategic module on real-world data provided by the railway company CN North America. The one-week horizon problem considered, from March 29 to April 5 (1994), involves the scheduling of 1,249 locomotives to cover 1,988 train segments, in addition to 164 output horsepower demands and 18 maintenance shop demands. The locomotives are divided into 26 different builder classes and 171 of them are considered critical and require maintenance during the scheduling horizon. Locomotives can be reassigned and rerouted at 26 power change points in the rail network.

The above problem is solved over a rolling horizon. It is divided into six successive two-day problems, with a one-day overlap for problems 2-3 through 6-7. The optimization period is identified by *Time Period* in the table. The number of *Train Segments* for each of these problems is given in the next column. The solution to the linear relaxation of the master problem at the first node of the search tree is denoted Z_{LP} while Z_{IP} is the cost of the best integer solution found. The integrality gap between Z_{LP} and Z_{IP} is given by GAP while the portion of this gap due to the under-covering of certain train segments (*penalty costs*) is denoted GAP^*. Identifiers for under covered trains, that is, with less than 90% of their requested number of locomotives, horsepower or tonnage demand, are provided in the column labeled UCT. BB gives the number of branch-and-bound nodes in a depth first search strategy and COL is the total number of columns generated in $ITER$ master iterations. Finally, the total processing time in minutes, CPU, is provided in the last column. All problems were solved on a HP9000/735 workstation computer using GENCOL 4.1, an optimizer developed at GERAD, an inter-university research center in Montréal. The last line of the table provides average values where appropriate.

Table 2.1: Experiments of the Strategic Module for a one-week horizon problem

Time Period	*Train Segments*	Z_{LP}	Z_{IP}	GAP	GAP^*	UCT	BB	COL	$ITER$	$CPU^{(1)}$
1-2	558	110,949	115,226	3.86%	2.32%	765	54	23,631	319	8.5
2-3	551	91,296	97,647	6.96%	6.44%	216	59	31,001	371	12.9
3-4	589	83,888	90,799	8.24%	7.34%	216	51	33,145	438	20.3
4-5	578	34,934	36,314	3.95%	2.86%	-	45	30,950	349	18.0
5-6	674	40,421	41,886	3.62%	2.85%	-	35	33,482	403	36.6
6-7	563	37,616	39,834	5.90%	4.10%	-	47	32,715	379	20.0
Average	586	66,517	70,284	5.42%	4.53%	-	48	30,821	377	19.4

(1) CPU times in minutes, on a HP9000/735 workstation

As can be seen from Table 2.1, the large cost values for the periods at the beginning of the scheduling horizon, as compared to the periods at the end, is due to the fixed cost associated with the utilization of a locomotive. Similarly, the increase of the integrality gap in the last period is due to penalty costs associated with horsepower target demands at all the 26 power change points, at the end of the scheduling horizon. This is necessary to ensure a good distribution of the motive power at the beginning of the next scheduling period; these penalties are not active in any other period.

A large proportion of GAP is due to penalty costs incurred for the under-covering of some train segments: the ratio GAP^*/GAP accounts for more than 80% of the gap, on average. Uncovered train segments mostly occur, in the solution process, within the branch-and-bound when, for example, a locomotive shared by two trains in a fractional solution is fully assigned to one of these trains. Column *UCT* of Table 2.1 identifies two uncovered trains, numbered 765 and 216 with the latter appearing in two successive periods because of the one-day overlap.

The optimization procedures described in the next sections provide tools to find train delays that reduce these penalty costs incurred for under-covering of certain train segments. In addition, they help to reduce the integrality gap.

3 Feasible Solutions with Train Delays

The optimization method presented in this paper is based on the following observations which are addressed separately in the next two subsections:

- It is possible to postpone the departure of an uncovered train segment until there is a sufficient number of locomotives.
- In the case of an uncovered express train, for which no delays are permitted, one can instead postpone the departure of a train that precedes the departure of this express train.

3.1 Departure Delays for Uncovered Train Segments

The first part of the method consists in modifying the train departure of an uncovered train segment in such a way that the arrival of additional locomotives will provide that train segment with sufficient motive power. This kind of modification is illustrated in Fig. 3.1. Assume that uncovered train segment T1 in the solution of the strategic module should depart at 07:00 from power change point PCP-1 and arrive at 08:15 at PCP-9. The proposed method suggests delaying the departure of train segment T1 until 08:00, i.e., until train segment T2 arrives with additional locomotives.

However delaying the departure of a train segment can create a domino effect as illustrated in Fig. 3.2 which starts from the information of Fig. 3.1. Suppose that train segment T3, departing from PCP-9 at 08:30, requires two locomotives that were already available from the solution of the strategic module. On the other hand, the three-locomotive requirement of train segment T1 is not fulfilled, the strategic module solution providing only two of them. At PCP-1, by delaying the departure of train segment T1 until the arrival of train segment T2, we cause train segment T1 to arrive at 08:50 at PCP-7 and this may create a shortage of locomotives for train segment T3. If this is the case, we have deferred the scheduling difficulty until 08:50. In this manner, it is possible to postpone the situation until it takes place outside the peak hours. Since a train is subdivided into smaller portion, called train segments, if the schedule of a train segment is modified, so must all the

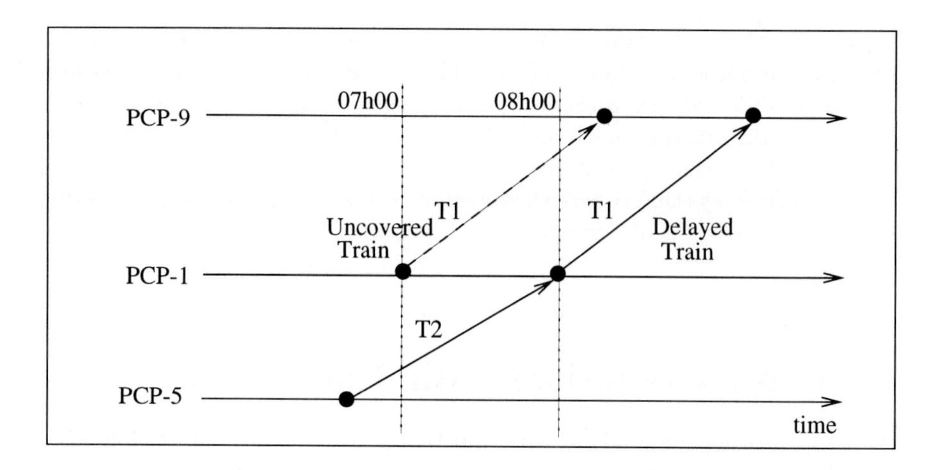

Figure 3.1: Departure delay for an uncovered train segment

segments that follow. Otherwise, locomotives as well as cars might not be in the right place at the right time.

The reader should notice that there is no guarantee that this procedure will produce a feasible solution without uncovered trains. However, we have observed that in practice, it almost always fulfills the train demands, sometimes with a single missing unit. Since we re-optimize the problem with the proposed new schedules, we have a good chance of satisfying all the train demands. The process can be repeated as needed.

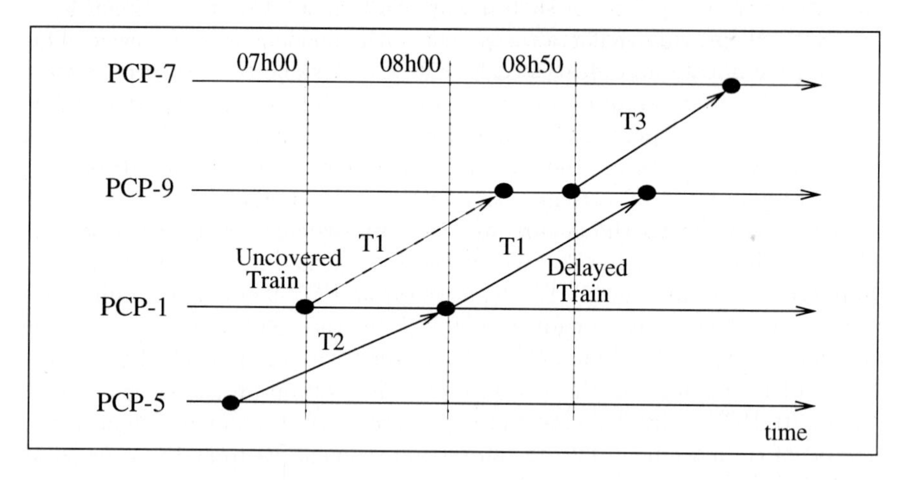

Figure 3.2: Domino effect for train delays

Computational Experiments: Table 3.1 presents some additional details of uncovered trains. The first column, *UCT Segment*, provides uncovered train segment identifiers. The reader can observe that, for both trains 765 and 216, a sequence of two train segments must be rescheduled. Hence delays are applied to a total of four train segments. This is a fairly small number compared with the total number of train segments involved in the strategic module solution process but indeed, without rescheduling, the routing is infeasible and consequently costly for the railway company. The next three columns show the *Departure Time*, the *Departure Date* and the *Trip Time* of the postponed train segment. The proposed *Delay* appears in the last column.

Table 3.1: Details of uncovered train segments

UCT Segment	*Departure Time*	*Departure Date*	*Trip Time*	*Delay*
765-30;1	06:10	March 30	48h40m	15h20m
765-30;2	09:20	April 02	13h10m	12h01m
216-01;1	22:15	April 01	5h35m	5h15m
216-01;2	13:55	April 02	68h30m	5h00m

Table 3.2 presents the solution with the uncovered train segment delays of the strategic module. With the proposed delays, which total less than 38 hours (see Table 3.1), all uncovered train segments have been covered adequately. Compared with the results of Table 2.1, the integrality gaps, given in the *GAP* column, have decreased substantially for the first three periods: this is mainly due to the reduction of GAP^* which is now below 2%. This reduction occurs because the train segment delays ensure that no penalty costs are incurred for uncovered trains. The average gap over the six periods was reduced from 5.42% to 3.65%. Problems with delays differ from the original ones and, for these experiments, *CPU* times (in minutes) refer to optimizations from scratch. They show a slight increase of 13% on average.

Table 3.2: Results of the strategic module, with train delays

Time Period	*Train Segments*	Z_{LP}	Z_{IP}	*GAP*	GAP^*	*BB*	*COL*	*ITER*	$CPU^{(1)}$
1-2	558	103,109	105,749	2.56%	0.50%	52	20,938	340	9.5
2-3	551	49,322	50,557	2.50%	0.82%	45	25,087	320	13.8
3-4	589	43,444	44,680	2.85%	1.80%	45	25,028	346	22.0
4-5	578	35,937	37,644	4.75%	2.80%	44	24,485	344	20.9
5-6	674	39,937	41,451	3.79%	3.33%	41	28,205	394	42.4
6-7	563	37,675	39,738	5.48%	4.12%	44	26,085	379	23.4
Average	586	51,570	53,303	3.65%	2.22%	45	24,971	353	22.0

(1) CPU times in minutes, on a HP9000/735 workstation

3.2 The Case of Express Trains

In the railway industry, express passenger trains exist for which no delays are acceptable. Given the very large penalty cost associated with the under-covering of such trains, it is almost certain that all the corresponding train segments will be covered adequately. However, we must be able to deal with undercover.

Suppose that in Fig. 3.3, train segment T1 is an uncovered express train resulting from the solution of the strategic module. To cover it, we must supply enough motive power at PCP-1 for its departure at 07:00. One way to do this is to postpone the departure of a train that precedes the departure of this express train, at the same power change point. In this way, we can use the motive power that is already available to increase the coverage of the express train. In Fig. 3.3, the departure of train segment T4 can be postponed from 06:00 to 08:00, that is, until the arrival of another train which can supply new motive power. Clearly the preceding train segment cannot itself be an express train. For practical purposes, one must also avoid choosing a train segment that is the first of a long train sequence; otherwise, this choice could create a long sequence of uncovered train segments.

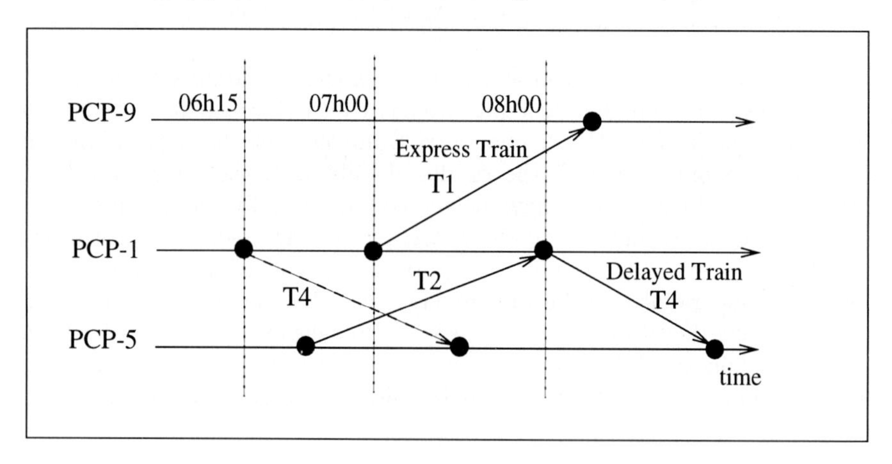

Figure 3.3: Delay example for an uncovered express train

Computational Experiments: Penalty costs for uncovered train segments differ by train type; i.e., the penalty for an express is higher than for a freight train. In the solution of the strategic module presented in Table 3.2, the segments numbered 1 and 2 for train 216-01 have both been delayed, by 5 hours 15 minutes and 5 hours, respectively. Assume now that train 216-01 is an express train and hence the company cannot delay its departure. We have conducted another computational experiment in which the penalty cost per minute has been increased for the corresponding train segments. Details of uncovered train segments are presented in Table 3.3. There we can see that, instead of express train 216-01, non-express train segments 5665L-01;1 and B5604L-02;2 should be delayed by 17 hours 5 minutes and 16 hours 55 , respectively.

Table 3.3: Express train case: details of uncovered train segments

UCT Segment	*Departure Time*	*Departure Date*	*Trip Time*	*Delay*
765-30;1	6:10	March 30	48h40m	15h20m
765-30;2	9:20	April 02	13h10m	12h01m
5665L-01;1	3:30	April 02	24h50m	17h05m
B5604L-02;2	4:20	April 03	29h10m	16h55m

Table 3.4 provides the computational results for the solution of the strategic module corresponding to the results presented in Table 3.2. The average integrality gap is 3.52%, about 2% below the value given without any delays in Table 2.1.

Table 3.4: Express train case: results for the strategic module

Time Period	*Train Segments*	Z_{LP}	Z_{IP}	*GAP*	*GAP**	*BB*	*COL*	*ITER*	$CPU^{(1)}$
1-2	558a	103,003	105,074	2.01%	0.05%	48	20,015	303	11.1
2-3	551	48,700	49,593	1.83%	0.12%	46	23,082	281	14.9
3-4	589	42,876	44,321	3.37%	2.05%	50	23,575	364	27.6
4-5	578	35,592	37,050	4.10%	1.98%	45	23,736	359	26.8
5-6	674	40,278	42,009	4.27%	2.50%	37	26,923	394	51.5
6-7	563	38,105	40,215	5.54%	3.85%	47	24,267	357	27.0
Average	586	51,426	53,044	3.52%	1.76%	46	2,600	343	26.5

(1) CPU times in minutes, on a HP9000/735 workstation

4 Computation of a Lower Bound

The proposed method can find feasible solutions by making schedule changes for express as well as non-express trains, but we do not yet have a measure of the quality of the solution found. The cost function does not include any penalty for delaying the departure of certain train segments. This kind of penalty, called *delay cost*, generally depends on train segment types as well as on trip durations.

By adding the delay cost to the cost of a feasible solution found using train delays, we obtain the total cost of this solution. We now need to evaluate a lower bound on this value so that an integrality gap can be measured.

The linear relaxation of the strategic module, without delays, is not a valid lower bound on a feasible solution with delays. On the one hand, the solution cost includes penalty costs for under-covering of certain train segments, while on the other hand, the problem solved is more constrained since delays are not allowed. In the same way, the linear relaxation of the model with train delays does not provide a valid lower bound on a feasible solution which covers all train segments because this model *a priori* imposes the train delays. Indeed it is a lower bound on any solution with the imposed train delays but there might be better solutions, with different delays.

To determine a valid lower bound, we consider an augmented optimization model in which the objective function comprises three parts:

- *fixed* and *routing* costs of the locomotives;
- *penalty* costs for under-covering of train segments;
- *delay* costs for train schedule changes.

The following paragraph describes an augmented network constructed to compute the lower bound while a numerical example shows the feasibility of the proposed approach.

An Augmented Network: The model described in this paragraph allows a train segment to be covered, at the price of a delay cost, by locomotives that arrive later than its departure time. The time-space network is modified to include, at power change points, *backward* arcs (see Fig. 4.1). For a non-express train segment T1, the head of such an arc is at the departure time of the train segment considered while the tail comes from the arrival time of a later train segment (see Wren 1972). Since express trains cannot be rescheduled, no backward arcs can point to them.

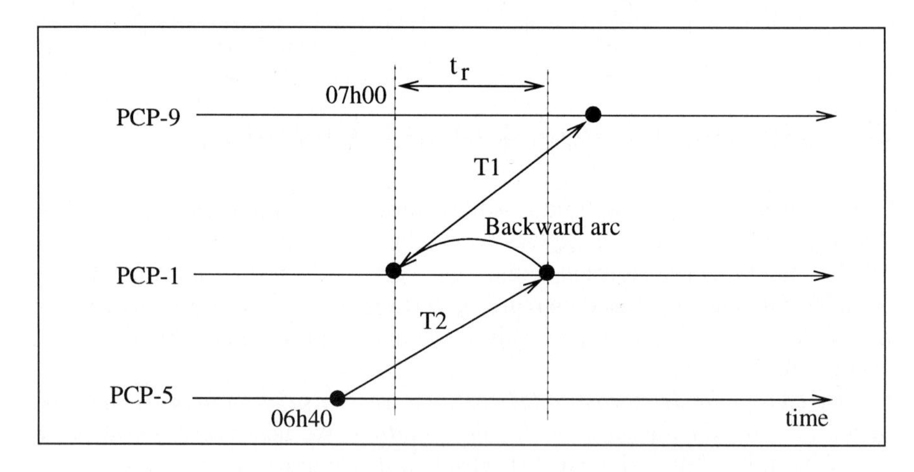

Figure 4.1: A backward arc in the augmented network

For a backward arc r, the following expression is used to compute the delay cost, defined as c_r^k for a k-type locomotive: $c_r^k = c^k t_r$. In this expression, c^k is the unit delay cost per minute for a k-type locomotive and t_r is the difference between the two departure times of the train segments considered.

For a given non-express train, many backward arcs can be created by selecting alternative later train segments. For practical reasons in the solution of the shortest path subproblems, one should be careful not to create cycles in the network.

Computational Experiments: The computation of a lower bound has been done for a single problem. The augmented network presented above has been constructed for the first 2-day period problem (558 train segments) and the results are shown in Table 4.1 which comprises three main result rows. The first row provides results taken from Table 2.1. The first line in the second row comes from Table 3.2; whereas the second line takes the delay costs into account: here 9200 is added to both Z_{LP} and Z_{IP}, computed as follows. The solution with penalties covers only half the demand for train 765 in terms of the number of locomotives – two are required. A delay of 15h20m (920 minutes) at a unit cost of \$10 per minute totals \$9200. The third row gives the lower bound: this is done by solving only the linear relaxation of the model on the augmented network. Any integer solution found on the augmented network is not a feasible one: real locomotives cannot take backward arcs!

Two values in Table 4.1 appear in bold characters: these are comparable. The optimization of the linear relaxation problem on the augmented network constitutes a relaxation of the scheduling problem with delays. Therefore, the value $Z_{LP} = 109,752$ is a lower bound on the feasible solution found with delays, that is, on the value $Z_{IP} = 114,949$. Hence the integrality gap is 4.73%. While this gap may seem large at first sight considering the average gap found for integer solutions *which are not feasible* which is over 5% in Table 2.1, this result is quite promising and shows the viability of the approach. Of course additional computational experiments need to be performed to adjust the relative values of the various costs used.

Table 4.1: Bound comparisons

OPTIMIZATION ...	Z_{LP}	Z_{IP}	GAP	GAP^*	BB	COL	$ITER$	$CPU^{(1)}$
with Penalties	110,949	115,226	3.86%	2.32%	54	23,631	319	8.5
with Delays	103,109	105,749	2.56%	0.50%	52	20,938	340	9.5
(2)	112,309	**114,949**	2.35%	0.46%				
on Augmented Network	**109,752**	–	–	–	–	–	–	7.1

(1) CPU times in minutes, on a HP9000/735 workstation

(2) delay costs: 9200 units added

5 Conclusion

The proposed method for finding feasible solutions to problems with a slight shortage of motive power appears to be satisfactory. In experiments on real data provided by CN North America, we showed that a total delay of about 38 hours was necessary to obtain a solution satisfying all train segment demands for a problem involving close to 2000 train segments. The method can deal with express trains for which no delays are permitted as well as non-express trains, such as freight trains. We have also discussed a tool for computing a valid lower bound on a feasible solution. This requires the construction of an augmented network with backward arcs. The computational results presented in the different sections show that, in the railway context where a small percentage improvement corresponds to millions of dollars a year, this solution approach with train delays is of practical interest.

Acknowledgments. This research was partly supported by the Natural Sciences and Engineering Council of Canada. The authors would like to thank CN North America which provided the data for this research.

References

Ahuja, R.K./Magnanti, T.L./Orlin, J.B. (1993): Network flows: theory, algorithms, and applications. (Prentice Hall) Englewood Cliffs, New Jersey.

Benders, J.F. (1962): Partitioning procedures for solving mixed variables programming problems. in: Numerische Mathematik 4, 238 - 252.

Booler, J.M. (1980): The solution of railway locomotive scheduling problem. in: Journal of Operational Research Society 31, 943 - 948.

Cordeau, J.-F/Toth, P./Vigo, D. (1998): A survey of optimization models for train routing and scheduling. Les Cahiers de GERAD, Ecole des Hautes Etudes Commerciales, Montréal, Canada. In preparation.

Fischetti, M./Toth, P. (1989): An additive bounding procedure for combinatorial optimization problems. in: Operations Research 37, 319 - 328.

Florian, M./Bushell, G./Ferland, J./Guérin, G./Nastansky, L. (1976): The engine scheduling problem in a railway network. in: INFOR 14, 121-138.

Forbes, M.A./Holt, J.N./Watts, A.M. (1991): Exact solution of locomotive scheduling problems. in: Journal of Operational Research Society 42, 825 - 831.

Ribeiro, C./Soumis, F. (1994): A column generation approach to the multiple depot vehicle scheduling problem. in: Operations Research 42, 41 - 52.

Wren, A. (1972): Bus scheduling: an interactive computer method. in: Transportation Planning and Technology 1, 115 - 122.

Wright, M.B. (1989): Applying stochastic algorithms to a locomotive scheduling problem. in: Journal of Operational Research Society 40, 187-192.

Ziarati, K. (1997): Affectation des locomotives aux trains. Ph.D. dissertation, École Polytechnique de Montréal, Montréal, Canada. In French.

Ziarati, K./Soumis, F./Desrosiers, J./Gélinas,S./Saintonge, A. (1997): Locomotive assignment with heterogeneous consist at CN North America. in: European Journal of Operational Research 97, 281 - 292.

Optimal Real-Time Control Strategies for Rail Transit Operations During Disruptions

Susan W. O'Dell[1] and Nigel H.M. Wilson[2]
[1]The SABRE Group, 23 Third Avenue, Burlington, MA 01803 Email: susan_odell@notes.sabre.com
[2]Massachusetts Institute of Technology, Department of Civil and Environmental Engineering, Room 1-240, Cambridge, MA 02139 Email: nhmw@mit.edu

Abstract: Rail transit systems are frequently subject to short term disruptions which temporarily block traffic, leading to increased passenger waiting times and overcrowding of trains. This paper focuses on the development of a real-time decision support system for rail transit operations. We develop a deterministic model of a rail system, and mixed integer programming formulations for several holding and short-turning strategies. We apply the formulations to problem instances based on the MBTA Red Line and demonstrate that passenger waiting time can be significantly reduced (on the order of 15-50%) by applying the resulting controls. We show that the great majority of benefits can be realized by applying only limited control actions on a small set of trains. Our formulations determine optimal control strategies, typically in less than 30 seconds.

1 Introduction

When a rail transit system is operating close to capacity, a relatively minor service disruption can lead to serious degradation in system performance if appropriate control actions are not taken immediately. Examples of disruption sources include a faulty door which won't close, a malfunctioning signal, or a medical emergency on board. It is unlikely that any of these incidents would delay a train at the point of occurrence for more than 10-15 minutes; however, operations on the line may be negatively affected long after the initial source of the disruption has been rectified. The dependence of a train's dwell time on its preceding headway causes the long headways ahead of the blockage to lengthen further, while the headways following the blockage are shortened. The problem is amplified in an uncontrolled, or poorly controlled, system.

It is extremely difficult, if not impossible, for even an experienced controller to assess the situation and quickly determine the best actions from a system-wide perspective. Fortunately, recent advances in automatic vehicle identification, control, and information technologies make it feasible to design computer-based real-time decision support systems for rail transit operations. This paper focuses on the development of such a system for control of a rail transit line during and following minor service disruptions.

Our work is partly motivated by the development of a new Operations Control System now being used by the Massachusetts Bay Transportation Authority (MBTA) to control its rail system. We use the MBTA Red Line as a case study; however, the methods we develop can be applied to any rail line for which train location information is available centrally. Furthermore, our formulations can be easily modified for use in routine operations control.

We study train operations during, and immediately following, a disruption which temporarily blocks traffic in one or both directions. We address incidents of relatively short duration (i.e. 10 to 20 minutes) which occur frequently in many transit systems. We do not consider longer incidents which would necessitate a systematic re-routing of trains according to an alternative operations plan.

We investigate two of the most commonly used control strategies: holding and short-turning. Our primary objective is the minimization of passenger waiting time. Since we are analyzing real-time control strategies, operating costs such as wages and benefits, electric power, and equipment depreciation are virtually fixed.

We develop a deterministic model of a rail line which closely approximates the behavior of a rail transit system. Based on this model, we present mathematical programming formulations which can be used to determine optimal holding and short-turning strategies, and test the formulations on several problem instances using data from the MBTA Red Line.

2 Model Description

In this section, we describe the model in detail, including features, assumptions, data requirements, the "impact set" which bounds the problem, variable and parameter definitions, the objective function, and the dwell time function.

2.1 Features and Assumptions

The model on which we base our formulations incorporates the following features:

- The transit system need not be a simple loop line; it may contain branches.
- The passenger arrival rate and alighting fraction, train running times, and minimum safe headways between trains are all station-specific parameters.

- Dwell time of trains at stations is modeled as a linear function of the number of passengers boarding and alighting, as well as the crowding conditions on the train. The dwell time function can also be station specific.
- The order of trains need not remain constant. Trains may enter or leave service, or may change order as in the case of short-turning strategies.
- The capacity of the train is considered; during peak hours, passengers may be left behind on the platform.

The model is subject to the following assumptions and limitations:

- The number of passengers boarding and alighting, and the load on the trains at the time the disruption occurs are estimated from existing passenger flow data which may vary by time of day.
- Stochastic elements, such as actual running and dwell times, are approximated by their expected values in order to make the problem more tractable.
- The dwell time function, which may be non-linear, has been approximated by a linear function.
- The problem formulations and control strategies consider only a limited "impact set" of trains and stations (see discussion of the impact set in Section 2.3 below). As a result, the effects of the disruption and control strategies at stations beyond the impact set are approximated by a function of the headway distribution at the last station in the impact set.
- Perhaps the greatest limitation of the model is the treatment of the running time and minimum headways between stations. The movement of trains over a rail transit line can be quite complex, depending on the signal system and control lines. In order to make the problem more tractable, we do not allow a train to depart from a station until it can travel to the next station at the maximum permissible (free-running) speed, without stopping in between. Inter-station stopping or deceleration due to the position of the preceding train is prohibited. Since many transit systems, including the MBTA Red Line, do not operate this way, the model uses holding at stations in order to meet this requirement. If the controls output by the model are applied, then the trains will run at maximum permitted speed between stations until they leave the impact set of stations.

2.2 Data Requirements

The following input data is required to find optimal control strategies using the model formulations:

- Track configuration including branches and crossover locations.
- Train schedule information, including when and where trains enter and leave service, routes, and any specific schedule constraints.

- Passenger arrival rate and alighting fraction at each station, by time of day and day of week.
- Minimum safe headway between trains at each station.
- Minimum (free-running) time between stations, including acceleration and deceleration time.
- Dwell-time function parameters, which may be station specific.
- Passenger capacity of train.
- Real-time information on train locations, from which the last station departure time for each train may be calculated. This information can also be used to estimate the current load on each train.
- Location and estimated duration of the blockage, and corresponding restrictions on train movements.
- Short-turning parameters, such as the time to cross a train between the tracks, time to operate the switch, time to notify an operator of a short-turn, and minimum safe headways around the short-turning train.

2.3 Impact Set

The problem formulations and control strategies developed below consider only a limited "impact set" of trains and stations in front of and behind the disruption. The station set is determined by the train set: the first (last) station in the impact set is the station from which the first (last) train departs next, immediately following the disruption. The number of trains in the set is governed by several factors. First, the problem must be solved in real time. As the number of variables and constraints is increased, the problem size and solution time also increase. Second, the number of trains to be considered should be no larger than can feasibly be controlled, given the existing signal and communication systems. Finally, it is unrealistic to project train movements far into the future, due to the stochastic nature of any transit system. Each of these factors presents an upper bound on the size of the impact set. It is intuitively obvious, however, that larger impact sets result (up to some point) in more effective control strategies, and a decrease in passenger waiting time. It is desirable, therefore, to find a set size which balances these conflicting requirements. The appropriate impact set size is explored in Section 4.3.1 below.

2.4 Variable, Parameter, and Set Definitions

The following variables are used in the formulations:

$d_{i,k}$	=	departure time of train i from station k
$l_{i,k}$	=	load on train i departing station k
$p_{i,k}$	=	people left behind on platform by train i at station k
i'	=	predecessor of train i at a particular station (see definition of Set *IatK* below)

i'_b	=	predecessor of train i on branch portion of the line
$h_{i,k}$	=	preceding headway of train i at station k
$\Delta h_{i,k}$	=	absolute value of the difference between headways of train i and its predecessor train
$z_{i,k}$, $z'_{i,k}$	=	variables used in quadratic objective function approximation (see Section 2.5)
$v_{i,k}$	=	1, if train i is loaded to capacity when it departs station k
	=	0, otherwise
$w_{i,k}$	=	1, if train i is held at station k
	=	0, otherwise

The following parameters are used in the formulations:

A_k^o	=	arrival rate at station k, for passengers with a destination on the trunk portion of the line
A_k^b	=	arrival rate at station k, for passengers with a destination on branch b
Q_k	=	passenger alighting fraction at station k
R_k	=	free running time from station $(k-1)$ to station k, including acceleration and deceleration time
H_k	=	minimum safe headway at station k
C_j	=	dwell time function parameter
$Sch_{i,k}$	=	scheduled departure time of train i from station k
M	=	"sufficiently large" coefficient, used in IP constraints
U_j	=	weighting factor in the objective function
$L_{\max}$	=	passenger capacity of train
T_{BL}	=	earliest time at which the blocked or disabled train can leave the next station
k_{if}	=	first station reached by train i after disruption occurs
i_{BL}	=	blocked or disabled train
k_l	=	last station in impact set

The following sets are used in the formulations:

I	is the set of trains in the impact set.
I_b	is the set of trains from branch b in the impact set.
K	is the set of stations in the impact set.
K_o	is the set of stations on the trunk portion of the line which are beyond k_l but which are included in the objective function.
K_b	is the set of stations in branch b which are beyond k_l, but are included in the objective function.
$IatK$	is the train/station impact set which contains one element corresponding to each decision variable $d_{i,k}$. An element (i, i', k) consists of train i, station k, and the predecessor train i' of train i at station k.

2.5 Objective Function

During short-term service disruptions, our chosen objective in determining optimal control strategies is to minimize passenger waiting time. An alternate objective would be to minimize total travel time (waiting plus on-board time); this alternate objective is discussed in Section 4.3.3 below.

The passenger waiting time at stations within the impact set, assuming no passengers are left behind by an overloaded train, is given by:

$$\frac{1}{2} \sum_{IatK} \left(A_k^o (d_{i,k} - d_{i',k})^2 + A_k^b (d_{i,k} - d_{i'_b,k})^2 \right)$$

The first term is the waiting time for passengers whose destinations are on the trunk portion of the line. These passengers may board any train. The second term is the waiting time for passengers whose destinations are on one of the branches. Note that the sub/superscript b in this term corresponds to the destination branch b for train i.

For stations beyond the impact set, we can approximate the passenger waiting time by using the headways at the last station in the impact set. The set of stations included in the objective function is bounded by several considerations. First, if there is "sufficient" layover time at the dispatching terminal, then uneven headways will be corrected there, and stations beyond the terminal should not be included in the objective function. Second, it is unrealistic to include stations far beyond k_l because the headway distribution at k_l will not be a good approximation of the headways at those stations. For stations in the trunk portion of the line, we have:

$$\frac{1}{2} \sum_{K_o} \left(A_k^o \sum_{I} (d_{i,k_l} - d_{i',k_l})^2 + \sum_{b} A_k^b \left(\sum_{I_b} (d_{i,k_l} - d_{i'_b,k_l})^2 \right) \right)$$

and for stations in the branched portion of the line, we have:

$$\frac{1}{2} \sum_{b} \left(\sum_{K_b} A_k \left(\sum_{I_b} (d_{i,k_l} - d_{i'_b,k_l})^2 \right) \right)$$

The waiting time terms can be multiplied by weighting factors U_j based on the distance of the stations from k_l; as the distance from k_l increases, the approximation of the headway distribution becomes less accurate and the weighting factor should decrease.

There is additional waiting time for passengers who are left behind at a station when they are unable to board an overloaded train. This waiting time is equal to the number of passengers left behind by each train, multiplied by the headway of the applicable *following* train.

The waiting time terms, for passengers who are able to board the first train departing for their destination, are quadratic functions. Since the formulations developed below are for large mixed integer programs, and are intended for real-time use, we developed a piecewise linear approximation for these functions. This approximation is of the form

$$h_{i,k}^{2} = (d_{i,k} - d_{i',k})^2 \approx z_{i,k} = \max_{n=1..N} [f_n \times h_{i,k} - g_n] \qquad (2.1)$$

where N is the number of piecewise segments used in approximating the quadratic function.

There is an inherent problem with the piecewise linear approximation. For a given mean headway, the passenger waiting time is minimized when the headway variance is minimized, i.e. all headways are equal, if possible. With the piecewise linear approximation, however, the passenger waiting time is minimized for an infinite number of headway combinations chosen along one piecewise segment. Although the passenger waiting time will be close to optimal, erratic headway patterns may result. To remedy the situation, we added a term to the objective function which penalizes headway variance:

$$PENALTY = U_{pen} \sum_{IatK'} A_k \left| h_{i,k} - h_{i',k} \right| = U_{pen} \sum_{IatK'} A_k \Delta h_{i,k} \qquad (2.2)$$

Even a small weighting factor U will achieve the desired result of eliminating unnecessary headway variance.

The additional waiting time for passengers left behind by an overloaded train is also described by a nonlinear function. Unfortunately, the function is the inseparable product of two variables, $p_{i,k}$ and $h_{i+1,k}$, and is not readily approximated by a linear function. Therefore, in our formulations we simply minimize the number of passengers left behind, multiplied by a weighting factor. This factor depends on whether the passengers are left behind by a train which is in front of the blocked train, where headways are longer than scheduled, or behind the blocked train, where they are shorter. The solutions which result from this simplified objective function should be close to optimal for the minor disruptions we are considering.

After we incorporate the piecewise linear approximation and penalty terms, and eliminate the constant, the objective function is:

$$\min \sum_{IatK} \left(A_k^o z_{i,k} + A_k^b z'_{i,k} \right) + U_o \sum_{K_o} \left(A_k^o \sum_I z_{i,k_l} + \sum_b A_k^b \sum_{I_b} z'_{i,k_l} \right) +$$

$$\sum_b U_b \sum_{K_b} A_k \sum_{I_b} z'_{i,k_l} + U_{pen} \sum_{IatK'} A_k \Delta h_{i,k} +$$

$$U_{ahd} \sum_{\substack{IatK \\ ahd}} p_{i,k} + U_{bhd} \sum_{\substack{IatK \\ bhd}} p_{i,k} \tag{2.3}$$

where

$$z_{i,k} = \max_{n=1..N} \left[f_n \times h_{i,k} - g_n \right]$$

$$z'_{i',k} = \max_{n=1..N} \left[f_n \left(d_{i,k} - d_{i'_b,k} \right) - g_n \right]$$

$$\Delta h_{i,k} = \left| h_{i,k} - h_{i',k} \right|$$

and "*ahd*" and "*bhd*" refer to the trains ahead of, and behind, the blocked train, respectively.

2.6 Dwell Time Function

To apply the model, we need a credible dwell time function. Although the dwell time at each station could perhaps be approximated by its mean during routine operations, the dwell time variance will be larger during the service disruptions for which this model was developed. The true dwell time function may be non-linear; however, we use a linear function in the formulations in order to make the problem more tractable. When the passenger load is well below the train capacity, crowding is not an issue and we use a dwell time function of the form:

$$dwelltime = C_o + C_1 A_k (d_{i,k} - d_{i',k}) + C_2 Q_k l_{i,k-1} + C_1 p_{i',k} \tag{2.4}$$

where C_o is the constant term, and C_1 and C_2 are the marginal boarding and alighting times that apply in uncrowded conditions. The number of passengers boarding the train equals $A_k(d_{i,k} - d_{i',k}) + p_{i',k}$; the number alighting equals $Q_k l_{i,k-1}$. As the passenger loads approach the train capacity, we model the dwell time by a pair of functions. When train i will depart from station k with a load less than its capacity, we use Equation (2.4) for its dwell time. However, if train i will be loaded to capacity when it departs, we use:

$$dwelltime = C_3 + C_4 (L_{\max} - (1 - Q_k) l_{i,k-1}) + C_5 Q_k l_{i,k-1} \tag{2.5}$$

where C_3 is a constant term, and C_4 and C_5 are the marginal boarding and alighting times that apply in crowded conditions. In this case, the number of passengers who board is equal to the number of available spots in the train after the alighting process. We can simplify Equation (2.5), writing it as:

$$dwelltime = C_3 - C_4 l_{i,k-1} + C_5 Q_k l_{i,k-1} \quad (2.6)$$

3 Problem Formulation

We developed mathematical programming formulations for problems in which holding and short-turning strategies are used to mitigate the effects of a short-term blockage.

3.1 Holding Formulations

We developed a series of formulations in which one of several holding strategies is used to mitigate the effects of a short-term blockage. The three holding strategies we considered differ in the extent of required control actions, and potentially in their effectiveness. The first strategy permits holding each train at any of the stations in the impact set ("Hold All"). Although this strategy is the most effective, it is infeasible unless there exists an efficient mechanism for controlling train departure times. An example of such a mechanism would be a holding signal at each station which indicates when a train should depart. The second, and simplest, strategy is to hold each train at the first station it reaches after the disruption occurs ("Hold at First"). The third strategy is to hold each train at only one optimally chosen station in the impact set ("Hold Once"). The three formulations differ in their complexity and in their solution time, as described in Section 4.

We developed formulations for both the Fixed Order and Variable Order problems. In the Fixed Order problem, the order in which trains from separate branches enter a junction is given. Any problem on a loop line is a Fixed Order problem. In the Variable Order problem, the order in which trains from separate branches enter a junction is a decision variable.

We distinguish between formulations in which the train capacity is constrained, and those in which it is not. Formulations without capacity constraints apply when the passenger load on the trains is well below the train capacity. This might be a realistic assumption for off-peak operations. During peak hours, however, capacity constraints are typically required.

3.1.1 Fixed Order Holding Problem With Capacity Constraints

We present below a formulation for the Fixed Order Holding Problem with Capacity constraints (FOHPC). We will assume that the impact set does not include any stations on the trunk for which there is significant boarding of passengers with destinations on a branch, i.e. A_k^b is negligible for trunk stations in the impact set. If we relax this assumption, then we replace the term $A_k(d_{i,k}-d_{i',k})$ in the constraints below with $A_k^o(d_{i,k}- d_{i',k}) + A_k^b(d_{i,k}- d_{i'_b,k})$. To be strictly correct, we should also consider that Q_k will vary between trains, depending on the ratio of $A_k^o(d_{i,k} - d_{i',k})$ to $A_k^b(d_{i,k} - d_{i'_b,k})$. However, this is a second order effect that can typically be neglected. We first develop the formulation for the "Hold All" strategy, and then extend it to the "Hold at First" and "Hold Once" strategies.

$$\min \sum_{IatK} A_k z_{i,k} + U_o \sum_{K_o} \left(A_k^o \sum_{I} z_{i,k_l} + \sum_{b} A_k^b \sum_{I_b} z'_{i,k_l} \right) +$$

$$\sum_{b} U_b \sum_{K_b} A_k \sum_{I_b} z'_{i,k_l} + U_{pen} \sum_{IatK'} \Delta h_{i,k} + U_{ahd} \sum_{\substack{IatK \\ ahd}} p_{i,k} + U_{bhd} \sum_{\substack{IatK \\ bhd}} p_{i,k} \qquad (3.1)$$

subject to:

Piecewise A:	$z_{i,k} \geq f_n \times h_{i,k} - g_n \quad \forall (i,i',k) \in IatK, \ \forall n \mid 1 \leq n \leq N$
Piecewise B:	$z'_{i,k} \geq fn\ (d_{i,k} - d_{i'_b,k}) - g_n \quad \forall (i,i',k) \in IatK, \ \forall n$
Piecewise C:	$h_{i,k} = d_{i,k} - d_{i',k} \quad \forall (i,i',k) \in IatK$
Piecewise D:	$\Delta h_{i,k} \geq h_{i,k} - h_{i',k} \quad \forall (i,i',k) \in IatK'$
Piecewise E:	$\Delta h_{i,k} \geq h_{i',k} - h_{i,k} \quad \forall (i,i',k) \in IatK'$
Load A:	$l_{i,k} \leq A_k\ (d_{i,k} - d_{i',k}) + (1 - Q_k) l_{i,k-1} + p_{i',k} \quad \forall (i,i',k) \in IatK$
Load B:	$l_{i,k} \leq L_{\max} \quad \forall (i,i',k) \in IatK$
Load C:	$l_{i,k} \geq A_k\ (d_{i,k} - d_{i',k}) + (1 - Q_k) l_{i,k-1} + p_{i',k} - M_1 v_{i,k} \quad \forall (i,i',k) \in IatK$
Load D:	$l_{i,k} \geq L_{\max} \times v_{i,k} \quad \forall (i,i',k) \in IatK$
Left Behind:	$p_{i,k} = A_k\ (d_{i,k} - d_{i',k}) + (1 - Q_k) l_{i,k-1} + p_{i',k} - l_{i,k} \quad \forall (i,i',k) \in IatK$
Headway:	$d_{i,k} \geq d_{i',k+1} + H_{k+1} - R_{k+1} \quad \forall (i,i',k) \in IatK \mid k \neq k_l$

RunTime A: $d_{i,k} \geq d_{i,k-1} + R_k + C_o + C_1A_k(d_{i,k} - d_{i',k}) +$
$C_2Q_kl_{i,k-1} + C_1p_{i',k} - M_2v_{i,k} \quad \forall (i,i',k) \in IatK$

RunTime B: $d_{i_{BL},k_{BL}} \geq d_{i,k-1} + R_k + C_3 - C_4l_{i,k-1} + C_5Q_kl_{i,k-1} -$
$M_3(1 - v_{i,k}) \ \forall (i,i',k) \in IatK$

Blockage: $d_{i_{BL},k_{BL}} \geq T_{BL}$

$d_{i,k},\ l_{i,k},\ z_{i,k},\ z'_{i,k},\ h_{i,k}\ ,\ \Delta h_{i,k}\ ,\ p_{i,k} \geq 0 \quad \forall i,k$

$v_{i,k} \in \{0,1\} \quad \forall i,k$

M_1, M_2, and M_3 are "sufficiently large" coefficients which prevent Load C, RunTime A and RunTime B from binding when $v_{i,k}$ equals 0 or 1, as applicable.

The Piecewise constraints are used to approximate the quadratic objective function. Note that the absolute value and maximum functions in Equations (2.1) and (2.2) have been replaced by linear inequalities.

The four Load constraints calculate the number of passengers on board train i as it departs station k, as a function of the load on the train as it enters the station and the number of passengers boarding and alighting. The load is equal to the *minimum* of the load capacity and the number of passengers desiring to board. When the train is full, $v_{i,k} = 1$, and the Load B and Load D constraints are binding. When the train is not yet full, $v_{i,k} = 0$, and the Load A and Load C constraints are binding. Load A and Load C include variable $p_{i',k}$, the number of passengers who were unable to board the previous train.

Variable $p_{i,k}$ is calculated in the Left Behind constraints as the sum of the passengers who arrived at the station since the last train departed, the load on train i when it arrived, and the number left behind by the previous train, minus the fraction who alighted at station k and the load on train i as it departs.

The Headway constraints ensure that the minimum safe headway between trains is maintained. These constraints prevent train i from departing station k until it can run to the next station ($k + 1$) without being blocked by the preceding train i'. Thus, inter-station stopping is prohibited.

The two RunTime constraints indicate that the earliest time of departure for train i from station k is equal to the departure time from the previous station, plus the running time to station i, plus the dwell time of train i at station k. If train i is not full as it departs station k, then $v_{i,k} = 0$, and RunTime A applies. When the train is full, however, $v_{i,k} = 1$, and constraint RunTime B is the binding constraint. RunTime B reflects the increased dwell time due to congestion effects, as was discussed in Section 2.6.

Finally, the Blockage constraint(s) restrict the movement of the train(s) directly affected by the disruption, based on the estimated duration of the blockage. Blockage constraints may be written for multiple trains at multiple stations.

The above formulation presents the most basic version of the problem. For transit system geometries which include branches, special constraints must be written to reflect the discontinuities in station numbering which occur at junctions.

There may be special headway constraints at intermediate terminals, or station-specific dwell-time functions in the RunTime constraints. There may also be schedule constraints which prevent trains from departing ahead of schedule, of the form $d_{i,k} \geq Sch_{i,k}$.

As formulated, FOHPC is a mixed integer program. Note that when the passenger load on the trains is well below the train capacity (i.e. off-peak operations), then the above formulation is greatly simplified. The binary variable $v_{i,k}$ is eliminated, along with constraints Load B, Load C, Load D, and RunTime B. Load A is replaced by an equality, and the last term is eliminated from RunTime A. The formulation for the Fixed Order Holding Problem *without* capacity constraints (FOHP) is a linear program, and may be solved very quickly using a commercial software package such as CPLEX, as described in Section 4.2 below.

"Hold at First" As discussed above, it may not be feasible or desirable to hold a train at multiple stations. Therefore, we consider a control strategy in which each train can be held at only the first station it reaches after the disruption occurs. At all other stations, the train must depart as soon as it is able, i.e. either one of the RunTime constraints or the Headway constraint must be binding. In keeping with the assumptions of our model, trains are held for sufficient duration at the first station to prevent inter-station stopping or deceleration due to blocking at any point within the impact set. Thus, a RunTime constraint must be the binding constraint. The "Hold at First" strategy will not be as effective as the "Hold All" strategy; we will compare their effectiveness for several problem instances in Section 4.2.2.

For the "Hold at First" strategy in FOHPC, the formulation is the same as for the "Hold All" strategy with the exception of the RunTime constraints. We must use inequalities to ensure that *either* RunTime A or RunTime B is satisfied with equality when $k \neq k_{if}$:

RunTimeG A: $d_{i,k} \geq d_{i,k-1} + R_k + C_o + C_1 A_k(d_{i,k} - d_{i',k}) + C_2 Q_k l_{i,k-1} + C_1 p_{i',k} - M_2 v_{i,k} \quad \forall\ (i, i', k) \in IatK$

RunTimeG B: $d_{i,k} \geq d_{i,k-1} + R_k + C_3 - C_4 l_{i,k-1} + C_5 Q_k l_{i,k-1} - M_3(1 - v_{i,k}) \quad \forall (i,i',k) \in IatK$

RunTimeEq A: $d_{i,k} \leq d_{i,k-1} + R_k + C_o + C_1 A_k(d_{i,k} - d_{i',k}) + C_2 Q_k l_{i,k-1} + C_1 p_{i',k} + M_3 v_{i,k} \quad \forall (i,i',k) \in IatK \mid k \neq k_{if}$

RunTimeEq B: $d_{i,k} \leq d_{i,k-1} + R_k + C_3 - C_4 l_{i,k-1} + C_5 Q_k l_{i,k-1} + M_2(1 - v_{i,k}) \forall (i, i', k) \in IatK \mid k \neq k_{if}$

The "Hold at First" formulation, like the "Hold All" formulation, can be written as a linear program for the problem without capacity constraints.

"Hold Once" The "Hold Once" strategy entails holding each train at only one station, similar to the "Hold at First" strategy above. However, the holding station for this strategy need not be the first station, but instead can be any station in the impact set. The formulation for this strategy is the same as for "Hold All" or "Hold First", with the exception of the RunTime constraints and the addition of binary variable $w_{i,k}$, which equals 1 if train i is held at station k.

For the "Hold Once" strategy in FOHPC, the RunTime constraints are:

RunTimeG A: $d_{i,k} \geq d_{i,k-1} + R_k + C_o + C_1 A_k(d_{i,k} - d_{i',k}) + C_2 Q_k l_{i,k-1} + C_1 p_{i',k} - M_2 v_{i,k} \; \forall (i,i',k) \in IatK$

RunTimeG B: $d_{i,k} \geq d_{i,k-1} + R_k + C_3 - C_4 l_{i,k-1} + C_5 Q_k l_{i,k-1} - M_3(1-v_{i,k}) \quad \forall (i,i',k) \in IatK$

RunTimeEq A: $d_{i,k} \leq d_{i,k-1} + R_k + C_o + C_1 A_k(d_{i,k} - d_{i',k}) + C_2 Q_k l_{i,k-1} + C_1 p_{i',k} + M_3 v_{i,k} + M_4 w_{i,k} \quad \forall (i,i',k) \in IatK$

RunTimeEq B: $d_{i,k} \leq d_{i,k-1} + R_k + C_3 - C_4 l_{i,k-1} + C_5 Q_k l_{i,k-1} + M_2(1-v_{i,k}) + M_4 w_{i,k} \quad \forall (i,i',k) \in IatK$

$$\Sigma_k w_{i,k} \leq 1 \quad \forall i$$

$$w_{i,k} = 0 \quad \forall i,k$$

M_4 is a sufficiently large coefficient to prevent the RunTimeEq constraints from binding when $w_{i,k} = 1$. Thus, either RunTime A or RunTime B must be satisfied with equality at all stations except the one holding station for each train. Note that because of binary variable $w_{i,k}$, the "Hold Once" formulation is a mixed integer program, even for the problem without capacity constraints.

3.1.2 Variable Order Holding Problem

The Variable Order Holding Problem (VOHP) applies only when a blockage occurs on one of the branches of the line. In VOHP, we allow trains to deviate from their scheduled order as they merge at a junction, i.e. we allow trains from the unblocked branch to "skip ahead" of one or more trains from the blocked branch. The impact set must, of course, include the section of the line where the branches enter the trunk.

We introduce the following binary variable for the Variable Order Holding Problem:

$y_{j,i}$ = if train j precedes train i through the junction and at station s on the trunk portion of the line
= 0, otherwise

The formulation of VOHP is very similar to that for FOHPC. However, for stations on the trunk portion of the line, each constraint for train i in FOHPC is replaced by a set of constraints - one for each possible predecessor train j. The term $M \times (y_{j,i}-1)$ is added (subtracted) to each constraint so that it is only binding when train j precedes train i. Note that we can decrease the size of the problem by limiting the number of trains which may skip ahead of, or slip behind, train i, thereby limiting the number of possible predecessor trains j.

Constraints are added to ensure that each train has only one predecessor and one successor on the trunk portion of the line.

All versions of the Variable Order Holding Problem are formulated as mixed integer programs because of the binary variable $y_{j,i}$.

3.2 Short-Turning Formulations

We developed mathematical programming formulations for the short-turning problem, in which short-turning is used in conjunction with holding to mitigate the effects of a minor disruption. To motivate these formulations, we first consider a simple loop line. Assume that we have a central business district (CBD) near the middle of the loop, and that peak traffic flow is in and out of the CBD in the morning and evening, respectively. When a one-track blockage occurs, there are three possibilities for the short-turning of trains: (1) one or more trains may be short-turned from behind the blockage, (2) one or more trains may be short-turned from the opposite track into the gap in front of the blockage, or (3) trains are short-turned both from behind the blockage and into the gap in front of it. If a train which has not yet reached the CBD is short-turned during the morning peak hours, then it is likely that many people will be forced to alight, which is undesirable. If a train which has not yet reached the CBD is short-turned in the evening peak hours, then few people will benefit. Thus, the decision whether to short-turn from behind the blockage, or into the gap in front of the blockage, should be made so that the short-turned train(s) are already past the CBD.

The discussion above suggests several strategies for which we could develop formulations, including strategies in which none, one, or multiple trains are short-turned from behind or in front of the blockage. The final train order (after the short-turn) may be predetermined, or may be determined as part of the optimization.

3.2.1 Short-Turning Problem with Predetermined Train Order

The formulation for the Short-Turning Problem with Predetermined Train Order (STPP), using the "Hold All" strategy, is quite similar to the "Hold All" formulation for FOHPC given in Section 3.1.1, with the following exceptions:

- The train/station impact set *IatK* and the constraints reflect the revised ordering of trains after the short-turn.
- Constraints are added which incorporate the time required to throw the switch at the cross-over, before and after the short-turn.
- The runtime constraint for the short-turning train, i_{st}, is replaced by two constraints. One states that train i_{st} must arrive at the short-turning station k_{st} and unload all of its passengers before it can depart. The other states that train i_{st} must first be notified that it will be short-turned and must then unload all of its passengers, including any who boarded at k_{st} prior to the notification.
- The headway constraint for the short-turning train is replaced by a constraint which prevents i_{st} from departing station k_{st} until it can travel over the cross-over and into position at the opposite platform without being blocked by its predecessor after the short-turn.
- Minor modifications are required to handle the passengers left behind by train i_{st} when it short-turns, and to initialize the passenger load on train i_{st} to 0 as it completes the short-turn.

3.2.2 Short-Turning Problem with Undetermined Train Order

Consider a situation in which we might short-turn a train from behind the blockage. In the Short-Turning Problem with Undetermined Train Order (STPU), the decision of whether, and in front of which train, we should short-turn is determined during the optimization. A similar problem occurs when we consider short-turning a train into the gap ahead of the blockage. For this problem, we must decide which (if any) train to short-turn.

For STPU, elements (i, i', k) of *IatK* are generated assuming that no short-turning takes place; *IatK* is the same set as for FOHPC. We introduce the binary variable y_i, which equals 1 if we short-turn i_{st} in front of i. We also introduce the set S, which includes all trains in front of which we can short-turn. Then, the constraint $\sum_S y_i \leq 1$ requires that train i_{st} can be short-turned in front of no more than one train in S. The train order for the short-turning train and the trains in S, at stations beyond the short-turning station, are determined as part of the optimization. For these trains and stations, the constraints in FOHPC are replaced by sets of constraints; terms My_i and $M(1-y_i)$ are used to control which constraint in each set is binding.

The formulation for STPU is rather long and complicated. More importantly, the number of constraints, problem complexity, and solution time increase. Note that we could simply compare the solutions of several versions of STPP, in lieu of solving STPU.

4 Model Application

We applied our formulations to several problems using data from the MBTA Red Line during morning peak-hour operations. In this section, we discuss the problem setting, solution methods and results, and some implementation issues.

4.1 Problem Setting

We describe the MBTA Red Line, the methods we used to generate the input data, and some complications we encountered when comparing the control strategies.

4.1.1 The MBTA Red Line

The MBTA Red Line is a heavy rail system with a branching structure. Schedules allow for layovers of approximately five to ten minutes at the terminals under routine operations. There are five, six and twelve stations on the two branches and the trunk, respectively. With the exception of the terminals, the northbound and southbound platforms are considered to be separate stations for modeling purposes; thus, we have a total of 43 stations in the Red Line model.

Each of the problem instances we tested is based on a disruption occurring at 8:15 AM, when passenger volumes are high and operating headways are short. Under these conditions, a relatively minor disruption can have serious consequences if appropriate control actions are not taken. At 8:15 AM, there are 27 trains operating on the Red Line. Each six car train has a capacity of approximately 1200 people. The trains are dispatched from the two branches at headways of approximately 8 minutes and 6 minutes, resulting in a headway of approximately 3.4 minutes on the trunk. Since we tested the model formulations off-line, we did not use "real-time" information on train locations, but instead located the trains based on the dispatching schedule and nominal running times.

In addition to the constraints given in the formulations in Section 3, a number of additional system-specific constraints were required to model the Red Line, such as special constraints at the junction and at the terminals.

4.1.2 Input Data

The data requirements for the model were described in Section 2.2. In order to apply the formulations, we needed actual data or good estimates for each of the items.

To estimate the arrival rates and alighting fractions at each station, we used an extensive data set collected by Massachusetts' Central Transportation Planning Staff (CTPS). The data set consists of detailed counts of passengers arriving and alighting at each station, for 15 minute intervals throughout the day. The boarding rates and alighting fractions were used in conjunction with the scheduled headways to estimate the current load on each train at the time of the disruption.

A small set of dwell time measurements was provided by the MBTA. This was used, along with the CTPS data, to estimate a crude dwell time function in which boarding and alighting rates are constrained to be equal:

Uncrowded Conditions: $dwelltime = 0.44 + 0.00150 * num$
Crowded Conditions: $dwelltime = 0.44 + 0.00195 * num$

where *num* equals the total number of passengers boarding and alighting. This function applies at all typical stations, but was modified for one station at which doors open on both sides of the train.

After we obtained detailed control line information for the Red Line, a simulation model was used to determine minimum safe headways and free running times. The short-turning parameters we used are estimates based on discussions with MBTA personnel.

4.1.3 Comparison of Control Strategies

We solved each of four problem instances using the "Hold All", "Hold Once", and "Hold at First" strategies of the Fixed Order Holding Problem with Capacity constraints (FOHPC), and as a Short Turning Problem with Predetermined train order (STPP). We also solved the problems for a "Do Nothing" strategy in which no active controls were applied.

Note that to compare the effectiveness of various control strategies for a particular problem instance, we must take care to compare the waiting times for the same set of passengers, i.e. the passengers arriving at each station k during the same fixed time interval. The objective function includes the waiting time for passengers arriving at each station k between the last train departure prior to the disruption, and the departure time of the last train in the impact set. For the problems we considered, the departure times of the last train in the impact set were different for Hold Once and Hold at First than they were for the other strategies, so the objective functions cannot be directly compared (see Section 4.2.2).

4.2 Solution of FOHPC and STPP

We used the modeling software package GAMS (V. 2.25) (Brooke/Kendrick/ Meeraus (1992)) to enter the model formulations and data sets, and the MIP solver in CPLEX (V. 3.0) (CPLEX Optimization, Inc. (1994)) to find solutions using branch-and-bound, for FOHPC and STPP. The problems were solved on a Sun

SPARC 20 workstation. The solution process was terminated when the objective function was guaranteed to be within 0.1% of the optimal value.

We tested several different piecewise linear approximations for the objective function, varying N from 4 to 7. Since there was no significant penalty in running time, we selected the more accurate 7 segment approximation, which overestimates the true waiting time function by no more than 4% for headways between 1.5 and 21 minutes. At headways of 21-27 minutes, however, the approximation underestimates the true function by 0-11.5%.

4.2.1 Incidents

We considered four problem instances, based on delays of 10 and 20 minutes at each of two incident locations.

Incident 1 delays a train which is heading away from the central business district on the trunk portion of the line. The impact set of 13 trains includes eight trains ahead of the blocked train, and four trains behind. For the Short-Turning Problem, a train from behind the blockage is short-turned; the location of the blockage and the cross-over require that it skip nine stations in the impact set.

Incident 2 causes a delay on one of the two branches, for a train that has just left the dispatching terminal. The impact set includes 11 trains; seven are ahead of the blocked train, and three are behind. The merge at the junction is included in the impact set. For the Short-Turning Problem, a train is short-turned into the gap in front of the blockage, skipping only a few stations.

For stations beyond the impact set, weighting factors of 0.5 and 0.25 are used for stations close to, and far from, the impact set, respectively. We use $U_{pen} = 0.1$ to weight the penalty for headway variation, and $U_{ahd} = 10$ and $U_{bhd} = 5$ for the people left behind on the platforms by overcrowded trains.

4.2.2 Results

Tables 4.1-4.4 summarize the results for the four problem instances. "Total Waiting Time" is the passenger waiting time which corresponds to the departure times output by the model; it is calculated using the quadratic objective function. Times are given in passenger-minutes. "Time Ahead" is the waiting time for the trains up to and including the blocked train, and "Time Behind" is the waiting time for the trains behind the blocked train. For Hold Once and Hold at First, the sets of passengers boarding the trains behind are not the same as for the other strategies, therefore the "Total Waiting Time" and "Time Behind" values can not be compared and have been omitted from the tables. The "Savings" entries give the passenger-minutes saved, and the percentage savings, over the Do Nothing strategy. "Maximum Load" is the maximum number of passengers wishing to board a train in the impact set; any number higher than 1200 indicates that a train is unable to clear a platform, and passengers are left behind. The "Problem Size" is measured by the number of $d_{i,k}$ variables in the problem (train/station pairs in the impact set). "CPU Time" is the solution time in seconds, using the software and hardware described above.

As indicated in the tables, the active control strategies we considered result in significant savings of passenger waiting time, for each of the four problem instances. For three of the four instances, the savings are realized almost entirely by passengers who board either the blocked train or one of the trains ahead of it. There are virtually no savings for passengers boarding trains behind, for either the Hold All or the short-turning strategy. Since Hold Once and Hold at First can be no more effective than Hold All, there can be no savings behind for these strategies either. There are small savings for the trains behind in Incident 1 with a ten minute delay.

Table 4.1 Incident 1, ten minute delay

		FOHPC			STPP
	Do Nothing	Hold All	Hold Once	Hold at First	Hold All
Total Waiting Time	15993	13627			14609
Time Ahead	11202	8863	8931	8961	9997
Savings (min.)		2338	2270	2240	1204
Savings (percent)		15%	14%	14%	8%
Time Behind	4791	4763			4753
Savings (min.)		28			39
Savings (percent)		0%			0%
Maximum Load	988	603	614	666	603
Problem Size ($d_{i,k}$)		95	95	95	88
CPU Time (seconds)		22	37	21	16

Table 4.2 Incident 1, twenty minute delay

		FOHPC			STPP
	Do Nothing	Hold All	Hold Once	Hold at First	Hold All
Total Waiting Time	46087	24767			21523
Time Ahead	36868	16934	17306	17385	16836
Savings (min.)		19934	19563	19483	20032
Savings (percent)		43%	42%	42%	43%
Time Behind	9218	7833			6842
Savings (min.)		1386			2377
Savings (percent)		3%			5%
Maximum Load	1646	666	759	805	651
Problem Size ($d_{i,k}$)		95	95	95	88
CPU Time (seconds)		25	82	27	17

Table 4.3 Incident 2, ten minute delay

		FOHPC			STPP
	Do Nothing	Hold All	Hold Once	Hold at First	Hold All
Total Waiting Time	38088	28421			28419
Time Ahead	32495	23101	24465	25327	23016
Savings (min.)		9394	8031	7186	9479
Savings (percent)		25%	21%	19%	25%
Time Behind	5593	5320			5404
Savings (min.)		273			189
Savings (percent)		<1%			<1%
Maximum Load	1336	1137	964	985	776
Problem Size ($d_{i,k}$)		69	69	69	78
CPU Time (seconds)		17	274	23	12

Table 4.4 Incident 2, twenty minute delay

		FOHPC			STPP
	Do Nothing	Hold All	Hold Once	Hold at First	Hold All
Total Waiting Time	94977	55102			44208
Time Ahead	88204	48978	52620	55487	38244
Savings (min.)		39226	35584	32717	49960
Savings (percent)		41%	37%	34%	52%
Time Behind	6773	6124			5964
Savings (min.)		649			809
Savings (percent)		<1%			<1%
Maximum Load	1653	1422	1343	1307	1200
Problem Size ($d_{i,k}$)		69	69	69	78
CPU Time (seconds)		25	2458	763	62

As shown in the tables, the short-turning strategy is more effective for the twenty minute delays than for the 10 minute delays. A 10 minute delay is not much longer than the time required to short-turn a train (approximately seven minutes, for the MBTA Red Line). As a result, with only a 10 minute delay, trains are blocked while they wait for the short-turning train to get into position ahead of them. The short-turning strategy is more effective for Incident 2 than for Incident 1. In Incident 1, the number of stations (and passengers) outside the short-turning loop is large, and the number of passengers benefiting from the short-turn is small. In contrast, only a few low-demand stations are skipped by the short-turning train in Incident 2, and many passengers benefit.

For Incident 1, the passenger waiting time for trains ahead is nearly equivalent for the three holding strategies. For Incident 2, the Hold Once and Hold at First are somewhat less effective than Hold All. The impact set for Incident 2, unlike that for Incident 1, includes trains which merge from the two branches. The Hold All strategy can be more effective than the other holding strategies because the headways for trains can be continuously adjusted both before and after the merge. In contrast, if trains can be held only once, it is not always possible to have even headways both before and after the merge. We re-ran the Incident 2 problem instances with a reduced impact set including only the trains ahead of the blockage. In this case, the trains from the unblocked branch which have not entered the junction prior to the disruption are not held until they pass the junction, i.e. the "first" station in Hold at First is the first station on the trunk. For the ten minute delay, the passenger waiting time savings for the trains ahead (with the reduced impact set) are then 22% for Hold at First and Hold Once, as compared to 25% for Hold All. For the twenty minute delay, the savings for Hold Once and Hold at First are 40% - nearly equivalent to the Hold All savings.

The Hold Once and Hold at First strategies require far less holding - only once per train - than the Hold All strategy. Since there is little or no benefit to holding trains behind the blockage, the impact set should be reduced to the blocked train and the trains ahead, in which case the majority of benefits can be realized with a minimum of controls for the Hold at First and Hold Once strategies.

For Incident 1 with a twenty minute delay, and Incident 2 with a 10 minute delay, the trains are always able to clear the platforms of passengers when any of the active control strategies is used. In contrast, there are passengers left behind at several stations for the Do Nothing strategy. For Incident 2 with a 20 minute delay, the short-turning strategy is the only one for which the trains could always clear the platforms. However, the maximum number of people left behind with any of the holding strategies is less than half that for Do Nothing.

The active control strategies result in far more even headway distributions than the Do Nothing strategy, particularly at the stations just beyond the blockage.

With the implementation used here, the execution times for several cases (in particular, for the Hold Once strategy) are too long to be used in a real-time decision support system. Execution time issues are discussed in Section 4.3.2.

4.3 Implementation Issues

4.3.1 Impact Set Size

As discussed in Section 2.3, several factors can be used to define an upper bound on the impact set size, including the necessity of a real-time solution, the infeasibility of controlling a large set of trains, and the stochastic nature of the system. However, larger impact sets should result in more effective control strategies. Therefore, we investigated the effect of varying impact sizes on the passenger waiting time savings for Incident 1, using the "Hold All" strategy of the Fixed Order Holding Problem. Since we demonstrated in Section 4.2.2 that there is little or no benefit to holding trains behind the blockage, we considered holding only the trains ahead.

We varied the number of trains held ahead of the blockage from 0 to 8. Under each scenario, we calculated the passenger waiting time for these same eight trains, plus the blocked train, at stations beyond the blockage, for both ten and twenty minute delays. The results showed that the majority of benefits is realized if only a few trains are held. The waiting time savings for two trains and for four trains are approximately 75% and 90% of the savings for eight trains, respectively. The results suggest that the marginal return for holding more than eight trains is small.

4.3.2 Execution Time

Tables 4.1-4.4 in Section 4.2.2 indicate the execution times for each problem instance using GAMS and CPLEX on a Sun SPARC 20 workstation. When we first solved the problems, we directly implemented the formulations given in Section 3, and used the default settings in CPLEX. Unfortunately, the execution times were often much longer than those shown in the tables. To speed up the execution, we made two modifications as described below.

First, we set the CPLEX MIP branch strategy parameter to -1. This parameter controls which branch (up or down) is taken first at each node in the branch-and-bound. By setting the parameter to -1, the down branch is taken first. There are only two sets of integer variables in FOHPC and STPP: $v_{i,k}$, which is used in each of the strategies, and $w_{i,k}$, which is used only in the Hold Once strategy. Recall that $v_{i,k} = 0$ unless train i is loaded to capacity as it departs station k, and $w_{i,k} = 0$ unless station k is the holding station for train i. Thus, these two binary variables are equal to zero far more often than they are equal to one, and we typically reach optimality more quickly by branching on zero first.

Second, we used a simple front-end algorithm to fix many of the $v_{i,k}$ variables before performing the optimization. The idea behind the algorithm is that for a given delay, we can identify a number of train/station pairs such that train i will not be loaded to capacity as it departs station k, no matter which control strategy - including Do Nothing - is applied. We iteratively calculate an upper bound for each $l_{i,k}$ as:

$$l_{i,k}^{\max} = A_k\,(h_{i,k}^{\max}) + (1 - Q_k)\,l_{i,k-1}^{\max} + p_{i',k} \qquad (4.1)$$

where $h_{i,k}^{\max}$, the maximum possible headway for train *i,* is equal to the (estimated) length of the delay at the blockage plus the scheduled operating headway. If the upper bound $l_{i,k}^{\max}$ is less than the train capacity L_{max}, we set $v_{i,k} = 0$ before optimizing. Thus, we effectively reduce the number of binary variables, to which the MIP execution time is extremely sensitive.

The execution times given in Tables 4.1-4.4 in Section 4.2.2 reflect the two modifications described above. Although 10 of the 16 cases given in Tables 4.1-4.4 ran in less than 30 seconds, several ran too slowly to be used in a real-time decision support system. The Hold Once formulation, which has the most binary variables, ran particularly slowly. However, the problems we ran are very large; the impact sets of 11 - 13 trains result in 69 - 95 $d_{i,k}$ variables. Since we demonstrated above that there is little benefit to holding trains behind the blockage, we reduced the impact sets to include only trains ahead of the blockage, and re-ran the six cases for which the execution time exceeded 30 seconds. The execution times were reduced to less than 30 seconds in all but one case (Hold Once for Incident 2 with a 20 minute delay) which ran in 53 seconds. For this most difficult case, we were able to reduce the execution time to 34 seconds by using a refined version of the front-end algorithm to fix several additional $v_{i,k}$ variables. (In the refined version of the algorithm, we consider that the sum of the headways for train i and its predecessor train i' at station k can be no greater than $h_{i,k}^{\max}$ plus the scheduled operating headway.)

4.3.3 On-Board Time

The objective in the formulations is to minimize the passenger waiting time. An alternate objective would be to minimize the total passenger travel time, including the on-board time. We chose to minimize only waiting time for several reasons. First, the effect of holding and short-turning on on-board time is relatively small, when compared to the effect on passenger waiting time. With the exception of the time spent by passengers who are on-board trains being held, the effects of the control strategies on on-board time are generally second order. Second, on-board time is generally assumed to be less onerous than waiting time. Finally, the on-board time terms must be calculated as the (typically) non-linear product of load ($l_{i,k}$) and time, which greatly reduces the tractability of the problem.

Fortunately, if we neglect second order effects, we can include the on-board time in our objective function when the Hold at First strategy is used. The load on each train at the time of the disruption is a constant, given as input data to the problem. We multiply this constant by $(1 - Q_k)$ and the holding time at the first station to calculate the on-board time as a linear function of the holding times of each train.

For the Hold at First formulation of FOHPC, we solved each of the four problem instances with two different objective functions: (1) the minimization of passenger waiting time (PWT), and (2) the minimization of total passenger travel time (PTT), including the on-board time. In each case, we calculated the resulting passenger waiting, on-board, and total travel time. Since waiting time is considered more onerous than on-board time, the latter was weighted by a (widely-used) factor of 0.4 in the objective function. Our results are summarized in Table 4.5; the total travel time is calculated as the weighted sum of the waiting and on-board time. Note that the impact set was reduced to include only trains ahead of the blockage.

When the on-board time was included in the objective function, the total weighted passenger travel time was reduced by less than 5%. Although the on-board time was reduced by 40 - 80%, the larger passenger waiting time was increased by 1 - 9%.

Table 4.5 Passenger travel time vs. waiting time

Incident	Delay	Objective Function	Passenger Time		
			Waiting	On-Board	Total
1	10 Min.	PWT	8961	1543	9578
		PTT	9074	271	9182
1	20 Min.	PWTT	17385	2372	18334
		PTT	17659	806	17982
2	10 Min.	PWTT	23411	8666	26877
		PTT	23702	5920	26070
2	20 Min.	PWT	50018	17617	57065
		PTT	51201	10488	55396

5 Conclusions

In this paper, we presented mathematical programming formulations for the Fixed Order Holding Problem with Capacity constraints, using three different holding strategies: Hold All, Hold at First, and Hold Once. We discussed how the problem is simplified when the capacity constraints are removed, and described how to modify the formulations for the Variable Order Holding Problem, the Short-Turning Problem with Predetermined train order, and the Short-Turning Problem with Undetermined train order. We tested the formulations on four problem instances based on the MBTA Red Line. For each problem, we determined the passenger waiting time using five different control strategies: the Hold All, Hold Once, and Hold at First version of OHPC; the Hold All version of STPP; and a Do Nothing strategy.

The results of our tests are very encouraging; the active control strategies result in significant passenger waiting time savings. For the two incidents we considered, we were able to reduce waiting time by 15-25% for a 10 minute delay, and by more than 40% for a 20 minute delay. The vast majority of the time was saved by passengers boarding either the blocked train, or one of the trains ahead of it. Therefore, we recommend that the impact set include only trains ahead of the blockage.

For trains ahead of a blockage, the Hold Once and Hold at First strategies are virtually as effective as the Hold All strategy, and require significantly fewer control actions. Among holding strategies, we recommend that the simplest Hold at First strategy be used. Short-turning should be considered if (1) the length of the blockage is significantly longer than the time required to execute the short-turn, and (2) the number of stations outside the short-turning loop is small, so that the number of passengers who will benefit from the short-turn is large in comparison to the number who will be inconvenienced.

We experimented with various impact set sizes, and found that the majority of the waiting time savings can be realized if only a small set of trains is held. For the incidents we studied, there is little marginal benefit to holding more than eight trains; in fact, the waiting time savings for two trains is approximately 75% of the savings for eight trains.

We investigated an alternate objective function in which on-board time is minimized along with waiting time. Such an objective function is non-linear for all problems unless the Hold at First strategy is used. For the four problems we tested, the total weighted passenger travel time was reduced by less than 5% when the on-board time was included in the objective function, thus there is little penalty for omitting it. However, since the Hold at First strategy is the recommended strategy, the alternate objective function can be easily used.

When we solved a direct implementation of the four test problems using the default settings in CPLEX, several of the cases ran too slowly to be considered real-time. However, by reducing the impact sets to a reasonable size, and using a simple front-end algorithm to fix some of the binary variables in advance, we were able to reduce the execution time of each problem to approximately 30 seconds or less. It appears that the mathematical programming formulations we developed may be solved quickly enough to be used as part of a real-time decision support system for rail transit systems. We believe this paper represents a significant step towards the development of such a system.

References

Brooke, A./Kendrick, D./Meeraus, A. (1992): Gams: A User's Guide, Release 2.25. (Scientific Press). South San Francisco, California.

CPLEX Optimization, Inc. (1994): Using the CPLEX Callable Library, Version 3.0.

Modeling Real-Time Control Strategies In Public Transit Operations

Xu Jun Eberlein[1], Nigel H. M. Wilson[2], and David Bernstein[3]
[1] Caliper Corporation, 1172 Beacon Street, Newton, MA 02162
[2] Department of Civil and Environment Engineering, MIT, Cambridge, MA 02139
[3] Department of Civil Engineering and Operations Research, Princeton University, Princeton, NJ 08554

Abstract: This paper presents recent research to model and solve real-time transit operations control problems. Types of control strategies studied include deadheading, expressing, and holding, both single and in combination. We formulated mathematical models for these control problems assuming real-time vehicle location information available. Important properties of the models are proved and illustrated. Efficient algorithms are developed for solving the control models. The advantages, disadvantages, and conditions for each type of control strategy are investigated. We show that while station-skipping strategies such as deadheading and expressing work in a totally different manner from holding, a coordinated combination of them results in the most effective and least disruptive control policies. Using AVI data collected for the MBTA Green Line (Boston, MA), the computational results show the control models are generally quite effective. The algorithms are also tested in situations where there are significant stochastic disturbances. Under stochastic conditions model performance is found to decline only moderately.

1 Introduction

Public transit operations are subject to various influences and conditions that often result in variable headways, long waiting times and generally poor performance. To improve service, transit operating agencies use a variety of real-time control strategies. In the most general sense, these control strategies can be divided into three categories: station control, interstation control, and others. The first category contains two main classes of strategies: holding and station skipping. The second includes speed control, traffic signal pre-emption, etc. The third includes strategies such as adding vehicles, splitting trains, etc.

In the US transit industry, real-time control decisions are often made by field inspectors at various points along the routes, mostly based on information on vehicles that have passed the point. The effectiveness of such manual control decisions vary by the personal experience and judgment of field inspectors. In fact, due to the complexity of transit operations, it is very difficult to make systematic and effective control decisions in such a manual control process. On the other hand, there has been surprisingly little research on the theory and implementation of real-time control strategies.

Motivated by this, and also by the greatly increased availability of automated information systems such as Automatic Vehicle Location (AVL) and Automatic Vehicle Identification (AVI) systems, we recently conducted research to develop theoretical models and solution algorithms for real-time transit operations control problems. Considering that strategies in the first control category are by far the most common, our research focuses solely on station control strategies, namely holding, deadheading and expressing. A common characteristic of the three strategies is that they do not typically involve changes in the order of vehicles. Another station-skipping strategy, short-turning, which often results in changes in the vehicle sequence, is not addressed in this research.

This paper focuses on combinations of the three station control strategies, as compared to using each of them alone. It is organized as follows. In Section 2, we define the transit system in question and present two important properties. Sections 3 and 4 present mathematical formulations and properties for the holding, deadheading and expressing problems. In Section 5 we combine the three strategies with a heuristic algorithm, and report computational results and findings. In Section 6 we test the algorithm under various stochastic conditions. Section 7 concludes the paper.

2 System Definition and Properties

The transit network considered is a one-way loop as shown in Fig. 2.1, and we assume passengers arrive at a station k randomly at a rate r_k. The vehicles are operated in a first-in first-out order with an evenly scheduled headway. The following equations define arrival time $a_{i,k}$, dwell time $s_{i,k}$, departure time $d_{i,k}$ and departure load $L_{i,k}$ of vehicle i at station k respectively without control:

$$a_{i,k} = \max(d_{i,k-1} + R_k + 2\delta,\ d_{i-1,k} + h_0) \tag{2.1}$$

$$s_{i,k} = c_0 + c_1 r_k h_{i,k} + c_2 q_k L_{i,k-1} \tag{2.2}$$

$$d_{i,k} = a_{i,k} + s_{i,k} \tag{2.3}$$

$$L_{i,k} = L_{i,k-1} + r_k h_{i,k} - q_k L_{i,k-1} = r_k h_{i,k} + (1-q_k) L_{i,k-1} \quad (2.4)$$

where $R_k+2\delta$ is the expected free vehicle running time between stations k-1 and k (including the extra time for acceleration and deceleration, 2δ), h_0 is the minimal safe headway, r_k is passenger arrival rate at station k, $h_{i,k}$ is the headway between vehicle i-1 and i at station k, and q_k is passenger alighting rate at station k. c_0, c_1 and c_2 are estimated parameters for the dwell time function (2.2).

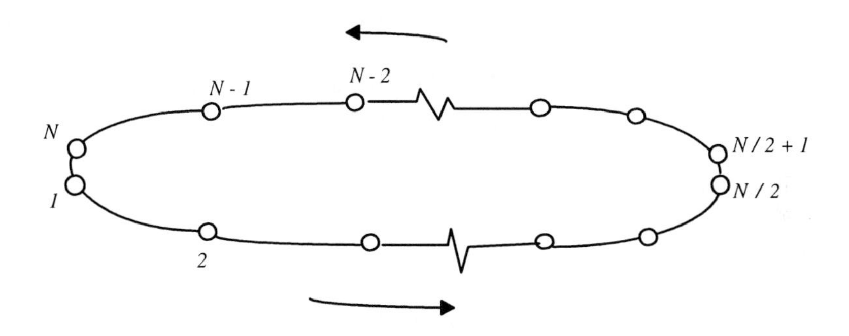

Fig. 2.1 The Network

Given the real-time, dynamic nature of transit operations control, we adopt a *rolling horizon* formulation of the control problems. This means that each time we solve a deterministic optimization problem considering only m consecutive vehicle trips i, i+1,..., i+m-1, with vehicle trip i+m as the "boundary vehicle trip" which is not controlled. The resulting optimal control policy is then applied to the first vehicle, i , only. Such a set of m +1 vehicles is called the *impact set* and is denoted by $\mathbf{I}_m$, where m is called the *rolling horizon size*. Such a horizon is rolled forward and the control process is repeated for each vehicle trip i during the entire study period.

We now present two important properties of the defined transit system.

Dispatching headway effects. If a vehicle i's dispatching headway is smaller than the minimal headway of its preceding vehicle i-1, i's headway will be monotonically decreasing along the route until it is blocked or reaches the terminal. If i's dispatching headway is larger than the maximal headway of i-1, its headway will be monotonically increasing along the route. As a consequence of the dispatching headway effects, when i's dispatching headway is in the midrange of the maximum and minimal headway of i-1's, i's headway variation along the route will be small. The significance of this property is its independence from the demand pattern and hence it is a general result. It provides a convenient and accurate way to forecast vehicle headway patterns down the line regardless of demand pattern, and hence helps to develop good control decisions early on.

Trajectory change diminishing effect. In a transit system with $c_1 r_k < 0.5$, where c_1 is the average time each passenger spends to board a vehicle and r_k is the expected passenger arrival rate at station k, the impact of a change in trajectory of vehicle i on the following vehicles diminishes quickly. In particular, after a small number of vehicle trips, m, the impact of i's trajectory change becomes insignificant. The value of m depends on the value of $c_1 r_k$, which can be easily estimated. This property is most useful for determining the length of rolling horizon, which is important for modeling dynamic systems such as the one studied in this paper. For the MBTA Green Line case presented later, the appropriate value for m was estimated to be 3.

For detailed proofs the reader is referred to Eberlein (1995).

3 Formulation and Properties of the Holding Problem

Even with a transit route scheduled at even headways, the headways will inevitably become uneven due to fluctuation in demand and randomness in the service, not to mention incidents and disturbances. In this case, *holding* can be used to shorten long headways of following vehicles and reduce those passengers' waiting time.

More precisely, the *holding problem* can be defined as follows: when a vehicle is ready to depart from a station after its normal loading and unloading process, it may be held for an amount of time in order to even out the headways between its preceding and following vehicles. The holding problem is to decide at a given time at a control station, which vehicle is to be held and for how long, such that the total passenger waiting time is minimized.

We formulate the holding problem as follows:

[HP] Minimize

$$f_{i,k}(\mathbf{d}) = \sum_{j \in \mathbf{I}_m} \sum_{k'=k}^{k_t} r_{k'}(d_{j,k'} - d_{j-1,k'})^2 \tag{3.1}$$

Subject to:

$$d_{j,k} - a_{j,k} - s_{j,k} \geq 0, \forall j \in \mathrm{I}_m, \; d_{j,k'} - a_{j,k'} - s_{j,k'} = 0, \forall j,k': k'>k \tag{3.2}$$

$$d_{i+m,k} - a_{i+m,k} - s_{i+m,k} = 0 \tag{3.3}$$

$$a_{j,k'} - d_{j-1,k'} \geq h_0, \forall j,k',k'>k \tag{3.4}$$

$$d_{j,k_c} \leq \max(t_{j,c}, d^0_{j,k_c}), \forall j \in \mathbf{I}_m \tag{3.5}$$

where i is the vehicle considered for control, k is the holding station, and h_0 is the required minimal headway between vehicles. $d_{i-1,k'}$ $\forall k' \geq k$ are departure times of vehicle i-1 at each station k and are known at the dispatching point of vehicle i. $t_{j,c}$ is the scheduled departure time at the next dispatching terminal k_c minus minimal layover time.

The objective of control is to minimize total waiting time of passengers randomly arriving at stations k' during headway $h_{i,k'}$, as stated in (3.1), and the decision variables are the vehicle departure times $d_{j,k}$. Constraint (3.2) says that each vehicle j can leave the holding station k either when finished normal unloading and loading or later, while at other stations it leaves right after the last passenger boards. Constraint (3.3) is a boundary condition, which says the last vehicle, $i+m$, in the impact set is not controlled at k (though its departure time, $d_{i+m,k'}$, may change due to control on preceding vehicles). Inequality (3.4) is the safe headway constraint. (3.5) is the terminal schedule constraint incorporated at the next dispatching station k_c, which eliminates any further delay in scheduled dispatching time for the next trip. In other words, if a vehicle is already going to be late for its next trip, it will not be held. This constraint may be of practical importance to most transit agencies in the U.S., because vehicle operators do not like to have their relief times delayed. On the other hand, depending on how tight the schedule (or layover time) is, constraint (3.5) can greatly decrease the effectiveness of holding, because once a vehicle is late, vehicles behind it are also more likely to be running late and hence won't be held.

Program [HP] is a nonlinear quadratic program. It is non-convex in general due to the vehicle arrival time equation (2.1). The implication of this nonlinearity in reality is that vehicles may pair or bunch together, which causes interstation stopping of the blocked vehicles. We developed an efficient heuristic algorithm to solve this problem, see Eberlein (1995) for more detail. When holding is used alone, our analysis and computational tests shows the following properties:

- A holding solution depends mainly on the vehicle headway pattern at the holding station, and is largely independent of passenger demand pattern along the transit line.
- As a consequence of the above property, real-time information on vehicle headways may be sufficient for determining holding policies, and real-time passenger demand information may not be necessary.

- The impact of holding a vehicle on the following vehicles diminishes quickly. This indicates the rolling horizon size can be very small, which permits fast implementation of the holding algorithm in real-time.
- The most effective holding station is the dispatching terminal at each end of a transit line. A second holding station may not be effective because the headway pattern down the line largely depends on the dispatching headway pattern, and early control benefits more stations.
- Holding is very effective in saving passenger waiting time. In the computational tests using realistic data from the MBTA, the average daily reduction of passenger waiting time is 31% without the terminal schedule constraint (3.5), and is 20% with the constraint.

The most interesting property of holding is that it is insensitive to demand patterns. Unlike the station skipping control policies which are very demand sensitive, optimal holding policies just reconcile dispatching headways of vehicles considered with the midrange headway of the preceding vehicle.

4 Formulations and Properties of the Deadheading and Expressing Problems

Deadheading and *Expressing* are two station skipping control strategies with both similarities and differences. When a vehicle is deadheaded, it runs empty from the dispatching terminal through a number of stations, in order to save time and thus reduce preceding headways at later stations. The *real-time deadheading problem* is to decide, at any given time, which vehicles should be deadheaded and how many stations should be skipped by each deadhead vehicle. Expressing is similar to deadheading in the sense that an express vehicle also skips stations. The difference is that expressing can start at any station, not just the terminal, and the vehicle generally does not run empty. In other words, passengers may get on an express vehicle but not a deadhead vehicle. The *real-time expressing problem* is to decide, in addition to which vehicles should be expressed and how many stations should be skipped, at which station expressing should start.

We formulate the real-time deadheading problem as follows.

[RTDP] Minimize

$$f(\mathbf{h}) = \sum_{i \in \mathbf{I}_m} [\sum_{k \in \mathbf{K}_c} (r_k h_{i,k}^2/2 + P_{i-1,k} h_{i,k}) + u_c \sum_{i \in \bar{\mathbf{K}}_c} r_k h_{i,k_t}^2/2] \quad , 0 \le u_c \le 1 \tag{4.1}$$

Subject to:

$$a_{i,k}(\mathbf{y}) - d_{i-1,k}(\mathbf{y}) \geq h_0 \quad \forall(i,k) \tag{4.2}$$

$$y_{i,k} - y_{i,k+1} \leq 0 \quad \forall(i,k) \tag{4.3}$$

$$\sum_{k \in K_c} y_{i,k} < N/2 \quad \forall_i \in I_m \tag{4.4}$$

$$y_{j+m,k} = 1 \ \forall k ; \tag{4.5}$$

where j is the first vehicle in $\boldsymbol{I}_m$; $P_{j-1,k}$, $d_{j-1,k} \forall k$, d_{i,k_0-1} $\forall i$ and k_0 are given. The arrival time, dwell time, and departure load of each vehicle is defined as in equations (2.1)-(2.4).

The objective function (4.1) is again to minimize passengers waiting time. However, we can no longer use the departure times as decision variables, due to the fact that deadhead vehicles must skip stations. Instead, the decision variables in RTDP are now binary variables, denoted as $y_{i,k}$, defined as follows:

$$y_{i,k} = \begin{cases} 1 & \text{if vehicle } i \text{ stops at station } k \\ 0 & \text{otherwise} \end{cases} , \quad y_{i,k} \in \{0,1\} \ \forall(i,k) \tag{4.6}$$

Inequality (4.2) is the deadhead feasibility constraint. (4.3) ensures that the skipped stations are consecutive . Since a deadheaded vehicle always starts to skip stations from k_0 , the skipped stations always have smaller index values than the other stations. Also, since the control variable has a smaller value (i.e., $y_{i,k}= 0$) at a skipped station than at a non-control station (i.e., $y_{i,k} =1$), if a non-control station k is followed by a skipped station k+1, we would have $(y_{i,k} = 1) > (y_{i,k+1} = 0)$. Thus (4.3) holds only if any skipped stations are consecutive. (4.4) says that the number of skipped stations can be at most $N/2$-1, which implies that a deadhead segment can not change direction. (4.5) imposes the boundary and initial conditions that (i) vehicle $i+m$ is not controlled; and (ii) the trajectory of vehicle i-1, the leftover passengers from vehicle i-1, and the departure times of all vehicles at station k_0-1 are all given. Note that by our station index convention k_0-1 is station N when k_0=1.

The real-time expressing problem is formulated as follows.

[RTEP] Minimize (4.1)

Subject to:

$$a_{i,k}(\mathbf{y}) - d_{i-1,k}(\mathbf{y}) \geq h_0 , \ \forall(i,k) \in \mathbf{I}_m \times \mathbf{K}_c \tag{4.7}$$

$$\sum_{k=k_0+1}^{k_t} (y_{i,k} - y_{i,k}y_{i,k-1}) \leq 1, \ \forall i \tag{4.8}$$

$$y_{i,k_0} = y_{i,k_t} = 1, \forall i \in \mathbf{I}_m \tag{4.9}$$

$$y_{j+m,k} = 1, \forall k \in \mathbf{K}_c \tag{4.10}$$

$$y_{i,k} \in \{0,1\}, \forall (i,k) \in \mathbf{I}_m \times \mathbf{K}_c \tag{4.11}$$

where j is the first vehicle in $\mathbf{I}_m$; $y_{i,k}$ is defined by equation (4.6); and, $P_{j\text{-}1,k}$, $d_{j\text{-}1,k} \forall k \in \mathbf{K}_c$, $\forall i \in \mathbf{I}_m$, as well as k_0, are given. The arrival time, dwell time, and departure load of each vehicle is defined as in equations (2.1)-(2.4).

While this program looks a lot like [RTDP], there are a couple of major differences between them. In [RTDP], constraint (4.3) prevents any intermediate station from being an initiation station of a deadhead segment; this constraint is not present in [RTEP]. Instead, constraint (4.8) requires that there be at most one express segment for each vehicle in a control direction, and constraint (4.9) rules out the possibility that either terminal station in the control direction will be skipped.

Both RTDP and RTEP are nonlinear integer programs. We developed efficient algorithms for solving them and analyzed their properties (see Eberlein/ Wilson/Bernstein/Barnhart (1998) and Eberlein (1995)). Our findings include:

- The station skipping strategies are much more sensitive than holding to passenger demand pattern across stations.
- When each is used alone, deadheading and expressing result in quite similar passenger waiting time savings. In our computational tests using realistic data, the average daily reduction in passenger waiting time using either strategy is about 14%.
- The station skipping strategies result in significant total passenger riding time savings, with expressing outperforming deadheading in this respect.
- While deadheading is not suitable for a route starting with high demand stations, expressing is particular effective in reducing both total passenger waiting and riding times when the optimal initial station is a high demand station.
- In terms of waiting time reduction alone, deadheading is often more effective in a high headway variation situation because a deadhead vehicle is subject to control earlier and more time is saved at the dispatching terminal compared to expressing. Deadheading may also be less frustrating since no passengers are forced to alight.

5 Combining Control Strategies

While deadheading and expressing have some similarity, they work in a totally different way from holding. Our research shows that holding is more smooth and often more systematically effective than deadheading or expressing (Eberlein, 1995). Furthermore, holding is easier to implement and has much less negative effects. This suggests that holding should be used more often than the other two strategies. What if we combine the strategies? How much better can we do than using each of them alone? One observation we had is that, when holding is used alone, some vehicles were held much longer than others. Intuitively, in such a situation deadheading and expressing may play a role to supplement holding. For example, if vehicle i has a much shorter preceding headway than its following vehicle $i+1$'s, we can either hold i for a time t, or deadhead/express $i+1$ to skip a number of stations n, or hold i for a time $t' < t$ and also deadhead/express $i+1$ to skip a number of stations $n' < n$. The last strategy may do better than holding alone because the holding time of i is reduced and deadheading/expressing $i+1$ is also beneficial. This motivates the study of the *combined control problem.*

Although a much more complicated problem than a single type of control, the combined control problem can be simplified by the fact that at most one of the three strategies would be actually applied on a particular vehicle. The station skipping strategies and holding strategy are mutually exclusive because of their opposite functionality: the former shorten the preceding headway while the latter lengthens it. The two station skipping strategies, deadheading and expressing, will not be both used on the same vehicle trip because it is unnecessary and will increase passenger frustration. Hence, it is reasonable to assume that a vehicle trip can have at most one skipped segment in a direction, and can be controlled by using at most one type of strategy. Therefore, the combined control problem reduces to the problem of choosing which (if any) strategy is appropriate for each vehicle.

What complicates the combined control problem is that different strategies may be considered for different vehicle trips within the same impact set. Fortunately, based on the research results on each single control strategy, the development of an efficient and effective heuristic for this problem becomes quite straightforward. Next we develop such a heuristic for the combined control problem (CCP).

5.1 Heuristic Algorithm for CCP

The logic of this algorithm is basically a hierarchical comparison structure. Since holding is a "smooth" control strategy which can be used most often with the least negative effects, for a vehicle set $\mathbf{I}_m$ we first try the holding strategy alone. We then try deadheading and expressing separately for i while holding other vehicles. The skipping strategy which gives lower total cost will be chosen to compare with holding-only solution. The best solution is then chosen. The efficiency and effectiveness of this hierarchical comparison algorithm is grounded in the efficiency and effectiveness of each individual strategy algorithm.

Algorithm CCP

For vehicle set $\mathbf{I}_m$:

Step 1. Apply algorithm HP alone to the set of vehicles i, $i+1,\ldots,i+m$. Set $d^* = d_{i,k_0}$. The resulting set cost is f^h.

Step 2. Check the headway condition of vehicle i to decide whether station skipping is feasible. If it is feasible, go to step 3; otherwise terminate and i is held until d^*.

Step 3. Try deadheading i, starting from $n=1$, $k_e = k_0+1$.
Iteration n: Let i skip n stations from k_0 to k_e-1. Compute the new trajectories of vehicles i to $i+m$, and then apply algorithm HP to vehicles $i+1,\ldots,i+m$. Compute cost $f^{(n)}$. If $f^{(n)} > f^{(n-1)}$, stop; set $n^d = n-1$, $f^d = f^{(n-1)}$. Otherwise increase n and k_e by 1, go to next iteration.

Step 4. Try expressing i, starting from $k_s = k_0$, $k_e = k_0+2$.
While $k_s < k_t-2$ **Do**
  While $k_e < k_t$ **Do**
   Let i skip n stations from k_s +1to k_e-1.
   Compute the new trajectories of vehicles i to $i+m$.
   Apply HP to vehicles $i+1,\ldots,i+m$.
   Compute $f^{(n)}$, where $n = k_e - k_s - 1$.
   If $f^{(n)} > f^{(n-1)}$, $f^* = f^{(n-1)}$, $n^* = n-1$, break;
    else $k_e = k_e+1$.
  End inner while loop
  $k_s = k_s+1$, $k_e = k_0 + n^*$.
  $f^e = \min_{k_s}(f^*)$.
End outer while loop

Step 5. If $\min(f^h, f^e, f^d) = f^h$, i is held until d^* and then dispatched. If $\min(f^h, f^e, f^d) = f^e$ and $n = k_e - k_s - 1 > 0$, i is expressed from k_s to k_e. If $\min(f^h, f^e, f^d) = f^d$ and $n^* > 0$, i is deadheaded to k_e.

We now discuss the computational results from running algorithm CCP on the MBTA Green Line data sets.

5.2 The Test Data

We obtained one week of vehicle location data by time of day from the AVI system of MBTA Green Line B line (in Boston, Massachusetts). The MBTA Green Line is a light rail transit system, which has network and operational characteristics similar to the type of system of interest here. There are a total of 52 stations in the two directions of the B line, with 13 AVI detectors located along the route. The inbound (IB) direction goes from station 1 to station 26, and the outbound (OB) direction is from station 27 to 52. The AVI data for morning peak hours for a week were used in our computational tests. The scheduled headway during this period was 5 minutes.

The demand data used here was compiled by Macchi (1990) from a survey of 1985 Green Line passengers and is illustrated in Fig. 5.1. There is layover time at the end of direction 1, but not direction 2. We use a Green Line train dwell time function which was estimated by Lin/Wilson (1992) in the form of equation (2.2).

5.3 Efficiency of Algorithm CCP

The algorithm was coded in C and implemented on a 25 MHZ 486 PC. The average execution time of the procedure is about 2.5 seconds for a vehicle set of size m=3 and 26 stations per direction in the Green Line data. The highest set average is about 5 seconds. Such a small computation time clearly meets the needs of a real-time control system.

5.3.1 Comparison between Single and Combined Strategies

Table 5.1 summarizes the computational results of CCP and Fig. 5.2 shows cost reduction by control strategy for all data sets. For the station skipping strategies, the results with no adjacent deadheading/expressing are used and for holding the results with the terminal schedule constraints are used in the comparison. Among the single strategies, the effectiveness of the two station skipping strategies, deadheading and expressing, is quite similar in terms of waiting time reductions, while expressing does slightly better in direction 1 and deadheading slightly better in direction 2. The station skipping strategies are less effective in direction 2 due to the demand pattern. On the other hand, holding is more effective in direction 2 than the station skipping strategies, and is also more effective than holding in direction 1. This is due to the dispatching headway patterns. However, when the terminal schedule constraint is tight, holding can be much less effective, as shown in the extreme case in Fig. 5.2 of data set "th1". In all cases, combined control shows clear advantages. The effectiveness of combined control is close to holding-only in direction 2, because the station skipping strategies are not very effective here. The cost reduction by combined control is as high as 37%. The marginal benefit of combined control, compared to the most effective single strategy in each data set, varies between 0.4% (m1) and 7% (w1), with an average of about 4% (or about 4000 passenger-minutes) per data set. The average standard deviation of vehicle headways is also the lowest in the CCP solution. We will see that this improvement is mainly due to the complementarity between station skipping and holding strategies.

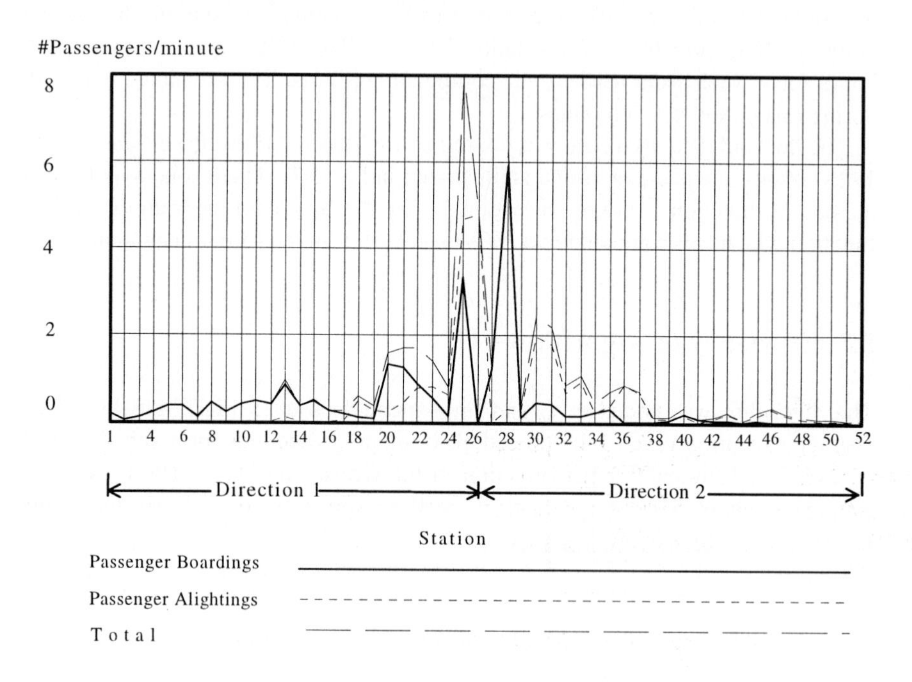

Fig. 5.1 Demand Pattern of MBTA Green Line B line.

While the marginal benefits of the combined control may not seem very high, it does have some structural advantages as we will see.

5.3.2 Solution Structure of Combined Control

Considering all data sets together, about 76% of vehicle trips (263 out of 346) are controlled. Among the controlled vehicles, about 27% are subject to station skipping strategies, and 73% are held. Holding is far more frequently used in the solution, and the portion of deadhead vehicle trips (11.8%) is slightly lower than the portion of express vehicle trips (15.6%) among the controlled vehicles. A number of characteristics in the solution structure are discussed below.

Table 5.1 CCP Results (m=3)

		No Control			with Combined Control							
Set	M	Cost (Psg. min)	Headway (H)		# Skipped Stations	# Vehicles Controlled			Change in Cost		Held Time (min.)	hdw StDev (min.)
			Mean	StDev		DH	Exp	Held	psg.min	%		
m1	36	62,467	4.6	1.8	5	2	2	23	-15,738	-25.2	25.0	0.8
tu1	35	62,701	5.0	1.7	9	1	6	20	-14,280	-22.8	18.7	0.6
w1	34	67,757	5.0	2.2	13	2	6	12	-15,889	-23.5	15.5	1.0
th1	37	66,083	4.9	1.9	11	2	5	25	-13,760	-20.8	22.5	0.7
f1	31	52,427	4.9	1.5	4	2	2	22	-10,745	-20.5	16.2	0.4
m2	36	38,820	4.6	5.7	10	5	3	21	-14,493	-37.3	48.3	2.6
tu2	35	37,792	5.1	5.3	13	3	7	18	-13,840	-36.6	43.1	1.9
w2	34	38,119	4.9	5.6	11	7	2	9	-7,364	-19.3	17.4	4.1
th2	37	38,640	4.9	5.4	13	4	4	22	-10,381	-26.9	41.2	2.5
f2	31	29,939	5.0	4.8	9	3	4	19	-9,459	-31.6	33.6	1.8
Avg		98,949	4.9	3.6					-25,189.8		1.5	1.6
Tot.	346	494,745			98	31	41	191	-125,949	-25.5	281.5	

H=5min. % Change in Cost = Change in Cost/Cost*100%. Average cost and change in cost are per morning peak. Average holding time is per held vehicle. M - # vehicle trips. m1 - Monday Direction 1

The use of station skipping strategies increases the feasibility and effectiveness of holding. It is not surprising that in the combined control solution holding is used most frequently. Its "smooth" and nonrestrictive nature makes it the most feasible and effective strategy. Interestingly though, total holding time has increased from 267 minutes to 281 minutes, and the number of vehicles held is increased from 171 to 191, compared to holding alone. This is because the number of vehicles which are feasible for holding has increased due to station skipping control of other vehicles. For example, to hold vehicle i may be infeasible in HP because it is blocked by vehicle i-1 and therefore i's arrival time at the terminal is binding on the terminal schedule constraint. However, after deadheading/expressing vehicle i-1, i may no longer be blocked and may arrive at the terminal earlier than scheduled. In this case holding i becomes feasible. Thus in combined control, station skipping strategies can increase the feasibility and effectiveness of holding.

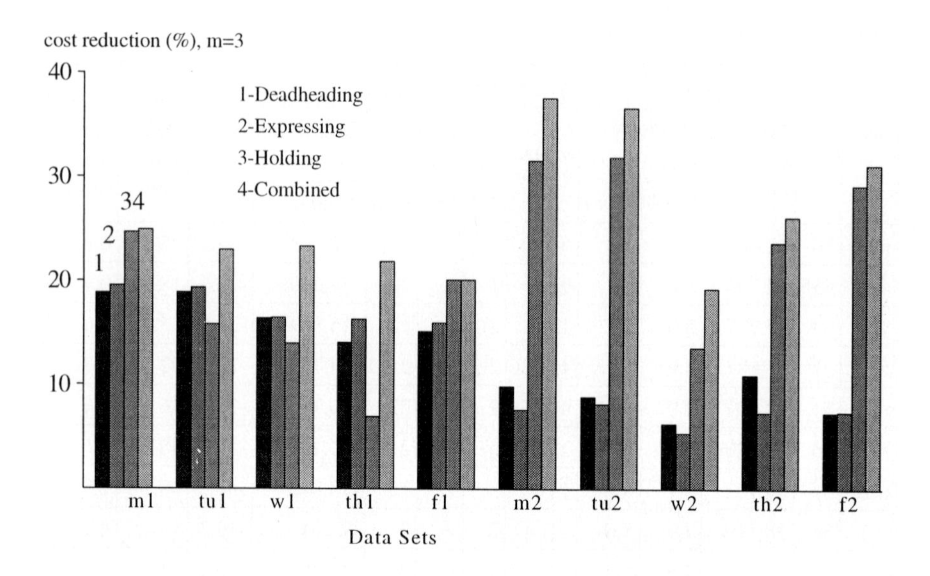

Fig. 5.2 Comparison between Single and Combined Control

The increased frequency of holding also slightly decreased average holding time per held vehicle from 1.6 to 1.5 minutes.

The use of holding decreases the frequency and side effects of station skipping. Compared to deadheading or expressing alone, the frequency of station skipping decreases dramatically in the CCP solution. While 118 vehicles are deadheaded and 193 stations skipped in the RTDP solution, and 143 vehicles are expressed and 252 stations skipped in the RTEP solution, there are only 72 vehicles expressed or deadheaded, and 98 stations skipped in the CCP solution. That is, both the frequency and magnitude of station skipping control are substantially decreased. This is no doubt good news to passengers, because no one who waits at a station likes to see the bus or train passing-by without stopping.

In addition, the side effects of station skipping almost all disappeared in the CCP solution due to the use of holding. For example, in a RTDP or RTEP solution, vehicle i might "over-skip" stations in order to "push back" the trajectory of a following vehicle. As a result, passengers who wait for vehicle i

and i+1 may be benefited less or not at all. In combined control, since trajectories of following vehicles can be pushed back by holding them, over-skipping is no longer needed. An additional consequence of this is that the cost curves are more often quasi-convex than in RTDP/RTEP. This, needless to say, contributes to the performance of the solution algorithm.

Frequent control actions are needed for maximum effectiveness. As we have seen from the CCP solution, 76% vehicles are controlled. This implies that for real-time control to be fully effective, it should be a continuous process. Less frequent control may result in more severe control actions, and increase passenger frustration induced by those control actions.

6 Effects of Randomness

In the above analysis we have taken a deterministic modeling approach to attack the real-time control problems, which assumes known trajectory of vehicle i-1 and perfectly predictable trajectories of vehicles in the impact set . In reality, however, these assumptions only partially hold. At the point a control decision is to be made for vehicle i, i should either approach or have arrived at the dispatching station, and vehicle i-1's trajectory should have at least partiallybeen realized and hence recorded by an AVI or AVL system. While this realized information can be treated as the deterministic part of input to our control models, future movement of vehicles in the impact set, as well as future passenger demand at each station, must be forecast. In our control models, this forecasting is done using the deterministic equations (2.1)-(2.4) . As stochasticity and forecast errors do exist in reality, an important question is how well our models do under such conditions. We expect that when vehicle running time and passenger demand variances and forecast errors are not too large, the deterministic models will still work well. In this section we test the CCP model performance under various stochastic scenarios.

In performing the tests, we introduce different degrees of randomness by changing the extent to which variables will change around their deterministic values. Thus we can gain an understanding of the impact of randomness on the models and algorithms. Since passenger waiting and travel times are the elements for which random disturbances will be most important, we run the same deterministic procedures with random disturbances added to the passenger arrival rates and vehicle running times.

With this randomness in place we run simulations in which the control decisions are made based on known past information and estimated future information, but in a stochastic environment. More specifically, we assume that we know past vehicle movement and passenger demand, but future demand and vehicle running times are just estimates at the time the control decision is made. In this way, we have a fairly good idea of how sensitive the control policies developed are to the important sources of randomness in the real system.

Since the travel time randomness and passenger demand randomness may have different impacts, we considered different stochastic scenarios as given in Table 6.1. As for the range of randomness, the Green Line AVI records show the coefficient of variation for composite travel times is between 0.05-0.1. In the interest of "worst case" testing, we chose to look at the range of 10-30% variation around the mean. Because we have no empirical data for the variances of passenger arrival rates, the same range of variances are assumed for the vehicle travel times $R_{i,k}$ and passenger arrival rates r^j_k. In addition, to consider systematic biases in vehicle running times due to driver behavior or mechanical conditions, we tested two more scenarios 4A and 4B, where a vehicle dependent but not station dependent random disturbance around estimated mean $R_{i,k}$ is assumed.

Table 6.1 Sensitivity Test Scenarios

Scenario	1A	1B	2A	2B	3A	3B	4 A	4B
C.O.V. for r^j_k	0.0	0.0	0.1	0.3	0.1	0.3	0.3	0.3
C.O.V. for $R_{i,k}$	0.1	0.3	0.0	0.0	0.1	0.3	0.3	0.3
C.O.V. for mean $R_{i,k}$	0.0	0.0	0.0	0.0	0.0	0.0	0.05	0.1
Note: C.O.V. = Coefficient of Variation								

We denote the deterministic scenario as Scenario 0, where both C.O.V.'s for $R_{i,k}$ and r^j_k are zero, and the expected values R_k and r_k are assumed.

We again use the Green Line data sets for these tests. The deterministic values are regarded as the expected values. Using the random parameters given in Table 6.1, 8 random data sets of no-control vehicle trajectories and passenger demand are generated for each direction each morning. This gives a total of 80 data sets for the 5 days.

6.1 Forecast Information

After generating the "true" free travel times and passenger demand incorporating randomness, we perform the tests as follows. For each control problem considered for vehicle set $\mathbf{I}_m = \{i,...,i+m\}$, we set $t_0 = a_{i,k_0}$, the known arrival time of vehicle i at the first station in the control direction. When evaluating the cost function for the vehicle set, all vehicle trajectories and passenger demand up to t_0 are computed using their "true" values as generated, since they are already realized. For forecast information on future vehicle trajectories and passenger demand, we consider the following two cases:

Case *m*: All vehicle trajectories and passenger demand after t_0 are predicted using the mean interstation travel time across vehicles (varying by segment), and expected passenger arrival rates over time (varying by station). This is the same as in the deterministic model. Here "*m*" stands for "mean".

Case *p*: All vehicle trajectories and passenger demand after t_0 are predicted with perfect information. That is, the true random values of $R_{i,k}$ and r^J_k are used as generated. Here "*p*" stands for "perfect".

Here Case *m* represents a realistic forecasting method. Case *p*, on the other hand, plays two different roles in the tests. By comparing it with Case *m*, we can gain insights on the impact of forecasting information. By comparing it with the deterministic case, we can learn about the impact of randomness on model performance. The logic of such comparisons are explained below.

6.2 Impacts to be Tested

The first impact of randomness to be tested is on the validity of the system properties. The system properties presented in Section 2 may be changed by the stochasticity. The proofs of the properties were based strictly on the assumptions that free interstation running times are constant across vehicles, and also that passenger arrival rates do not change with time. The CCP algorithm is developed based on these system properties. Because the random disturbances violate the deterministic conditions used in the proofs, the difference between the solutions in the deterministic scenario and a random scenario with Case *p* will reflect the model sensitivity to the property changes. Hence we want to compare Case *p* in the stochastic scenarios with the deterministic scenario.

Second, using Case *p* to compare with Case *m* in each stochastic scenario will reflect two types of impacts: in addition to the impact from system property changes there is also the impact from forecast information accuracy. A control policy for vehicle *i* is generated by the CCP model using forecast information on future vehicle trajectories and demand. In a test with Case *m*, this forecast is based on the mean or expected values. The control policy based on this information will then be "performed" on *i* in the "true" stochastic environment. Since we obtain and perform the control decisions in two different environments, one without and the other with accurate information, such a test will further show the model sensitivity to forecast information errors. By comparing Cases *p* and *m* for each random scenario, we can estimate how much impact forecast accuracy has on algorithm performance.

6.3 Results of Sensitivity Tests

For 6 unbiased stochastic scenarios and 2 forecast information cases, 12 simulations runs are performed for each input data set. Another 4 simulation runs are performed for scenarios 4A and 4B with systematic biases in vehicle travel times. Table 6.2 and Fig. 6.1 show the test results (week totals).

A number of important results from the randomness tests are summarized below.

Worse system performance in absence of control. Compared to the deterministic Scenario 0, all but one (1A) stochastic scenarios have higher costs and all have higher standard deviation of headways in the no-control situation. Also, the costs and variances increase as the randomness increases. With the same variance, the no-control system performance with a single source of randomness is better than with both sources of randomness. Thus, the system performance is worst in scenario 3 among the unbiased random scenarios. With the 0.05 COV level of biases in mean vehicle travel times and other things held equal, the system performance in scenario 4A is no worse than 3B. However, with the 0.1 COV level of biases in mean vehicle travel times, the system performance degrades sharply: the total waiting time has increased 10% compared with the deterministic scenario. All this is as expected.

Lower cost reductions after control, but not much lower in unbiased random scenarios. All cost reductions in the stochastic scenarios are lower, which is also expected. In all "A" unbiased scenarios (where the COV is 10% or 0 for either type of randomness), however, the difference between the highest and the lowest cost reduction is only 1.1%. The lowest cost reduction is only 1.2% lower than that of the deterministic scenario . This shows the control models perform very well at the 10% COV level. At the 30% COV level, the model performance is both worse and less consistent. For unbiased scenarios, the difference between the highest and lowest cost reductions is about 5 percent which translates into about 5,000 passenger-minutes per morning. Though this is a significant amount, the model generated controls still decreases the total passenger waiting time by (21%-25%) relative to the no control situation.

With systematic biases in vehicle travel times, however, the performance declines as the level of bias increases. At the 0.05 COV level of bias, the total cost reduction has been reduced to 19% , and to only 15% at the 0.1 COV level of bias. This decline is quite significant. This shows that the models developed here are quite sensitive to systematic biases, which is as expected since we did not consider these biases in our modeling.

These results show that the deterministic models have an acceptable performance in a stochastic environment with coefficients of variation of up to 0.3, in either vehicle travel times or passenger arrival rates, or both, and with low

(<0.05) systematic biases. Performance degrades as the amount of randomness and systematic bias increases, and this degradation is most pronounced for variation in vehicle running time. We should also note that a 0.3 COV for free interstation travel times and the 0.1 COV for travel time bias are probably unrealistically high. The tests we have performed indicate that the control models developed here will probably work well in most transit systems.

System properties insensitive to randomness. Compared to Scenario 0, Case *p* in all random scenarios show insignificant differences. The range of such differences is from only 0.2% to 1.1% in cost reduction. This means that randomness in interstation trip time and passenger demand has little impact on the properties of the models, which in turn indicates the robustness of the control models. This is to say, with accurate forecast information, the control models perform well under all stochastic scenarios considered here.

Table 6.2 Results of Randomness Tests

				No control		After control			
Forecast information						Case *m*		Case *p*	
	C.O.V for $R_{i,k}$	C.O.V for r^J_k	C.O.V. for mean $R_{i,k}$	psg.min/day	hdw. StDev	% cost reduction	hdw. StDev	% cost reduction	hdw. StDev
unbiased $R_{i,k}$									
0	0.0	0.0	0.0	98,948.8	3.6	25.5	1.5		
1A	0.1	0.0	0.0	98,797.2	3.6	24.3	1.7	25.2	1.7
1B	0.3	0.0	0.0	101,650.1	3.6	21.0	2.0	25.0	1.9
2A	0.0	0.1	0.0	99,475.0	3.6	25.4	1.6	25.4	1.6
2B	0.0	0.3	0.0	101,056.7	3.7	25.3	1.7	24.6	1.8
3A	0.1	0.1	0.0	99,776.1	3.6	24.7	1.7	24.3	1.8
3B	0.3	0.3	0.0	105,078.3	3.7	22.3	2.0	24.4	2.1
baised $R_{i,k}$									
4A	0.3	0.3	0.05	103,655.7	3.7	19.1	2.1	23.3	1.9
4B	0.3	0.3	0.1	109,627.5	3.9	15.6	2.4	24.1	2.1

Larger impact from vehicle travel time variance. Free interstation vehicle travel time variation has a much larger impact than passenger arrival rate variation at any given value for their respective coefficients of variation. This is reflected most clearly by the lower cost reduction in scenario 1 relative to scenario 2.

There are a number of explanations for this difference: first, vehicle travel times in most segments are much larger quantities than dwell times. Second, the variation of dwell times around the mean at each station across vehicles is much smaller than the variation of mean dwell times across stations. Since the mean dwell times are already captured by the deterministic model, the small added randomness apparently does not have a significant impact. Furthermore, the time dependent randomness in passenger arrival rates may, to some degree, cancel each other out over a long period, and this is probably why there is almost no difference between Case *m* and Case *p* in Scenario 2. Finally, the constant term of the dwell time function (2.2) is not affected by any randomness in the test.

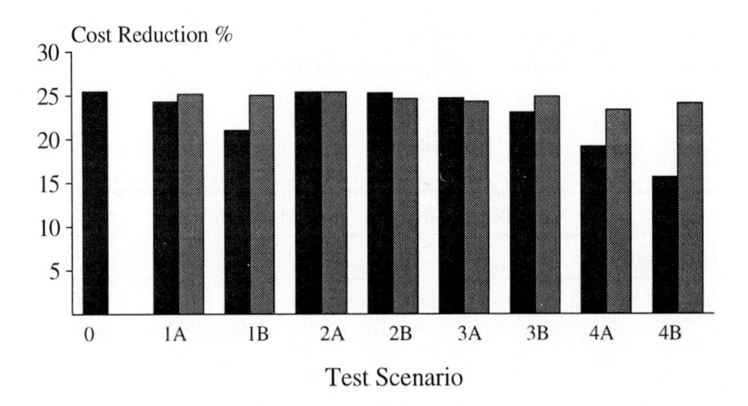

Fig. 6.1 Cost Reduction after Control under Randomness

Greater sensitivity to forecast information accuracy for vehicle travel times. Apparently, forecast information accuracy has larger impact in scenarios 1B, 3B, 4A and 4B, where the coefficient of variation of vehicle travel times are the highest (0.3). In all other stochastic scenarios, such impact is small. This indicates that forecast accuracy for vehicle travel times is more important than for passenger arrival rates. This is good news since real-time vehicle information is easier to obtain than passenger information. On the other hand, this also shows a limitation of the deterministic models: in an operating environment in which vehicle speed variance is high, and forecast information for vehicle interstation travel times is not accurate, the models will have worse performance.

7 Conclusions

In this paper we presented original mathematical programming formulations for three real-time control problems: holding, deadheading and expressing. Based on these formulations, we developed a heuristic algorithm for combining all three control strategies. Computational results show that the effectiveness of combined control is higher than any single type of control, although the marginal benefits are limited. Furthermore, combined control leads to less severe control actions. It also increases control feasibility, and as a consequence more vehicle trips were controlled than in any single type of control. This tendency is welcome because it brings, at the same time, more benefits and less frustration to passengers. One important conclusion from the results is that for maximum effectiveness frequent control actions are needed. In the tests using Green Line data sets, over three quarters of vehicle trips were subject to some form of control.

Our analysis on each of the problems shows that holding is the most effective single control strategy. When holding is used continuously, headway variance decreases, and the chance for beneficial station skipping control also decreases. This indicates that combined control is most effective where holding is tightly restricted by schedule constraints.

Because in reality various degrees of stochasticity affect all transit systems, it is important to test the deterministic models under randomness. We designed tests to evaluate the models when various degrees of randomness exist in the system. The computational results for the combined control model show that, while system performance without control was worse with stochastic disturbances, the change in effectiveness of control was not large when systematic biases were small. The system properties presented in section 2 are insensitive to randomness. The impact from randomness in vehicle interstation travel times is larger than the impact of randomness in passenger arrival rates. The effectiveness of control is more sensitive to systematic biases in vehicle travel times. The model performance is also more sensitive to forecast information accuracy on vehicle travel times than on passenger demand. In conclusion, the control models developed here are likely to work well in many transit systems.

References

Eberlein, X. J. (1995): Real-Time Control Strategies in Transit Operations: Models and Analysis. Ph.D. dissertation, Department of Civil and Environment Engineering, Massachusetts Institute of Technology.

Eberlein, X.J./Wilson, N. H.M./Bernstein B./Barnhart C. (1998): The Real-Time Deadheading Problem in Transit Operations Control. In: Transportation Research 32B, 77-100.

Lin, T./Wilson, N.H.M. (1992): Dwell Time Relationships for Light Rail Systems. In: Transportation Research Record 1361.

Macchi, R.A. (1990): Expressing Vehicles on the MBTA Green Line. Master thesis, Civil Engineering, Massachusetts Institute of Technology.

Knowledge-Based Decision Support System for Real-Time Train Traffic Control

Alexander Fay and Eckehard Schnieder
Technische Universität Braunschweig, Institut für Regelungs- und Automatisierungs-technik, Langer Kamp 8, D - 38106 Braunschweig, Germany
email: a.fay@tu-bs.de, phone: +49 (0) 531 391 3329, fax: +49 (0) 531 391 5197

Abstract: Modern public transport and train traffic systems have to fulfill increasing demands on service reliability and availability. Train operators can only satisfy these requirements by quickly developing an efficient action in case of traffic disturbances. This paper describes a dispatching support system which is intended for use in railway operation control systems, but could also be useful for other kinds of public transport. The system consists of a knowledge-based decision support system, a simulation tool and a graphical user interface. The assistance consists of conflict detection, display of relevant information, prediction of certain dispatching measures' impacts, and proposals for appropriate dispatching actions. Thus, the decision support system should significantly improve traffic performance, reliability, and customer satisfaction.

1 Introduction

1.1 Meeting Increasing Demands with Computer Assistance

Public transport providers face increasing demands concerning

- competitive transport markets,
- need for effective and efficient use of resources,
- reduction of personnel for train operation and control,
- changing and increasing customer requirements.

Aspects like profitability, customer satisfaction and quality assurance are becoming more and more important. High transport capacity should be obtained and maintained, but the transport system has to be attractive to passengers. Irregular services - and especially train delays or cancellations - are a severe obstacle to these aims. The challenge of increasing train speeds, tight time schedules and higher traveler demands force train operators to improve the punctuality and reliability of their train services.

The use of computer-aided systems in both planning and operations control of train traffic is becoming more and more important because of effectiveness and efficiency. Possible areas of application range from strategic and operational planning to operational control and monitoring. One aim is an integrated system in which the computer support of operations control forms a central part (see Daduna (1993)).

The paper is organized as follows: Section 1.2 develops a hierarchical method to structure traffic management tasks. Section 1.3 explains the role of the dispatcher. Section 1.4 reveals the influence of dispatching on traffic quality and economic performance. Section 1.5 gives an overview of the state-of-the-art of dispatching with special focus on the role of the computer. In Sect. 1.6, the combination of planning and control is discussed. Section 1.7 presents the two main approaches to dispatching support: analysis and heuristics. The train traffic control loop is presented in Sect. 1.8 as a starting point for the description of our dispatching decision support system (DSS) in Sect. 2. Section 2.1 starts with the overall system structure. The different modes of dispatching support are explained in Sect. 2.2. Sections 2.3 to 2.5 focus on the three main components of the system: graphical user interface, simulation tool, and decision support system, respectively. In Sect. 2.6, the state of implementation is discussed, including an example of the functionality of the dispatching support system. Concluding remarks are given in Sect. 3.

1.2 Hierarchy of Traffic System Management Tasks

Various groups in society impose requirements on traffic: users, providers, manufacturers, ecologists, regional politicians etc. Hence, these requirements tend to be quite different and conflicting, for example:

- high availability in time and space,
- low resource consumption (space, time, money, energy),
- high comfort,
- high safety and security.

Not only can these requirements be conflicting, they are usually only vaguely described and not comparable. They concern various aspects which have to be taken into account in the planning, development and control of traffic systems. To meet them all would involve a complex multicriteria optimization problem.

For several reasons, control of public transport can not be satisfactorily tackled by means of the usual methods of control theory alone. These reasons are:

- the variety of requirements (see above),
- the complexity of the system to be controlled:

- the dimensions in space range from 10^0 m to 10^5 m,
- the number of system objects (vehicles, track elements, signals, ...) can be in the range of 10^4,
- processes are mixed discrete and continuous (hybrid),
- time constants range from milliseconds to several years,
- the influence of actions is neither easy to predict nor to measure.

To handle this kind of system appropriately, the problem has to be divided with respect to the time scales of both action and system reaction. Therefore, we distinguish four main levels (which have been named according to process engineering convention) in the task hierarchy. These range from long-term planning, scheduling and dispatching to operation (Table 1.1).

Table 1.1: The task hierarchy

Level	Tasks	Typical time horizon
Strategic (planning)	Changes in infrastructure and traffic supply	Years
Dispositional (scheduling)	Scheduling of lines, trains, vehicles and crews; duty rostering	Weeks - 1 year
Tactical	Dispatching / control to avoid and/or handle conflicts	Minutes - hours
Operational	Propulsion control, interlockings	Seconds - minutes

Of course, this is only a rough structure which can be refined if necessary. In this hierarchy, each of the task levels sees the results of the higher levels as fixed, which act as constraints on the lower level tasks.

1.3 The Role of the Dispatcher

The schedule is the basis for all dispatching activities. It is the dispatchers' task to ensure optimal train traffic performance according to the schedule and to minimize the impacts of the schedule deviations even in presence of unforeseeable disturbances. The latter might be due to

- planning mistakes (like poorly calculated headways or service times),
- technical reasons (engine breakdowns, signaling failures, track closures, and the like),
- organizational problems (late or absent staff members, extra trains for urgent transportation demands).

These disturbances can cause traffic conflicts. One can distinguish between

- connection conflicts for passengers,
- resource conflicts for trains and personnel,
- availability conflicts for track sections,
- delays (usually in combination with one of the above).

In general, traffic conflicts result in situations where not all of the technical and operational requirements can be met. A conflict can be resolved by relaxing some of the requirements. Only operational requirements, such as arrival times or predefined platforms, can be relaxed, whereas technical requirements, such as maximum speed restrictions, can not be changed.

The answers to the following questions contain the most important information about existing anticipated conflicts:

- Where is the conflict?
- Which resources (tracks, trains, staff) are affected and for how long?
- Which resources remain?

Possible actions might include

- prolonged or additional stops in stations,
- crossing and overtaking,
- shifts and detours,
- canceled or added trains.

Dispatching is a complex conflict solution and traffic optimization process. The dispatcher has to adapt various conflict resolution measures to the actual traffic situation and select one which resolves the conflict with a minimum of negative side effects, considering the different objectives of the traffic provider and the customers. These objectives, such as service cost minimization or connection possibilities, can be mutually exclusive and must be weighed against each other and combined in an optimal way.

Before discussing possibilities of supporting the dispatcher, the crucial significance of dispatching for the transport business is discussed.

1.4 Influence of Dispatching on Traffic Quality

Passengers have to be seen as customers who have the choice between different means of public and private transportation. They will always opt for the one which offers the largest benefit for them, i.e. the "product" on the "traffic market" whose features are most similar to their desires. If the expected benefit can be realized, the customers are satisfied ("customer satisfaction") and will probably choose the same means of transportation again in future. Products which offer a high amount of benefit can be seen as "quality products". This quality is not an objective measure or feature of the product but a subjective attribute from the point of view of the customer.

Important quality factors of a trip by train are to arrive on time and to be able reliably to change trains at connecting nodes of the network. Traffic provider's failure in assuring these criteria will result in significant future revenue losses. Punctuality and connections, however, are subject to the actions of the dispatchers.

The correct estimation of all consequences of the dispatching action is crucial for optimal use of resources. Thus, effective and efficient incident management is important to regain the normal level of service as soon as possible and to reduce the loss of profit.

Dispatching actions thus have an enormous impact on the customer, the perception of the service quality, and, ultimately, on the economic success of the traffic provider in a competitive market.

1.5 Dispatching in the Past and in the Future

In the past, train supervision and control was performed by numerous decentralized operators, each of them being responsible for a small and distinct part of the network. Communication among them typically consisted of the frequent exchange of routine messages by phone. To decide about major deviations from schedule, dispatchers had to collect data manually to gain an overview of the overall traffic situation to be able to develop decisions.

Fig. 1.1 shows that the main part of these tasks is the responsibility of humans, and only a small part is assigned to the computer. The computer may visualize schedule deviations and the present and probable future process data through a time-space-diagram and/or a network overview.

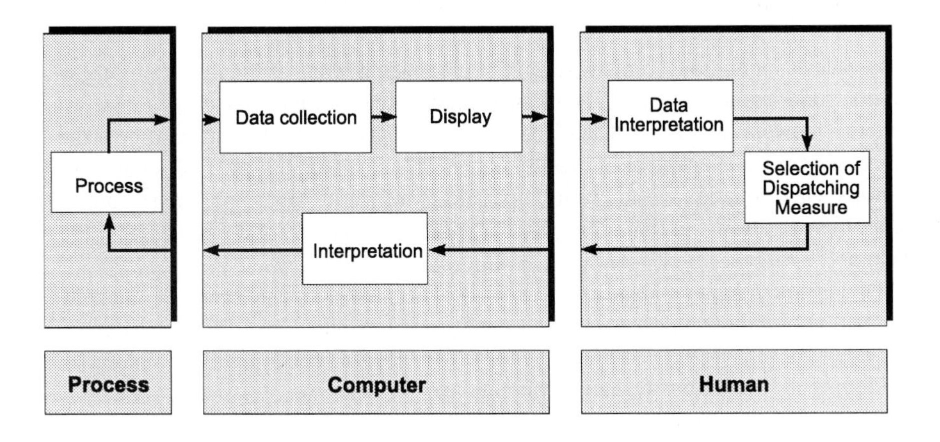

Fig. 1.1 Burden sharing between dispatcher and computer

The dispatcher has to collect data from both diagrams, to derive the relevant information from it and to evaluate it. At the same time, an optimization has to occur in the human mind. The dispatcher compares the actual and probable future process values with the timetable, recognizes functional dependencies for a set of events and tries to minimize the difference. After having completed this calculation, the dispatcher has to find a dispatching method to implement the solution in practice. High traffic density and high train velocities result in stress for the dispatcher with negative impacts on the quality of these decisions.

This problem can be ameliorated by shifting the human/computer interface to the right, for example by passing the task of data collection and information representation to the computer (Fig. 1.2). As a consequence, the total job is divided so that each part of the system does the job it is best qualified for.

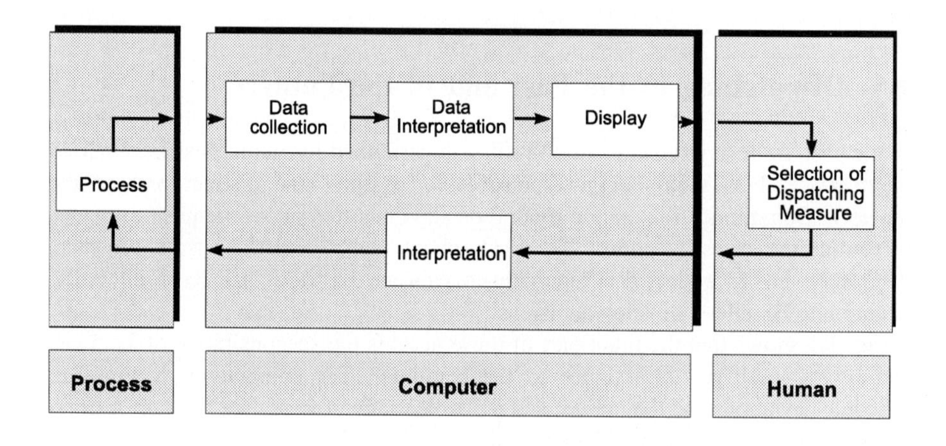

Fig. 1.2 Shift of data interpretation tasks towards the computer

Nowadays, operations centers largely replace local control facilities. In these centers, train supervision and operation functions are concentrated and carried out for large network areas (see Oser/Thorwarth/Biernatzki (1994)). Furthermore, the supervision and management of maintenance works is integrated in these centers to optimize coordination with train operation. Modern signaling and control technologies allow automation of large-scale systems and processes (see Daduna (1993)).

The increased requirements imposed on traffic providers can only be met by automation of tedious routine jobs and complicated calculations. Thus, the human dispatcher can concentrate on those tasks which require human creativity, such as developing new solutions for new problems.

Because information about the position and state of all vehicles, tracks and other system components is always available at the control center, many of the dispatching tasks usually performed manually in conventional train systems can be automated. This provides additional time for the more important and difficult dispatching problems.

Based on the problem characteristics, conflicts can be solved in one of the following ways:

- fully automated (if the conflict can be detected automatically and only one solution is applicable or is clearly dominant),
- partially automated (if several solutions exist and a selection by the dispatcher is desirable),
- manually (if the conflict can not be detected automatically or no solution can be found by the computer).

To ensure an optimal dispatching process in an environment of crowded tracks and reduced personnel, the dispatcher has to be supported by improved tools which focus on the most important conflicts, present all the necessary information, and offer effective solutions.

1.6 Trends in Operations Control of Train Traffic Systems

Regarding the development of train traffic control, two main trends can be observed:

- To optimize train traffic systems in the sense of an optimal use of resources, more and more sophisticated methods and algorithms are used in traffic planning. The greater the success of these methods, the greater is the desire to make use of them also for short-term problems such as traffic control. As explained in Sect. 1.2, planning and control have a similar problem structure. The difficulty in transferring the effective planning methods results from the shorter time available for decision-making: most algorithms developed for planning tasks are too time-consuming to yield immediate answers to traffic control problems. Therefore, heuristics have to be found and employed, not to replace the algorithms, but to adapt them to reduce the solution space significantly.
- To further improve traffic safety, quality and profitability, most functions on the level of train, track and signaling control have already been automated. This development is to be repeated in the area of traffic operation and control for the same reasons, but this is much more difficult because the technical and organizational processes are less formalized on this level. The desired solution consists of gradual automation with "burden sharing" between human and computer: each performs the tasks most qualified for. Computer-assisted traffic analysis and computer-aided development of conflict resolution proposals offer valuable support.

So while train traffic control problems are similar to planning problems, the approaches have to be developed further given the special operations control characteristics. Decision support systems can offer sound solutions. In the following section, different approaches are sketched about how one could influence the system with mathematical and technical methods.

1.7 Algorithmic and Heuristic Approaches

The numerous approaches in international research and application to tackle train traffic planning and control can be divided into two main branches:

- algorithmic optimization approaches, which stem from mathematics, classical computer science and operations research and can find the global optimum with respect to the goal function chosen,
- heuristic approaches, which try to find good (not necessarily optimal) solutions fast by employing AI methods and expert knowledge.

These approaches are not necessarily mutually exclusive but can be combined in a useful manner.

Examples of algorithmic optimization approaches for real-time tasks are

- a decision support system for routing trains through railway stations (see Zwaneveld/Kroon/Ambergen (1996)),
- train scheduling on a single-track railway (see Nachtigall (1996)),
- optimal regulation for metro lines (see Fernandez/deCuadra/Garcia (1996)).

Most optimization approaches stemming from operations research show NP-complexity and can, therefore, not be used for practical applications. They can be transformed to algorithms with polynomial time complexity, but only by making severe simplifications of the real problem. Hence, the search space has to be reduced significantly to achieve implementable results. This reduction can be performed with sophisticated search strategies or heuristics. Heuristics, which include knowledge of dispatching experts, can lead to straight-forward dispatching decisions. Examples employing heuristics for dispatching support are

- ESTRAC-III, an expert system for train traffic control (see Komaya/Fukuda (1989)),
- PETRUS, an underground railway traffic expert system (see Moirano/Rossi/Sassi/Sissa (1989)),
- an expert system for public transport control (see Adamski (1993)),
- an expert system for real-time train dispatching (see Schaefer/Pferdmenges (1994)).

The system developed by the authors (described in Sect. 2) is also based on heuristics in the form of expert rules. The paramount importance of rules derived from expert knowledge can be seen in the success of Fuzzy Control. In a similar way, knowledge of dispatching experts can be collected and used for traffic operations control.

Yet heuristics alone will not necessarily always yield satisfactory solutions: the knowledge base is always limited, and not all situations can be covered in advance by appropriate rules. On the other hand, dispatching actions can have large impacts on traffic behavior and service quality. Therefore, it is desirable to estimate these effects in advance before implementing a dispatching strategy. For this purpose, simulation plays an important role. Though it is difficult to build an appropriate model of the traffic system, the most important features can be captured in simulation models, which then offer an inexpensive and riskless means for evaluating dispatching alternatives. Section 2 shows how a simulation tool can be integrated in a dispatching decision support system.

1.8 Train Traffic Control Loop

As explained in Sect. 1.2, control theory alone is unable to solve most traffic operations control problems. Nevertheless, the representation of train dispatching as a control loop is very helpful as it reveals possibilities for improvement (compare Fig. 1.3 and Fig. 1.4).

Without dispatching, the traffic follows the planned time schedule. Unavoidable random disturbances cause deviations from schedule. State-of-the-art dispatching compares the traffic process data with the schedule (Fig 1.3).

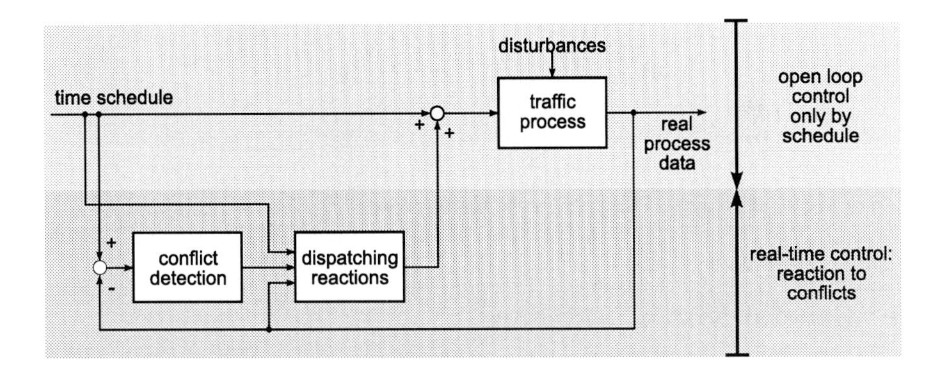

Fig 1.3 Reactive control

The dispatcher tries to detect differences and to find appropriate dispatching strategies to minimize the difference and to reduce the negative consequences of the conflict. The dispatching process can significantly be improved by the use of a fast simulation tool (Fig. 1.4).

This simulation tool yields predicted process data and, in comparison with the schedule, allows the forecast of upcoming conflicts. The simulation approach also provides sufficient time for the dispatcher to generate dispatching measures thoroughly which either prevent the conflict or minimize its negative impacts. This approach is implemented in the structure described in the following section.

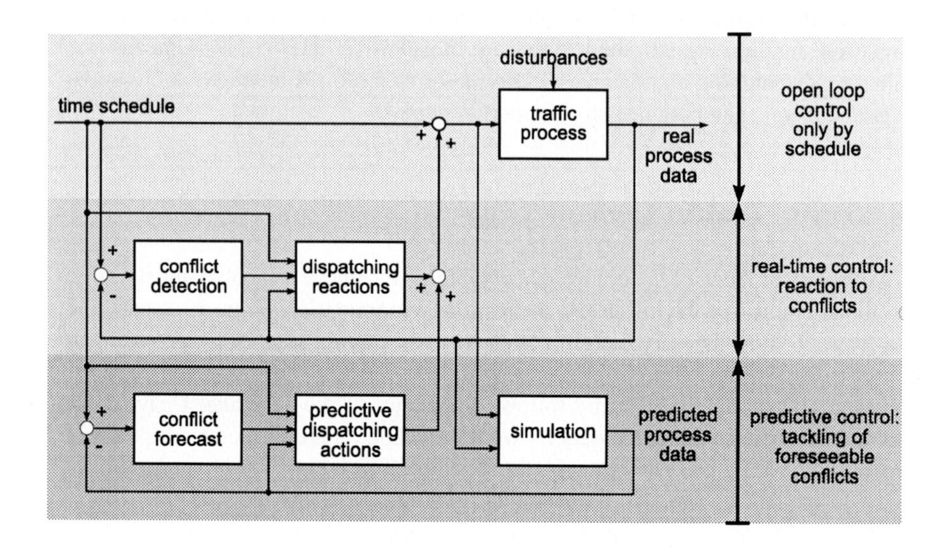

Fig. 1.4 Predictive Control

2 Dispatching Support System

2.1 General System Structure

To meet the requirements outlined in Sect. 1 and to support the dispatcher with immediate and sound assistance, an enhanced dispatching support system has been developed. The remainder of Sect. 2.1 describes the overall system structure. The different modes of dispatching support are explained in Sect. 2.2. Sections 2.3 to 2.5 focus on the three main components of the system: graphical user interface, simulation tool, and decision support system, respectively. In Sect. 2.6, the state of implementation is discussed, including an example of the functionality of the dispatching support system.

The system consists of a human-computer-interface for the visualization of actual and future traffic situations and for effective human-computer-communication, a simulation tool for the prediction of the short-term traffic conditions, and a decision support system for the development of dispatching proposals (Fig. 2.1).

The real process is controlled by the dispatched schedule. In addition, it is influenced by external stochastic events which cause random disturbances. The traffic process states are displayed to the dispatcher through the human-computer-interface. The need for dispatching activities can be deduced directly from the state of the process (e.g. in case of reservation conflicts) or from a comparison with the schedule (e.g. delays or connection conflicts).

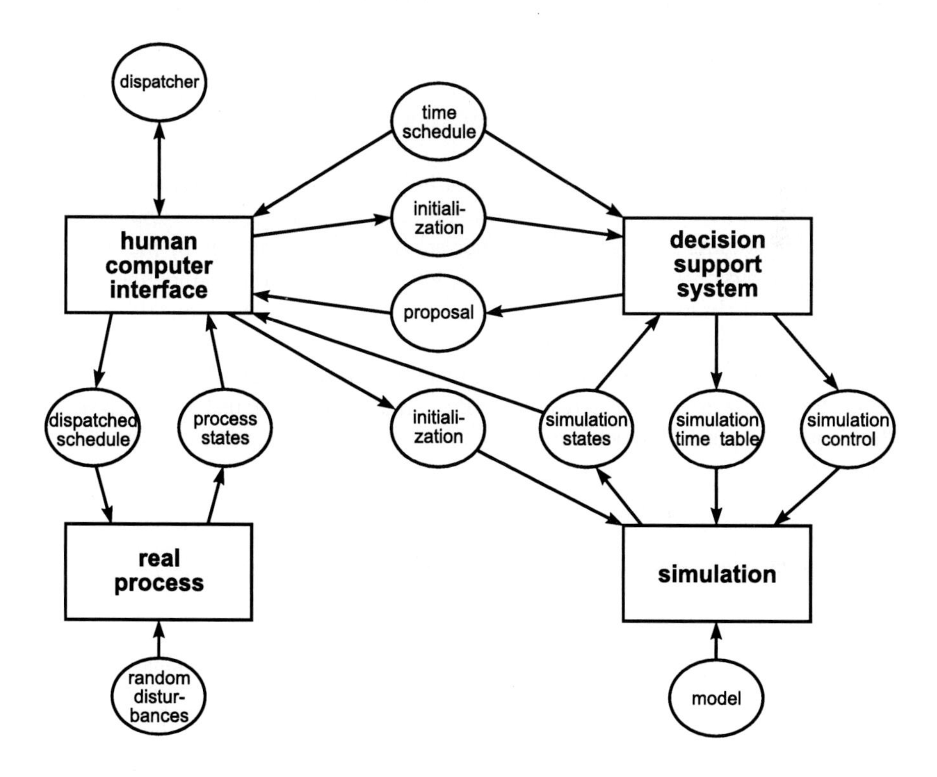

Fig 2.1 System structure

The integration between the human-computer-interface, the simulation and the decision support system is explained in Sect. 2.2. in combination with the functions provided for the user.

2.2 Different Modes of Dispatching Support

The dispatching support system assists the dispatcher in finding appropriate dispatching strategies to resolve conflicts in an effective way. The available assistance features are:

- early conflict detection,
- extraction and display of only the relevant information for the specific conflict,
- prediction of dispatching measures' impacts,
- proposal development of dispatching actions.

These four assistance functions are described in Sections 2.2.1 to 2.2.4, respectively.

2.2.1 Conflict Detection and Prediction

A crucial task in dispatching is to detect conflicts as soon as they occur or even before they occur, if possible. The traffic process is continuously scanned for existing and upcoming conflicts. The former are detected by analysis of the actual process values and comparison with the schedule and other service guidelines, the latter by periodic simulation of the future traffic conditions (approximately thirty minutes into the future). The simulation also allows prediction of the probable effects of the conflict, i.e. which trains are affected and how the conflict will spread over the network with time. The results of the simulation and detection are shown to the dispatchers, and their attention is drawn to the most urgent traffic conflicts by visual means.

2.2.2 Extraction and Display of Relevant Information

The process values of the train and its control system were classified into four layers according to their relevance for train dispatching, and different diagrams for their visualization were developed, following a multiple hierarchical information layer concept. On the highest layer, displayed by a large projection, the general state of the network is shown. On subsequent layers, smaller sections of the network are shown with those displays containing more and more detail, according to the task to be performed.

2.2.3 Prediction of the Impacts of Dispatching Actions

As mentioned in previous sections, the dispatching support system contains a simulation tool which simulates the real process as described in more detail in Sect. 2.4. The probable effects of dispatching measures which the dispatcher considers can thus be predicted. An example of this kind of support is shown in Fig. 2.2 (which is a screenshot from the implemented graphical user interface).

VORGESCHLAGENE UMLEITUNG

FAHRT-NR.: 21
FHRZG.-NR.: 200

STATION	ANKUNFT	ABFAHRT
HAMBURG-HBF	-	18:35:00
HAMBURG-BILLWERDER	18:50:00	18:55:00
SCHWERIN	19:55:00	20:00:00
BERLIN-SPANDAU	21:00:00	21:00:00

AKZEPTIEREN

ABBRUCH

Fig. 2.2 Dispatching proposal screenshot

Due to blocking of track sections, a train has to be diverted between Hamburg and Berlin over an alternative track. This detour will take more time, and the probable arrival times at the subsequent stations can be predicted by applying the simulation tool (the results are shown in Fig. 2.2). The delays (in comparison with the original arrival times) are a measure of the quality of the dispatching actions. This allows the dispatcher to assess and compare alternative dispatching possibilities.

2.2.4 Proposed Dispatching Strategy

In case of difficult conflicts which can neither be resolved by the underlying automation nor by the dispatcher alone, the dispatcher can ask the decision support system for advice. Furthermore, the decision support system simultaneously analyses the actual traffic situation as well as the one predicted by the simulation tool in order to detect conflicts as soon as possible. The function of the expert system is described in more detail in Sect. 2.5.

2.3 Human-Computer Interaction

The one developed by the authors is based on a newly designed human-computer-interface (HCI). For its development, the tasks of the dispatcher and the necessary information have been examined in a user-centered design approach. This approach led to a complete redesign of the dispatchers' work area, of the large projection displays, of their I/O-devices, and of the graphical user interface on the dispatcher's computer screen. The working conception and the visualizations increase the dispatchers' satisfaction and working quality. Several improvements were obtained in the dispatchers' working area, including a large screen projection showing a projection of the network which can be viewed simultaneously by all members of the working group. A detailed description of these strategies can be found in Müller/Fay/Schnieder (1996). This section will focus on the HCI.

State-of-the-art dispatching environments rely heavily on time-space-diagrams and a network overview. These tools are simple transformations of traditional displays and drafts onto the computer. The capabilities and power of modern computers remain largely unused, and human mental capacity and abilities are not considered in these state-of-the-art approaches.

In the development described in this paper, however, the question of how information should be presented in general and how it can be graphically transformed for easy perception and understanding by the human's mind was addressed. For efficient design, two important questions had to be addressed:

- What information does the dispatcher need to control trains in the network?
- Which process values contain this information?

To reduce the amount of data, all process information was classified according to its importance for the dispatcher.

An example is shown in Fig. 2.3 which is intended to give a quick overview of the train network and the conflicts. For this purpose, actual train positions are not necessary. Instead, the tracks between two major stations are divided into several blocks, which are again divided into parallel slices. Each of the slices represents the conflict state of this block at a certain time. The conflict state (in this case the sum of deviations from schedule) is color-coded from red to yellow, according to the colors of glowing metal or fire. Thus, the dispatcher can perceive the development of conflicts in time and space at a single glance.

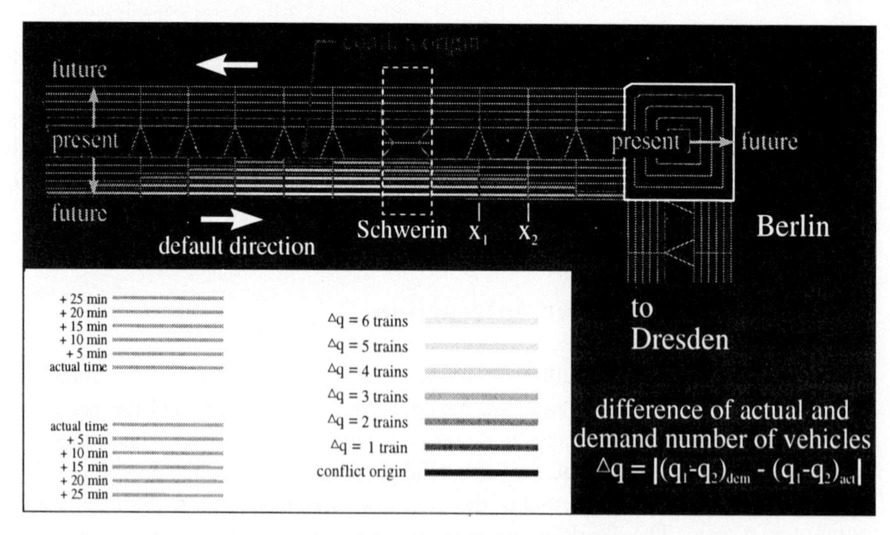

Fig. 2.3 Conflict overview

2.4 Simulation

The simulation component of the system allows a fast estimation of the future traffic situation. A simulation is always a compromise between simulation speed and the precision of the results. To assess the effects of a conflict or of a dispatching action, the traffic development over the next sixty minutes (approximately) has to be forecast. Results of the simulation should be available in less than a second, recognizing that the expert system may need several iterations, and therefore several simulations, to define a suitable dispatching strategy. Hence, a speedup factor of approximately 10,000 has to be achieved compared with real time. Nevertheless, the simulation results have to be calculated precisely enough to yield reliable results. Special attention has to be given to the train headways due to their significant effects on traffic behavior.

The execution of simulation experiments is controlled by the initialization and a set of control commands. The initialization can be either an actual state of the real process or an intermediate stage or final stage of a previous simulation experiment. With this concept, the expert system is able to reject decisions while looking iteratively for an effective dispatching strategy. It can be restarted from a good intermediate stage to try other dispatching actions.

During the course of this project, two alternative approaches have been followed to build a simulation tool that fulfills these requirements. The first is based on Petri Nets, the second is object-oriented programming.

2.4.1 Simulation Based on Petri Nets

The use of Petri Nets for a description language allows structured design of complicated and complex systems. Thus, a mathematically based proof of safety is possible in safety relevant applications (e.g. public transport). The Petri Net model completely describes the static and dynamic behavior of the train network and the control system. The system description designed within the period of system development is reused for consistency and effort reduction reasons and adapted for the simulation.

Petri Nets are a discrete event orientated description language. Every transition is checked to see whether its input and output arc inscriptions are fulfilled so that the transition is enabled to "fire". If the transition "fires", the tokens are taken from the input places and are set onto the output places, and then the test for enabling starts again. Obviously, many transitions are checked without "firing" so that the computing effort is very high for large nets, reducing overall simulation speed.

The solution to this problem is the calculation of the reachability graph for all net components, which is a complete calculation of the system states which implies the tests for enabling. It is very similar to the structure of state machines. Using this method, an efficient discrete event simulation model can be obtained. Regarding the input- / output- signals of separated partial reachability graphs as input- / output-arcs of the Petri Net submodels, a network of state machines (Mealy automata) is obtained. This way, a library of automata is developed. Some examples for automata are:

– automata for track reservation (controlling the state of the guideway),
– automata for handing over the vehicle to the next section.

Using a graphical editor a complete simulation model consisting of automata based building blocks (with the Petri Net features inside) can now be constructed.

2.4.2 Object-Oriented Simulation

Following accepted software development practice, an object-oriented simulation tool has been developed (in C++). Due to its intrinsic modularity, this allows fast simulation for different train systems and control systems. The tool can easily be adapted to the traffic system to be examined. The level of detail in the simulation can be adjusted to the desired accuracy of the result.

Distributed event-driven programming concepts are ideal to represent the distributed traffic system structure. Every train objects calculates its own movement data. The coordination of conflicting trains is performed by an event-driven timer. All train interactions and deviations are collected and stored for further evaluation and display.

An example of the capacity of the simulation is given in Sect. 2.6.2. As is shown by these (and other) examples, the C++ simulation tool provides an ideal means to predict traffic developments in the time available for an immediate system response.

2.5 Decision Support System

In Sect. 1.6, it was shown that the complexity of problems to be dealt with in real-time traffic control requires sophisticated heuristics. The knowledge of expert dispatchers, collected over many years and condensed to rules-of-thumb, forms a valuable resource for dispatching problem solving. Hence, the approach followed in this dispatching support system is to acquire this knowledge in cooperation with dispatchers and to derive a rule base from this knowledge which can automatically be used for the development of dispatching possibilities.

2.5.1 Expert Knowledge Modeling

The expert system consists of application-specific knowledge and the ability to combine this knowledge and to apply it to the problem to be solved. To ensure maintenance and adaptability of the system, the knowledge is stored in a knowledge base apart from the inference machine (Fig. 2.4); unlike with conventional algorithmic programming.

An important feature of the dispatchers' expertise is the vagueness of the rules as stated by human experts. A few rules can always be applied and are agreed by every dispatcher, but most of the rules hold only to a certain degree. This vagueness is often faced in knowledge acquisition. To achieve the desired system performance, it must not be ignored. To cope with the vagueness, two fuzzy concepts are made use of (see Fay/Schnieder (1997)):

- Whether a certain situation or condition is true is represented by a fuzzy number, e.g. a train which is 5 minutes late has a "minor delay" to the degree of 0.8 and a "severe delay" to the degree of 0.2.
- A "credibility factor" is attached to every rule to represent the dispatcher's belief in the rule (the higher the dispatcher's belief, the higher the credibility factor). For example, the rule "In case of a minor delay of an important train, let other important trains wait" might have a credibility factor of 0.7. Of course, this credibility is subject to change, according to the dispatchers priorities etc.

Together, the two fuzzy concepts allow for selection of dispatching actions which closely resemble the dispatching expert's decision process. To enhance the development and maintenance of the rule base further, a graphical description means is desirable. Therefore, the authors developed a concept of Fuzzy Petri Nets (FPN), which covers all the required features. This concept is described in detail in Fay/Schnieder (1997)).

2.5.2 Criteria for Dispatching Decisions

Dispatching decisions can not be taken with regard to only one or two main aims of the traffic provider, but with respect to a variety of criteria. Without claiming completeness, the criteria for dispatching decisions on tracks are:

- following the actual schedule as closely as possible,
- consuming little energy (for both cost and ecological reasons) by reduced maximum speeds, idle running, avoidance of braking and stops for heavy trains etc.,
- reduction of wear and tear by electrical instead of mechanical breaking,
- timing of departures to achieve optimal headways,
- reduction of overlaps by speed restrictions to free conflicting track sections,
- improvement of track usage.

The criteria mentioned above are only examples to show the complexity of the decision process. For dispatching at network nodes and partitions, additional criteria have to be taken into account. Each of them represent a complex optimization problem which is dealt with in special literature. Therefore, the criteria have to be described explicitly, and priorities between them have to be defined.

2.5.3 Development of Dispatching Proposals

Conflicts are first classified by a diagnostic tool. Then, the knowledge base is scanned for rules which are suitable for tackling or resolving the conflict with regard to overall traffic objectives and strategies. By application of the appropriate rules to the conflict situation, a set of promising dispatching actions is derived (Fig. 2.4).

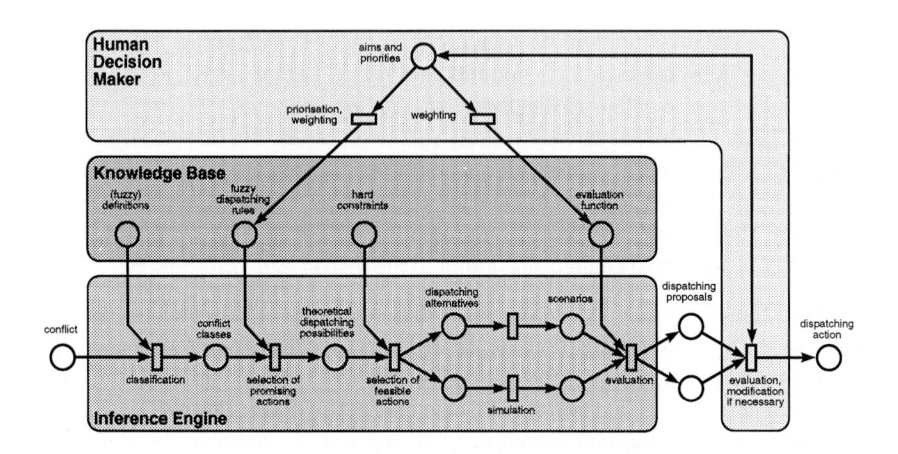

Fig. 2.4 Development of dispatching proposals

The resulting strategies are checked for satisfying the hard constraints (e. g. for overtaking, the parallel track has to be long enough). Only the strategies that satisfy all hard constraints are considered further.

To predict the probable effects of each strategy, they are applied to the actual traffic situation in parallel simulation runs, again making use of the simulation. The emerging scenarios are evaluated automatically with regard to different quality criteria, taking into account the various desires of all persons and organizations involved in the traffic process. To find out the aspects important for the customers, a conjoint measurement analysis approach has been carried out.

After having defined values that represent the most important criteria and are easy to measure in practice, in any conflict case the actual situation as well as the various dispatching alternatives can be evaluated with respect to these values.

The single evaluations alone can give hints to which dispatching alternative seems to be favorable. An effective comparison, however, requires a mathematical combination of the single values in an overall optimization function. Here, again fuzzy approaches like the ordered weighted averaging (OWA) operators can be applied (see Yager (1993)).

If the dispatching strategies seem to be inadequate to solve the conflict, or if additional conflicts arise during simulation, the selection and assessment of dispatching actions is repeated iteratively to seek further improvement. A complete search in the set of all possible solutions of dispatching problems results in a combinatorial explosion and can not be performed in reasonable time. Therefore, heuristic strategies to reduce the solution space are applied.

The dispatching measures which result in the best conflict solutions are proposed to the dispatcher, together with explanations of how they were achieved, to what extent they contribute to the solution of the problem and the general traffic optimization, and which other solutions may exist. The dispatcher can either follow the advice and accept the proposal, modify it or try out his own solutions. In any case, the dispatcher still has the final control over the executed dispatching measures.

2.6 Prototype Implementation

To prove the effectiveness of the system described above, a prototype has been developed at the authors' institute. Instead of implementing the complete dispatching support system as a whole, the key components of the system have been identified and implemented as a prototype. This is described in Sections 2.6.1, 2.6.2, and 2.6.3 for the human-computer-interface, the simulation, and the decision support system, respectively. In Sect. 2.6.4, conclusions are drawn concerning the state of implementation and the applicability in practice.

2.6.1 Implementation of the Human-Computer-Interface

As described in Sect. 2.3, the graphical user interface consists of several graphical representations on different layers. These graphics have been adapted and implemented for the special needs of the German TRANSRAPID maglev system. The development took place in close cooperation with the TRANSRAPID authorities and was supported by the German Federal Ministry of Research and Technology. All graphics have been designed conforming to the layout of the first TRANSRAPID line from Berlin to Hamburg, which is intended to start operation in the year 2005. The graphical user interface has been realized on Hewlett-Packard workstations using the graphical development kit SL-GMS©. The communication telegrams are in accordance with the ones described by the manufacturer of the TRANSRAPID control system.

2.6.2 Implementation of the Simulation

Both simulation approaches described in Sections 2.4.1 and 2.4.2, respectively, have been examined for implementation. The Petri Net simulation approach offers the advantage of directly reusing the existing Petri Net models, but at the moment, this approach still suffers from deficiencies in the Petri Net tools commercially available. Therefore, the main focus has been put on the C++ simulation. The simulation has been implemented on HP©-workstations. The simulation input (traffic network, train characteristics, time schedule, ...) and output (traffic scenarios) data interface is provided in text files. Thus, the simulation system can easily be configured and parametrized for any train traffic system to be simulated.

In a typical application, 6 trains were run on a net consisting of 23 nodes and 43 vertices. The total time interval covered approximately 2 hours. During the program run, 3306 events had to be calculated. The complete simulation run took only 0.32 s on an HP 9000/715-100L workstation (CPU PA7100LC, 100 MHz). Hence, the simulation tool is capable of computing the necessary simulation runs as fast as required.

2.6.3 Implementation of the Decision Support System

The crucial steps in the development of the knowledge-based dispatching support component (as of any expert system (see Puppe (1993)) are:

- the knowledge acquisition process,
- the appropriate modeling of the (fuzzy) knowledge,
- the structuring of the rule base for evaluation purposes.

Therefore, these steps have been carried out to prove the feasibility of our approach.

The Knowledge Acquisition Process: The knowledge has been acquired at the dispatchers' control centers during several sessions. Thus, the gained knowledge could be worked off and structured afterwards, and gaps could be closed and ambiguities and contradictions resolved in the following session. Altogether, about 100 rules have been accumulated this way, some of which have been further refined later.

Modeling of the Fuzzy Knowledge: The knowledge has been modeled in fuzzy rules to respect the vagueness of the knowledge and the uncertainty of the fulfilment of the conditions of the rules (see Sect. 2.5.1). This concept is described in more detail in Fay/Schnieder (1997). An example is given below.

Structuring of the Rule Base: The rules have been sorted according to several conflict classes. Each rule is based on two to eight conditions, which partly overlap in each conflict class. Some rules are structured hierarchically. During this procedure, even a small set of rules (three to ten rules) in one conflict class can become difficult to handle. To keep the rule base clear, however, is very important for several reasons:

- During rule base compilation, errors can be found more easily.
- During the solution of an actual conflict, the user can trace the rule evaluation - an important feature for the user's acceptance.
- Parallel to the practical use of the system, the user can maintain and enhance the rule base by adding rules, modifying rules, and deleting rules to adjust the system to his aims and ideas.

To enhance the clarity of the rule base, the rules are represented in a graphical manner by the use of Fuzzy Petri Nets (see Fay/Schnieder (1997) for details). An example is given in the following section.

An Example: The following example is based on a simplified selection from the real problem domain. It deals with the following situation:
In a certain station, train B waits for train A to arrive to allow passengers to change from train A to train B. Train A is late.

In this situation, the following alternatives may be considered:

- Train B waits for train A to arrive. In this case, train B will depart late.
- Train B departs on time. In this case, passengers from train A have to wait for a later train.
- Train B departs on time, and an additional train is employed for the passengers who wanted to change from train A to train B.

To make a decision, several side conditions have to be taken into account: the length of the delay, the number of changing passengers etc. An optimum has to be found regarding the divergent aims listed in Sect. 2.5.2. In the following handling of the example, the connection between the aims and the rule credibilities have been omitted for simplification. Concerning this traffic conflict, the following four rules have been collected from domain experts:

Rule 1: IF *many* passengers would like to change for train B
AND the delay of train A is *small*
THEN let train B wait for train A.

Rule 2: IF the delay of train A is *large*
THEN let train B depart according to schedule.

Rule 3: IF there is an *urgent* need for the track of train B
THEN let train B depart according to schedule.

Rule 4: IF *many* passengers would like to change for train B
AND train B departs according to schedule
AND train B was the last train heading for this direction on this day
THEN employ an additional train (in direction of train B).

The preconditions printed *in italics* are fuzzy and are, therefore, represented by fuzzy numbers.

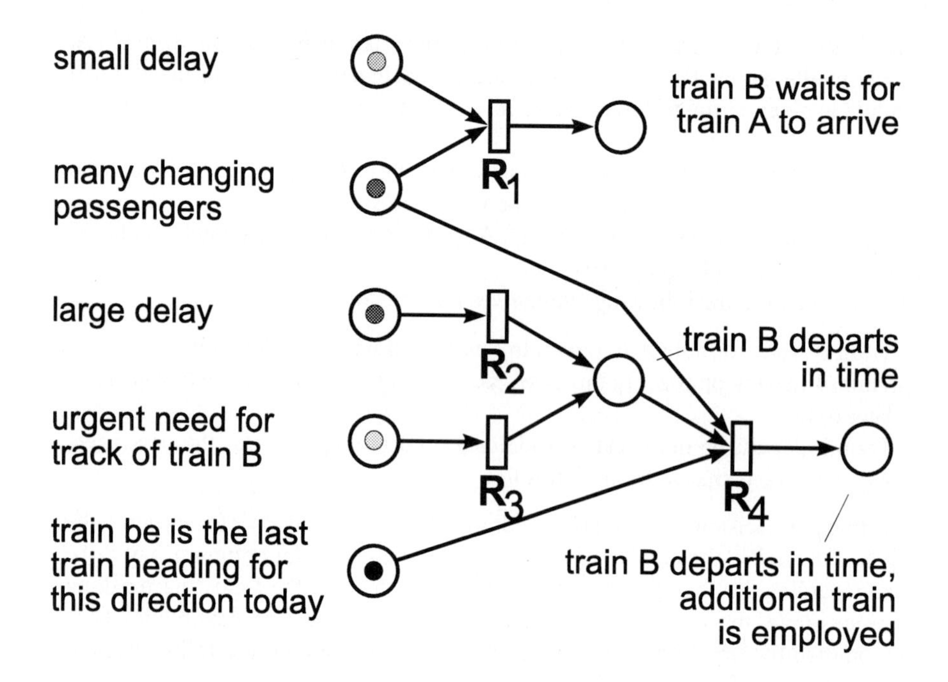

Fig. 2.5 Example of Fuzzy Petri Net for representation of fuzzy rules for train dispatching

These fuzzy numbers state

- to which degree the delay is small or large,
- in which way the estimated number of changing passengers equals the expert's idea of "many",
- how urgent the track is needed for other purposes.

In Fig. 2.5, the rules are shown together in one FPN. Note that the fuzzy preconditions are shown as shaded tokens in this graphic. In this example, "last train today" is the only crisp precondition. By means of evaluation of the preconditions, the rules which can fire are determined. Firing of these rules allows the computation of the support for the alternatives in question. In this way, the possible alternatives are ordered with respect to the preference they achieve from the knowledge base.

2.6.4 Conclusion Concerning the State of Implementation

As described in Sections 2.6.1 to 2.6.3, all three main system components (the human-computer-interface, the simulation and the decision support system) have been implemented as a prototype. These prototypes have proven the feasibility of the dispatching support system approach described in this paper. The next step towards a commercial application will be the integration of the system into a specific train traffic control system.

3 Conclusions

The intelligent assistant system for dispatching support presented in this article can be integrated in a control center. Due to its flexible and modular structure, the system is the core for the development of dispatching support systems for various public transport systems.

By making use of the achievements of modern computer science and technology, assistance can now also be provided for short-term train traffic control and management, problem domains which seemed to be rather difficult to tackle so far.

The system can assist the dispatcher by the proposal of sound, validated and traceable dispatching measures. Thus, it can contribute to an improvement in traffic performance, reliability, and customer satisfaction.

References

Adamski, A. (1993): Real-time computer aided adaptive control in public transport from the point of view of schedule reliability. Proceedings of the CASPT '93 in Lisbon, 6th Workshop on Computer-Aided Scheduling in Public Transport, 278 - 295 (Springer) Berlin.

Daduna, J.R. (1993): The integration of computer-aided systems for planning and operational control in public transport. Proceedings of the CASPT '93 in Lisbon, 6th Workshop on Computer-Aided Scheduling in Public Transport, 347 - 357 (Springer) Berlin.

Fay, A./Schnieder, E. (1997): Fuzzy Petri Nets for knowledge representation and reasoning in rule-based systems. ISFL '97, Proceedings of the 2nd International ICSC Symposium on Fuzzy Logic and Applications, 146 - 150, (ICSC) Zurich.

Fernandez, A./de Cuadra, F./Garcia, A. (1996): SIRO - an optimal regulation system in an integrated control centre for metro lines. Proceedings of the COMPRAIL '96 in Berlin, 299 - 308, Computational Mechanics Publication, Southampton.

Komaya, K./Fukuda, T. (1989): ESTRAC-III: an expert system for train traffic control in disturbed situations. Control, Computers, Communications in Transportation, 147 - 153. IFAC/IFIP/IFORS Symposium (Pergamon Press) Paris.

Moirano, M./Rossi, C./Sassi, C./Sissa, G. (1989): An underground railway traffic expert system. Proceedings of the International Conference on Artificial Intelligence in Industry and Government, 330 - 335, Hyderabad.

Müller, J.O./Fay, A./Schnieder, E. (1996): DISPOS - maglev train traffic supervision with a new human-computer interface. INFORMATIQUE MONTPELLIER '96, Proceedings of the 5th International Conference, Interface to Real & Virtual Worlds, 219 - 227 (EC2 & Cie) Montpellier.

Nachtigall, K. (1996): Train scheduling on a single-track railway. Proceedings of the SOR Symposium on Operations Research (DGÖR/IFIP) Braunschweig.

Oser, U./Thorwarth, W./Biernatzki, F. (1994): Betriebszentralen - eine neue System-technische Lösung für das Netz der Deutschen Bahn AG. Eisenbahningenieur 45 (1994) 12, 859 - 866 (in German).

Puppe, F. (1993): Systematic Introduction to Expert Systems - Knowledge Representations and Problem-Solving Methods. (Springer) Berlin..

Schaefer, H./Pferdmenges, S. (1994): An expert system for real-time train dispatching. Proceedings of the COMPRAIL '94. in: Madrid, 27 - 34, Computational Mechanics Publication, Southampton.

Yager, R.R. (1993): Families of OWA operators. in: Fuzzy Sets and Systems 59 (1993), 125 - 148.

Zwaneveld, P.J./Kroon, L.G./Ambergen, H.W. (1996): A decision support system for routing trains through railway stations. Proceedings of the COMPRAIL '96 in Berlin, 219 - 226, Computational Mechanics Publication, Southampton.

Requirement for, and Design of, an Operations Control System for Railways

Leena Suhl and Taïeb Mellouli
Decision Support & OR Laboratory, Department of Business Computing,
University of Paderborn, Germany.

Abstract: In this paper, we discuss computer-based systems that support operations control for railways. Specifically, we focus on the design of decision support tools for dispatchers. We have studied requirements of systems supporting operations control processes from the expert/user point of view for the German railway, "Deutsche Bahn AG." The main goal is to help dispatchers ensure passenger traffic with the best possible quality in a dense network with more than 30,000 trips daily. We suggest that such a computer-based system must support the dispatcher in recognizing forthcoming conflicts as early as possible, rescheduling of passenger connections taking into account customers' acceptance and cost, and reallocating vehicles/crews in a convenient way.

We represent a knowledge-based, object-oriented system architecture that includes components for information management, simulation, and optimization. Besides the scheduled state of a transportation system, our model also covers the actual (until now) and expected (in the future) states. The decision support system automatically recognizes conflict situations, such as missed passenger connections due to trains being delayed. What-if analyses can be performed to support the estimation of network-wide effects of a dispatcher's decision. Expert's knowledge rating possible decisions may be formalized and stored in the system to help in generating later decision proposals. Remote Java-based clients provide service point employees and passengers with station overviews and alternative connections based on the actual state of the network. Vehicle rescheduling is currently in progress.

1 Operations Control and its Interaction with Planning Phases

Providers of passenger and public transportation systems, such as airlines and railway companies, are faced with a very complex planning and control process consisting of several phases, including demand estimation, timetable construction, resource allocation, and operations control. In the planning phase, the schedule and resource allocation are determined. We distinguish three main phases of the planning process: Product Planning, Production Planning and Scheduling, and Resource Allocation. These phases are followed by the Operations Control phase including rescheduling of passenger connections as well as shifts in the allocation of resources such as vehicles and crews. Fig. 1.1 shows the main components of this planning, scheduling, and control process (see Suhl (1995) for further details).

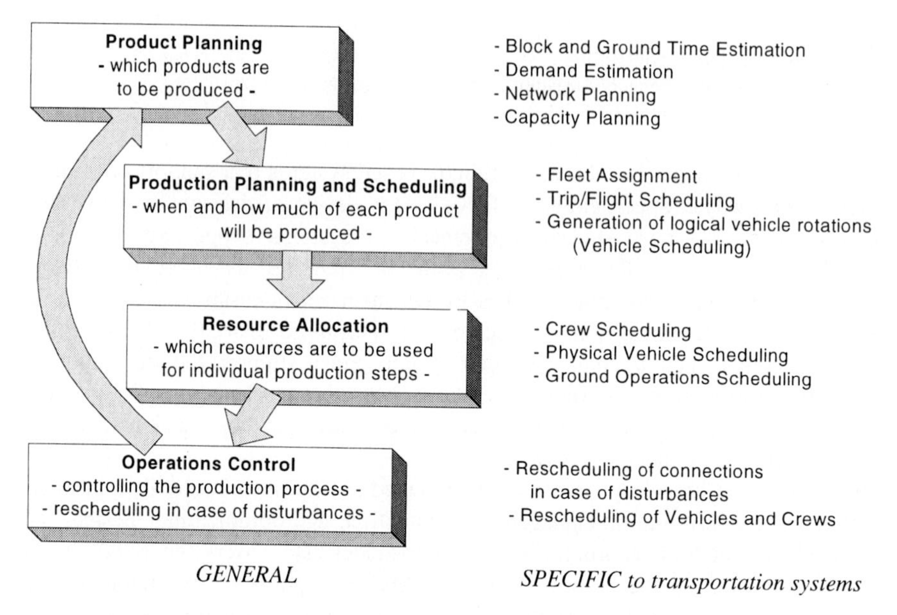

Fig. 1.1 Phases in Production Planning, Scheduling, and Control

This paper discusses computer-based systems to support the decision making of an expert in the last phase of this cycle. Such a system should be able to help a dispatcher ensure passenger traffic with the best possible quality. In the case of a large, dense network, as in the case of Deutsche Bahn with about 30,000 trips daily, it is a challenge to design ways to support a dispatcher in his/her tasks related to operations control. In passenger traffic, short-term changes of a given schedule are required because of disturbances, such as technical failures, or accumulated vehicle delay. For example, trips have to be delayed or new trips introduced to enable passengers to make their connections or take alternative routes to reach their destinations. The former case implies rescheduling of passenger connections and the latter rescheduling of vehicles and crews.

Deutsche Bahn is currently in the process of being privatized and simultaneously split into nine independent companies. This means that in the future new requirements for decision making will have to be taken into account. It will be important to know, to a much greater extent than today, who caused a certain disturbance and to whom a decision is going to be disadvantageous as well as the costs of a certain decision for each independent company involved.

For two years, the authors have been involved in projects dealing with design and development of computer-aided systems to support vehicle scheduling and operations control for railways (see Suhl/Mellouli (1997), Mellouli (1997a)).

We suggest that these two components interact with each other as shown in Fig. 1.1, and propose a tight integration of both components within a Scheduling and Control System as will be discussed below. The product of a commercial transportation provider is a transportation service defined by a time period (e.g., summer 97), day of the week (e.g., daily, except Sundays), departure and arrival times, the terminal stations, and capacity (expressed in number of passengers, possibly divided into several service classes). One special instance of a product is called a trip. The construction of rotations (sequences of trips) of a vehicle schedule involves trip times (block time) and minimum turnaround times between two consecutive trips (ground time).

Results of planning, that is, automatically generated schedules, are forwarded to operations control. In the case of railways, a schedule for trains consists of periodically workable vehicle rotations which have to be feasible with respect to maintenance. A maintenance strategy could be that each train has to be checked every three days for about four hours in one of a number of maintenance bases. Large-scale vehicle scheduling tasks of this type were efficiently solved using a mathematical network flow model for vehicle scheduling with maintenance routing proposed in Mellouli (1997b). During operations, a given vehicle schedule is subject to changes if disturbances occur. These changes are performed within the vehicle rescheduling component of the operations control sub-system. Crew schedules may be forwarded and managed analogously to vehicle schedules.

For a second component dealing with rescheduling of passenger connections, we also need the intermediate stations of the trips together with their arrival and departure times. As an example, train number ICE577 stops in Kassel at 1:40 p.m. and leaves this station at 1:41 p.m. on its way to the next station Frankfurt/Main en route to Stuttgart. In Kassel, there are connections such as train IR2550 starting at 2:00 p.m. to Düsseldorf serving Paderborn, Dortmund, and Essen. A main goal of this system component is to support the dispatcher in performing fast what-if analyses. Thus, he or she may consider network-wide effects of possible decisions and provide suggestions to reschedule affected connections so that factors such as customers' acceptance and costs are taking into account.

The control components of a computer-based support system may be used, except in the control phase, also to improve planning quality. First, rescheduling experience and what-if analyses by dispatchers help to improve planning in the future by detecting bad connections in the planned timetable/schedule. Second, having constructed a timetable/schedule in the planning phase, the operations control component may be used to test its robustness in the event of disturbances. Therefore, distributions of disturbances, for example for lateness, have to be constructed in order to simulate a given schedule.

In the next section, we discuss the main problems in operations control of transportation systems. Requirements for an operations control system for railways are then presented within a general framework in Section 3. We then design a system architecture integrating various problem solving components. Finally, Section 5 presents a decision support tool for dispatchers with an interface for expert users, which has been partially implemented within a prototype.

2 The Operations Control Problem in Transportation Systems

Because traffic is always subject to many external factors, many types of disturbances cannot be avoided which make short-term changes to a given schedule necessary. When a disturbance such as a technical defect, accumulated vehicle lateness, missing crew members, or congestion, occurs, a dispatcher has to react within a few minutes, sometimes seconds, and decide about changes in the schedule and reallocation of resources. It often takes several days to reconstruct a consistent schedule after ad hoc changes.

In a control center, especially for airlines and railway companies, there is usually a team of dispatchers working 24 hours a day 7 days a week. To ensure fast and good decisions, permanent connections are needed between moving vehicles, stations expecting the vehicles to arrive, other control centers, and crews ready to be called to work at any time.

Normally, a dispatcher first notices the symptoms of a disturbance and then has to find out the underlying reasons. Frequent symptoms are, for example,

- delay of a vehicle
- delay of a crew supposed to start working
- a closed station or airport
- unavailability of part of the network.

The reasons leading to various symptoms can be, for example,

- bad weather, such as fog or ice
- vehicle has technical problems
- too much traffic waiting for takeoff or landing at an airport: inadequate capacity available
- a train (or aircraft) waits for a late connecting train (or aircraft)
- security problems.

After analyzing the situation, the dispatcher has to decide about

- canceling flights or trains
- rerouting vehicles
- exceptions to waiting rules between trains
- temporary changes in the given schedule
- temporary train or flight connections beyond the given schedule.

The decisions of a dispatcher will usually affect many passengers and, therefore, have to be made very carefully. A train arriving late at a station may imply delays to other trains waiting at the station to ensure passenger connections. Since the dispatcher has great responsibilities, he or she should be supported in an optimal way. Current support systems may be improved, especially, if exact data about numbers of connecting passengers is available.

Our experience shows that dispatchers are able to make good decisions, if there are only a few (less than three) disturbances occurring simultaneously. However, if there are many occurring simultaneously, the quality of the decisions made by a dispatcher without computer support deteriorates. Often it takes several days to recover from the decisions made in such a situation. In this area, the available computer support is not yet optimal. In the following, we discuss in more detail the requirements for, and design of, a computer-based decision support system for dispatchers, as well as the reasons why there are currently no such systems generally available.

There are several types of decisions that a dispatcher has to make. Some of them are fixed based on given rules within the company, others require ad hoc decisions by a human dispatcher. Since it is not possible to anticipate or model all types of decisions of a dispatcher, it is also not possible to automate them. However, they should be supported in an optimal way. In this area, it is our belief that computer-based support systems do not yet exploit available technologies in an optimal way, for several reasons:

- Dispatchers work under great time pressure. If there are five telephones ringing simultaneously, there is no time to position a mouse on a screen and search for data in a computer-based information system; he/she has to react immediately.
- Some decisions are difficult or impossible to formalize: The type of decisions to be made varies from well-structured to completely unstructured ones and many decisions can only be made based on the personal experience of a dispatcher.
- We are considering a highly distributed system: information is usually required simultaneously from the moving vehicle and other stations as well as other control centers.
- Large amounts of information are required: a decision can affect the whole network since often connections cannot be guaranteed without further changes of the schedule which again imply further changes and so on.
- Dispatchers are often used to make decisions without computer support and may not see the advantages of a decision support system at first glance. This may lead to a situation where a support system has been installed but is not used in an optimal way, if at all.
- The information technology infrastructure of transportation companies is often very heterogeneous: several incompatible systems were installed at various times and are in parallel use.

It is of critical importance that the decisions of a dispatcher are communicated in time to the passengers, who are affected by the given changes. However, in today's complex railway networks it is not possible for a passenger to have up-to-date information about all connections in situations where significant changes were caused by short term decisions. Normally, all passengers with questions have to contact a service point employee who often receives information about current changes by telephone. Passengers themselves as well as service point employees would benefit greatly from reliable, actual information about connections considering changes in the schedule. The information should be provided in an easily usable format allowing flexible queries. Since a large transportation network contains thousands of stations with different levels of equipment, this is a complex task not yet solved optimally.

3 Requirement for an Operations Control System for Railways

In this and the following sections, we describe methods and solutions designed to improve the quality of decision support for dispatchers, specifically in the railway context.

A necessary requirement for a dispatcher support system in a large network is the ability to cope with an extremely large amount of data. Basic data of an operations control system involves information about the scheduled trips and the actual trips. The data handling system has to monitor all differences between these two. In advanced transportation systems, especially airlines and railway companies, the real position of a vehicle is transmitted automatically from a vehicle control system to the information system of the control center. For example, the most frequently used lines of Deutsche Bahn have a measurement device every three kilometers automatically sending a signal to the control center when a train passes. Real-time data of this type is shown graphically on a screen for the dispatcher.

Our discussions with dispatchers have shown that most poor rescheduling decisions during operations result from the actual state not being fully processed by humans in a short time interval. It is impossible for a dispatcher to estimate correctly all the effects of a decision because of the extremely large amount of information in a dense network consisting of more than 30,000 trips daily. Thus, the system should be able to represent this information in such a way that a human dispatcher can recognize the essential problems needing to be solved.

A dispatcher support system should also be able to support decisions of very different types. The problem types vary from well-structured (formalizable) to extremely ill-structured (non-formalizable). Ideally, the system would be able to recognize the structure of a given problem and provide automatic solutions to the well-structured problems as well as optimal support to the ill-structured ones. Besides that, dispatchers have their individual ways of decision making and would like the system to support their own methods. This means, that even for one special problem, different solution techniques and methods have to be provided.

The question of whether a train should wait for a late connecting train is among the most important decisions that a dispatcher makes, and serves as an example of a well-structured as well as of an ill-structured problem. Small delays can be handled according to general waiting time rules expressing a maximum waiting time for a special connection, but larger delays with many passengers involved are always treated as special cases requiring human intervention.

At Deutsche Bahn, there is a corporate rule system ("Wartezeitvorschrift") which defines for each type of train for which other types it has to wait and for how long. Each train belongs to a given type, including ICE (InterCityExpress), IC (InterCity), EC (EuroCity), and IR (InterRegio). There are also rules given for special situations: for the last connection of a day, a train is supposed to wait longer to ensure that passengers arrive at their destinations that day. Local special rules may be defined to overwrite the global, general rules.

Note that not each deviation from the schedule constitutes a conflict situation. If a connection train waits according to given waiting time rules, then passengers are supposed to make their connections, and the induced deviation of the schedule is assumed not to have further ramifications. Thus, the dispatcher need not react in this case. A conflict situation arises, if the connection train would have to wait for passengers longer than the regular waiting times, including local exceptions, imply. If the train does not wait, passengers may have to wait too long for the next connection train. If the dispatcher lets the train wait, perhaps other more complicated conflicting situations may be produced in other stations inducing inherent costs. This is the case particularly if extra trains or crews have to be introduced.

The Deutsche Bahn waiting time rules are expressed in a book with more than one hundred pages containing general rules and local exceptions. These rules consider minimum transit times (MTT) needed by passengers to proceed from one railway platform to another within a station as well as regular waiting times (RWT) depending on the types of connecting trains. Examples of these times are:

- For the case of minimum transit times: If the connection train leaves the same platform or that on the opposite side from the arriving train, MTT is respectively 2 and 3 minutes. Otherwise the passengers have to change platform by using a tunnel, and the MTT is 5 minutes.
- For the case of regular waiting times: StadtExpress-trains (SE-trains) for local traffic wait for EC-trains up to 10 minutes but EC-trains do not wait for SE-trains. ICE/EC/IC/IR-trains wait for each other up to 5 minutes. SE-trains and other local trains also wait for each other up to 5 minutes.

One of the main difficulties is that the given times may differ from one station to another. For example, Munich has a large railway station with long distances within the station and many types of possible connections. This is why there are a lot of exceptions to general rules in Munich: some particular trains do not have to wait for certain others, e.g., IR of line 25 does not wait for ICE of line 4, contrary to regular waiting times. The minimum transit times in Munich may be as high as 15 instead of 5 minutes, if the railway platforms of connecting trains are far apart.

A dispatcher may decide that a train waits longer than the waiting time rules allow, if this helps many passengers to make their connections and if the implied conflicts at other stations (and implied delays of other trains at the same station) are manageable. A decision of this type depends, among other things, on the number of transit passengers for each connection as well as on the network structure and amount and severity of further conflicts which may be created. Since this may be a very complex task, it often cannot be performed manually in an optimal way, especially under time pressure. A decision support system should help the dispatcher to make the right decision, especially by processing the possible network-wide effects of each decision.

We suggest that a computer-based system for operations control must support the dispatcher in:

1. Modeling the given schedule and the actual state of the network in such a way that differences can easily be recognized (the difficulty here is to handle the overwhelming amount of data involved),
2. Managing corporate rules, such as waiting time rules with local exceptions,
3. Automatically recognizing and locating forthcoming conflicts as early as possible,
4. Rescheduling passenger connections taking into account factors such as customers' acceptance and cost,
5. Reallocating vehicles and crews so as to minimize the effort to regain the initial scheduled state or another consistent state.

Dealing with this complexity - related to rescheduling of passenger connections caused by train delays - an operations control system also has to differentiate between several types of disturbances and to take advantage of special methods to solve conflicts. In the case of railways, we distinguish between: (1) train delay, (2) train technical problems, (3) track damage between two stations, and in some exceptional cases, (4) an unavailable station. To solve each of these conflict types, the dispatcher has to perform one or more of the following steps:

- Detect the disturbance and recognize its cause,
- Estimate its network-wide effects, especially, recognizing implied conflicting situations at other stations,
- Decide about delays of connection trains considering conflicting situations at stations,
- Organize maintenance for trains with technical defects,
- Introduce new trips with new trains and/or new crew,
- Determine alternative routes for trains,
- Inform passengers about alternative connections to reach their destinations.

4 A System Architecture for Operations Control

We emphasize that a computer-based system for operations control should neither be a black box which solves all the problems caused by different types of disturbances, nor a mere information system telling the expert which steps should be performed to solve the conflicts. Rather, we suggest that the system should, in the first place, provide solution components for specific problems, such as

- Simulating the network-wide effects of disturbances automatically. This is possible with a reasonable representation of the actual state of the network.
- Updating the actual state of the network in the system including integration of new and old trains.
- Automatically detecting conflict situations affecting passenger connections.
- Checking all corporate rules, especially the waiting time rules taking into account local deviations of regular waiting times and minimum transit times.
- Performing what-if analyses to estimate network-wide effects of a decision.
- Automatically determining alternative routes for passengers to reach their destinations based on the actual state of the network, taking into account expected train delays.
- Updating the actual state of vehicle schedules and crew schedules as well as of the numbers of vehicles and crew members available or on-call at stations.
- Automatically determining possible alternative routes for trains in case of partial unavailability of the railroad based on the actual network state.
- Recording statistics for causes of disturbances for future analysis.

Each of these problem solving components can be used to solve particular problem types. If a conflict is detected by the support system, it must be determined whether some of these components are suited to solve or to support solving the problem. This can partially be carried out automatically by a Dispatching Assistant using problem classification knowledge (see Fig. 4.1) or by the expert who can work interactively and activate some specific what-if analyses using the Dispatching Workbench. The results of analyses carried out by the problem solving components are communicated to the dispatcher through a graphical interface displaying relevant information. This supports the expert in his/her decision making process. In addition to this, a knowledge-based component may be used to rate possible alternative decisions based on formalized global/local rules of the expert and on the results of what-if analyses. The best possible choices are provided to the user as proposals, letting the user make the final decision.

Conflicts which are to be resolved by the expert with system support may either be special disturbances, e.g., track damage between two stations reported by a train driver, or conflict situations detected by the operations control system (see below). Especially, in case of special disturbances, the classification of the problem is complex and an expert should decide in the first place what kind of work and which analyses are needed to reach a decision. To support this, the system may provide the expert (through a case-based support component) with descriptions of similar disturbances that occurred in the past together with the decisions made by experts at that time. For example, one of the tasks to be carried out in the case of track damage is to find alternative routes for passing trains and to inform passengers in nearby stations about alternative routes. For such cases, problem-solving components, for example using shortest path algorithms based on the actual network, may be called to provide solutions for the current problem.

The proposed system architecture consists of three main levels (Fig. 4.1). The Dispatching Workbench and the knowledge-based Dispatching Assistant (discussed below) constitute the higher level of the system.

The middle level of the support system contains several problem-solving components together with case-based support and statistics components which are called up by the Dispatching Assistant. In addition, a conflict detection component automatically checks global and local rules and passes detected conflict situations, especially violations of waiting time rules, to the Dispatching Assistant.

The problem solving components and the conflict detection component work on the basis of the actual state of the network which is managed by an object-oriented component constituting the main part of the system's lower level. The object-oriented data management component is capable of modeling the actual state and the computed expected state of the transportation network as well as the original schedule. Whenever vehicles are delayed, the delay is propagated over the entire network. Information on train delays may be transmitted either automatically from measurement devices or manually by users. Through Monitoring and Information System Components interacting with object-oriented data management, dispatchers, service point employees, and passengers have access to the actual state of the schedule concerning passenger connections and/or vehicle and crew schedules.

Therefore, the proposed system supports the dispatcher in processing the extremely large amount of information about disturbances and predicting their network-wide effects. This enables the system (and hence the expert) to be informed about the current state and to predict conflicts before they occur based on reliable data. Thus, the system provides the necessary time and information to reschedule connections in an optimal way.

To recognize conflicting situations automatically, we modeled the waiting time rules of Deutsche Bahn in our prototype system. As discussed earlier in this paper, these rules consider the minimum transit times for passengers and the regular waiting times depending on the types of connection trains. The difficulty of having more than one hundred pages of exceptions was overcome by the object-oriented architecture, since local exceptions of each station can be managed within an object modeling this "static" station. Each trip served by a train in the network is modeled as a list of objects, say "dynamic" stations, inheriting the scheduled arrival and departure times from the corresponding static station. Thus, several dynamic stations of different trains interact with one static station where information on local minimum transit times and waiting times can be called, as needed. Whenever

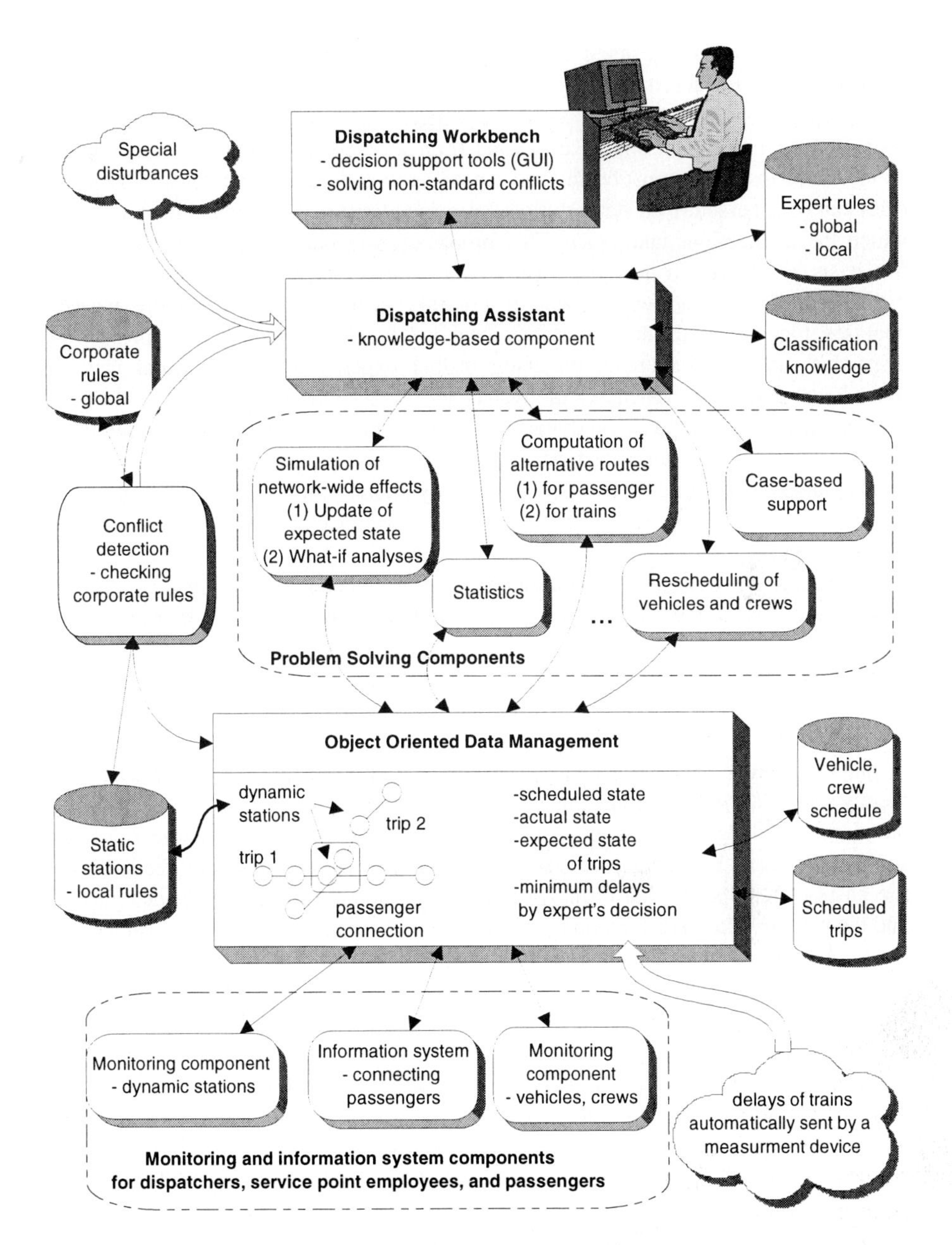

Fig. 4.1 System architecture of a decision support system for operations control

lateness of trains is registered by the system, the actual values of arrival and departure times are stored in the dynamic stations.

A propagation mechanism updates the expected arrival and departure times at subsequent dynamic stations of a delayed train as well as at dynamic stations of connection trips which must be delayed according to general or local rules. The latter rules can be accessed from static stations corresponding to those dynamic stations at which the connections take place. This propagation mechanism can also be run as the basis of a what-if analysis. Thus, our system supports the dispatcher in estimating the network-wide effect of possible decisions by performing what-if analyses together with an automatic check of local constraints and rules. The dispatcher can then concentrate on using his/her experience in an optimal way to make better decisions. Whenever the dispatcher decides that a train is to be delayed, this information is stored in the dynamic station of the connection train separately as a minimum delay (and not merely handled as expected departure time). The reason for this is that expected times may be overwritten by updates induced by the propagation of certain other delays. However, this is to be avoided in order not to lose the expert's input.

To support the task of rescheduling passenger connections, our system starts a what-if-analysis for the case that the connection train waits in a conflicting situation. The result of this analysis is used as a factor to support making the right rescheduling decision. Other factors may be the number of passengers taking the connection train, the necessary waiting time for the next connection, and also the induced delays of other trains using the same platform or the same rail parts. Besides displaying results of what-if analyses based on reliable data (actual schedule, not merely scheduled information), our system may generate *proposals* for rescheduling based on knowledge-based rules previously formulated by experts. These rules are supposed to formalize the expert's rating process of different alternative decisions on the basis of results of what-if analyses and other factors. Since the relevant factors and their rating criteria differ among experts, this component has to be developed in tight collaboration with the experts.

The task of reallocating vehicles and crews may be supported in two phases. As a first step, a graphical interface for the planned and actual schedules allows the dispatcher to make changes easily. As a second step, optimization techniques initially developed for the planning phase can also be used. However, the goal is no longer to compute an optimal global schedule, but to minimize the effort of reconstructing the planned state or another consistent schedule state. For this, we proceed as follows. Given an initial optimal schedule, i.e., set of rotations for a number of vehicles, say V_1, V_2,..., V_p. If a vehicle, say V_i, is delayed in such a way that the next planned trip of its rotation cannot be operated, either the trip must be delayed or another vehicle is used to serve the next trip of the delayed vehicle V_i. If an additional unused vehicle V_{p+1} is available in the same station or in a nearby station, simply use this vehicle V_{p+1} instead of V_i to serve next trips of V_i's rotation and let V_i stand instead of V_{p+1}. If no additional vehicle is available in the station or in a nearby station, we have to make changes in the current vehicle schedule. The idea is to change parts of some vehicle rotations by inserting some deadhead trips (non-profit trips) to serve next trips of V_i's rotation by another vehicle. One way to do this is first to suppose that a vehicle V_{p+1} is available. The unrealistic vehicle schedule with p+1 vehicles is considered, and the residual network for the not yet performed part of this substitute schedule is constructed. After inserting "promising" deadhead trips, a *shortest path* from the sink, (see Mellouli (1997a)), which has *to go back through* the flow introduced by V_{p+1}, corresponds to the minimum deadhead to resolve the conflict, i.e., to get a p-vehicle schedule. Aspects of assuring maintenance for rescheduled vehicles and minimum schedule changes are being studied.

5 Decision Support Tools

In this section, we present some interactive, graphical decision support tools which are partly developed in a prototype within our laboratory and have achieved experts' acceptance. During operations, the expert is interested, in the first place, to get a clear representation of the actual state of trips as well as of vehicle and crew rotations and their deviation from the scheduled state. This may be supported by Monitoring and Information system components (see Fig. 4.1).

Often, the expert wishes to analyze a situation at a special station of the network. A representation device which can be used to illustrate passenger connections at stations and also vehicle/train connections to trips at terminal stations is given in Fig. 5.1. This representation is useful for many components of the operations control system and may be used by dispatchers, service point employees, and vehicle and crew (re)schedulers. In some cases, the behavior of a single trip is considered. This view may be easily supported by a list of stations with the necessary information on scheduled, actual as well as expected arrival and departure times. Within a support system, we may have two menu points from which the scheduled state and the actual (optionally together with the expected) state of the network may be accessed, respectively. From each of these menu points, information on passenger connections, train rotations, and so on may be called. Information may be requested locally for a station or for a special line. Fig. 5.1 outlines this idea.

Support tools for resolving conflicts discussed below use corporate knowledge, including waiting time rules, minimum transit times, and numbers of available vehicles and crews at stations, as well as formalized expert knowledge. This knowledge may be accessed from two other menu points (see Fig. 5.2). For example, the expert may easily access a special local exception to minimum transit times by clicking a mouse and specifying the station considered (without consulting a book with more than one hundred pages).

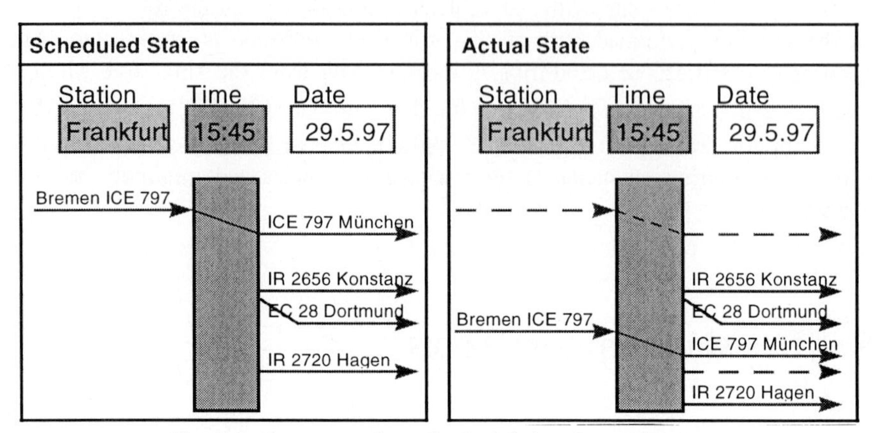

Fig. 5.1 Representation of scheduled and actual states

Whenever conflict situations are automatically recognized by the conflict detection component, the expert/dispatcher is informed where this conflict exists. This may be illustrated by an indicator on a graphic as shown in the following screen view of the prototype developed in our laboratory (see Goecke (1996)).

In this example, the ICE 797 coming from Hamburg was delayed shortly before Hannover on its way to Munich. After propagating this delay and checking waiting time rules automatically, a conflict situation is detected in Hannover, since passengers may miss their connecting trips at this station. The expert clicks on station Hannover to analyze the situation. The screen view of Fig. 5.3a then shows the current state of connections for this station. The train ICE 797 is expected to arrive in Hannover at 1:33 p.m. instead of 1:18 p.m. Passengers may miss their connection trains IR 2343 to Braunschweig and ICE 789 to Munich with scheduled departure times 1:26 p.m. and 1:27 p.m., respectively. The expert may analyze each of these connections by clicking the respective buttons on the right-hand side.

The screen view of Fig. 5.3b is divided into three parts. The first part shows results of a what-if analysis carried out by the system for the case that IR 2343 waits. The induced conflicts in this case may be shown in a tree form on request (see Fig. 5.4). The second part outputs a proposed decision of the system based on expert knowledge. The third part considers special solutions to the conflict, such as calculating alternative routes for passengers using other possible substitute trains or considering an exceptional stop of another passing train.

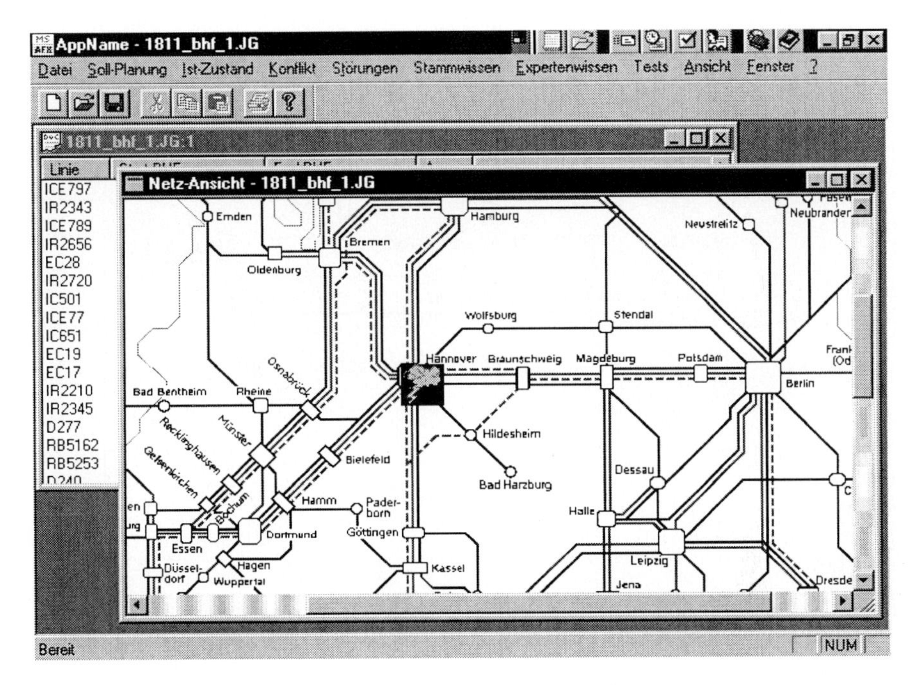

Fig. 5.2 A graphical representation of conflict location

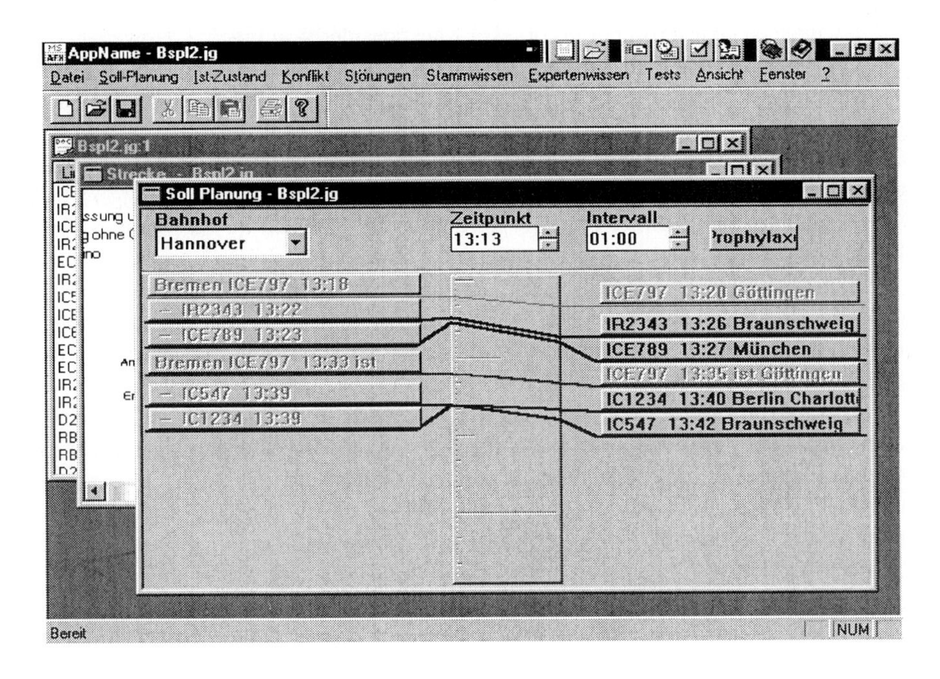

Fig. 5.3a Analysis of passenger connections for a conflict solution

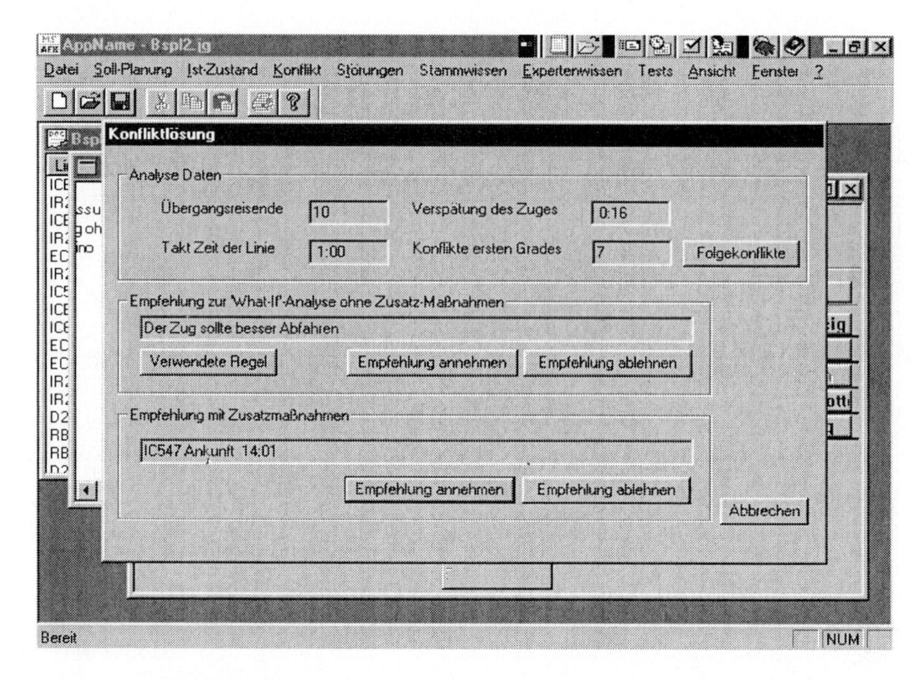

Fig. 5.3b Analysis of passenger connections for a conflict situation

Only induced conflict situations of level 1 are considered in Fig. 5.4. However, the expert may specify the mode of what-if analysis or click on a special connection as shown in Fig. 5.4 to analyze it further by clicking the button on the top right-hand corner.

In addition to this component supporting dispatchers in rescheduling passenger connections, monitoring and information system components for passengers and service point employees (see Fig. 4.1) have been implemented in the prototype (see Stelbrink (1998)). These components make use of the central object-oriented component containing the actual and expected state of the overall network via dynamic station objects (see Fig. 4.1). This subsystem is based on Intranet-technology implemented in Java and consists of a station-server connected to the dispatching data base as well as different clients.

Fig. 5.5a and Fig. 5.5b show screens of the client to be used by service point employees. They may ask for a particular connection for passengers or request an overview of passenger connections at certain stations. To determine a particular passenger connection, a shortest path problem is solved based on the actual state of the network. For example, in Fig. 5.5a a delay of train B43051 is expected to occur between Salzkotten and Paderborn. This causes a delayed departure of train IR02552 from Paderborn to Brackwede. The screen view in Fig. 5.5b illustrates the situation in Paderborn when train B43051 from Lippstadt has a delay of ten minutes.

Connecting trains are shown as a tree which can contain several branches corresponding to further conflicts and connections.

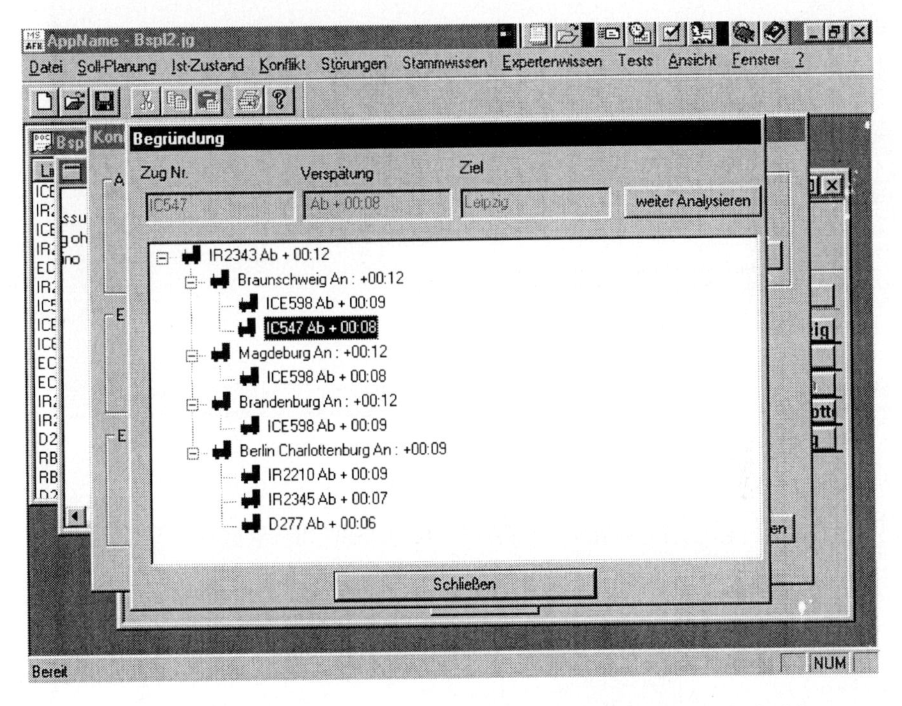

Fig. 5.4 Representation of induced conflict situations by an expert's decision

To illustrate overviews of passenger connections at certain stations, the graphical user interface illustrated in Fig. 5.1 and Fig. 5.3a may also be used as an alternative representation.

Fig. 5.6 shows the input screen of a self-service terminal for passengers. Letters of a station name can be typed on a touch-screen, and the system looks for corresponding stations on the server. The current time is provided as a default, because a passenger normally looks for his or her connection right now, but other times are also possible. In addition to passenger connections, self-service terminals may also be used to request other useful information, e.g., about station, city, and local traffic.

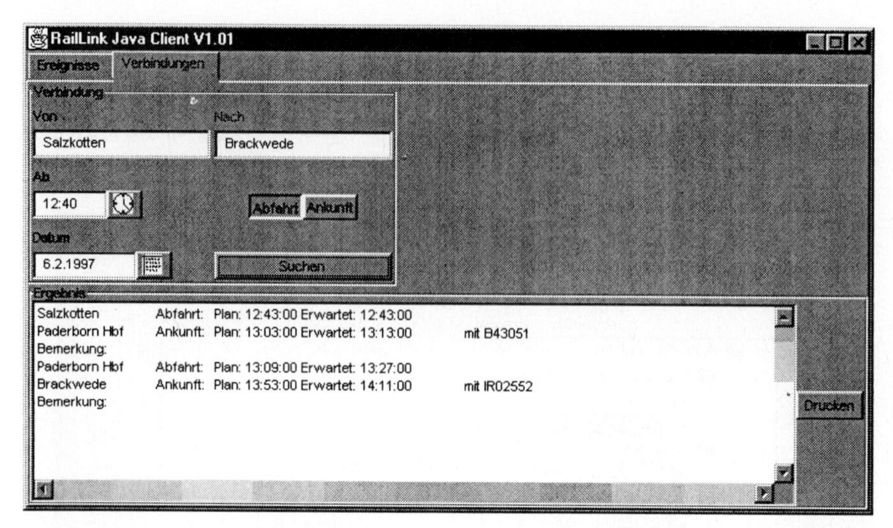

Fig. 5.5a Client for service point employees

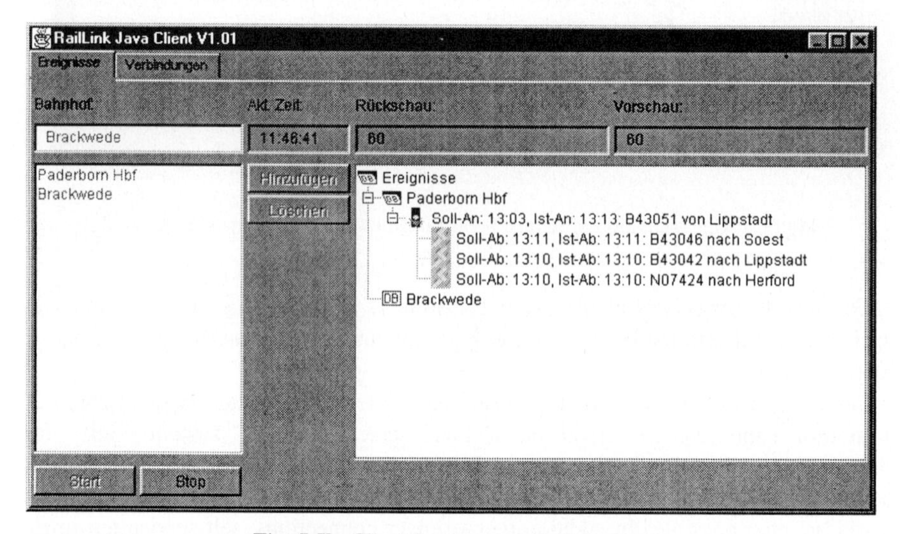

Fig. 5.5b Client for service point employees

6 Conclusions

In this paper, we discussed the operations control problem in transportation systems. Illustrating various facets of this problem in a railway context, we concluded that both dispatchers and service point employees are not yet optimally supported by existing information technology infrastructure, at least at Deutsche Bahn. After studying the requirements for an operations control system for railways, we presented an integrated knowledge-based, object-oriented system architecture for such a decision support system involving various problem solving components. The proposed architecture (see Figure 4.1) internally manages the scheduled, actual, and expected states of a dense network and supports both dispatchers and service point employees to ensure passenger traffic with the best possible quality.

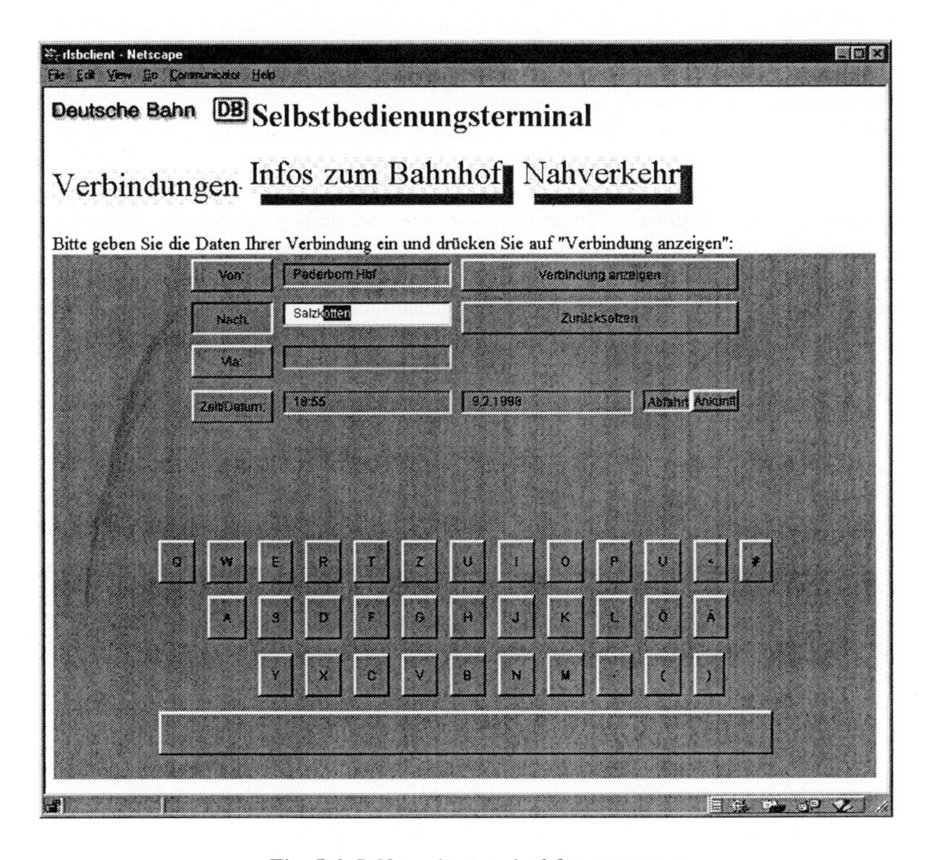

Fig. 5.6 Self-service terminal for passengers

The prototype has been implemented under Windows NT/95 using MS-Visual C++ and includes a component for propagating disturbances, recognizing conflicting situations, and performing what-if analyses to support estimating network-wide effects of a dispatcher's decision, as well as monitoring and information system components for passengers and service point employees. Another component currently under development deals with vehicle rescheduling.

The design and development are being carried out, together with the authors, by students within their diploma theses (see Goecke (1996), Stelbrink (1998), Kopp (1998)).

Currently, we are working with professional schedulers and dispatchers of Deutsche Bahn to test the ideas and search for an optimal system design to meet the different and difficult requirements that arise within this challenging task.

References

Suhl, L. (1995): Computer-aided scheduling: An Airline Perspective. (Gabler–DUV) Wiesbaden.

Suhl, L./Mellouli, T. (1997): Supporting planning and operation time control in transportation systems. in: Operations Research Proceedings 1996, 374-379. (Springer) Heidelberg.

Mellouli, T. (1997a): Improving vehicle scheduling support by efficient algorithms. in: Operations Research Proceedings 1996, 307-312. (Springer) Heidelberg.

Mellouli, T. (1997b): A network flow model for vehicle scheduling applied to maintenance for airlines and railways. Presented at the seventh international workshop on Computer-Aided Scheduling of Public Transport. MIT, Boston.

Goecke, J. (1996): Entwicklung eines graphisch-interaktiven Systems zur Unterstützung der netzweiten Konfliktlösung bei Zugverspätungen der Deutschen Bahn AG. Diploma thesis. Decision Support & OR Laboratory. University of Paderborn.

Kopp, A. (1998): Vehicle rescheduling with application to railways. Diploma thesis in progress. Decision Support & OR Laboratory. University of Paderborn.

Stelbrink, M. (1998): Konzeption und prototypische Implementierung eines verteilten, echtzeit-basierten Kundeninformationssystems bei der Deutschen Bahn AG unter Verwendung von Intranet-Technologie. Diploma thesis. Decision Support & OR Laboratory. University of Paderborn.

Telebus Berlin: Vehicle Scheduling in a Dial-a-Ride System

R. Borndörfer[1], *M. Grötschel*[1], *F. Klostermeier*[2], *and C. Küttner*[2]

[1] Konrad-Zuse-Zentrum für Informationstechnik Berlin, Takustraße 7, 14195 Berlin, Germany, Email: [surname]@zib.de, URL: www.zib.de
[2] Intranetz Gesellschaft für Informationslogistik mbH, Klopstockstr. 9, 14163 Berlin, Germany, Email: info@intranetz.de, URL: www.intranetz.de

Abstract: *Telebus* is Berlin's dial-a-ride system for handicapped people who cannot use the public transportation system. The service is provided by a fleet of about 100 mini-buses and includes assistance in getting in and out of the vehicle. Telebus has between 1,000 and 1,500 transportation requests per day. The problem is to schedule these requests onto the vehicles such that punctual service is provided while operation costs are minimized. Additional constraints include pre-rented vehicles, fixed bus driver shift lengths, obligatory breaks, and different vehicle capacities.

We use a *set partitioning* approach for the solution of the bus scheduling problem that consists of two steps. The first *clustering* step identifies segments of possible bus tours ("orders") such that more than one person is transported at a time; the aim in this step is to reduce the size of the problem and to make use of larger vehicle capacities. The problem of selecting a set of orders such that the traveling distance of the vehicles within the orders is minimal is a set partitioning problem that can be solved to optimality. In the second step the selected orders are *chained* to yield possible bus tours respecting all side constraints. The problem to select a set of bus tours such that each order is serviced once and such that the total traveling distance of the vehicles is minimum is again a set partitioning problem that is solved approximately.

We have developed a computer system for the solution of the bus scheduling problem that includes a branch-and-cut algorithm for the solution of the set partitioning problems. A version of this system has been in operation at Telebus since July 1995. Its use made it possible for Telebus to serve about 30% more requests per day with the same resources.

1 Handicapped People's Transport in Berlin

Better accessibility of the public transportation system has become an important political goal for many municipalities, partially met by introducing

Figure 1.1: A Telebus picks up a customer

low-floor buses, installing lifts in subway stations, etc. But many handicapped and elderly people still have problems because they need additional help, the next station is too far away, or the line they want to use is not yet accessible. Berlin, like many other cities, offers these people a special transportation service. The system, called *Telebus*, provides door-to-door transportation with assistance at the pick-up and the destination. The system is financed by the Senate of Berlin's department for Social Affairs (SenSoz) and operated by the Berliner Zentralausschuß für Soziale Aufgaben e.V. (BZA), an association of charitable organizations. Fig. 1.1 shows a Telebus vehicle picking up a customer.

Telebus is a *dial-a-ride system*: every entitled user (currently about 25,000 people) can order up to 50 rides per month through the BZA's telephone centre. If the order is placed one day in advance, Telebus guarantees to service the ride as requested. Later "spontaneous" requests are serviced if possible. The advance orders, about 1,500 during weekdays and 1,000 on weekends, are collected and scheduled into a fleet of mini-buses that are rented on demand from charitable organizations and commercial companies. These buses will pick-up the customers at the desired time (within a certain tolerance) and transport them to their destinations; if required, the crew will provide assistance in leaving the apartment, entering the vehicle, etc. This service is available every day from 5 am to 1 am.

Telebus was established 15 years ago and since then the number of customers and orders have increased continuously. Until recently, the *vehicle scheduling* was done manually by experienced planners who could work out a feasible schedule in about 16 man-hours. But when East Berlin's handicapped people also started to use the system after the reunification of Germany, it was clear that the traditional way of scheduling would not be able to cope

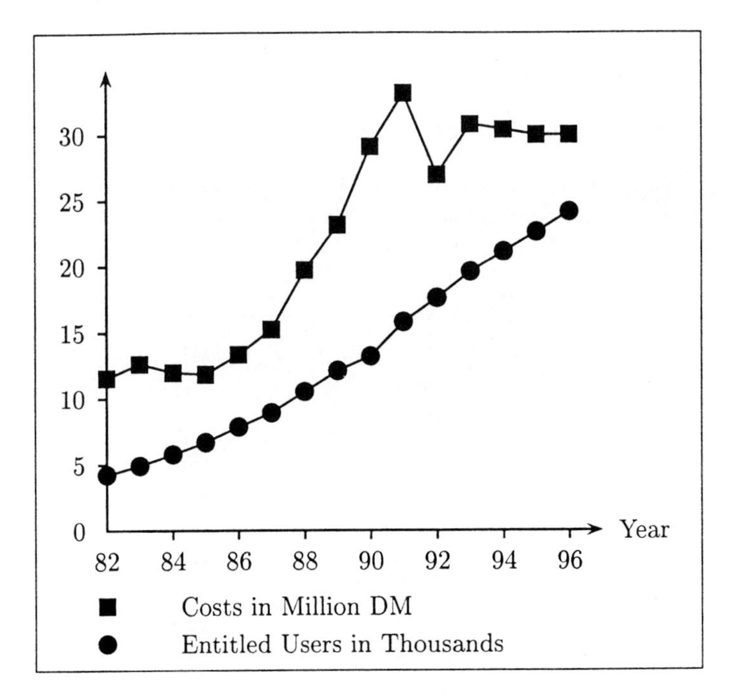

Figure 1.2: Development of Telebus

with the projected additional demand. The problem was not only to come up with a *feasible* schedule: more requests in an area of service which had doubled in size led to rising costs and put the system under heavy pressure to be more efficient. Fig. 1.2 illustrates the explosive growth of the system; the numbers for the years up to 1993 are taken from report T 336 of Berlin's audit division for the year 1994, the other data were provided by the BZA.

Modern computer hard- and software was needed to solve the scheduling problems of the BZA and the *Telebus project*, involving cooperation between the Konrad-Zuse-Zentrum für Informationstechnik Berlin (ZIB), the BZA, and the SenSoz, was started to develop it. The result of the project is a new *Telebus-computersystem*, that supports and integrates the complete sequence of operations at the BZA: ordering, vehicle scheduling, radio communication, accounting, controlling, and statistics. The system which consists of a tool box of software modules, running on a network of 20 MacIntosh computers, has been operating at the BZA since 1995. Its use, together with a simultaneous reorganization of many parts of the Telebus service, led to

(i) improvements in service: for example, a reduction of the advance ordering period from three days to one day (needed for vehicle renting) and increased punctuality of the computer schedule in comparison to the result of manual planning,

(ii) cost reductions, such that today about 30% more requests can be serviced with the same resources, and

(iii) simplifications of the work in the Telebus centre.

A comparison of the number of requests and the costs for a month before and after the installation of the system is shown in Fig. 1.3.

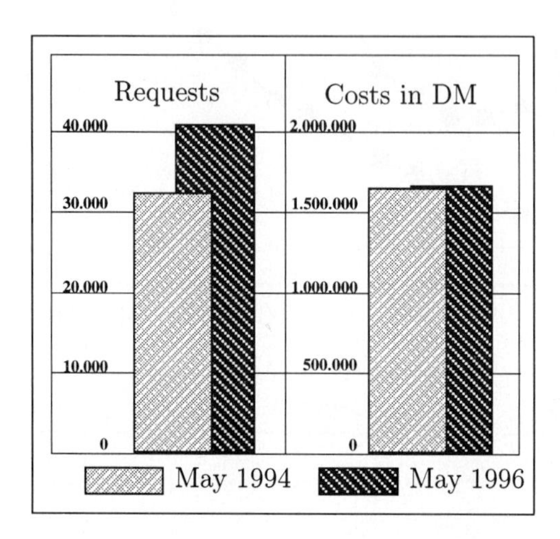

Figure 1.3: Results of the Telebus Project

The heart of the Telebus-computersystem is the *vehicle scheduling module.* This module is based on mathematical optimization techniques that are described in this paper. Our aim is to show that methods of this kind can make a significant contribution to the solution of real world transportation problems: the results at Telebus are of interest for similar dial-a-ride systems. It goes without saying, however, that optimization at Telebus did not only consist of better vehicle scheduling, but involved many other important factors: restructuring of the operation of the centre, negotiations with vehicle providers, and personal dedication (F. Klostermeier and C. Küttner, in particular, worked for more than a year in the Telebus centre, drove Telebuses, etc.). More details on this *consulting aspect* of the Telebus project can be found in the (German) articles Klostermeier/Küttner (1993) and Borndörfer/Grötschel/Herzog/Klostermeier/Konsek/Küttner (1996).

2 Vehicle Scheduling at Telebus

The most important task at Telebus is the daily construction of the *vehicle schedule* which determines both *operational costs* for vehicles and crews and *customer satisfaction* in terms of punctual service. The vehicle scheduling problem (VSP) at Telebus can be stated in an informal way as follows:

(VSP) Given a number of requests and a number of available vehicles, rent a suitable set of vehicles and schedule all requests to them such that a number of constraints like punctuality and labour regulations are satisfied and operational costs are minimized.

The aim of this section is to describe the VSP precisely and to introduce our set partitioning approach for its solution. We start with a discussion of the VSP's data, its constraints, and objectives.

The basis for vehicle scheduling are some number ν of *vehicles* of different types. Actually, a vehicle is in this context not just a car, but also a crew for a shift of operation: the BZA does not rent vehicles, but shifts of operation of a car and a crew. Such a (manned) vehicle b, $b = 1, \ldots, \nu$, is characterized by the following data:

	c_b	type (class): Teletaxi, 1- & 2-bus (small or large)
(V)	$A_b = (A_b^{\text{wchair}}, A_b^{\text{seat}})$	capacity: no. of wheelchair places and seats
	g_b	group: type, depot location, shift

There are approximately 100 buses available for renting. Vehicles can be distinguished by a type (or class) and a group. There are five types: Teletaxis, small buses with one driver (1-bus), large 1-buses, small 2-buses, and large 2-buses. The type is important for deciding whether a request can be serviced by a particular vehicle: Teletaxis can transport only ambulatory customers and those with folding wheelchairs. Non-folding wheelchairs require a bus. Staircase aid require a bus with a crew of two. The type of a vehicle also determines its capacity: Teletaxis can transport one handicapped customer and one non-handicapped companion, small buses have a capacity of $(2, 3)$, large buses of $(3, 4)$. Capacity is a sub-parameter of the type, but is given a symbol of its own for convenience of notation. Vehicles of the same type fall into groups, that play a role for the construction of tours: a group contains vehicles that are indistinguishable in the sense that they have the same type, are stationed at the same depot, and can be rented for identical shifts.

The vehicles will be used to service some number m of transportation *requests*. The following data are associated with each request $i = 1, \ldots, m$:

	$v_i^{\text{pick}}, v_i^{\text{dest}}$	pick-up and destination node
	$p(v_i^{\text{pick}}), p(v_i^{\text{dest}})$	pick-up and destination point
	$T(v_i^{\text{pick}}) = [\underline{t}(v_i^{\text{pick}}), \overline{t}(v_i^{\text{pick}})]$	interval of feasible times to arrive at pick-up point
(R)	$T(v_i^{\text{dest}}) := [\underline{t}(v_i^{\text{dest}}), \overline{t}(v_i^{\text{dest}})]$	interval of feasible times to arrive at destination point
	$t^{\text{service}}(v_i^{\text{pick}}), t^{\text{service}}(v_i^{\text{dest}})$	service time at pick-up and destination
	C_i	set of feasible vehicle types
	$a_i = (a_i^{\text{wchair}}, a_i^{\text{seat}})$	no. of wheelchairs and seats needed

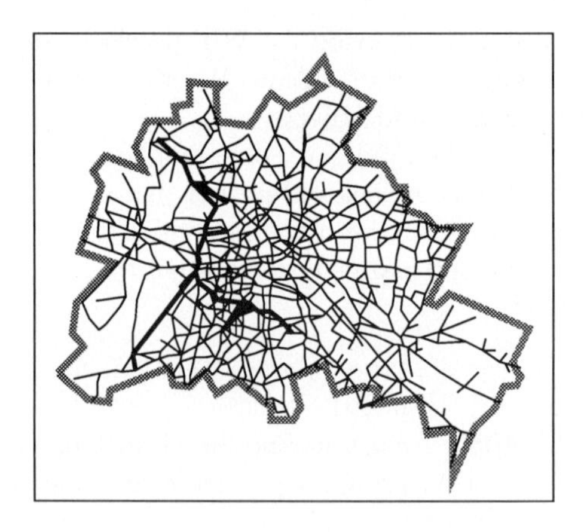

Figure 2.1: Graph of Berlin

There is a *pick-up node* v_i^{pick} and a *destination node* v_i^{dest}, that correspond to the pick-up and delivery *events* of a request. The pick-up and destination locations or points[1] $p(v_i^{\text{pick}})$ and $p(v_i^{\text{dest}})$ of a request are stored as nodes of a graph of Berlin that is shown in Fig. 2.1. The 2,510 edges of this graph are labelled with average travelling times and distances that we use to compute shortest routes between its 828 nodes. In addition to this spatial information, a request bears temporal data that is measured in units of 5 minutes. There is an interval of feasible pick-up times $T(v_i^{\text{pick}})$ that is computed according to Telebus specific rules. The rules try to find a compromise between punctual service and more degrees of freedom for the vehicle scheduling process. Currently, most requests have $T(v_i^{\text{pick}}) = t^\star(v_i^{\text{pick}}) + [-3, 3]$, where $t^\star(v_i^{\text{pick}})$ is the time desired by the customer, i.e., the vehicle is allowed to arrive $3 * 5 = 15$ minutes early or late. Similar, but more complex rules are used to determine a feasible time interval $T(v_i^{\text{dest}})$ to arrive at the destination; here, the shortest possible travelling time and a maximum detour time play a role.

[1] We distinguish between *nodes* that belong to space-time networks and locations or *points* that correspond to geographical data.

Finally, some service time $t^{\text{service}}(v_i^{\text{pick}})$ and $t^{\text{service}}(v_i^{\text{dest}})$ is needed at the pick-up and the destination point. The amount of assistance, the wheelchair, and other factors determine what kinds of vehicles C_i (Teletaxi, 1-bus, or 2-bus) can or must be used, and the final load data a_i gives the number of wheelchair places and seats needed. Fig. 2.2 shows a typical distribution of Teletaxi, 1-bus, and 2-bus demand.

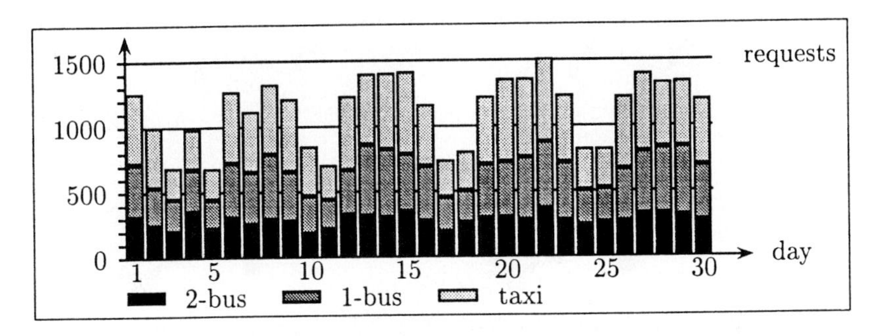

Figure 2.2: Telebus request pattern for June 1995

Rules for feasible vehicle tours arise primarily from bus rental contracts and labour regulations for bus drivers. Most rental contracts are for shifts of $8\frac{1}{2}$ or $10\frac{1}{2}$ hours, although some vehicles can be rented by the hour to cover demand peaks. The majority of renting is done on a daily basis on demand, but vehicles can also be rented on a long term basis. Labour regulations prescribe maximum driving hours and rules for obligatory breaks. The current rule at Telebus is that a break of 30 minutes has to be taken between the fourth and sixth hour of a shift. Two other rules state that a feasible vehicle tour must start and end at the vehicle's depot, and that it is not allowed to wait or take a break with a customer "on board".

The *objective* of the VSP is to minimize operational costs, but the BZA seldom uses this criterion in its pure form. The reason is that the planned schedule and the one that is really executed on the next day differ significantly because of cancellations of requests, spontaneous requests, vehicle break-downs, and other unpredictable events. The BZA must safeguard against every day's emergency situations and does so by preferring "safer" plans at some extra cost. The main tool to do this is to introduce components into the objective that aim at schedules of a safer type; we will come back to this point in the discussion of the set partitioning model.

Our solution approach for the VSP is based on the concept of a *cluster* of requests. A cluster or, in BZA terminology, an *order*, consists of a set of requests that are advantageously serviced simultaneously. It corresponds to a maximal subtour such that the vehicle is never empty: the subtour starts with an empty vehicle picking up a first customer, services the requests of the cluster, and becomes empty for the first time when the last customer leaves the vehicle at his/her destination. This results in "simultaneous service" of the requests in the cluster in the sense that, while one customer is transported, at least one other person is picked up or transported to his/her destination. Fig. 2.3 shows a number of clusters: collections, insertions, simple and continued concatenations.

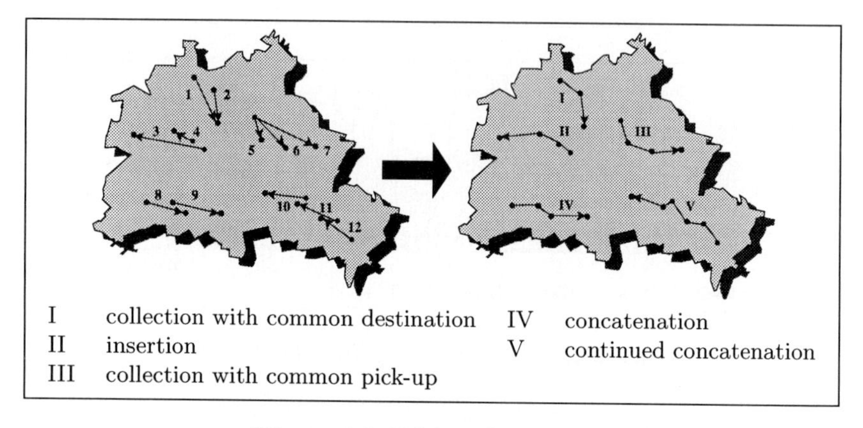

Figure 2.3: Telebus cluster types

Clusters can be used to *decompose* the vehicle scheduling process into two phases: a *clustering* phase that combines requests into clusters and a subsequent *chaining* phase that builds tours as sequences of clusters. The flavour of clustering is that of a local optimization to make use of larger vehicle capacities, while chaining must deal with constraints for the feasibility of complete tours, such as depot locations, breaks, and shift lengths. The advantage of this approach is that it gets easier to construct tours from a comparably smaller number of orders in a non-overlapping way. The disadvantage is that a hierarchical planning process will generally yield suboptimal solutions.

To use this approach, it makes sense to describe a cluster c as follows:

	$S_c := (v^1, \ldots, v^l)$	sequence of pick-up and destination nodes
	$T_c^{\text{pick}} := [\underline{t}_c^{\text{pick}}, \overline{t}_c^{\text{pick}}]$	interval of feasible times to arrive at first pick-up point
(C)	$T_c^{\text{dest}} := [\underline{t}_c^{\text{dest}}, \overline{t}_c^{\text{dest}}]$	interval of feasible times to end service at last destination point
	t_c	total time to service cluster completely
	C_c	set of feasible vehicle types

The subtour corresponding to a cluster is given by a sequence of pick-up and destination nodes S_c that will be serviced in this order. More precisely, if $v^j = v_i^{\text{pick}}$ is a pick-up node, the vehicle will drive to the corresponding location and pick-up the customers complete with service. If $v^j = v_i^{\text{dest}}$ is a destination node, the vehicle will go to the destination location and service the customer. A cluster sequence S_c must, of course, satisfy several constraints: the initial node v^1 must be a pick-up node, the terminal node v^l a destination node, each node can appear at most once, each destination node must be preceded by the pick-up node of the same request and vice versa, and the sequence must describe simultaneous service, i.e., the service of each request except the first/last overlaps with the service of a succeeding/preceding request in the sense that the customers share the vehicle. An important observation is that the cluster sequence completely determines the operation of the vehicle: since it is not allowed to wait with a customer "on board," the vehicle either drives to the next node or the crew provides service. This means that the total time t_c to service the cluster is constant and that the service of the complete cluster can be shifted as a block over some feasible interval of time. Thus, there is a maximal interval T_c^{pick} of feasible times to arrive at the first pick-up node of the cluster and a corresponding interval of feasible end times, and these have the property

$$t_c = \underline{t}_c^{\text{dest}} - \underline{t}_c^{\text{pick}} = \overline{t}_c^{\text{dest}} - \overline{t}_c^{\text{pick}}.$$

The sequence of serviced requests also determines the possible types of vehicles C_c: these depend on the most "demanding" vehicle type of the requests and the maximum number of occupied wheelchair places and seats needed.

Vehicle tours are the last structure that is needed, and just as a cluster can be described as a sequence of request nodes, a tour k can be seen as a sequence of clusters:

	$S_k := (c^1, \ldots, c^l)$	sequence of serviced clusters
	$T_k^{\text{pick}} := [\underline{t}_k^{\text{pick}}, \overline{t}_k^{\text{pick}}]$	interval of feasible times to start service of the first cluster
(T)	$T_k^{\text{dest}} := [\underline{t}_k^{\text{dest}}, \overline{t}_k^{\text{dest}}]$	interval of feasible times to end service of the last cluster
	t_k	total time to service tour completely
	k	vehicle

A tour k consists of a sequence of clusters S_k that are serviced in the given order. To deal with depot locations, breaks, and shift lengths we also allow for additional pull-in, break, and pull-out clusters. Pull-in clusters will prescribe a starting location and time of a tour, break clusters an obligatory break between the fourth and sixth hour of service of a tour, and pull-out clusters again model depot locations and maximum shift lengths. Pull-in and pull-out clusters will fix the possible times to begin and end a tour, but we nevertheless introduce the time windows T_k^{pick} and T_k^{dest} for later use in our tour construction algorithm. Additional parameters give the total time to service a tour, i.e., the shift length, and the vehicle.

What is a good way to do the clustering? In principle, one would like to construct a set of clusters which will later result in the construction of a good set of tours. We try to approximate this goal using secondary criteria such as the *travelling distance* or *time* within the clusters. This leads to the *clustering problem* to construct a set of clusters, such that each request is contained in exactly one cluster and some objective, such as the sum of the internal travelling distances, is minimal. Given a decision for a set of clusters, the *chaining problem* can be stated in a similar way. This time, we want to construct a set of tours, such that each cluster is serviced by exactly one tour, such that there are enough vehicles of the required types and groups, and such that operational costs or a similar objective becomes minimal.

Both questions can be modeled as a *set partitioning problem*

$$\min c^T x \quad Ax = \mathbb{1},\ x \in \{0,1\}^n, \tag{SPP}$$

where $A \in \{0,1\}^{m \times n}$ is a 0/1-matrix and $c \in \mathbb{R}^n_+$ is a positive cost function.

In the clustering case, row i of the matrix A corresponds to request i, and each column $A_{\cdot j}$ of A to a feasible cluster: the entry a_{ij} is equal to one if cluster j services request i and zero; otherwise, the objective c_j denotes, for example, the internal travelling distance or time within the cluster. Then, the feasible solutions x of the integer program (SPP) are in one-to-one correspondence to sets S of clusters such that each request is contained in exactly one cluster via the relation $x_j = 1 \iff j \in S$ and the optimum solution $x^\star$ of (SPP) corresponds to the best such combination.

In the chaining case, the rows correspond to the clusters selected in the clustering step, the columns to tours, and the objective is some cost criterion associated with a tour, for example, operation costs. The only additional point to consider is that the model as stated does not respect *vehicle availability*. The tour matrix A contains for each vehicle all possible tour-columns that this vehicle can service, and it is possible that a solution of (SPP) will use a vehicle more than once. To prevent this additional constraints are included of the form

$$\sum_{j \in J(k)} x_j \leq 1 \quad \text{or} \quad \sum_{j \in J(k)} x_j = 1,$$

where $J(k) \subseteq \{1, \ldots, n\}$ denotes the set of tours serviced by vehicle k. These inequalities fit into the set partitioning model. They give rise to additional rows that correspond to vehicles instead of requests (possibly introducing additional columns as well, that correspond to slack variables).

A set partitioning model is well suited for the VSP, because it allows correct treatment of constraints and objectives that do not arise from individual components of a tour, but from a tour as a whole. Break rules, for instance, are observed by constructing only such tour-columns for the chaining SPP that correspond to tours with feasible breaks. If operation of a vehicle at night incurs additional costs, we can modify the objective accordingly. We also can penalize "packed tours" that operate at capacity because delays are more likely and try to produce safer schedules at some additional cost. A second advantage is that a correct tour matrix A already guarantees that all feasibility constraints are satisfied such that the selection of the best set of tours can be done in a second step on an abstract level. If the rules for feasible tours change, the cluster or tour matrix changes, but a solver for set partitioning problems will still be useful. This makes the approach particularly useful to analyze different operating scenarios.

Our clustering and chaining approach to the VSP using set partitioning can now be stated as follows:

- *Clustering*

  (i) *Cluster generation* to construct all possible clusters and set up the clustering SPP.

  (ii) *Cluster selection* to solve the clustering SPP to select a best set of orders such that each request is contained in exactly one order.

- *Chaining*

  (iii) *Tour generation* to construct a set of feasible tours and set up the chaining SPP.

  (iv) *Tour selection* to solve the chaining SPP and thus choose a best set of tours.

The approach requires an implementation of three modules: a *cluster generator*, a *tour generator*, and a *set partitioning solver*. Our cluster generator is based on complete enumeration. It turned out that there are usually about 100,000 to 250,000 legal clusters in a typical VSP that can be produced in a couple of minutes. The corresponding set partitioning problems are of a size that can be solved to near or proven optimality using branch-and-cut algorithms and it is possible to do this in the Telebus case. The number of possible tours in the chaining problem is, however, much larger, and we can neither compute nor store all of them. We have nevertheless chosen to use the same branch-and-cut algorithm as for the clustering problems in the chaining instances, and we must thus restrict the set of considered tours to a (small) subset of, say, 50,000 possible tours that we construct heuristically. It turned out that the chaining SPPs are computationally much harder than the clustering ones, and we cannot solve them to optimality. But our tour optimization still yields significant savings in operational costs of about 10% in comparison to what we can achieve with heuristic chaining methods.

Our set partitioning clustering and chaining approach is a *static variant* of the methods discussed in Cullen/Jarvis/Ratliff (1981), that solve a sequence of dynamically generated set partitioning problems in both the clustering and the chaining phase using column generation techniques, or Ioachim/Desrosiers/Dumas/Solomon (1991), that use dynamic programming techniques in a column generation algorithm for the clustering problem. An overview of related techniques and pointers to the extensive literature on vehicle routing can be found in the survey articles Desrochers/Desrosiers/Soumis (1984) and Desrosiers/Dumas/Solomon/Soumis (1995) and in the annotated bibliography of Laporte (1997).

3 Cluster Generation

The aim of the cluster generation step is to enumerate all possible clusters. As was pointed out in Sect. 2, we will ignore feasibility conditions for complete tours like breaks, depot locations, and shift lengths for the moment, i.e., we ignore all information related to vehicle groups. Different vehicle types (Teletaxi, small and large 1- and 2-bus), however, give rise to different possible clusters. We can deal with this parameter by enumerating the clusters for each of the five types separately. For ease of exposition, we can thus assume that there is only one type of vehicle that can service all requests.

A way to enumerate all possible clusters in a systematic way is to consider the operation of the vehicle in a cluster as the result of a sequence of decisions to pick-up or deliver a next customer, or, in other words, to add a next node to the cluster sequence. Each time this is done, the vehicle must drive to the corresponding node and pick-up or deliver the customer, before the next decision can be taken.

The possible *states* of the vehicle can be recorded in terms of cluster subsequences $S = (v^1, \ldots, v^l)$, where each node v^j denotes a pick-up or delivery node of some request. We adopt the convention that a vehicle in state S has just serviced the last pick-up or delivery node v^l. More information on the vehicle can be derived from this basic state description. First, there is the set of yet unserviced pick-up nodes

$$R(S) := \{v_i^{\text{pick}} : \exists v^j = v_i^{\text{pick}}, \nexists v^j = v_i^{\text{dest}}\}.$$

The customers of the unserviced requests are sitting in the car that has at state S a total load of

$$a(S) := \sum_{v_i^{\text{pick}} \in R(S)} a_i.$$

Since it is forbidden to wait with a customer on board, the total time since service of the sequence S began is independent of the precise starting time and amounts to

$$t(S) := \sum_{j=1}^{l} t(v^{j-1}, v^j) + t^{\text{service}}(v^j),$$

where $t(v^{j-1}, v^j)$ denotes the time to drive from node v^{j-1} to node v^j and where $v^0 := v^1$ such that $t(v^0, v^1) = 0$. Depending on the time intervals associated with the nodes in S, the service of the complete sequence S may be shifted back or forth over a certain feasible time interval. This results in intervals of feasible times $T^{\text{pick}}(S)$and $T^{\text{dest}}(S)$ to start service of the sequence and to end service at the current last node v^l. Since the total service time $t(S)$ is a constant, these intervals have the same length and, in fact,

$$T^{\text{pick}}(S) + t(S) = T^{\text{dest}}(S).$$

We will discuss shortly how $T^{\text{pick}}(S)$ and $T^{\text{dest}}(S)$ can be computed iteratively.

With this terminology, we can devise a simple algorithm to enumerate all possible clusters. We start by setting S to an *initial state*

$$S := (v_i^{\text{pick}}).$$

Then:

$R(S) = \{v_i^{\text{pick}}\}$	request i is not yet serviced
$a(S) = a_i$	customers and companions of request i are in the car
$t(S) = t^{\text{service}}(v_i^{\text{pick}})$	the total time spent to service the cluster was used to pick-up request i
$T^{\text{pick}}(S) = T(v_i^{\text{pick}})$	service of the cluster can start whenever i is eligible for pick-up
$T^{\text{dest}}(S) = T(v_i^{\text{pick}}) + t(v_i^{\text{pick}})$	service of the cluster ends in the same interval shifted back the serviced time $t(v_i^{\text{pick}})$.

We can now decide the next node to service and this decision will lead to a transition to a new state. In general, a *state transition* from a state S to a state S' servicing an additional node v^{l+1} servicing request i is as follows:

$S' := (v^1, \ldots, v^{l+1})$ — the new node v^{l+1} is added to the cluster subsequence

$$R(S') = \begin{cases} R(S) \cup \{v^{l+1}\} & \text{if } v^{l+1} = v_i^{\text{pick}} \\ R(S) \setminus \{v^{l+1}\} & \text{if } v^{l+1} = v_i^{\text{dest}} \end{cases}$$

a request is serviced or there is another customer to be serviced

$$a(S') = \begin{cases} a(S) + a_i & \text{if } v^{l+1} = v_i^{\text{pick}} \\ a(S) - a_i & \text{if } v^{l+1} = v_i^{\text{dest}} \end{cases}$$

customers and companions of request i enter/leave the car

$t(S') = t(S) + t(v^l, v^{l+1}) + t^{\text{service}}(v^{l+1})$ — total time to service the cluster goes up by time to drive from v^l to v^{l+1} and to service v^{l+1}

$$T^{\text{dest}}(S') = ((T(S) + t(v^l, v^{l+1})) \cap T(v^{l+1})) + t^{\text{service}}(v^{l+1})$$

possible times to complete service of v^{l+1} are as follows: service at v^l ends in $T(S)$, the vehicle arrives at v^{l+1} in $T(S)+t(v^l, v^{l+1})$, but feasible times are in $T(v^{l+1})$, time $t^{\text{service}}(v^{l+1})$ passes until the request is serviced

$T^{\text{pick}}(S') = T^{\text{dest}}(S') - t(S')$ — the time interval to start service of the cluster is possibly reduced

We will denote this state transition by

$$S' := S \leftarrow v^{l+1}.$$

Not all states that we can produce in this way are feasible or correspond to a cluster. Conditions for a feasible state S for some vehicle k are

$a(S) < A_k$	the load does not exceed the vehicle's capacity
$T^{\text{pick}}(S) \neq \emptyset$	all customers can be picked up in time
$T^{\text{dest}}(S) \neq \emptyset$	all customers can be delivered in time

Other feasibility conditions are that a state S must contain a node only once and that each destination node must be preceded by the corresponding pick-up node and vice versa. A state that does not satisfy all of these conditions is called *infeasible.* The state corresponds to a cluster c when $R(S)$ becomes empty; such a state is called *terminal.* In this case, we can set

$$S_c := S, \quad T_c^{\text{pick}} := T^{\text{pick}}(S), \quad T_c^{\text{dest}} := T^{\text{dest}}(S), \quad \text{and} \quad t_c := t(S).$$

(The vehicle type was fixed at the beginning of this section by assumption.)

A simple algorithm to enumerate all possible clusters is to consider all possible initial states and, starting from these, to do all possible state transitions recursively. The recursion stops when a terminal or infeasible state is reached, the terminal states are returned.

```
void dfs (state S, digraph D) {
    if (infeasible (S)) return;
    if (eliminated (S)) return;
    if (terminal (S)) output (S);
    // service next request
    for all transitions v^{k+1} ∈ γ^+(v^k)
        dfs (S ← v^{l+1}, D);
}

void cluster (digraph D=(V,A)) {
    for all pick-up nodes v_i^pick ∈ V
        dfs (initial (v_i^pick), D);
}
```

Figure 3.1: Generic cluster generation

Most state transitions, however, will immediately lead to infeasible states, and some effort must be spent to filter these out. We do this using a *transition digraph* $D = (V, A)$, whose vertices are the pick-up and destination nodes. There is an edge from node u to v if

$$(T(u) + t^{\text{service}}(u) + t(u, v)) \cap T(v) \neq \emptyset,$$

that is, if it is possible to arrive at u, service u, and arrive at v at a feasible time. Since the destination time interval $T^{\text{dest}}(S)$ of some state S with terminal node v^l is always a subset of $T(v^l) + t^{\text{service}}(v^l)$, only the heads $\delta^+(v^l)$ of the arcs that go out from v^l qualify as candidates for feasible transitions.

Other states that must turn infeasible contain unserviced pick-up nodes v_i^{pick} such that the corresponding destination nodes can no longer be reached in time. An easy criterion to detect this is

$$\max T(v_i^{\text{dest}}) < \min T(S) + t(v^l, v_i^{\text{dest}}),$$

that is, when it is impossible to arrive at the destination node of the unserviced request i in time even if we go there immediately. One can work out more elaborate *state elimination criteria*, but for Telebus this one proved to be efficient enough.

C-type pseudocode for our generic recursive procedure to enumerate all clusters (for a fixed vehicle type) is given in Fig. 3.1. The procedure searches in a depth first way starting from all possible initial states. `digraph` is a data structure to store the transition digraph, and `D=(V,A)` is this digraph as produced somewhere else. `state` is a data structure for cluster subsequences that contains the data items discussed in this section. `infeasible`, `eliminated`, and `terminal` are boolean functions that check a state for infeasibility, whether it can be eliminated, or is terminal as described above. `initial` is a function that returns an initial state corresponding to a pick-up node, `output` saves a cluster sequence to some medium.

Our procedure for cluster enumeration at Telebus is very simple: we do not use a dynamic program, and our state space elimination criteria are straightforward. There are two reasons why this algorithm is successful for the Telebus instances. One is the ratio of service time, transportation time, and maximum detour time at Telebus. Service of a request takes about 30 minutes on average: 5 minutes pick-up service, 20 minutes driving, and another 5 minutes of service at the destination. Since a customer is not satisfied if his transportation takes more than, say, 15 minutes longer to pick-up or drop somebody else, it is often just not possible to service more than two requests simultaneously. A second reason is that BZA rules do not accept all clusters as produced by the above generic cluster generation routine. In fact, there is a catalogue of "legal" clusters at Telebus, consisting of collections, insertions, concatenations, and continued concatenations of a maximum "depth" (currently at most 3). We use more restrictive derivatives of the generic routine to produce the legal clusters and these are, of course, less than what the generic routine would yield.

The cluster generator routines usually produce, depending on the requests, the complete set of 100,000 to 250,000 legal clusters in a couple of minutes. The resulting set partitioning problems are large scale, but computationally not difficult in the sense that one can find near or proven optimal solutions in about the same time. Optimizing the internal travelling distance of the vehicles within the clusters, one obtains a reduction of about 20% in comparison to individual transportation, while the number of clusters is up to 40% less than the number of requests. Fig. 3.2 illustrates these reductions.

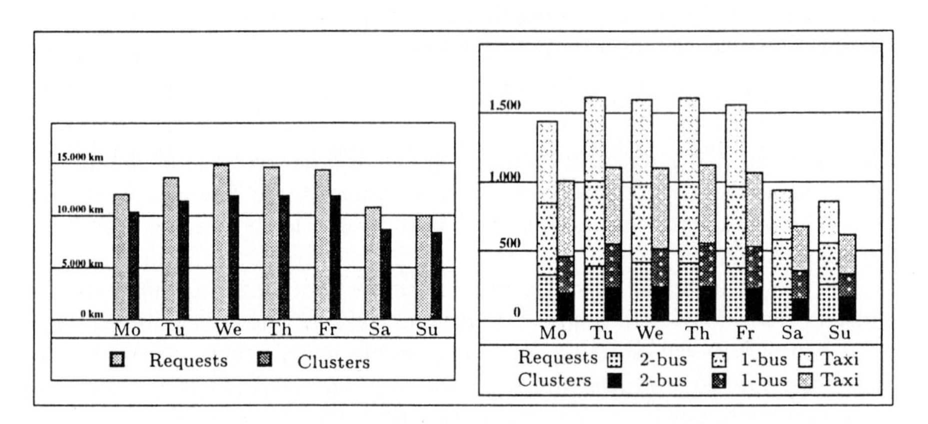

Figure 3.2: Clustering requests of September 16–22,1996

4 Tour Generation

The aim of the tour generation step is to produce feasible vehicle tours as sequences of clusters. The basic flavour is similar to cluster generation where service nodes are replaced by complete clusters. But where clustering had an eye on local optimization and ignored *tour feasibility* conditions, vehicle group information like depot locations, break rules, and shift lengths must be considered in tour construction. Another difference is that while the service of clusters cannot be interrupted, it is not only legal, but often advantageous to wait *between* service of two clusters.

We deal with different *vehicle groups* by constructing the tours for vehicles of each group separately and will assume in the remainder of this section that we have fixed a depot location, the shift length, and the vehicle type similar to what we did in cluster generation. We can then also assume that all clusters can be serviced by the vehicles of the group under consideration.

Again analogous to cluster generation, our approach to chaining is to build tours iteratively as sequences of clusters, but with an additional eye on tour feasibility criteria, and represent the possible *states* of a vehicle in terms of a tour subsequence of serviced clusters

$$S = (c^1, \ldots, c^l);$$

the interpretation of state S is that the vehicle has just completed service of the terminal cluster c^l.

The main difference between clustering and chaining is the additional consideration of driver breaks and shift lengths. Both criteria are in terms of *total elapsed time* since the start of the tour: the shift length simply sets an upper bound to this value, the break rule prescribes an obligatory break of 30 minutes between the fourth and sixth hour of work. Our approach to control the total time is simply to consider all possible times when a tour can start. All possible times means in this case every quarter of an hour, because 15 minutes is the minimum accounting unit of the vehicle providers.

We can model the different possibilities of *pull-in* times t^{pullin} to start a tour by means of a "pull-in" cluster c^{pullin} with

$$\begin{aligned}
S_{c^{\text{pullin}}} &= (v^{\text{pick}}_{\text{pullin}}, v^{\text{dest}}_{\text{pullin}}) && \text{pull-in cluster (starts and) ends at depot}\\
T^{\text{pick}}_{c^{\text{pullin}}} &= [t^{\text{pullin}}, t^{\text{pullin}}] && \text{pull-in time of tour}\\
T^{\text{dest}}_{c^{\text{pullin}}} &= T^{\text{pick}} && \text{pull-in time of tour}\\
t_{c^{\text{pullin}}} &= 0 && \text{no service}
\end{aligned}$$

that represents the start of a vehicle tour and will be used to initialize the cluster sequence of the tour. The pull-in cluster contains two service nodes with service time zero, that point to the depot location. There is a unique feasible pick-up time such that the pull-in cluster fixes the starting time of a tour. An analogous *pull-out* cluster is supposed to terminate the tour. Its service time intervals are chosen to model the shift length, i.e., for an $8\frac{1}{2}$ hour shift we would have

$$T^{\text{pick}}_{c^{\text{pullout}}} = T^{\text{dest}}_{c^{\text{pullout}}} = T^{\text{pick}}_{c^{\text{pullin}}} + 8.5 * 12 \qquad (1 \text{ hour} = 12 * 5 \text{ minutes}).$$

When the starting time t^{pullin} of the tour is fixed, breaks can be modelled by a *break cluster* c^{break} with

$$\begin{array}{ll} S_c = \emptyset & \text{no pick-up and destination node} \\ T^{\text{pick}}_{c^{\text{break}}} = t^{\text{pullin}} + [4, 5.5] * 12 & \text{feasible time interval to start break} \\ T^{\text{dest}}_{c^{\text{break}}} = t^{\text{pullin}} + [5.5, 6] * 12 & \text{feasible time interval to end break} \\ t_{c^{\text{break}}} = 6 & \text{duration of break } (6 * 5 = 30 \text{ minutes}) \end{array}$$

that has to be serviced by the tour. We adopt here the convention that an empty cluster sequence results in the vehicle standing at its current location. Our goal is to construct all cluster sequences that start at a fixed pull-in cluster, contain a feasible break cluster, and end at the corresponding pull-out cluster.

An algorithm for this must derive and update only a single data item from a state S, the interval

$$T^{\text{dest}}(S)$$

of feasible times to end service of the last cluster in the tour subsequence, and even here only the earliest such time $\underline{t}^{\text{dest}}(S)$ is relevant, because one can always wait arbitrarily long to service the next cluster.

The algorithm starts in a (fixed) *initial pull-in state* $S = (c^{\text{pullin}})$ with

$$T^{\text{dest}}(S) = T^{\text{dest}}_{c^{\text{pullin}}} = [t^{\text{pullin}}, t^{\text{pullin}}].$$

We can now decide the next cluster to service, add this to the tour cluster subsequence, and so on. In general, we will be in a state $S = (c^1, \ldots, c^l)$ and decide to service a next cluster c^{l+1}. This results in a *state transition* to the new state S' with

$$S'_c = (c^1, \ldots, c^{l+1}) \quad c^{l+1} \text{ is the new terminal cluster}$$

$$T^{\text{dest}}(S') = (T^{\text{dest}}(S) + [t(c^l, c^{l+1}, \infty]) \cap T^{\text{pick}}_{c^l} + t_{c^l}$$

feasible times to end service of cluster c^{l+1} are as follows: service of cluster c^l ends in $T^{\text{dest}}_{c^l}$, $t(c^l, c^{l+1})$ is needed to drive from c^l to c^{l+1}, one possible waits, feasible times to start service of C^{l+1} are $T^{\text{pick}}_{c^{l+1}}$, it takes another $t_{c^{l+1}}$ to service c^{l+1},

where $t(c^l, c^{l+1})$ is the time needed to drive from the terminal node of c^l to the initial node of c^{l+1}. We denote this state transition by

$$S' = S \leftarrow c^{l+1}.$$

Feasibility conditions for a state are $T^{\text{dest}}(S) \neq \emptyset$ and that each cluster is contained only once. A feasible state that contains the pull-in cluster under consideration as the initial cluster, the corresponding break cluster c^{break}, and the terminal cluster c^{pullout} is called *terminal*.

The aim of tour generation is to enumerate all terminal states. A simple algorithm to do this is to consider all possible initial pull-in states, to examine all feasible state transitions recursively, and to output all encountered terminal states.

To make this approach work we want to consider only transitions that do not immediately lead to infeasible states because of incompatible service times. A necessary condition for the existence of a feasible transition from some cluster u to another cluster v is

$$(T_u^{\text{dest}} + [t(u,v), \infty)) \cap T_v^{\text{pick}} \neq \emptyset,$$

i.e., it is possible to service u, drive to the initial node of v, possibly wait, and start service of v at a feasible time. We can store this set of possible follow-on clusters in another *transition digraph* $D = (V, A)$ that has an arc from cluster u to v if this condition holds. Then, $\gamma^+(u)$ is the set of possible follow-on clusters for a cluster u. But different from the situation in cluster generation, the number of possible follow-ons is very large: an hour in the future every cluster is eligible!

Elimination criteria for states that cannot lead to a terminal state focus on the break and pull-out cluster. If it is no longer possible to make a feasible break because

$$\min T^{\text{dest}}(S) > t^{\text{pullin}} + 6 * 12$$

or pull-out is no longer possible because

$$\min T^{\text{dest}}(S) + t(c^l, c^{\text{pullout}}) > \bar{t}_{c^{\text{pullout}}},$$

we can forget about state S.

The generic program for tour enumeration that results from these considerations is so similar to the cluster generation routine that we refrain from giving the pseudocode here.

As we have already pointed out, the combinatorial situation for tour generation differs from the clustering scenario because the number of possible follow-on clusters is much higher. In fact it is not possible to produce all possible vehicle tours in this way, and the reason is not that the routine wouldn't work fast enough, but that the output is simply so large that there is no hope of even storing it. Also, the majority of tours obviously consist of rather inefficient tours, such that an optimal plan will contain only a few of them — which of course does not release us from trying to find "the right ones".

Since our set partitioning solver is a branch-and-cut code, we decided to reduce the solution space by producing only a "promising" set of tours that hopefully combine to a good vehicle schedule. Our tour generation routines are modifications of the above generic procedure that produce tours along *heuristic* strategies that we have developed in cooperation with the BZA. All these heuristics work very fast and together they can also be used as a stand-alone vehicle scheduling module (in fact, this was a first stage of installation of the Telebus-computersystem at the BZA).

The x *best neighbors* heuristic tries to produce "good" tours by applying the generic enumeration algorithm to a restricted transition digraph where the outdegree of each cluster, i.e., the number of follow-on clusters, has been limited to some value (we use $x = 2$ and $x = 3$). The x surviving neighbors of each cluster are chosen with respect to local criteria, like "nearest clusters".

The *tour-by-tour greedy* heuristic tries to work in a slightly more global way by iteratively producing a feasible tour. It selects an initial pull-in state and adds "best fitting" clusters (including the break) until the pull-out state is reached. The serviced clusters are removed from the transition digraph, the next tour is started, and so on. This heuristic tends to produce "good" tours at the beginning and yields unsatisfactory results at the end when only far-out or otherwise unattractive clusters are left. Tour-by-tour produces complete vehicle schedules.

Time sweep also constructs a complete schedule by scanning the clusters in some order. At every step, the next cluster is assigned to a best fitting tour (that is eventually created), until all clusters are scheduled. We use the natural orderings in time (from morning to evening and from evening to morning), and a "peaks first" variant, that tries to smooth out peaks of demand and link the resulting subtours.

A *hybrid* time sweep greedy heuristic performs a time sweep, but always adds not only one, but some x best neighbors to a tour.

Of a similar flavor is the *assignment* heuristic, that subdivides the time interval into slots of half an hour, and constructs an assignment of the subtours (possibly starting new ones) to the follow-on clusters of the next slot.

Another set of methods imitates the *hand planning* methods that were in use at the BZA earlier. These methods partition the requests by hour and city districts. Doing a time sweep from morning to evening, one looks at densities of requests in districts and hours and tries to concentrate vehicles in or near regions of high demand.

These methods can produce vehicle schedules that are already significantly superior to comparable hand planning. We use them in this way to set up chaining set partitioning problems with up to 100,000 columns. These IPs turned out to be computationally much harder than the clustering instances. A possible explanation is that clusters have a local nature and do not interact much, while tours extend over much longer time periods and larger service areas and thus exert more influence on each other. So we cannot solve the chaining set partitioning problems to optimality, but we nevertheless obtain significant reductions in operational costs of about 10% in comparison to what we can achieve by only using the chaining heuristics. There is, of course, even more potential for cost reductions if a better column generation method is used.

5 Set Partitioning

The third module of our vehicle scheduling system for Telebus consists of a branch-and-cut algorithm to solve large scale set partitioning problems. High-level pseudocode for the algorithm is shown in Fig. 5.1. We will now quickly state our branch-and-cut terminology and then discuss some aspects of our implementation.

The algorithm uses a *branch-and-bound enumeration scheme* for solving set partitioning problems that is based on considering *subproblems*

$$\min c^T x \quad Ax = \mathbb{1},\, l \leq x \leq u,\, x \in \{0,1\}^n, \qquad (\mathrm{SPP}(l,u))$$

of the original problem, where the lower and upper bounds l and u are 0/1-vectors. The original problem reads in this notation $\mathrm{SPP}(0,\mathbb{1})$, and a subproblem is formed by setting some of the upper bounds to zero, such that the corresponding variables are fixed to zero, and some of the lower bounds to one.

The scheme computes for each subproblem $\mathrm{SPP}(l,u)$ a *lower* and an *upper bound*

$$\underline{z}(l,u) \leq z^\star(l,u) \leq \overline{z}(l,u) = c^T \overline{x}(l,u)$$

on the *optimal objective value* $z^\star(l,u)$: the lower bound is derived from the LP relaxation $\mathrm{QSPP}(l,u)$, the upper bound and a corresponding feasible solution $\overline{x}(l,u)$ are computed by a heuristic to be discussed later; when the heuristic fails, we have $\overline{z}(l,u) = +\infty$ and $\overline{x}(l,u)$ is "undefined".

Subproblems are useful to search the solution space of $\mathrm{SPP}(0,1)$ in a divide-and-conquer way. The technique involves a rooted binary *searchtree* T, whose nodes are subproblems $\mathrm{SPP}(l,u)$. The tree is *initialized* to consist of only the root node $\mathrm{SPP}(0,\mathbb{1})$ and by setting $\underline{z}(0,\mathbb{1}) := -\infty$ and $\overline{z}(0,\mathbb{1}) := +\infty$, i.e., no lower and upper bounds for $\mathrm{SPP}(0,\mathbb{1})$ are known in the beginning. The algorithm *works* the root node by improving $\underline{z}(0,\mathbb{1})$ and $\overline{z}(0,\mathbb{1})$ and *labels* the root as being processed. If this step results in $\underline{z}(0,\mathbb{1}) = \overline{z}(0,\mathbb{1})$, the problem is solved and $\overline{x}(0,\mathbb{1})$ is the optimal solution. Otherwise, a *branching step* is taken to subdivide the problem into two subproblems $\mathrm{SPP}(l_1,u_1)$ and $\mathrm{SPP}(l_2,u_2)$, that become the sons of the root node. The subdivision must be done in such a way that the optimal solution for the root problem is contained in one of the two subproblems:

$$\min\{z^\star(l_1,u_1), z^\star(l_2,u_2)\} = z^\star(0,\mathbb{1}).$$

Since the subproblems are restrictions of the father problem, their lower bounds are at least as large and we can initialize them

$$\underline{z}(l_1, u_1) := \underline{z}(l_2, u_2) := \underline{z}(0, \mathbb{1})$$

with the father's lower bound. In general, the algorithm picks an unlabelled node v, works, and labels it.

```
// initialization
read problem;
initial preprocessing;
set up searchtree;

// branch-and-bound loop
while (∃ unlabelled subproblem) {
    select unlabelled subproblem;
    label it;

    // LP plunging heuristic
    set-up local LP relaxation;
    do {
        solve LP relaxation;
        if (integral) {
            update z̄(T);
            break;
        }
        set some fractional variables
            to integer values;
        do out pivoting;
        preprocessing;
        do in pivoting;
    }
    while (!infeasible);

    // cutting plane loop
    set-up local LP relaxation;
    do {
        solve LP relaxation;
        if (integral) {
            update z̄(T);
            break;
        }
        if (fathomed) break;
        out pivoting;
        preprocessing;
        in pivoting;
        separation;
        LP management;
    }
    while (progress);
    branch;
}

output z̄(T);
```

Figure 5.1: A branch-and-cut algorithm.

Either the node can be solved, or a branching step is taken adding two new unlabelled subproblems as the sons of v to the tree. To guarantee finiteness of this process, the branching process is done in such a way that each subproblem has at least one stricter bound than its father. This results in one more variable being fixed, and after a finite number of steps all variables are fixed and the subproblem is trivial to solve.

To save work, the algorithm maintains a *global upper bound*

$$\overline{z}(T) = \max_{\text{SPP}(l,u) \text{ unlabelled node of } T} \overline{z}(l, u),$$

which is the value of the best solution encountered in any of T's subproblems. The bound can be used to *fathom* subproblems that cannot contain a better solution than the currently best know because

$$\underline{z}(l,u) \geq \overline{z}(T);$$

such nodes can be labelled immediately and are not considered any further.

This standard branch-and-bound algorithm leaves a lot of freedom to implement its generic subroutines. We will explain some aspects of our algorithm in the following subsections.

5.1 Searchtree

The generic branch-and-bound algorithm does not specify the rule to choose the next unlabelled node. We use the so-called *best first* rule, that chooses the node with the smallest lower bound, i.e., the node that has most potential for possible improvement of the global upper bound. The smallest lower bound is also called the *global lower bound*

$$\underline{z}(T) = \min_{\text{SPP}(l,u) \text{ unlabelled node in } T} \underline{z}(l,u).$$

The best first choice potentially raises the global lower bound and thus decreases the *duality gap*

$$\overline{z}(T) - \underline{z}(T),$$

which is a measure of the global progress of the algorithm.

Best first requires that we can jump from one problem in the searchtree to any other. Our code uses a *local setup procedure* to do this, that simply generates the complete LP relaxation of a subproblem from scratch. This looks like a time consuming operation at first sight, but the method has advantages when additional cutting planes are used and redundant parts of the problem are removed by preprocessing: redundant parts for one subproblem are not necessarily redundant for others such that removed parts have to be restored, and similar actions are necessary if different sets of cutting planes are used in the subproblems. But removing and reinserting parts of a subproblem's description takes about the same time as a set-up from scratch.

The method to derive lower bounds $\underline{z}(l,u)$ for the subproblems of the branch-and-bound tree is to solve the LP relaxation

$$\min c^T x \quad Ax = \mathbb{1},\, l \leq x \leq u, \tag{QSPP(l,u)}$$

of the integer program (SPP) and a crucial point is that this need not be done from scratch every time. Rather, the *dual simplex method* allows using the optimal solution of the father's LP relaxation as a *dual feasible starting basis* for the LP relaxations of its sons and often only a few iterations are needed to recover primal feasibility and thus optimality. To benefit from this favorable behavior, we store this optimal basis for later use as starting basis.

A last point to specify is the *branching rule* that we use to subdivide a subproblem into two smaller problems. We mainly use Ryan/Foster (1981)'s rule and *strong branching*, see CPLEX (1995), that perform similarly in our instances.

5.2 Cutting Planes and LP Management

The LP relaxations of the subproblems can be strengthened by adding various types of globally valid cutting planes (see, e.g., Balas/Padberg (1976)). We use *clique inequalities* and *simultaneously lifted odd cycle inequalities* of the associated set packing polytope (see Padberg (1973)), and a class of set covering inequalities that arise from an associated set packing problem via "complementing" and "aggregating" variables (see Borndörfer (1998)). Clique inequalities are separated both heuristically and by an exact branch-and-bound algorithm, cycle inequalities are separated using the exact polynomial algorithm of Grötschel/Lovász/Schrijver (1988) and a Chvátal-Gomoroy simultaneous lifting procedure, and the covering inequalities by heuristic procedures. Details of these methods are discussed in Borndörfer (1998).

Working on a subproblem means to solve and strengthen the LP relaxation iteratively by adding violated cutting planes until the subproblem is either solved, fathomed, or some other stopping criterion is satisfied and we branch. In our implementation, we use the duality gap $\overline{z}(l,u) - \underline{z}(l,u)$ as a measure of progress of the cutting plane loop and continue as long as this gap is reduced by 10% in every three successive iterations.

We also remove rows from a subproblem's LP relaxation, because the time to solve LP relaxations of set partitioning problems increases with the number of rows of the constraints matrix. Another important point in a branch-and-cut framework is that more rows also tend to produce more fractional variables in the LP solution. To reduce running time and get a more integral solution, it is important to remove redundant cutting planes from a subproblem's description and we do this heuristically when the slack exceeds 10^{-3}. Each subproblem involves a different subset of all cutting planes that we have found throughout the course of the algorithm and if we want to be able to reproduce a subproblem exactly in the local set-up step, we must maintain a global *pool* of all cutting planes. An advantage of this method is that the computation on invocation of a subproblem becomes independent of the history of the branch-and-bound algorithm.

The LPs themselves are solved using the CPLEX dual steepest edge simplex algorithm, (see CPLEX (1995)).

5.3 Problem Reduction and Pivoting

Significant speed-ups for the solution of the LP relaxations of the subproblems can be achieved by removing redundant parts like columns of variables that are fixed to zero or one, or rows that intersect columns that are fixed to one. Such fixings do not only arise from branching decisions, but also from the logical structure of a set partitioning problem, and *preprocessing* is the use of simple techniques to detect such redundancies. Preprocessing techniques for set partitioning problems are know to be highly effective, and our code uses a concept of *repeated problem reduction* that applies preprocessing techniques after each individual LP solution. Repeated preprocessing of a similar type has been used by Atamturk/Nemhauser/Savelsbergh (1995) for a Lagrangian heuristic for SPPs, but the technique does not seem to have been tried in a branch-and-cut framework before.

The *preprocessing techniques* that we use include ones from the literature, like elimination of duplicate columns and rows, fixing of singletons, elimination of columns that are neighbors of a variable fixed to one, dominated rows, and some new ones. These procedures must be applied several times, because elimination of dominated rows can lead to more duplicate columns, etc. Our preprocessor performs another pass as long as it detects redundancies.

An important point in a dual simplex framework is the proper linking of preprocessing and LP solving: preprocessing must not destroy dual feasibility of the basis, because otherwise we would have to solve the LP essentially from scratch. The consequence is that we are not allowed to remove fixed *basic* variables and we cannot remove redundant *nonbasic* rows. The desire to remove such redundant parts of the problem nevertheless leads to some algorithmic consequences that we explain now.

Dual feasibility of the basis forces us to distinguish between *fixings* and *eliminations* of variables. Fixing is the setting of bounds of variables, elimination involves a real removal of data from memory. Our preprocessor works only with fixings, a subsequent elimination removes all fixed nonbasic variables and all detected redundant basic rows from memory. In this way, we combine a maximum of problem reduction with maintenance of the basis's dual feasibility. Nevertheless, one would like to remove all detected redundancies from memory, and this leads to the consideration of *pivoting techniques.* The aim of these techniques is simply to perform a number of (degenerate) pivots to move from one optimal basis to an alternative one, such that all fixed variables are nonbasic, all detected redundant rows are basic (have their slack/artificial variable in the basis), and all of these redundancies can be eliminated. A pivoting technique that is implemented in CPLEX is the *in pivoting* of rows with zero dual multiplier into the basis. Unfortunately it turns out that most of the (known) redundant rows have nonzero duals and the reason is that fixed variables tend to proliferate in the basis: often more than 30% of the basis consists of "junk" of this type, inhibiting removal of the same number of rows. Fixed variables can also be pivoted out of the basis using dual simplex steps, and we are grateful to Robert E. Bixby that we had access to a version of CPLEX that provides this novel out pivoting routine. Application of the procedure usually leads to a faster problem reduction, but out pivoting is not cheap: it requires one dual pivot for each fixed variable. One thus has to compare the benefits of eliminating large numbers of fixed variables by a consequently large number of pivots with the possibly few simplex iterations required to solve the next LP without prior pivoting. Eliminations, however, are inherited by all offspring problems and our computational experience is that out pivoting is worth its price.

5.4 Primal Heuristics

We use the popular *LP plunging* heuristic to generate upper bounds and feasible solutions for a subproblem in the searchtree. This heuristic solves the LP relaxation of a subproblem, fixes some fractional variables to integer values, and iterates, until the solution becomes integral or the problem infeasible. Our algorithm does not have a separate implementation of this routine, but simply uses the main cutting plane loop in a "primal mode" where separation is turned off. This results in particular in iterative preprocessing after each fixing decision, and this results in a fast reduction of the problem size. The heuristic is nevertheless expensive: a sequence of LPs has to be solved, and the elimination of (the largest) parts of the associated data forces a subsequent second local setup of the subproblem to initiate the cutting plane loop. For this reason we call the heuristic only once at the invocation of a new subproblem.

6 Computational Results

In this section we report on computational experiences with our vehicle scheduling system. Our aim is to discuss two complexes of *questions.* Our first and main goal is to evaluate the usefulness of our set partitioning approach for the solution of VSPs at Telebus. Does clustering lead to savings in internal travelling distance? Does tour optimization lead to better results than our heuristics? Second, we want to look at the performance of our software modules for Telebus instances. What is the size of the problems that we can solve in reasonable time? What is the quality of the solutions?

To answer the second question, we ran our branch-and-cut algorithm on a *test set* of Telebus clustering and chaining problems. It is not interesting to provide performance data for the cluster and tour generators, because there is no computational bottleneck in these procedures. Our branch-and-cut code is implemented in `C` and consists of about 1 MB of source code in 140,000 lines, the LP solver is the CPLEX Callable Library V4.0, CPLEX (1995). All test runs were made on a Sun Ultra Sparc 1 Model 170E, the code was compiled with the Sun `cc` compiler using the switches `-fast -xO5`, and we used a time limit of 7,200 CPU seconds. The format of the upcoming tables is as follows. Column 1 gives the name of the problem, columns 2-4 contain the size of the problem in terms of the number of rows, columns, and nonzeros. The next three columns give the number of rows, columns, and nonzeros after the initial preprocessing of the problem at the root node. Comparing columns 2-4 to columns 5-7 shows the performance of our preprocessing. The next three columns give solution values. $\underline{z}$ reports the value of the global lower bound. This number coincides with the global upper bound $\overline{z}$, when the problems is solved to proven optimality. Otherwise, we are left with a duality gap that we report in percent of the global upper bound, i.e., the gap is computed as $(\overline{z}-\underline{z})/\overline{z}$. The following five columns give information about the performance of the branch-and-cut algorithm. There are, from left to right, the number of in and out pivots (Pvts), the number of cutting planes (Cuts), the number of simplex iterations to solve the LPs (Itns), the number of LPs solved (LPs), and the number of branch-and-bound nodes (B&B). The next five columns show timings: the percentage of the total running time spent in problem reduction (PP), pivoting (Pvts), separation (Cuts), LP solution (LPs), and the heuristic (Heu). The last column gives the total running time in CPU seconds.

Our first set of test problems consists of 14 clustering problems for the weeks of April 15–22, 1996, (v0415–v0421) and the already well-known week September 16–22, 1996, (v1616–v1622) that we used to produce most of the diagrams in this article. The first five problems of each data set correspond to the weekdays Monday to Friday, the last two show a significantly smaller number of rows (= requests) and belong to the weekend. The two test sets were generated with different parameter settings of the cluster generator.

In April 1996, rules for legal clusters were very restrictive: continued concatenations and insertions were not allowed, maximum detour time was small, etc. The cluster generator thus found only relatively few feasible clusters and the clustering SPPs are small. Moreover, most of the legal clusters provide simultaneous service to very few requests: the average number of nonzeros per column is a little above two for four days of the week, the three larger instances have a higher average because they contain many clusters for a couple of large collective requests, but the remainder of the problem has the same characteristic. This means that individual clusters do not interact much, the problem sort of decomposes and becomes easy. And in fact the initial call to the preprocessor is very successful, in particular the number of rows is reduced by more than 50%. This trend continues in the branch-and-cut phase: all problems are solved to proven optimality in at most 3 minutes, and we can see from the pivoting (Pvts) and preprocessing (PP) columns that the problem is basically solved by iterative preprocessing. In particular, the high number of out pivots shows that variables could be fixed in large numbers and the sizes of the problems were reduced very fast.

	Original Problem			Preprocessed			Solutions			Branch-and-Cut					Times in %					
Name	Rows	Cols	NNEs	Rows	Cols	NNEs	$\underline{z}$	$\overline{z}$	Gap	Pvts	Cuts	Itns	LPs	B&B	PP	Pvts	Cuts	LPs	Heu	Total Time
v0415	1518	7684	20668	598	4536	10988	2429415	2429415	0.000	12774	70	755	36	9	32	32	4	12	3	5.68
v0416	1771	19020	58453	812	11225	33991	2725602	2725602	0.000	325151	1305	4677	1970	643	19	28	4	15	3	120.53
v0417	1765	143317	531820	715	55769	206131	2611518	2611518	0.000	61309	294	1360	171	41	35	21	8	10	3	174.07
v0418	1765	8306	20748	742	4957	11177	2845425	2845425	0.000	12203	81	941	25	7	29	31	6	15	3	5.72
v0419	1626	15709	52867	650	7852	25052	2590326	2590326	0.000	4106	55	801	4	1	29	17	11	17	5	3.99
v0420	958	4099	10240	417	2593	6124	1696889	1696889	0.000	2538	47	511	4	1	28	23	8	18	5	1.31
v0421	952	1814	3119	286	1134	1437	1853951	1853951	0.000	2304	34	317	9	3	32	18	4	18	3	0.72
v1616	1439	67441	244727	1230	52926	199724	1006460	1006460	0.000	1295605	11123	177084	4811	1605	6	30	8	41	8	4219.41
v1617	1619	113655	432278	1409	85457	336147	1102357	1102586	0.021	15257970	16169	67051	15661	3571	22	46	6	10	3	7200.61
v1618	1603	146715	545337	1396	90973	349947	1152989	1154458	0.127	2418105	5549	70533	1461	296	11	27	16	21	9	7222.28
v1619	1612	105822	401097	1424	85696	336068	1156072	1156338	0.023	5774346	9040	124824	4203	880	15	40	15	12	10	7205.74
v1620	1560	115729	444445	1365	89512	353689	1140604	1140604	0.000	7460098	20801	111073	19230	8161	19	29	14	11	2	5526.43
v1621	938	24772	76971	807	16683	54208	825563	825563	0.000	12214	130	1415	13	5	23	20	16	22	3	13.79
v1622	859	13773	41656	736	11059	35304	793445	793445	0.000	13325	99	1147	14	3	27	29	10	18	3	9.69
v1616	1439	67441	244727	1230	52926	199724	1006261	1006460	0.020	83501	828	6193	70	11	19	26	17	22	4	125.06
v1617	1619	113655	432278	1409	85457	336147	1101822	1103036	0.110	48690	426	2972	20	4	14	22	27	20	3	137.51
v1618	1603	146715	545337	1396	90973	349947	1152150	1156417	0.369	38494	436	2976	15	3	12	18	32	22	2	130.19
v1619	1612	105822	401097	1424	85696	336068	1155336	1157851	0.217	48584	528	3228	15	3	13	17	35	21	3	146.06
v1620	1560	115729	444445	1365	89512	353689	1140238	1142159	0.168	35910	377	2940	15	3	12	20	24	29	3	133.33
v1621	938	24772	76971	807	16683	54208	825563	825563	0.000	12214	130	1415	13	5	22	20	17	22	3	13.82
v1622	859	13773	41656	736	11059	35304	793445	793445	0.000	13325	99	1147	14	3	26	29	10	19	3	9.40
21	29615	1375763	5070937	20954	952678	3625074	31105431	31117511	0.039	32932766	67621	583360	47774	15258	15	34	12	18	6	32405.34
t0415	1518	7254	48867	870	3312	20592	5163849	5590096	7.625	1291268	2029	93675	724	167	5	11	14	17	53	7218.94
t0416	1771	9345	62703	974	3298	19692	5882041	6130217	4.048	1334745	2163	92796	641	144	5	11	14	17	54	7207.46
t0417	1765	7894	54885	897	3774	24186	5656886	6043157	6.392	1614510	994	51439	316	71	6	17	6	11	60	7310.58
t0418	1765	8676	66604	999	4071	29368	6185168	6550898	5.583	629332	1066	67551	399	87	3	9	12	21	56	7239.54
t0419	1626	9362	64745	904	3287	19990	5689134	5916956	3.850	1891101	1235	57831	429	100	6	16	9	11	59	7251.57
t0420	958	4583	27781	562	1872	10271	4036526	4276444	5.610	3989264	4440	135766	1507	362	10	16	11	11	52	7208.44
t0421	952	4016	24214	557	1691	9015	4113080	4354411	5.542	4238861	4581	134126	1594	375	10	16	12	11	51	7213.44
t1716	467	56865	249149	467	11952	61110	122408	161636	24.269	1230592	886	39379	296	69	2	7	3	11	76	7212.95
t1717	551	73885	325689	551	16428	85108	135539	184692	26.613	1021307	592	27888	183	41	2	7	2	10	77	7331.93
t1718	523	67796	305064	523	16310	83984	127040	162992	22.058	982755	606	28048	203	44	2	6	3	10	78	7238.72
t1719	556	72520	317391	556	15846	83893	139332	187677	25.760	992993	404	22565	169	37	2	6	2	8	80	7281.77
t1720	538	69134	310512	538	16195	84194	127225	172752	26.354	899591	688	30360	187	38	2	6	3	11	77	7349.28
t1721	357	36039	148848	357	9043	44106	104698	127424	17.835	1965867	1921	69853	765	174	3	9	5	12	69	7243.42
13	13347	427369	2006452	8755	107079	575509	37482926	39859352	5.962	22082186	21605	851277	7413	1709	4	11	7	12	65	94308.04

Table 6.1: Clustering and Chaining.

In September 1996, rules were much more liberal: the clustering problems contain, for example, continued concatenations of a depth of up to 6. Consequently, there are many more possibilities for feasible clusters, the instances are larger, contain about 4 NNEs per column, and there is more overlap. This time, the initial preprocessing is still successful, but the number of rows is reduced far less than in the first test set. And in fact, the instances turn out to be harder in the sense that we cannot solve them to proven optimality as fast: in fact, there are three instances that we cannot solve completely within our time limit of 7,200 CPU seconds. Looking at the performance of the algorithm, we see that pivoting and preprocessing need most of the time, but are successful (remember that every pivot indicates a fixed variable). However, even though we find a significant number of cutting planes, the quality of the cuts does not prevent the algorithm from extensive branching, as can be seen from the B&B column. All of this effort is, however, only spent in proving the optimality of a solution of very good quality. To show this, we have run the algorithm another time with a time limit of 2 minutes, and we see that satisfactory solutions can be obtained in this period.

The clustering results are satisfactory in the sense that more or less independent of the parameter settings clusterings of proven optimality or with very good quality guarantee can be computed in short time, considering the complete solution space of all legal clusters.

We have used the clusterings that we computed in the previous test runs to set up the corresponding chaining problems as well. The April instances (t0415–t0421) contain duplicate rows for clustered requests and have thus the same number of rows as the corresponding clustering instances, the optimization criterion was operational costs, in the September instances (t1716–t1721) only the bus clusters were chained, the optimization criterion was travelling distance. Chaining rules were again more strict for April and the resulting instances are not very large in terms of columns and NNEs. Looking at the preprocessed instances (with the duplicate rows removed), however, the average number of NNEs is already larger, indicating a more complicated combinatorial structure. This can also be observed for the September instances: here, our preprocessor cannot even remove a single row in any of the instances. Although the preprocessed instances are not large, they turn out to be computationally difficult. In contrast to what is usually reported about real world set partitioning problems, there is a large duality gap between the value of the LP relaxation and the best know integer solution. In fact, even the duality gaps on *termination* of the algorithm as reported in column Gap are significant, in the case of the September instances even large. Most of the computational effort is spent in the heuristic, because the iterative preprocessing doesn't reduce the problems a lot in the early rounding steps. But even if we subtract this time completely, the algorithm performs comparably few iterations: the number of LPs is rather small and the same holds for the size of the searchtree. The reason for this is that the LP relaxations of the chaining problems are harder to solve, as can be seen by looking at the average number of simplex iterations per LP (column Itns divided by column LPs).

Day	Requests	Heuristics				Integer Programming		
		Clusters		Tours		Cluster		Tours
	No.	No.	km	DM	DM	No.	km	DM
Mo	1439	1167	10909	66525	60831	1011	10248	55792
Tu	1619	1266	11870	71450	67792	1106	11291	62696
We	1603	1253	12701	74851	68166	1107	11813	61119
Th	1612	1276	12697	74059	68271	1121	11821	64863
Fr	1560	1242	12630	71944	63345	1080	11757	61532
Sa	938	748	9413	45842	47736	676	8561	41638
Su	859	703	8850	42782	44486	620	8243	38803
$\sum$	9630	7655	79070	447453	420627	6721	73734	386443

Table 6.2: Comparing Vehicle Schedules.

Although the chaining step does not provide near optimal solutions, tour optimization is still valuable. Table 6.2 shows the results of a comparison of different vehicle scheduling methods for the week September 16–22, 1996. Column 1 gives the day of the week and column 2 the number of requests. The next three columns show the results of a heuristic vehicle scheduling using our cluster and tour generators as a stand-alone optimization module. There are, from left to right, the number of clusters obtained from a heuristic clustering, the internal travelling distance within these clusters, and the costs of a vehicle schedule based on this clustering. Skipping column 6 for the moment, we can compare these numbers with the results that we obtained using the set partitioning approach. Column 7 gives the number of clusters obtained in this way, column 8 the corresponding internal travelling distance, and the last column the costs of the vehicle schedule that was obtained by chaining the optimal set of clusters and solving the resulting chaining SPP approximately. Column 6 that we just left out gives the costs of a vehicle schedule that was constructed heuristically from the optimal clustering. Roughly speaking, these numbers show that an optimal clustering reduces the number of requests about 10% more than what we can achieve heuristically. Heuristic chaining based on optimized clusters results in vehicle schedules that are about 5,000 DM per day cheaper than a pure heuristic approach, while chaining optimization can save another 5,000 DM per day.

7 Summary

In this paper we have presented a set partitioning approach to vehicle scheduling in a dial-a-ride system for handicapped people. The results show that it is possible to solve vehicle scheduling problems for systems of this size in a satisfactory way. In the Telebus case, the use of modern computer technology and mathematical programming techniques resulted in improvements in service quality and *simultaneous* significant cost reductions. We think that such results can lead to a renewed interest in dial-a-ride systems for use not only as a special purpose system for handicapped people, but as a component of the public transport to service areas or times of low demand.

References

Atamturk, A. / Nemhauser, G.L. / Savelsbergh, M.W.P. (1995): A combined lagrangian, linear programming and implication heuristic for large-scale set partitioning problems. Tech. Rep. LEC - 95-07, Georgia Inst. of Tech.

Balas, E. / Padberg, M. (1976): Set partitioning: a survey. SIAM Rev., 18, 710 - 760.

Ball, M.O. / Magnanti, T.L. / Monma, C.L. / Nemhauser, G.L. (eds.) **(1995):** Network Routing, Vol. 8 of Handbooks in Operations Research and Management Science. (Elsevier Sci. B.V.) Amsterdam.

Borndörfer, R. (1998): Aspects of set packing, partitioning, and covering. PhD thesis, Tech. Univ. Berlin.

Borndörfer, R. / Grötschel, M. / Herzog, W. / Klostermeier, F. / Konsek, W. / Küttner, C. (1996): Kürzen muß nicht Kahlschlag heißen — Das Beispiel Telebus-Behindertenfahrdienst Berlin. Preprint SC 96-41[1], (Konrad-Zuse-Zentrum) Berlin.

CPLEX (1995): Using the CPLEX Callable Library[2]. CPLEX Optimization, Inc., Suite 279, 930 Tahoe Blvd., Bldg 802, Incline Village, NV 89451, USA.

Cullen, F. / Jarvis, J. / Ratliff, H. (1981): Set partitioning based heuristics for interactive routing. in: Networks 11, 125 - 143.

Dell'Amico, M. / Maffioli, F. / Martello, S. (eds.) **(1997):** Annotated bibliographies in combinatorial optimization. (John Wiley & Sons Ltd) Chichester.

---

[1]Avail. at URL http://www.zib.de/ZIBbib/Publications/

[2]Inf. avail. at URL http://www.cplex.com

Desrochers, M. / Desrosiers, J. / Soumis, F. (1984): Routing with time windows by column generation. in: Networks 14, 545 - 565.

Desrosiers, J./ Dumas, Y./ Solomon, M.M./ Soumis, F. (1995): Time constrained routing and scheduling. in Ball / Magnanti / Monma / Nemhauser (1995), Chap. 2, 35 - 139.

Grötschel, M. / Lovász, L. / Schrijver, A. (1988): Geometric algorithms and combinatorial optimization. (Springer Verlag) Berlin.

Ioachim, I. / Desrosiers, J. / Dumas, Y. / Solomon, M.M. (1991): A request clustering algorithm in door-to-door transportation. in: Tech. Rep. G-91-50, École des Hautes Études Commerciales de Montréal, Cahiers du GERAD.

Klostermeier, F. / Küttner, C. (1993): Kostengünstige Disposition von Telebussen. Master's thesis, Tech. Univ. Berlin.

Laporte, G. (1997): Vehicle Routing. in: Dell'Amico/Maffioli/Martello (1997), Chap. 14, 223 - 240.

Padberg, M.W. (1973): On the facial structure of set packing polyhedra. Math. Prog. 5, 199 - 215.

Ryan, D.M. / Foster, B.A. (1981): An integer programming approach to scheduling. in: Wren (1981), 269 - 280.

Wren, A. (ed.) **(1981):** Computer Scheduling of Public Transport: Urban Passenger Vehicle and Crew Scheduling. (North-Holland) Amsterdam.

Advanced Technologies in the Design of Public Transit and City Information Systems

Joachim R. Daduna[1] and Kamen Danowski[2]

(1) Fachhochschule für Wirtschaft Berlin, Badensche Straße 50 - 51, D - 10825 Berlin (Germany), e-mail: daduna@fhw-berlin.de

(2) Fraunhofer Institut für Informations- und Datenverarbeitung, Zeunerstraße 38, D - 01069 Dresden (Germany), e-mail: danowski@eps.iitb.fhg.de

Abstract: Passenger information is an important factor in improving the acceptance of public transit services. Since mobility should be viewed in a more general context, information must not be limited to public transit, and information relevant to the various user groups and different trip purposes should be included. Therefore, a modular expert system has been developed, to fit the required functionality. The most important technical aspects are a detailed data structure, an efficient knowledge based search procedure to calculate connections between any origin and destination point in street networks and public transit networks, and a favourable processing speed. Beside these factors the design of the user interface forms a second focal point, especially concerning user friendly dialog, a uniform structure of screens, and the usage of common icons and pictograms. In addition, multimedia elements have been used to form an attractive product. A passenger information system based upon the presented concept has been implemented in Dresden and has been in operation since 1995.

1 Information Structuring in Public Transit Services

Passenger information is an important obligation of public transit companies and is a key factor for better acceptance of the services offered. It is known that the availability of sufficient information on a product is a basic precondition to get clients or potential clients interested in using the services. For these reasons it is very important for providers of public transit services to offer complete and comprehensive information.

The efforts to achieve this goal must cover the whole process beginning with choice of means of transport up to the completion of the trip (see, for example, Daduna (1997)). The set of requirements for a user friendly information structure includes not only the availability of public transit services, but also the connections within the line network and information on operations. Functional co-ordination between the different measures is absolutely necessary, because isolated measures will not lead to appropriate solutions. Finally, only joint action making use of the different components can result in the full range of advantages of an information concept.

Over the past few years researchers have been seeking new tools based on the developments in different areas of information technologies, to complement the established information media such as network maps, timetable booklets, etc. Passenger information systems have become an essential part of these new tools (see Mentz (1986) or Daduna (1997)). They give many possibilities for improvement in information provision. The evolution of these systems must reflect the changing of requirements in passenger information, especially due to the impact of the competition with the automobile.

The provided information should not be limited to public transit, because mobility has to be seen in a more general context. Contents should also be related to different travel goals and user groups. Users require horizontal interconnection of different information content related to the reasons for mobility and to timetables. Therefore, the appropriate information structure is a means to influence the public transit modal split.

Beside this, information systems represent an important interface between the service provider and clients. This imposes special requirements on the design and the quality of the product. First, these requirements concern the technical efficiency of the systems and second, the user-friendly design of interfaces. These two points are crucial for the acceptance of the systems and for removing psychological barriers in using information technologies.

Based on these considerations a multi-functional passenger information system has been developed as a modular expert system. In the following sections the conceptual basis for the system and the software design are described. Priority is given to data provision, the routing processes, and the design of user interfaces. Several application areas are shown and experiences from the first installation in the City of Dresden are reported. Finally an overview of the future developments in this information segment is given, in light of possible changes in information technologies.

2 Basic Characteristics of an Efficient Information System

The basic characteristics that are essential for the development of the system are the *functionality in its content*, the *routing procedure*, and the design of the *user interface*. The requirements listed below are the basic prerequisites for an efficient passenger information system which form the framework for development.

- *Functionality*

  Together with the provision of timetable data different kinds of information for travel-related areas such as tourism, culture, sports, entertainment, public authorities and services should also be provided. The aim is to create a *city information system* with a wide functionality, which can be used both by citizens and by visitors. An important aspect is the improvement of information quality associated with simplifying the access through combining existing separate systems.

- *Routing*

  There are two specific points relating to the routing process, the efficiency of the algorithms used, and the selection criteria being sensitive to the user's needs. Both are important for the efficiency and acceptability of these systems by the clients. Therefore, to cover the different clients requirements it must be possible to use various criteria for the routing process, which normally lead to different route suggestions. For example route selection should be possible based on the shortest time, the cheapest, or the most convenient path (see Fig. 2.1) together with the associated constraints on the transfer times at transfer points. Therefore, the information about possible transfers must be exact so that the client can rely on it.

- *User Interface*

  The user interface must first be simple to use, to help remove the psychological barriers to acceptance and use of the system. To achieve a user oriented dialog, uniform structure of the screen, understandable icons, pictograms and texts should be implemented. In addition to the well known presentation forms from the area of the print media (text, graphic, picture) various multimedia elements could also be used.

Beyond these three functional aspects important factors are the modelling of the transit system and data handling, including data provision. They play key roles in the design of passenger information systems because they present to the client the range of available services.

3 Software Concept

The following sections describe the basic structure of the software concept, as they are developed and implemented in the *Public Transit & City Information System Dresden.* They are based on the principles set out in Sect. 2, which affect not only the efficiency of the system but also the user requirements. The problems relating to the user are given high priority, because the acceptance of an information system and more widely of the conception for such systems depends on this. Three problems are considered below, data provision, routing procedure, and design of the interaction processes.

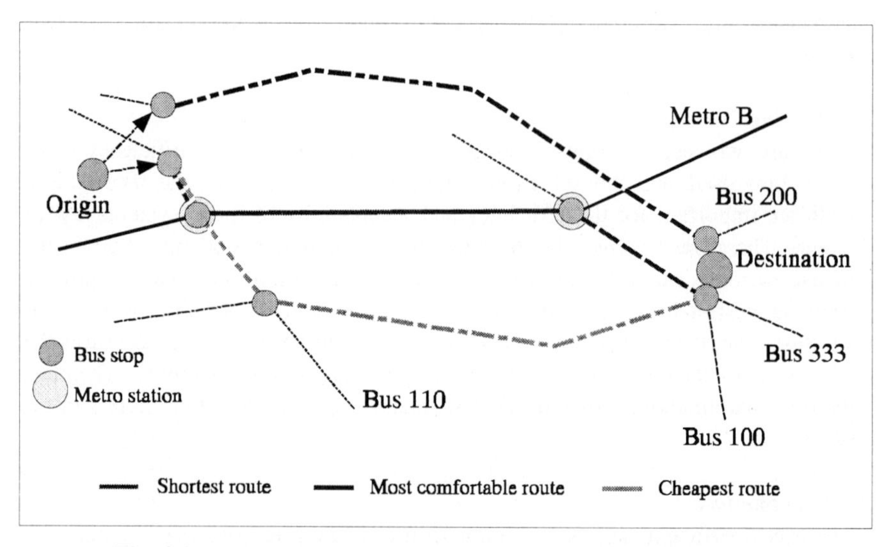

Fig. 2.1 Individual choice of the routes in public transit networks

3.1 Data Provision

Data provision concerns the supply of the system with planned data from relevant areas for different users. The incorporation of *dynamic data*, which are currently used in various applications of passenger information (see Daduna (1997)), is realized in the first steps. There are however limits to the functionality, because the necessary level of information depends on the interests of the user which will vary. For example, if an inquiry concerns an immediate trip, the currency of the data used is very important. But if an individual timetable has to be established for a trip to be made (in future), the current information on vehicle location is not relevant.

There are two types of data sources for planned data, which must be considered separately. First are the data on the passenger transit services offered (line networks, timetables) including two basic data sets. One deals with data used directly in the routing processes, such as data concerning local or regional public transit, or the long-distance rail service, which reflects the exact network connections throughout the whole system. Second are data of other means of transport, such as air timetables, giving the necessary information to the user to make easier transfers at specific nodes in the network.

Other data sources relate to the information concerning the purpose of a trip, which can be multilevel, and can vary depending on the different area of application (for example, see Daduna/Voss (1996)). So it is possible to include information about public facilities and public services, as well as about other services, including hotels, restaurants, etc. In addition, information about tourist sites of the area and about cultural, sports and entertainment events can be incorporated, as well as other commercial information.

It is obvious that the multilevel aspect of the data requires an efficient information management system, that can assure the processing and the provision of user oriented information services. Therefore, it is very important to establish an institution with defined responsibilities as an interface between the passenger information system and the different data sources. It must be responsible for the supply of data necessary for the system and for their updating over time.

This need must not be underestimated. Only after securing a stable and continuous updating of the data can a system work efficiently and realize its potential. The main tasks are the gathering, structuring, preparation and transmission of the needed data. Gathering includes the standards for data transmission and terminating of the procedures. Structuring concerns the selection of the available data, since they don't normally exist in the source systems in an appropriate form. This happens frequently with software tools used in public transit companies, which are designed as separate systems and do not give integrated solutions. Data editing is necessary to ensure the efficiency of the information system especially with respect to quasi real-time operations.

Together with the planned data it is possible to incorporate actual data in the form of dynamic passenger information, as mentioned above (see Daduna (1997)). This relates first to timetable information directly concerning a current trip. To be able to consider dynamic data, parallel data structures must be established, which have on one hand the planned data and on the other hand a continuously updated data set. No fundamental problems should arise here except a concern with the efficiency of the routing procedures. A basic requirement for the incorporation of dynamic data is the area-wide availability of an *Automated Vehicle Monitoring System* (AVM System), to avoid inconsistencies which could result from a mix of planned and updated data. Otherwise unacceptable routing suggestions could be generated. In the long term attention will have to be paid to the dynamic components, because they will give additional benefits to the clients.

3.2 Routing

The problem of routing within a given line network and timetable is to find the appropriate connections between a defined origin and destination. The criteria of searching can be to leave the origin or to arrive at the destination at a given (earliest or latest) time. Additional search criteria can be the travel time to reach the destination, the number of transfers and the price. On the basis of these requirements it is possible to formulate an optimization problem, which can be solved making use of procedures to determine routes in networks (see Mentz (1986)). The combinatorial nature of such problems requires for computer-aided solutions on one hand the development of efficient model structures that reflect the characteristics of the line networks and timetables of the transit system and on the other hand to have powerful algorithms to find the specific connections based on the data and restrictions given by the user (see Danowski (1997)).

There are various methods for finding connection possibilities in passenger information systems, which can be divided into three main categories:

- Making the choice from a set of already calculated and stored paths and connections.
- Multi-step calculation methods.
- One-step calculation methods.

The first category includes procedures which calculate in advance all possible connections in a transit network. Upon an enquiry the necessary data are selected directly from the computed set. There is no need to apply algorithms in the commonly used sense, only efficient data access and data selection methods are necessary. In timetable based public transit systems however such applications are exceptions. This method can be used only in very small transit networks or in special areas because of the large number of possible combinations, as compared with air transit networks. Public transit networks with a great number of stops/stations and short headways, as is normally the case in large cities, do not permit use of these methods, from either the technical or economic point of view. In addition, allowing the clients to specify requirements for routing further increases the range of possible connections.

The basic principle in multi-step methods is that in the first step one or more routes are defined as chains of nodes, only on basis of the existing network topology, taking into account the different objective functions. Then the possible routes are examined sequentially, starting with direct connections, then with one, two, etc. transfers. From this set of feasible connections a number of possible routes are selected making use of route-dependent preliminary weights. The criteria used are normally the estimated travel time or the number of transfers and the average transfer time. These depend on the modes of transport and on the route segments. At the beginning of the solution procedure the timetable information is not considered. In the following steps the expected travel times are calculated for all selected routes and among them, the shortest one is found. The necessity to incorporate at the first step preliminary weights that compensate for the absence of the timetable information could result in not finding the best connection in every case. This problem can appear in the following situations:

Timetable information in an area with a large number of line variants.

- Transition from one network to another, depending on daily time periods (e.g., day- and night-services in the network).
- Network with single trains or buses at the beginning or the end of the day.
- Transition between two operation days.

A proven optimal solution, corresponding to the given criteria can only be found on the basis of these framework requirements, if the travel times for all possible routes are included together with the information about the area. Such a method is called one step which considers in parallel the timetable and line/network characteristics.

The calculation of the possible connections, as implemented in the *Public Transit & City Information System Dresden* presented in Sect. 5, consists of a combined route-time-search procedure. A special step by step pass through the travel area is used, in which the transfer nodes are selected based on weights. The weight $B(x)$ for a node x or the route segment to it, is determined as the sum $g(x)$ of the calculated minimum time (for running and waiting time) to reach this node, from the origin, and the weighted time $h(x)$, needed to reach the destination from this node (see Fig. 3.1). The first part $g(x)$ can be calculated exactly from the running time for all the trip segments, the transfer times and waiting times within the transfer nodes on the route segment from the origin up to the analysed node x. The second part $h(x)$ is an estimate, which is based on a heuristic travel time matrix, established for this specific purpose (see Danowski (1994)).

Within the procedure of route calculation a search tree is developed whose branches are composed of trips or trip segments. Every reached node is checked to see if it has already been reached through other connections (on other routes with various time parameters). If this is the case, that means that a transfer node can be reached by two (or more) route segments in the search tree, the different segments are compared and weighted on the basis of defined criteria. There are different rules defined for this purpose in the system (see Danowski (1997)).

- Groups of rules for weighting single connections.
- Groups of rules for early recognition of "dead-end roads" and their preliminary exclusion.
- Procedures to avoid establishing meaningless route segments.
- Rules to handle exceptions and alterations in the standard timetable.

In accordance with these rules, every route segment is listed in a connection list or is excluded from further calculations. The order in which the route segments (branches) must be handled in the connection list is defined by priority rules. The procedure of step by step branching will be continued until the destination is reached. Finally defined route segments are included as meaningful connections or excluded.

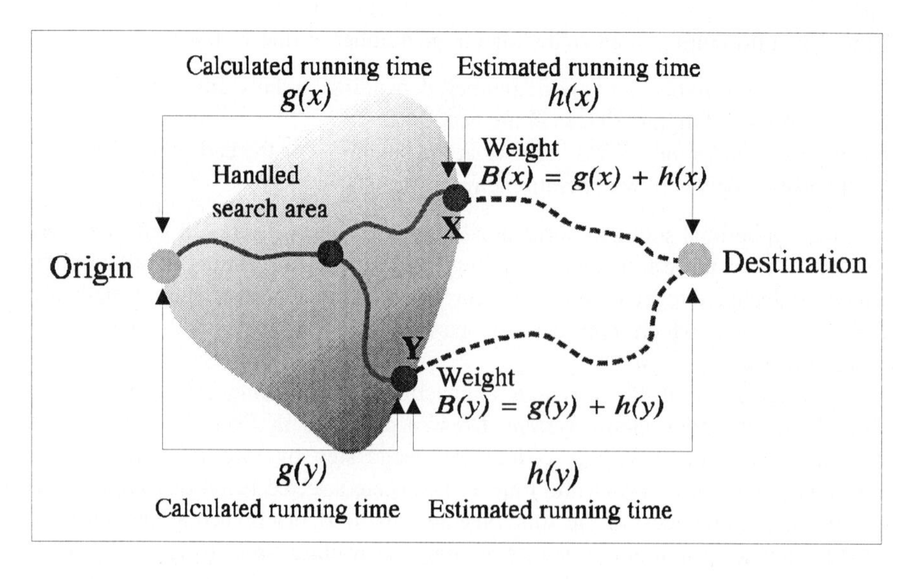

Fig. 3.1: Principle of the route search method to find possible connections

This procedure is computation intensive and therefore, different heuristics and decision rules are used (see Danowski (1994)) to limit the search area based on the various characteristics of the nodes. Because changes in modes of transport occur only at transfer nodes, all transit nodes between two neighboured transfer nodes are excluded from the branching process. So only transfer nodes are of significance for the connection tree. This simplification results from the fact that every transit node can be reached from either of two transfer nodes. If the origin is a transit node, exactly two neighbouring transfer nodes exist that allow access to the search network and these two are registered as start nodes for the search procedure. Applying this method, the calculations are made for the best route through one of the two nodes to the destination. If the destination is a transit node it can be reached through either of the transfer nodes assigned to it. An exception is the case, when the origin and the destination are transit nodes assigned to one and the same transfer node pair (e.g. both transit nodes are located on the same segment between the two transfer nodes (see Danowski (1997)).

The fact that the line network topology is considered at the same time as the timetable and the waiting time at the transfer nodes makes it possible to find a provable optimal solution for the model with the one-step method. A significant disadvantage of this method is the computation time required and the great number of accesses to the timetable data. This expenditure can be reduced through different heuristics (Danowski (1997)). Day-dependent alterations of the services (alterations of line network and timetable, etc.) are taken into consideration making use of time validity attributes, in the one-step as well as in the multi-step methods.

3.3 Design of Interactive Procedures

The acceptance of passenger information systems depends not only on the speed and the precision in finding solutions but also on how simple such a system is to use. Therefore, the areas of application (see Sect. 4) define different requirements for the user interface. This especially concerns the necessity to bring together the quite different views of adequate representation of the information and to provide the information in different languages. The acceptance of a system and therefore, its success depends substantially on the basic concept for the dialogues and the screens (see Danowski/Fünfstück (1997)).

Working on the principle that the users of public information systems have different experience in using computer systems, a number of groups can be defined. For each group the operation of the system must be simple, fast and if possible, without needing prior education. As a rule the user normally has little or no experience in dealing with such systems. This means that the information system must correspond in an appropriate way to the underlying mental model. This could be achieved by presenting the information directly in an easy-to-understand form, similar to common situations in the user's every day life. Thus, special procedures should not be imposed on the user to get the information, except if this is absolutely necessary for technical or conceptual reasons.

Important points for consideration are perception and structuring principles (see Hoffmann (1989) and Geiser (1990)) to achieve easy-to-learn systems. The user must be able to use the information terminal quickly and intuitively, and without a specific introduction. Self explanatory descriptions, clear metaphorical expressions, etc. are necessary. To avoid mistakes by the user and to facilitate the recognition of already known actions, a consistent design of the interface elements is needed, because some operations when repeated several times get automated (see Eberleh/Oberquelle (1994)).

The attractiveness of an information system from the user point of view depends mostly on the content of the information and on the type of presentation. At the same time the public information system is an advertising instrument for the institution that provides it. Therefore, it is important to prepare the information using multimedia elements to impress the user and encourage him to use the system. The implementation of sound, colored graphics, animations and videoclips can make the information system much more attractive, but the use of these elements must aim to improve the value of the system and should not be treated as an end in itself. Furthermore, overstated use of color and animation could tire or confuse the user (see Geiser (1990)). The features which are possible from a technical point of view, and the software-ergonomic characteristics have to be balanced, so that aesthetic impact and a clear vision are achieved.

4 Areas of Application

There are various areas of application for such information systems, while a differentiation has to be made between client advisory service systems and end-user systems. This distinction is necessary because it defines some specific differences in functionalty concerning the structuring of user interfaces and data handling and therefore also regarding the information spectrum. At the same time the overall decision-making process as it relates to mobility has to be considered because appropriate information provision can influence the choice of modes of transport potentially benefiting public transit (see Oppenheim (1995), Bergerhoff (1996) or Daduna/Voß (1996)). This concerns not only decisions before the start of a trip, but it covers the whole travel period and especially at transfer nodes.

Based on the distinction of the applications it is practical to consider the structuring of the user interface from two points of view:

❑ *Operation by Specialists*
In this case the client has no direct access to the passenger information system, it is used with the help of a "translator". The interface must be designed to make access to the information by a trained user fast and easy. These applications are common in service and communication centres and for telephone information services.

❑ *Operation by End-user*
The criteria for the structure in this case are quite different recognizing that the user may have no knowledge of using such systems. Here, it is necessary to make a further distinction based on the distribution channel used. When information terminals[1] are used, the requirements are much stronger than in the case of (individual) PCs for which the programs and data are on diskettes or CD-ROMs, or of direct access to the information system via network services, e.g. *WorldWideWeb* (www) on Internet. In the first application type the user is assumed to have no (specific) knowledge, whereas in the next two there is normally a certain level of familiarity assumed.

The quality of the information provided and the benefits for the users, as noted in 3.1, depend substantially on the currency of the data. For applications such as information terminals and access to network services, it is possible to assure continuous updating of timetable data through direct connection to a server. But when the data provision is made on data carriers, updating can not be easily achieved, except if they are significant alterations in the timetables which will be in effect for a long period. Only in these cases the preparation of updates makes sense.

---

[1] For the use of information terminals cf. e.g. Strobel (1993).

5 The Dresden Implementation

The multifunctional passenger information system described in the previous sections has been in operation for three years as the *Public Transit & City Information System Dresden* for the city Dresden (Germany) and its surroundings. In the foreground, as functional components of this system are timetable information with wide coverage of modes of transport and complete information about the city. The main functions as shown in the initial screen (Fig. 5.1) are under five menus:

- Connection search.
- Information about the city mass transit system and the regional transit systems.
- Information about long-distance transport.
- City map with search functions.
- (Actual) information about the city.

The core function of the system is the generation of client-oriented connections in the Dresden area public transit network (see various choices in Fig. 2.1), taking into account city rail, as well as tram and bus services. Special transit services are also included, such as the Airport-City-Liner, the Elbe River ferries, and a mountain railway. The different characteristics of the timetable and the line network, depending on the time of day and the type of day, achieve an exact representation of the services offered. Multidimensional knowledge-based description models are developed, to attain an appropriate repesentation of transit network structure and operations (see Danowski (1993)). The knowledge base contains the topology of the network, the timetables of the various modes of transport, running times for different segments of the network varying by time and type of day and the transfer times needed at stations or stops.

Efficient analysis of the knowledge base makes it possible to keep a large variety of continuously changing structures and operations of the transit system in a limited and controlled range (see Danowski (1994)). Origins and destinations in the route search procedure include not only the stops and stations of the Dresden public transit companies, but over 5,000 places, including names of streets (with numbers of houses) and addresses of hotels, shopping centers, schools and universities. Fig. 5.2 shows an example of timetable information.

Together with this ("classical") function, the *Public Transit & City Information System Dresden* gives actual information about more than 4,000 commercial objects within the city and in its surrounding area such as local governmental agencies, (specific) public facilities, tourist sites in the city and cultural facilities, sports and entertainment facilities, and shopping centers. In addition, the user receives, depending on the type of object a short description, historical data, addresses, telephone numbers, business hours, ticket prices and other data.

Fig. 5.1: Initial page of the Public Transit & City Information System Dresden

The provision of relevant client information is supported by a digital city map (Fig. 5.3) with two levels of enlargement that give the appropriate display during the search procedure. This map makes it possible to give a detailed view of the area around the origin and the destination. Together with the city and regional transit network in the information system are included the national and international long-distance connections to and from Dresden (air transport, long-distance rail transit, Elbe River transport). Current information on fares, operating buses and trams, on construction work, and detours for specific lines are available under a separate sub menu.

During the evaluation of the data handling concept for the *Public Transit & City Information System Dresden* consideration was given to both static information systems with fixed data, and dynamic systems with continuously updated data. At the moment the dynamic components are not in operation, because an area-wide AVM System is not yet operating. The linking of the different data sources, which are dependent on the applications, is of key importance and requires an exact definition of the data interface. Fig. 5.4 shows the present data sources for the schedule information, *Dresdener Verkehrsbetriebe AG*, *Deutsche Bahn AG*, *Sächsische Dampfschiffahrt-GmbH&Co. Conti Elbschifffahrt-KG* and the other information taken from the *Dresden-Werbung und Tourismus GmbH.*

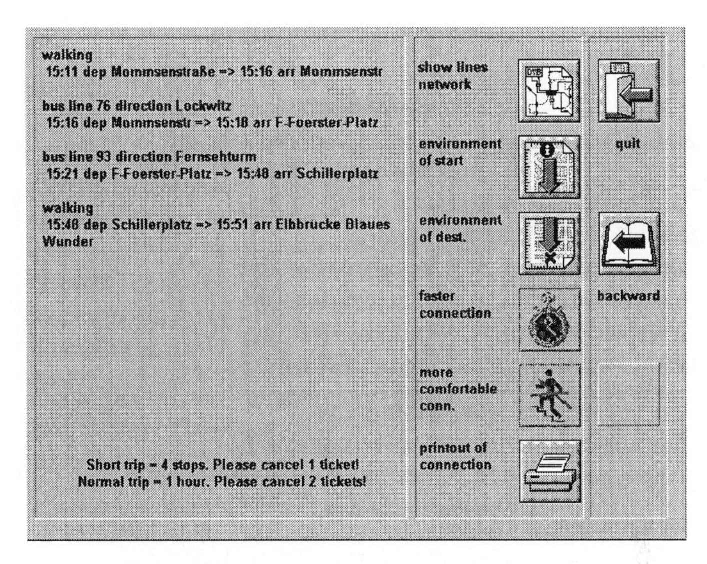

Fig. 5.2: An example of timetable information

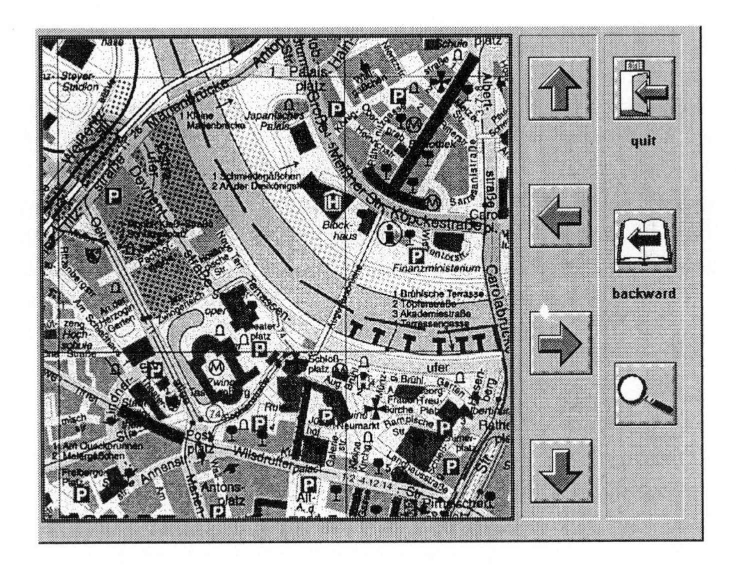

Fig. 5.3: The city map

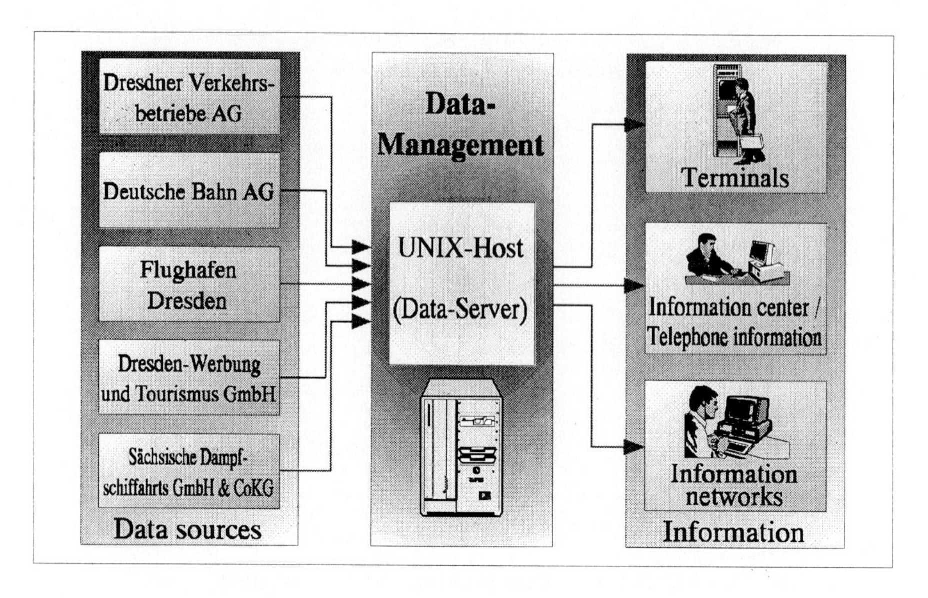

Fig. 5.4: System concept of the Public Transit & City Information System Dresden

To assure continuous data supply without interruptions in the operation of the system, a polling process is used, providing updated information from the computers in cycles via telephone network. The data are stored and managed on the server and when there is a request, they are transmitted to the information terminals. The implementation of different communications technologies (switched connections, leased lines, ISDN, X.25) is possible in principle and in every case the decisions are taken on the basis of the cost and the local framework requirements. This flexibility allows modular construction and extension of the system, up to data warehouse concepts with distributed data management and intelligence, based on fast communication media (see Danowski/Jung (1995)).

The *Public Transit & City Information System Dresden*, as explained above has been in operation since mid 1994 in the service centres of the *Dresdener Verkehrs-betriebe AG* and at public accessible (self-serve) information terminals. Much statistical data on the manner and frequency of use has been collected at these information terminals during these three years for the analysis of the acceptance of such systems and of the behaviour of the users. The dialog steps for input of data, whether the user spends too much time and quits, were analysed in depth and where necessary partly revised. In addition to this automatic information collection several interviews on the quality of the information provided were conducted and the resulting practical advise analysed.

The frequency of use statistics (Fig. 5.5) shows that daily between 260 and 550 information requests are processed (see Danowski (1997). During the observation period differences in the frequency of use between the introductory phase and the normal operation were found. Beside this, the significant increase of information requests at the beginning of new timetable period is of interest.

From the three years operation of the *Public Transit & City Information System Dresden* it can be concluded that this information system has very good acceptance by the clients and by the staff of the *Dresdener Verkehrsbetriebe AG*. The benefits from the application of this system are:

- Improvement of information quality by provision of current information.
- Improvement of the client services by showing the offered service level by *Dresdener Verkehrsbetriebe AG* and by other public transit companies.
- Incorporation of a large amount of information offered in a public transit environment.
- Relieving the staff of the service centres of the Dresdener Verkehrsbetriebe AG.

With the help of such information tailored to the individual client requirements on the transit offerings, public transit is taking the first step towards removing the psychological barriers to accessing services. The passenger information system contributes to a better public transit image and it is a marketing instrument to draw more clients to these services. It therefore becomes especially important in the environment of growing competition with the automobile (see Danowski (1997)).

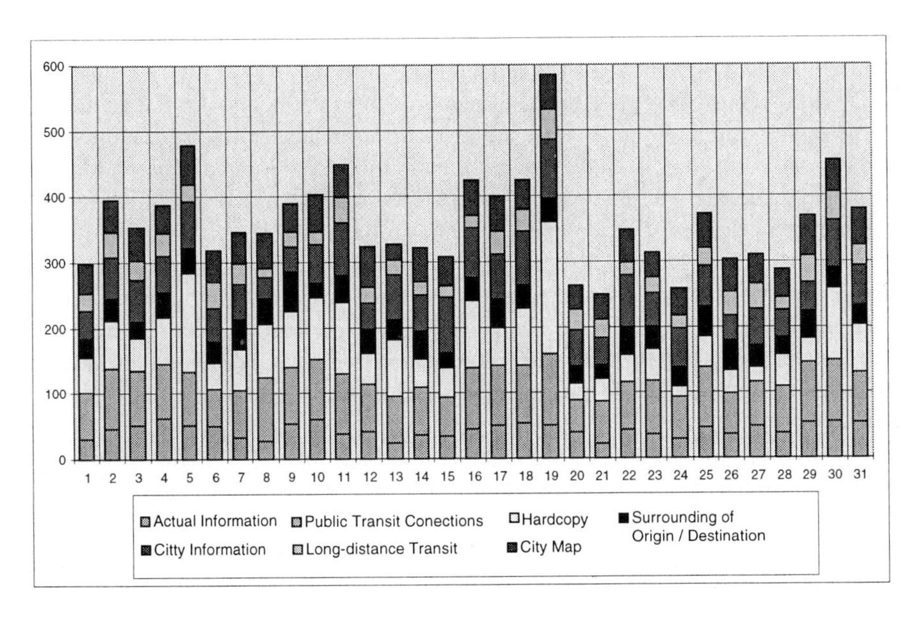

Fig. 5.5: Statistics on the information requests to a self-service machine (Location: Pirnaischer Platz / Dresden, Analysis interval: October 1996)

6 Conclusions

The experience gathered through the operation of the *Public Transit & City Information System Dresden* has shown that efficient information systems can make a significant contribution to the improvement of customer information in public transit. These systems will become much more important with structural changes, which may result from the emergence of supraregional service areas and from the framework requirements set by laws of the European Community concerning the deregulation of public transit (see Daduna (1995)). Another reason for its increasing importance is the separation of responsibiltity for planning and operational control on one hand and operational service provision on the other hand (see Daduna/Voß (1994)). A decisive point is the selective access to the data bases, which will become larger and larger. Making use of the classical print media will be less feasible for clients and eventually impossible.

With the aim of giving greater access to such systems and especially to allow clients to use it individually, since the beginning of 1996 a (modified) www version of the *Public Transit & City Information System Dresden* is available on the Internet (see Daduna (1997) and Schneidereit/Daduna/Voß (1998))[2]. This information system has, in addition to the above functions, an information module for the traffic volume on the street network of the city and in the region. This is very helpful to make a decision before starting a trip, to determine which mode of transport is best – public transit or automobile.

Despite these advantages it must be clear, that from the point of view of the use of these systems in form of either information terminals or Internet applications the current availability of these new information media is limited. The reasons lie in the access barriers for many clients (or potential clients) as a result of absence of experience in dealing with computers. This problem applies not only to public transit, it concerns nearly all areas of every day life, where computer-aided data management systems are becoming more and more common. On the other hand a significant reason is the limited availability of end-user infrastructure, concerning the installation of public accessible information terminals as well as the use of information services in the private life, which restricts the number of potential clients[3].

For the medium- and long-term, computer-aided information systems will clearly play a significant role in information provision in public transit. Two aspects are important: increased accessibility, for example the use of network services, and increased familiarity and knowledge by the public on how to use computer-aided systems. Considering technical developments it can be expected that within the

---

[2] The Internet address of the *Public Transit & City Information System Dresden* is http://dresden-info.fhg.de.

[3] c.f. the comments of Ebcinoglu / Ebert / Mattern (1997) concerning the use of computer-aided tools for passenger information.

next few years there shall exist supraregional information structures on the basis of interconnection among different information systems (see, e.g. Schneidereit/Daduna/Voß (1998)). Based on the functionality of such systems, their operation by providers of public transit services would be impossible in the long-term, because the changes in the legislative framework will lead to conflicts (see Daduna (1995)). Obviously in the future it is likely that commercial companies will be established, to serve as passenger information service providers.

References

Bergerhoff, J. (1996): Technologie und Fahrgastinformation. in: Public Transport International 6/96, 59 - 63.

Daduna, J.R. (1995): Organisationsstrukturen des öffentlichen Personennahverkehrs und ihre Einbindung in den komunalen Bereich. in: Zeitschrift für Verkehrswissenschaft 66, 187 - 206.

Daduna, J.R. (1997): Gestaltungsmöglichkeiten in der Fahrgastinformation im öffentlichen Personennahverkehr. (Working paper).

Daduna, J.R./Voß, S. (1994): Effiziente Leistungserstellung in Verkehrsbetrieben als Wettbewerbsinstrument. in: ZP Zeitschrift für Planung 5, 227252.

Daduna, J.R./Voß, S. (1996): Efficient technologies for passenger information systems in public mass transit. in: Pirkul, H./Shaw, M.J. (eds.): Proceedings of the First INFORMS Conference on Information Systems and Technology, Washington D.C., May 5 - 8, 1996, 386 - 391.

Danowski, K. (1993): Ein Beitrag zur wissensbasierten Modellierung von Entscheidungsprozessen in Verkehrsleit - und Informationssystemen. in: atp-Automatisierungstechnische Praxis, Heft 12/35, 677 - 682 .

Danowski, K. (1994): Ein wissensbasiertes, multimediales Fahrplan- und City Auskunftssystem. in: Mitteilungen aus dem Fraunhofer-Institut für Informations-und Datenverarbeitung '94, Fraunhofer-Institut für Informations-und Datenverarbeitung, Karlsruhe.

Danowski, K. (1997): Computergestuetzte Fahrtendisposition in Fahrplan-und bedarfsgesteuerten Verkehrssystemen: Ein Beitrag zur Integration wissensbasierter und algorithmischer Konzepte in Entscheidungsunterstützungssystemen. Dissertation Fakultät Verkehrswissenschaften "Friedrich List" der Technischen Universität Dresden.

Danowski, K/Fünfstück, F. (1997): Ein Beitrag zur Gestaltung von Benutzungsschnittstellen in Terminal-Informationssystemen am Beispiel des ÖPNV & City Informationssystems Dresden. in: Liskowski, R./Velichkovsky, B.M./Wünschmann, W. (Hrsg.): Software-Ergonomie '97, Usability Engeneering: Integration von Mensch-Computer-Interaktion und Software-Entwicklung. (B.G. Teubner) Stuttgart, 123 - 134.

Danowski, K/Jung, U. (1995): Zur Anwendung wissensbasierter Model-lierungsmethoden in Verkehrsinformationssystemen. in: Wissenschaftliche Beiträge zur Informatik. Technische Universität Dresden, Fakultät Informatik, Heft 1/8, 105 - 116.

Ebcinoglu, U./Ebert, H.M./Mattern, U. (1997): Integrierte Fahrgastinformation für die Region Hannover. in: Der Nahverkehr 4/97, 8 - 12.

Eberleh, E/Oberquelle, H. (1994): Einführung in die Software-Ergonomie: Gestaltung graphisch-interaktiver Systeme: Prinzipien, Werkzeuge, Lösungen. (de Gruyter) Berlin, New York.

Geiser, G. (1990): Mensch-Maschine-Kommunikation. (Oldenbourg) München, 1990.

Hoffmann, T. (1989): Handbuch zur softwareergonomischen Gestaltung von Bildschirmmasken. (VDI-Verlag) Düsseldorf.

Mentz, H.-J. (1986): Wegwahl in Raum und Zeit. in: v+t Verkehr und Technik 39, 21 - 28

Oppenheim, N. (1995): Urban travel demand modeling. (Wiley) New York/ Chichester/Toronto/Singapore/Brisbane.

Schneidereit, G./Daduna, J.R./Voß, S. (1998): Informations-distribution über Netzdienste am Beispiel des Öffentlichen Personenverkehrs (to appear).

Strobel, H. (1993): Intermodales Verkehrsleitsystem Ballungsraum "Dresden / Oberes Elbtal", Schlußbericht zu einer Studie erarbeitet mit Förderung des Bundesministeriums für Forschung und Technologie (Förderkennzeichen TV 9220), Interdisziplinäre Forschungsprojektgruppe INTER-VLUS-DRESDEN. c/o TCAC GmbH Dresden, June 1993.

An Overview of Models and Techniques for Integrating Vehicle and Crew Scheduling

Richard Freling[1], Albert P.M. Wagelmans[1], and José M. Pinto Paixão[2]
[1]Econometric Institute, Erasmus University Rotterdam, The Netherlands email: freling@few.eur.nl
[2]DEIO, Faculdade de Ciências da Uiversidade de Lisboa, Portugal

Abstract: In this paper, the problem of integrating vehicle and crew scheduling is considered. Traditionally, vehicle and crew scheduling have been dealt with in a sequential manner, where vehicle schedules are determined before the crew schedules. The few papers that have appeared in the literature have in common that no comparison is made between simultaneous and sequential scheduling, so there is no indication of the benefit of a simultaneous approach. In order to get such an indication before even solving the integrated problem, we propose a method to solve crew scheduling independently of vehicle scheduling. We introduce a mathematical formulation for the integrated problem, and briefly outline algorithms. The paper concludes with computational results for an application to bus scheduling at the public transport company RET in Rotterdam, The Netherlands. The results show that the proposed techniques are applicable in practice. Furthermore, we conclude that the effectiveness of integration as compared to a sequential approach is mainly dependent on the flexibility in changing buses during a duty.

1 Introduction

In this paper we consider an overview of models and techniques for the integration of bus and driver scheduling. Technical details are not considered here. For a more comprehensive description of the subject, including the technical details, we refer the readers to Freling (1997).

In Fig. 1.1 we show the relation between four operational planning problems which typically arise in public transport organisations. Decisions about which routes or *lines* to operate, and with what frequency, are based on the available infrastructure, service requirements for the passengers, and demand aspects. We assume that these are known for the operational planning phase. Also known are the travel times between various points on the route. These may differ for various parts of the planning period (e.g., the travel times may be longer during busy hours). Based on the lines and frequencies, timetables are determined resulting in *trips* with corresponding starting and ending times and locations. An example of a trip is departure from location 1 at 9:00 am and arrival at location 2 at 10:00 am.

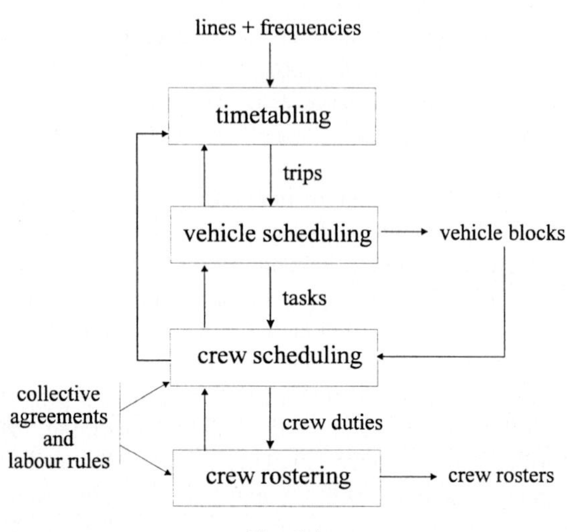

Fig. 1.1

The next two planning processes are vehicle and crew scheduling, which consist of assigning vehicles to the trips and crews to vehicles, respectively. Crew scheduling is short-term crew planning (e.g. one or several days), while the crew rostering process is long term crew planning (e.g. one month or half a year).

In most practical situations these processes interact with each other as shown in the figure. Presumably, the planning could be more efficient if an integrated approach were taken; however, because it is mathematically not feasible to consider the whole process at once, most theoretical studies deal with these processes in a sequential manner. Some approaches have been proposed in the operations research literature which deal with an integration of two of the planning problems. Several papers consider the vehicle scheduling problem with time windows, which is an example of the integration of timetabling and vehicle scheduling; that is, starting times of trips are not fixed but can vary within time windows (see Ferland/Fortin (1989)).

The integration of vehicle and crew scheduling is the topic of this paper, which is organized as follows. In Sect. 2, we discuss several approaches for vehicle and crew scheduling. Recent theoretical developments in the field of optimization, as well as rapid developments in computer technology in terms of power and speed at much lower cost, inspired us to tackle vehicle and crew scheduling in an integrated manner. One of our primary research objectives was to obtain insight into the effectiveness of integrating vehicle and crew scheduling compared with sequential approaches. In Sect. 3 we discuss the potential benefits of integrating vehicle and crew scheduling. The integration of vehicle and crew scheduling in general has received little attention in the literature. We provide an overview of the scarce literature in Sect. 4. In Sections 5 and 6, we outline models and algorithms. Another research objective was to test the practical applicability of the proposed techniques. In Sect. 7 we consider application of bus and driver scheduling at RET, the urban public transport company in Rotterdam, the Netherlands.

2 Problem Definition

Although in the early eighties several researchers recognized the advantage of integrating vehicle and crew scheduling, most of the algorithms published in the literature follow the sequential approach where vehicles are scheduled before and independent of crews. Algorithms incorporated in successful computer packages use this sequential approach as well, while sometimes integration is dealt with at the user level (see Darby-Dowman/Jachnik,/Lewis/Mitra (1988)). In the operations research literature, only a few publications address simultaneous approaches to vehicle and crew scheduling. None of those publications makes a comparison between simultaneous and sequential scheduling. Hence, they do not provide any indication of the benefit of a simultaneous approach.

In this paper, we consider vehicle and crew scheduling from a different angle. Besides a complete integration of vehicle and crew scheduling, we also consider the reverse sequential approach of scheduling crews before and independent of vehicles.

2.1 Traditional Sequential Approach

In the traditional sequential approach vehicles are scheduled before and independent of crews. The *single depot vehicle scheduling problem* (SDVSP) is defined as follows: given a depot and a set of trips with fixed starting and ending times, and given travel times between all pairs of locations, find a feasible minimum cost schedule such that (1) each trip is assigned to a vehicle, and (2) each vehicle performs a feasible sequence of trips. All the vehicles are assumed to be identical. A schedule for a vehicle is composed of *vehicle blocks*, where each block consists of a departure from the depot, the service of a feasible sequence of trips and the return to the depot. The cost function is usually a combination of vehicle capital (fixed) and operational (variable) cost. The SDVSP is known to be solvable in polynomial time.

For the sequential approach it is assumed that the vehicle scheduling problem has been solved when considering the scheduling of crews; that is, the set of vehicle blocks defining the vehicle schedule is known. The *Crew Scheduling Problem* (CSP) is defined as follows: given a set of *tasks*, find a minimum cost set of *duties*, such that (1) each task is assigned to one duty, and (2) each duty is a sequence of tasks that can be performed by a single crew. The vehicle schedule defines vehicle blocks which should be covered by duties at minimum cost. The blocks are subdivided at *relief points*, defined by location and time, at which a change of driver may occur. A task is defined by two consecutive relief points and represents the minimum portion of work that can be assigned to a crew. Each duty must satisfy several complicating constraints corresponding to work rules for crews. Typical examples of such constraints are maximum working time without a break, minimum break duration, maximum total working time, and maximum duration. The cost function is usually a combination of fixed costs such as wages, and variable costs such as overtime payment.

Crew scheduling has similarities to vehicle scheduling but is more complex due to several complicating constraints such as the aforementioned work rules for crews. Instead of assigning trips to vehicles, crew scheduling involves the assignment of tasks to crews or, better, to duties. A basic assumption is that every crew is the same, i.e. individual crew members are not considered. Since the type of constraints differ from application to application it is difficult to define a generic crew scheduling problem. Beasley/Cao (1996) propose the CSP with only spread time constraints as the generic CSP. In fact, this is the SDVSP with time constraints (see Freling/Paixão (1995)), which Fischetti/Martello/Toth (1987) have shown is NP-hard. The last group of authors have shown in Fischetti/Martello/Toth (1989) that the CSP with only working time constraints is NP-hard. Here we assume that the CSP has at least spread time or working time constraints, and is therefore NP-hard. Furthermore, we assume that at least one break must occur in a duty, which adds considerably to the complexity of the problem. We believe that, as long as one aims at developing techniques which are generally useful, a generic crew scheduling problem should at least have working or spread time and break constraints.

Since a duty is defined as a sequence of tasks and breaks, we can consider constraints which define the feasibility of a duty as resource constraints. Each task and break consumes a certain amount of a resource, and the total amount must be within the allowed interval. Below we present a summary of such *resource* or *local constraints* which define the feasibility of a single duty:

- *Time constraints*, placing limitations on the time in a duty. Examples are limited duration (i.e. *spread time*), limited working time, limited paid time, limited working/spread time without rest period or break.
- A minimum number of *breaks*. For example a minimum number of rest periods, coffee/tea breaks, meal breaks, overnight rests, etc.
- *Location constraints*, placing limitations on the locations visited during a duty. For example, a duty must originate and terminate at the same location.
- *Vehicle constraints*, defining links between crews and vehicles. For example, a limited number of changes of vehicle during a duty (*changeovers*), or *vehicle attendance* when a vehicle is stationed at a location other than the depot.
- *Crew dead-heading constraints*, placing limitations on crew transportation when the crew is not working on a vehicle.

Besides local constraints, *global constraints* also may exist which deal with groups of duties at once:

- *Time constraints*, e.g. limited average working time.
- *Location constraints*, e.g. limited number of crews available at a crew base.
- *Duty type constraints*, e.g. limited number of duties with overtime.

A particular application of the CSP is the Bus Driver Scheduling Problem (BDSP). For the BDSP the planning horizon is usually one day, that is, the BDSP consists of determining a set of duties or workdays which will then be assigned to individual bus drivers. An important notion here is the definition of a *piece of work*, which is a set of consecutive tasks in a vehicle block to be performed by a single driver. Sometimes a piece of work is defined more generally as a set of tasks without a break. A duty consists of one or more pieces of work. If a duty contains more than one piece of work, then these pieces are separated by breaks or free time periods. The duration of pieces of work and breaks in a duty is usually limited. Typical global constraints restrict the percentage of certain types of duties in the solution, such as split duties (i.e. with a large unworked period between pieces) and *trippers* (i.e. a single piece or short duty).

2.2 Integrated Approach

The *vehicle and crew scheduling problem* (VCSP) is the following: given a set of service requirements or trips within a fixed planning horizon, find a minimum cost schedule for the vehicles and the crews, such that both the vehicle and the crew schedule are feasible and mutually compatible. We make the following assumptions:

1. The vehicle scheduling characteristics correspond to the SDVSP as defined previously, that is, one depot, identical vehicles, fixed starting times of trips, and no time constraints.

2. The cost function for the VCSP is simply the summation of the vehicle and crew scheduling cost functions defined previously. The primary vehicle scheduling objective is to minimize the number of vehicles, while the primary crew scheduling objective is to minimize the number of crews.
3. A piece of work is defined as a sequence of tasks in a vehicle block which can be performed by one crew without interruption. This sequence of tasks is only restricted by its duration which must be within certain time limits.

The last assumption is not restrictive. For example, it can incorporate a restricted number of pieces, restricted spread time, and restricted working time. The three assumptions make our approach in principle applicable to bus and driver scheduling, although the crew scheduling characteristics are general.

We distinguish between two types of tasks, i.e., *trip tasks* corresponding to (parts of) trips, and *dh-trip tasks* corresponding to deadheading trips (*dh-trips*). All trip tasks need to be covered by a crew, while the covering of dh-trip tasks depends on the vehicle schedules and determines the compatibility between vehicle and crew schedules. In particular, each dh-trip task needs to be assigned to a crew if and only if its corresponding dh-trip is assigned to a vehicle. Note that one or more trip tasks may correspond to one trip, depending on the relief points along that trip. Similarly, one or more dh-trip tasks may correspond to one dh-trip. For example, a dh-trip between locations e_i and b_j which passes the depot corresponds to two dh-trip tasks, one from e_i to the depot and the other from the depot to location b_j. Here we assume that waiting at the depot is not a task because vehicle attendance at the depot is not necessary.

2.3 Scheduling Crews Independent of Vehicles

As an alternative to the crew scheduling problem, we consider crew scheduling independent of vehicle scheduling. The *independent crew scheduling problem* (ICSP) is the following: given a set of trip tasks corresponding to a set of trips, and given the travelling times between each pair of locations, find a minimum cost crew schedule such that all trip tasks are covered in exactly one duty and all duties satisfy crew feasibility constraints. When the ICSP is used as a method for determining crew schedules, vehicles need to be scheduled afterwards such that the vehicle and crew schedules are compatible. Therefore, the duties need to satisfy extra requirements in order to assure that a crew is available for each task induced by the vehicle schedule. In Sect. 3 we discuss the use of the ICSP for determining the potential benefits of integration.

3 Potential Benefits of the Integration of Vehicle and Crew Scheduling

Vehicle and crew scheduling problems often interact with each other: the specification of vehicle schedules will place certain constraints on the crew schedules and vice versa. Because vehicles are often much more flexible to schedule than crews, it may be inefficient to schedule vehicles without considering crew scheduling. Vehicle oriented characteristics of crew scheduling may affect the extent to which the integration of vehicle and crew scheduling is beneficial compared with the traditional sequential approach. Examples of such characteristics are the following:

- A restricted number of *changeovers*, that is, a crew is restricted in changing vehicles during a duty.
- Restricted crew dead-heading.
- Extra start-up time on a vehicle when a new crew is assigned to it.
- Compulsory continuous attendance when a vehicle is waiting.
- Minimum duration of a piece of work.
- Domination of crew costs over vehicle costs.
- Minimum break with the vehicle if a piece of work is longer than a certain duration.
- Crew relieves only occur at the depot, that is, changeovers are not allowed outside the depot.

If none of these characteristics exist, there is probably no need to integrate vehicle and crew scheduling because crews can move independently of vehicles. However, combinations of the mentioned characteristics appear frequently in practice. Integration may serve two purposes: feasibility and/or cost efficiency. We illustrate this with two examples.

Tosini/Vercellis (1988) consider the extra-urban bus driver scheduling problem, which is an example where integration is necessary for obtaining feasible solutions. In this situation, driver dead-heading is not allowed due to long distances, while driver relieves can occur only at crew bases. If buses are scheduled without attention to the driver scheduling, it may be that no feasible driver schedule exists, because a bus may be away from a crew base too long to be serviced by a driver. Therefore, the bus should pass by a crew base once in a while to change drivers. Figure 3.1 illustrates this case with six trips on one vehicle block, marked by their locations *L1*, *L2* and *L3*.

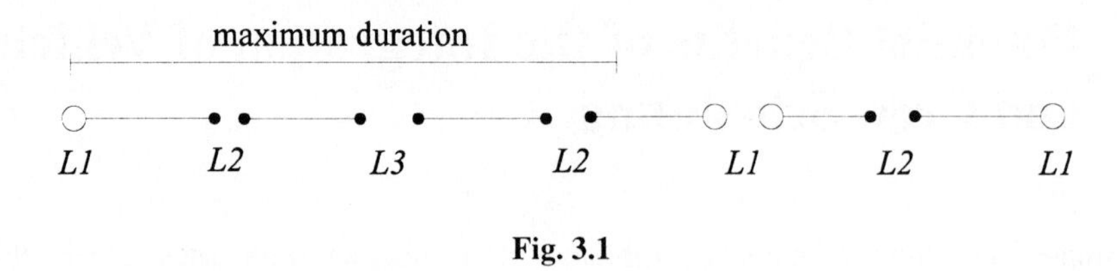

Fig. 3.1

Suppose that the only crew base is at location *L1*. Then, the problem is unfeasible if the duration of the first four trips away from the base exceeds the maximum duration allowed. Note that a possible approach for the VCSP is to consider crew rules for in the vehicle scheduling step. The example above can be tackled by determining vehicle schedules subject to a time constraint (see Freling (1995)).

An example of integration being more cost efficient is illustrated by two vehicle scheduling networks depicted in Figures 3.2 and 3.3, respectively. A directed vehicle scheduling network has nodes corresponding to a source s and a sink t and one node for each trip, where it is assumed that trips are numbered according to increasing starting time. An arc from the source to a trip corresponds to a vehicle leaving the depot to perform the trip, an arc from a trip to the depot corresponds to a vehicle entering the depot after performing the trip, while an arc between two trips corresponds to a vehicle performing both trips in sequence (see also Sect. 5). Only the arcs are drawn which correspond to two different vehicle schedules with three vehicles each. The idea is that the two different vehicle schedules may be served with crew schedules with different degrees of efficiency. Suppose that changeovers are not allowed and the maximum duty duration is such that at most two trips can be included in a duty. Then, at least five duties are necessary to cover the three vehicles in the vehicle schedule of Fig. 3.2. In Fig. 3.3 it is shown that it might be possible to save one duty by adjusting the vehicle schedule with the same number of vehicles. At least four duties are necessary to cover this schedule.

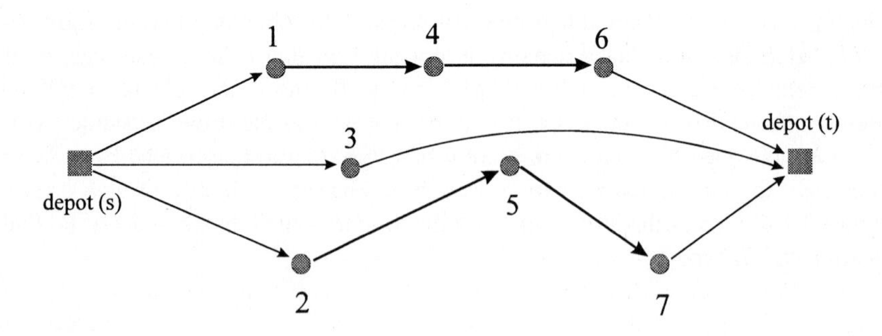

Fig. 3.2

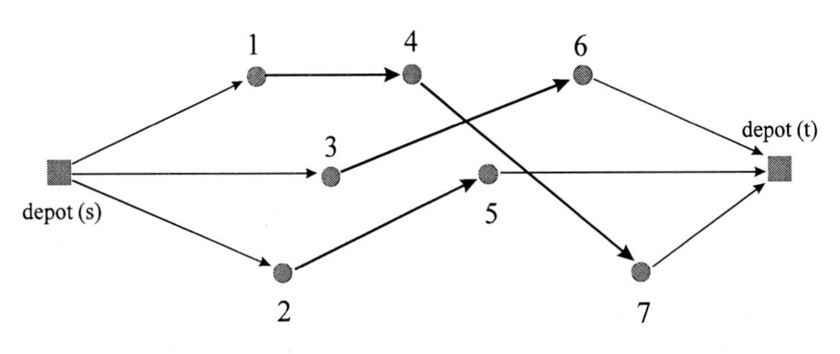

Fig. 3.3

It may be useful to have a measure of the potential benefit of integration with respect to cost efficiency, without needing to solve the VCSP. Consider the solution values of the following three approaches:

1. Traditional sequential approach (solution value v1): first solve the SDVSP and then the CSP.
2. Independent approach (solution value v2): independently solve the SDVSP and the ICSP.
3. Integrated approach (solution value v3): solve the VCSP.

The solutions of the independent approach are of no practical use since the resulting vehicle and crew schedules are usually not compatible. However, it is easier to obtain solutions for the first two approaches than for the third approach, and we know that $v2 \leq v3 \leq v1$. Thus, if v2 is significantly less than v1, it may be that the crew scheduling solution will improve significantly when considering integration compared with the sequential approach. On the other hand, we know that there is no need to integrate if v2 and v1 do not differ much.

4 Literature Review

Overviews of algorithms and applications for the SDVSP and some of its extensions can be found in Daduna/Paixão (1995) and in Desrosiers/Dumas/Solomon/Soumis (1995). Recent surveys on solution methods for the BDSP can be found in Odoni/Wilson/Rousseau (1994) and in Wren/Rousseau (1995). In this section, we discuss the literature on simultaneous scheduling of vehicles and crews. To our knowledge, this literature deals mainly with bus and driver scheduling, and all approaches proposed in the literature belong to one of the following two categories:

1. Schedule vehicles using a heuristic approach to crew scheduling.
2. Include crew considerations in the vehicle scheduling process, and schedule crews afterwards.

The traditional sequential strategy is strongly criticized by Bodin/Golden/Assad/ Ball (1983). This is motivated by the fact that in North American mass transit organisations the crew costs dominate vehicle operating costs, and in some cases reach as much as 80% of total operating costs. As noted before, although simultaneous vehicle and crew scheduling is of significant practical interest, only a few approaches of this kind are proposed in the literature. Most of the procedures are of the first category and are based on a heuristic procedure proposed by Ball/Bodin/Dial (1983). This procedure involves the definition of a scheduling network, which consists of vertices characterised by parts of trips called *d-trips* that have to be executed by one vehicle and crew, and two vertices *s* and *t* representing the depot. Several types of arcs can be grouped into two categories, those which indicate that a crew and vehicle proceed from one d-trip to another and those which indicate that only the crew proceeds from one d-trip to another (*crew-only arcs*). The solution procedure is decomposed into three components, emphasising the crew scheduling problem: a piece construction component, a piece improvement component and a duty generation component. All three components are solved using matching algorithms. The piece construction routine generates a set of pieces whose time duration is less than some constant T. Note that this corresponds to vehicle scheduling with time constraints. In the second step pairs of short pieces are combined into partial duties, while in a third step pairs of these duties and longer pieces are combined into two- and three-piece duties. Vehicle schedules are generated simultaneously by deleting the crew-only arcs and fixing arcs used by pieces in the solution. This procedure is applied to large VCSP instances which correspond to the entire physical network, i.e., all lines are considered at once, while no restrictions are placed on interlining, i.e., a crew may work on an arbitrary number of lines.

Similar heuristic approaches of the first category are proposed by Tosini/ Vercellis (1988), Falkner/Ryan (1992), and Patrikalakis/Xerocostas (1992). All these approaches use a similar crew scheduling network as in Ball/Bodin/Dial (1983). For the sake of illustration we briefly discuss a three phase procedure proposed in Patrikalakis/Xerocostas (1992). In the first phase, a set covering problem is solved to determine a set of crew duties which cover all timetabled trips. Because the vehicle movements are not known, the actual starting and ending times of crew duties and other parameters, such as idle time, are calculated approximately at this stage. In the second phase, a set of compatible vehicle schedules are built around the resulting duties by solving a minimum cost network flow problem. The compatibility of the crew and vehicle scheduling solutions is ensured by providing vehicles to all required crew movements. In the third phase, the crew duties are reconsidered using a restricted crew scheduling network to generate complete duties. Their conclusion is that the proposed approach can be more efficient than the traditional sequential approach when vehicle oriented crew constraints, such as a maximum number of changeovers and continuous attendance of crews, are not very important. However, we do not agree with this conclusion and believe that their conclusion is only based on the fact that such constraints affect the efficiency of the three-phase approach compared with the sequential approach which becomes more complex when vehicle oriented constraints are relaxed. In our opinion, and as discussed in the previous subsection, the potential benefit of a simultaneous approach increases when these kind of constraints are tightened. This statement is supported by computational results presented in Sect. 7.

Approaches of the second category are proposed by Darby-Dowman/Jachnik/ Lewis/Mitra (1988) as an interactive part of a decision support system, and by Scott (1985) who heuristically determines vehicle schedules which consider crew costs. An initial vehicle schedule is heuristically modified according to estimated marginal costs associated with a small change in the current vehicle schedule. The estimated marginal costs are obtained by solving the linear programming dual of the HASTUS crew scheduling model (see Rousseau/Blais/ HASTUS (1985)). Results obtained with public transport scheduling problems in Montréal, show a slight decrease in estimated crew costs which is mainly due to relatively high estimated marginal costs in the periods before the morning and evening peak hours.

5 Modelling

In Freling (1997) we propose several mathematical formulations for the VCSP and ICSP. We will briefly discuss a slightly simplified version the most important of these formulations. This formulation includes a formulation for the SDVSP. For the SDVSP, trips serviced by the same vehicle are linked by *deadheading trips* (*dh-trips*), that is, movements of vehicles without serving passengers. Dh-trips consist of travel time (or vehicle deadheading) and/or *idle time*. Idle time is defined as the time a vehicle is idle at a location other than the depot. Let b_i and e_i be the start and end locations, and let bt_i and et_i be the starting and ending times of a trip i, respectively. Two trips i and j are said to be *compatible* if the same vehicle can cover these trips in sequence, that is, if $et_i + trav(e_i,b_j) \leq bt_j$, where $trav(e_j,b_i)$ is the deadheading travel time from location e_j to location b_i. A sequence of trips is feasible if each consecutive pair of trips in the sequence is compatible.

Let $N=\{1,2,...,n\}$ be the set of trips, numbered according to increasing starting time, and let $E=\{(i,j) \mid i<j,\ i,j \text{ compatible},\ i\in N, j\in N\}$ be the set of arcs corresponding to dh-trips. The nodes s and t both represent the depot at location d. We define the vehicle scheduling network $G=(V,A)$, which is an acyclic directed network with nodes $V=N\cup\{s,t\}$, and arcs $A=E\cup(s\times N)\cup(N\times t)$. A path from s to t

in the network represents a feasible schedule for one vehicle, and a complete feasible vehicle schedule is a set of disjoint paths from s to t such that each node in N is covered. Let c_{ij} be the vehicle cost of arc $(i,j) \in A$, which is usually some function of travel and idle time. Furthermore, a fixed cost K for using a vehicle can be added to the cost of arcs (s,i) or (j,t) for all $i,j \in N$. For the remainder of this paper, we assume that the primary objective is to minimize the number of vehicles. This means that K is high enough to guarantee that this minimum number will be achieved. In Sect. 3, we have shown two examples of network G.

An important consideration when formulating the VCSP or ICSP is the way crew tasks are defined without knowing the vehicle schedules in advance. Recall that we consider two types of tasks, namely a set of trip tasks denoted by I_1 of which we can be sure that they have to be serviced by a crew, and a set of dh-trip tasks denoted by I_2 that need to be covered by a crew if and only if a vehicle traverses this dh-trip.

To our knowledge, only Patrikalakis/Xerocostas (1992) propose a formulation for the VCSP, but this model is only used for illustration and is computationally intractable. The mathematical formulation we propose for the VCSP contains a vehicle scheduling formulation based on network G=(V,A), which assures the feasibility of vehicle schedules. The remaining constraints in the formulation assure that each trip task is assigned to a duty and each dh-trip task is assigned to a duty if and only if its corresponding dh-trip is part of the vehicle schedule. Before providing the mathematical formulation, we need to introduce some notation. K denotes the set of all feasible duties, and K(p) is the set of duties covering trip task $p \in I_1$ or dh-trip tasks $p \in I_2$, and $I_2(i,j)$ denotes the set of dh-trip tasks corresponding to dh-trip $(i,j) \in A$. Decision variables y_{ij} and x_k are defined as follows: y_{ij} indicates whether a vehicle covers trip j directly after trip i or not, while x_k indicates whether duty k is selected in the solution or not. The VCSP can be formulated as follows (model VCSP1):

$$\min \sum_{(i,j)\in A} c_{ij} x_{ij} + \sum_{k \in K} d_k x_k$$

$$\sum_{\{j:(i,j)\in A\}} y_{ij} = 1 \qquad \forall i \in N, \tag{1}$$

$$\sum_{\{i:(i,j)\in A\}} y_{ij} = 1 \qquad \forall j \in N, \tag{2}$$

$$\sum_{k \in K(q)} x_k = 1 \qquad \forall p \in I_1, \tag{3}$$

$$\sum_{k \in K(q)} x_k - y_{ij} = 0 \qquad \forall (i,j) \in A, \forall q \in I_2(i,j), \tag{4}$$

$$x_k, y_{ij} \in \{0,1\} \qquad \forall k \in K, \forall (i,j) \in A.$$

The objective coefficients c_{ij} and d_k denote the vehicle cost of arc $(i,j)\in A$, and the crew cost of duty $k\in K$, respectively. The objective of this 0-1 linear programming problem is to minimize the sum of total vehicle and crew costs. The first two sets of constraints (1) and (2) correspond to a *quasi-assignment formulation* for the SDVSP, that is, each trip is assigned a predecessor and a successor in order to ensure that the nodes in network G are covered by a set of *s-t* paths which together include every trip node once. Constraints (3) ensures that each trip task *p* will be covered by one duty in the set K(*p*). Furthermore, constraints (4) guarantees the link between dh-trip tasks and dh-trips in the solution; that is, the constraints guarantee that each dh-trip task *q* is covered by a duty in the set K(*q*) only if the corresponding arc is in the vehicle solution. The model contains $\Theta(|A|+|K|)$ variables and $\Theta(|N|+|A|)$ constraints, which may already be quite large for instances with a small number of trips. This is probably the main reason why complete integration has not been considered previously. In Freling (1997) we show how to reduce the number of constraints considerably, that is, instead of $|A|$ being of order $|N|^2$ it is in practice of order $|N|$.

It is not obvious how to incorporate a restricted number of changeovers in the model. For the CSP, a restricted number of changeovers is often dealt with by restricting the number of pieces of work in a duty, that is, it is implicitly in the model due to the definition of the variables. This is valid since a piece of work corresponds to one vehicle, but also more restrictive because two pieces of work may correspond to one vehicle. However, this is not valid if the minimum number of breaks is larger than the maximum number of changeovers. For the VCSP, we can also restrict the number of pieces of work in a duty in order to deal with a restricted number of changeovers. A piece of work is here defined as a sequence of tasks in $I_1\cup I_2$ so that a piece of work corresponds to one vehicle. In Freling (1997) we propose a mathematical formulation for the particular situation where no changeovers are allowed, while at least one crew break is required, that is, restricting the number of changeovers by restricting the number of pieces of work is not valid. We also propose a formulation for the ICSP. Both are set partitioning formulations.

6 Algorithms

In Freling et al. (1997) we propose three new algorithms for the SDVSP: an auction algorithm, a two-phase approach in case of a special cost structure, and a core oriented approach dealing with a large number of arcs in network G. For one algorithm for the VCSP we solve the SDVSP up to hundreds of times using the auction algorithm. Therefore, the computational speed of such algorithm is very important. For real life data from the RET in Rotterdam, we have solved problems with up to 1328 trips within 17 seconds on a Pentium 90 PC. For the computational tests in the next section we consider problems with up to 148 trips. The auction algorithm solves each of these problems within 0.11 seconds.

The algorithms we developed and applied to the CSP, VCSP and the ICSP are all Lagrangian heuristics with a quality guarantee. That is, we do not attempt to solve the problems to optimality, but instead use optimisation based techniques in a heuristic algorithm which produces a guarantee of the quality of the solution in terms of the difference with a lower bound. Since the late eighties there has been increasing interest among operations researchers in exact optimisation techniques for crew scheduling applications (see Desrochers/Soumis (1989) and Paixão (1990)). From an application point of view, the main reason is a rapid changing market with increased competition. On the other hand, from a technical point of view, the main reasons are the application of column generation techniques, and the ability to solve large scale integer linear programs using improved computer hardware and software. We refer to Desrosiers/Dumas/Solomon/Soumis (1995), Carraresi/Girardi/Nonato (1995), and Caprara/Fischetti/Toth (1996) for examples of successful applications of column generation to routing and scheduling problems.

We developed and implemented algorithms which consist of column generation applied to Lagrangian relaxations. The column generation is necessary since the models usually contain a huge number of variables. For each model we have at least a set of variables which corresponds to the set of all feasible duties. The motivation for considering a huge number of variables in the formulation, is that this allows for considering complex resource constraints in the pricing problem of a column generation approach. Such an approach starts with a small set of initial variables (columns) and iteratively updates the set of variables by solving a master and a subproblem while keeping the number of variables small.

The master problem corresponds to a Lagrangian relaxation, and the subproblem to one or more shortest path type of problems. The aim of column generation is to end with the set of variables that contain the optimal solution. This is achieved when no *duties price out*, including those outside the set of variables currently contained in the master problem (see Freling (1997) for more details).

In case of the CSP, the subproblem is a constrained shortest path problem (see Desrochers/Soumis (1989)). For the VCSP and ICSP we developed an approach for the subproblem which is depicted in Fig. 6.1.

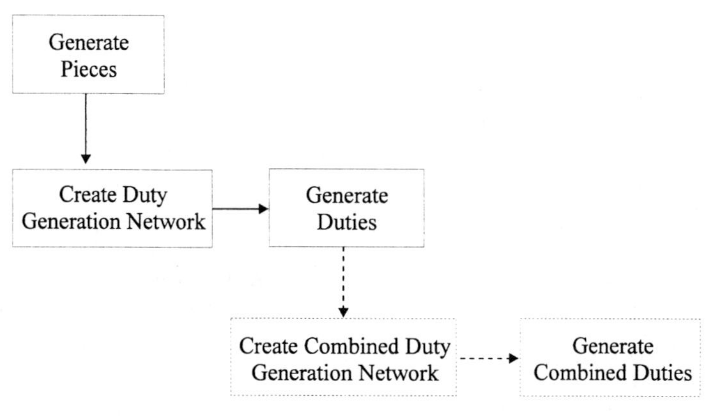

Fig. 6.1

The idea is that duties are generated in three or five steps:

1. Generate pieces: solve an all-pairs shortest path problem in an acyclic network.
2. Construct a network for duty generation.
3. Generate duties: solve a constrained shortest path problem in an acyclic network.
4. Construct a network for combined duty generation.
5. Generate combined duties: solve a sequence of shortest path problems.

Steps 4 and 5 are only necessary for the VCSP when changeovers are not allowed. In that case a duty is entirely assigned to one vehicle block, and a *combined duty* consists of one vehicle block and the crew duties assigned to it.

The structure of the algorithms for the CSP, VCSP and ICSP is depicted in Fig. 6.2. After convergence of the column generation algorithm, a heuristic algorithm is used to determine a feasible solution with the duties in the final master problem as input.

For the CSP, the ICSP and the VCSP without changeovers, we use a set covering heuristic. For the VCSP with changeovers, the heuristic consists of the following three steps:

1. determine a crew schedule using a set covering heuristic.
2. based on this crew schedule, determine a vehicle schedule which is possibly incompatible.
3. if the vehicle and crew schedules are incompatible, solve the CSP based on the vehicle schedule to get a compatible crew schedule.

Again, we refer to Freling (1997) for more details

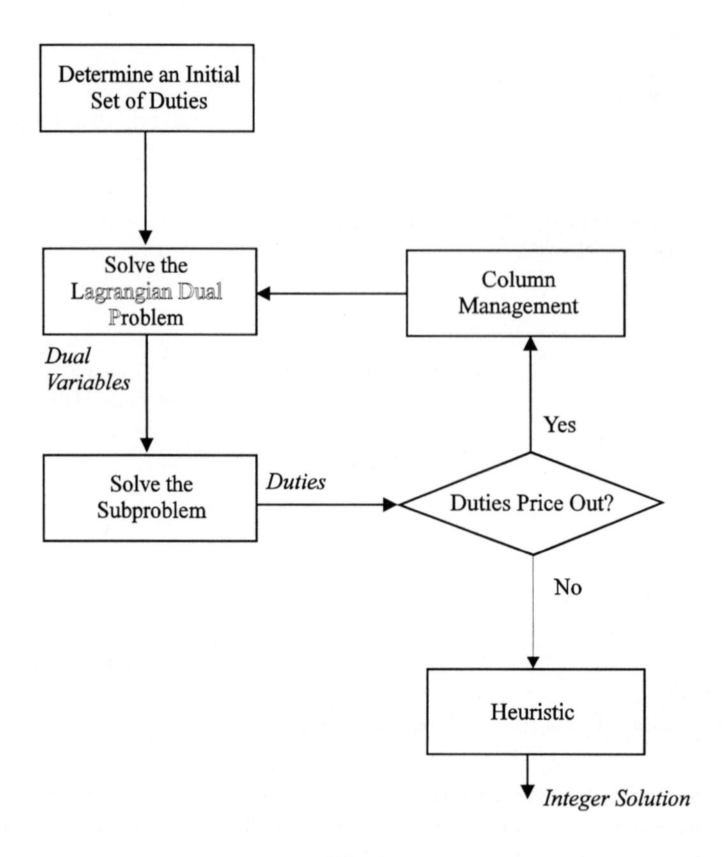

Fig. 6.2

7 Computational Experience at RET Rotterdam

We have tested the algorithms on a set of data provided by RET. This public transport company in Rotterdam provides passenger service with bus, metro and tram. The data corresponds to bus and driver scheduling for individual bus lines where the objective is to minimize the sum of the number of buses and drivers. Five duty types are allowed: a tripper, an early duty (start before 10:50 am), a normal duty (start before 3.15 pm), a late duty (start after 3.15 pm), and a split duty (break duration at least 3.5 hours). A tripper consists of one piece of work with a limited duration. The other duty types have limited duration of pieces of work, break, duration (spread time), and working time. At most two pieces of work are allowed in an early and normal duty, while at most three pieces of work are allowed in a late and split duty.

We present a summary of the computational experience reported in Freling (1997).

Table 7.1

	sequential				independent			integration			
trips	lower	upper	buses	drivers	lower	upper	drivers	lower	upper	buses	drivers
24	15	15	6	9	13	13	7	15	15	6	9
42	21	22	9	13	20	20	11	21	21	9	12
72	15	16	5	11	14	14	9	15	15	5	10
113	23	25	8	17	21	22	14	24	24	9	17
148	34	35	11	24	31	33	22	34	34	11	23

In Table 7.1 we show results obtained with five data sets, defined by the number of trips from 24 to 148. We compare solutions obtained with the sequential approach (columns 2-5), independent approach (columns 6-8) and the integrated approach (columns 9-11). Columns 2-3, 6-7 and 9-10 show the lower and upper bounds for the three approaches, respectively. In case of the sequential approach, these bounds are obtained by the summation of the minimum number of vehicles obtained by solving the SDVSP and the number of drivers corresponding to the solution obtained with the algorithm discussed in the previous section. Columns 3 and 4 show the number of buses and drivers in the solution. In case of the independent approach the bounds are obtained by solving the SDVSP and the ICSP independently. This solution is not useful in practice since bus and driver schedules are generally not compatible. However, it gives an indication of the potential benefit of integration as discussed in Sect. 3. The number of buses is the same as in column 4. The computation times on a Pentium 90 PC with 32Mb RAM are:

1. sequential approach: up to 54 seconds for the lower bounds, and up to 27 seconds for the upper bounds.
2. independent approach : up to 2814 seconds for the lower bounds, and up to 21 seconds for the upper bounds.
3. integrated approach : up to 3374 seconds for the lower bounds, and up to 63 seconds for the upper bounds.

The heuristic algorithms perform well, as can be seen from the table. All solutions for the integrated approach are optimal, and the gap between lower and upper bounds is at most two (8%) for the sequential and independent approach. Comparing the sequential with the independent approach, shows that the potential savings when integrating vehicle and crew scheduling are not very large. This is also clear from the results of the integrated approach. The upper bound is never below the lower bound of the sequential approach. In three out of five test cases, the integrated solution was better than the sequential solution due to the savings of one driver. However, it is very well possible that this difference is caused by the quality of the heuristics and not by the effectiveness of integrating vehicle and crew scheduling.

Table 7.2

	sequential				integration			
trips	lower	upper	buses	drivers	lower	upper	buses	drivers
24	22	22	6	16	16	16	7	9
42	28	30	9	21	22	22	11	11
72	18	22	5	17	15	15	5	10
113	30	33	8	25	30	34	10	24
148	46	47	11	36	39	46	14	32

In Table 7.2 we show results for the case where changeovers are not allowed. Computation times for the sequential approach are up to 10 seconds for the lower bounds and up to 27 seconds for the upper bounds, and for the integrated approach up to 1255 seconds for the lower bounds and up to 11 seconds for the upper bounds. The gap between lower and upper bounds for the integrated approach is up to seven (15%) which makes it more difficult to draw conclusions for the larger two instances. For the smaller three instances, the impact of integration is clearly more substantial compared with the results in Table 7.1. In four out five test cases, the integrated solution was better than the sequential solution due to the saving of up to 10 drivers at the cost of at most three extra buses. We can conclude that integration is better since heuristic solutions are below lower bounds of the sequential approach for three of the problems. This also confirms the expectation of the potential savings resulting from comparing the sequential with the independent approach. For example, in the case of 72 trips, the solution of the integrated approach saves seven drivers compared with the sequential approach, with the same number of buses. Interestingly, for this problem the effect of not allowing changeovers for the sequential approach is completely eliminated by considering the integrated approach.

Comparing Tables 7.1 and 7.2, it is clear that the impact of not allowing changeovers is huge.

8 Conclusion

Our first goal was to develop techniques which are applicable in practice. The results in the previous section show that we can get good solutions within reasonable computation times on a personal computer. Larger instances can be tackled by adding heuristic features to the column generation algorithm. Our second goal was to get an indication about when it is useful to integrate vehicle and crew scheduling. Although we can not draw general conclusions based upon one application, we can see that it is clear that the benefits of integration are greatest if changeovers are not allowed. In practice, it often occurs that either changeovers are not possible due to long distances or changeovers are not allowed for juridical or technical reasons In Freling (1997) we have performed tests for randomly generated data. The results support the conclusions obtained with the real life data.

For future research, we suggest to test and improve the algorithms for other, possibly larger sized, applications.

Acknowledgements: The authors would like to thank an anonymous referee for several helpful comments, and the RET company for providing the data used in the computational study.

References

Ball, M.O./Bodin, L.D./Dial, R. (1983): A Matching Based Heuristic for Scheduling Mass Transit Crews and Vehicles. in: Transportation Science 17, 4 - 31.

Beasley, J.E./Cao, B. (1996): A tree search algorithm for the crew scheduling problem. in: European Journal of Operational Research 94, 517 - 526.

Bodin, L.D./Golden, B./Assad A./Ball, M.O. (1983): Routing and Scheduling of Vehicles and Crews: The State of the Art. in: Computers and Operations Research 10, 63 - 211.

Caprara, A./Fischetti, M./Toth, P. (1996): A heuristic algorithm for the set covering problem, Working paper, DEIS, University of Bologna, Italy.

Carraresi, P./Girardi, L./Nonato, M (1995): Network Models, Lagrangean Relaxation and Subgradients Bundle Approach in Crew Scheduling Problems. in: J.R. Daduna/I. Branco/J.M. Pinto Paixão (eds.): Computer-Aided Transit Scheduling, Proceedings of the Sixth International Workshop. (Springer Verlag) 188 - 212

Daduna, J.R./Pinto Paixão, J.M. (1995): Vehicle scheduling for public mass transit - an overview. in: J.R. Daduna/I. Branco/J.M. Pinto Paixão (eds.): Computer-Aided Transit Scheduling, Proceedings of the Sixth International Workshop. (Springer Verlag) 76 - 90.

Darby-Dowman, K./Jachnik, J.K./Lewis, R.L./Mitra G. (1988): Integrated Decision Support Systems for Urban Transport Scheduling: Discussion of Implementation and Experience. in: J.R. Daduna/A. Wren (eds.): Computer-Aided Transit Scheduling: Proceedings of the Fourth International Workshop. (Springer Verlag) 226 - 239.

Desrochers, M./Soumis, F. (1989): A Column Generation Approach to the Urban Transit Crew Scheduling Problem. in: Transportation Science 23, 1 - 13

Desrosiers, J./Dumas, Y./Solomon, M.M./Soumis F. (1995): Time constrained routing and scheduling. in: M.O. Ball/T.L. Magnanti/C.L. Monma/G.L. Nemhauser (eds.): Handbooks in Operations Research and Management Science 8: Network Routing. (North-Holland) 35 - 139.

Falkner, J.C./Ryan, D.M. (1992): EXPRESS: Set partitioning for bus crew scheduling in Christchurch. in: M. Desrochers/J.M. Rousseau (eds.): Computer-Aided Transit Scheduling: Proceedings of the Fifth International Workshop. (Springer Verlag) 359- 378.

Ferland, J.A./Fortin, L. (1989): Vehicle scheduling with sliding time windows. in: European Journal of Operational Research 38, 213 - 226.

Fischetti, M./Martello, S./Toth, P. (1987): The fixed job schedule problem with spread-time constraints. in: Operations Research 35, 849 - 858.

Fischetti, M./Martello, S./Toth, P. (1989): The fixed job schedule problem with working-time constraints. in: Operations Research 37, 395 - 403.

Freling, R./Pinto Paixão, J.M. (1995): Vehicle scheduling with time constraint. in: J.R. Daduna/I. Branco/J.M. Pinto Paixão (eds.): Computer-Aided Transit Scheduling, Proceedings of the Sixth International Workshop. (Springer Verlag) 130 - 144.

Freling, R./Pinto Paixão, J.M./Wagelmans, A.P.M. (1995): Models and algorithms for single depot vehicle scheduling. Internal Report 9562/A, Econometric Institute, Erasmus University Rotterdam, Rotterdam, submitted for publication in Transportation Science.

Freling (1997): Models and techniques for integrating vehicle and crew scheduling. PhD Thesis, Tinbergen Institute, Erasmus University, Rotterdam.

Odoni, A.R./Rousseau, J.M. (1994): Models in urban and air transportation. in: S.M. Pollock/ M.H. Rothkopf/A. Barnett (eds.): Handbooks in Operations Research and Management Science 6: Operations Research and the Public Sector. (North-Holland) 129-150.

Paixão, J.M. (1990): Transit crew scheduling on a personal workstation. in: Operational Research '90, Selected Papers from the Twelfth IFORS International Conference on Operational Research, Athens, Greece.

Patrikalakis, I./Xerocostas, D. (1992): A New Decomposition Scheme of the Urban Public Transport Scheduling Problem. in: M. Desrochers/J.M. Rousseau (eds.): Computer-Aided Transit Scheduling: Proceedings of the Fifth International Workshop. (Springer Verlag) 407 - 425.

Rousseau, J.M./Blais, J.Y./HASTUS (1985): An interactive system for buses and crew scheduling. in: J.M. Rousseau (ed.): Computer Scheduling of Public Transport 2. (North-Holland) 473 - 491.

Scott, D. (1985): A Large Linear Programming Approach to the Public Transport Scheduling and Cost Model. in: J.M. Rousseau (ed.): Computer Scheduling of Public Transport 2. (North-Holland) 473 - 491.

Tosini, E./Vercellis, C. (1988): An Interactive System for Extra-Urban Vehicle and Crew Scheduling Problems. in: J.R. Daduna/A. Wren (eds.) Computer-Aided Transit Scheduling: Proceedings of the Fourth International Workshop. (Springer Verlag) 41 - 53.

Wren, A./Rousseau, J.M. (1995): Bus Driver Scheduling - An Overview. in: J.R. Daduna/I. Branco/J.M. Pinto Paixão (eds.): Computer-Aided Transit Scheduling, Proceedings of the Sixth International Workshop. (Springer Verlag) 173-187.

Appendix 1: List of Authors

Cynthia **Barnhart,** Center for Transportation Studies/Massachusetts Institute of Technology, Cambridge, MA 02139-4307 (United States)

David **Bernstein,** Department of Civil Engineering & Operations Research/Princeton University (United States)

Ralf **Borndörfer,** Konrad-Zuse-Zentrum für Informationstechnik, Berlin (Germany)

Alberto **Caprara,** DEIS/University of Bologna (Italy)

Avishai **Ceder,** Technion - Israel Institute of Technology, Haifa (Israel)

Joachim R. **Daduna,** Fachhochschule für Witschaft Berlin (Germany)

Kamen **Danowski,** Fraunhofer Institut für Informations- und Datenverarbeitung, Dresden (Germany)

Guy **Desaulniers,** École Polytechnique and GERAD, Montreal (Canada)

Jacques **Desrosiers,** École des Hautes Études Commerciales and GERAD, Montreal (Canada)

Xu Jun **Eberlein,** Caliper Corporation, Newton, MA (United States)

Nicolau D. **Fares Gualda,** Departamento de Engenharia de Transportes, Escola Politécnica, Universidade de São Paolo (Brasil)

Alexander **Fay,** Technische Universität Braunschweig, Institut für Regelungs- und Automatisierungs-technik, Braunschweig (Germany)

Matteo **Fischetti,** DMI/University of Udine (Italy)

Sarah **Fores,** School of Computer Studies, University of Leeds (United Kingdom)

Richard **Freling**, Econometric Institute, Erasmus University Rotterdam (The Netherlands)

Christian **Friberg,** Institut für Informatik und Praktische Mathematik, Christian-Albrechts-Universität zu Kiel (Germany)

Andrea **Gaffi,** MAIOR srl, Lucca (Italy)

Martin **Grötschel,** Konrad-Zuse-Zentrum für Informationstechnik, Berlin (Germany)

Pier Luigi **Guida,** Ferrovie dello Stato SPA (Italy)

Knut **Haase,** Institut für Betriebswirtschaftslehre, Christian-Albrechts-Universität zu Kiel (Germany)

Daeki **Kim,** (United States)

Fridolin **Klostermeier,** Intranetz Gesellschaft für Informationslogistik mbH, Berlin (Germany)

Christian **Küttner,** Intranetz Gesellschaft für Informationslogistik mbH, Berlin (Germany)

Raymond S.K. **Kwan,** School of Computer Studies, University of Leeds (United Kingdom)

Ann S.K. **Kwan,** School of Computer Studies, University of Leeds (United Kingdom)

Arielle **Lasry,** Numetrix Limited, Toronto (Canada)

Andreas **Löbel,** Konrad-Zuse-Zentrum für Informationstechnik, Berlin (Germany)

Taïeb **Mellouli,** Decision Support & OR Laboratory, University of Paderborn (Germany)

Marta **Mesquita,** Instituto Superior de Agronomia, Dep. De Matemática, Lisboa (Portugal)

Maddalena **Nonato,** Computer Science Department, University of Pisa (Italy)

Susan W. **O'Dell,** The SABRE Group, Burlington, MA (United States)

José **Paixão,** DEOI, Faculdade de Ciências de Lisboa (Portugal)

Margaret E. **Parker,** School of Computer Studies, University of Leeds (United Kingdom)

José M. **Pinto Paixão**, DEIO, Faculdade de Ciências da Uiversidade de Lisboa (Portugal)

Les **Proll,** School of Computer Studies, University of Leeds (United Kingdom)

Mohammad A. **Rahin,** School of Computer Studies, University of Leeds (United Kingdom)

Eckehard **Schnieder,** Technische Universität Braunschweig, Institut für Regelungs- und Automatisierungs-technik, Braunschweig (Germany)

Marius M. **Solomon,** Northeastern University, Boston, MA (United States) and GERAD, Montreal (Canada)

François **Soumis,** École Polytechnique and GERAD, Montreal (Canada)

Leena **Suhl,** Decision Support & OR Laboratory, University of Paderborn (Germany)

Ofer **Tal,** Technion - Israel Institute of Technology, Haifa (Israel)

Paolo **Toth,** DEIS/University of Bologna (Italy)

Daniele **Vigo,** DEIS/University of Bologna (Italy)

Albert P.M. **Wagelmans,** Econometric Institute, Erasmus University Rotterdam (The Netherlands)

Nigel H.M. **Wilson,** Department of Civil & Environmental Engineering/Massachusetts Institute of Technology, Cambridge, MA 02139-4307 (United States)

Anthony **Wren,** School of Computer Studies, University of Leeds (United Kingdom)

Koorush **Ziarati,** École Polytechnique and GERAD, Montreal (Canada)

Appendix 2: List of the Presented Papers Which are not Included in this Volume

Applying *CrewOpt* to "large" problems

Jean-Marc **Rousseau** / GIRO Enterprises Inc., Montreal, Quebec (Canada)

Vehicle Scheduling with Not Exactly Specified Departure Times

Joachim R. **Daduna** / Fachhochschule Konstanz, Konstanz (Germany) / Manfred **Völker,** / HanseCom-Gesellschaft für Informatik und Informationsverarbeitung mbH, Hamburg (Germany)

Bus route and route headway determination

Mark **Miller** / Yongchang **Zhao** / Trapeze Software Inc., Mississauga, Ontario (Canada)

Graphical Scheduling - A Powerful Tool for Service Planning and Control

Eric C. **Bruun** / National Transit Institute, Rutgers University, New Brunswick, New Jersey, USA / Vukan R. **Vuchic** / Yong-Eun **Shin** / Department of Systems Engineering, University of Pennsylvania, Philadelphia, Penn., USA

A New Network Model for Vehicle Scheduling Applied to Maintenance for Airlines and Railways

Taïeb **Mellouli** / Department of Business Computing, Decision Support & OR Lab, University of Paderborn, Paderborn (Germany)

Fully automated paratransit scheduling for large multi-contractor operations

Nigel **Hamer** / GIRO, Inc., Montreal, Quebec (Canada)

Use of Genetic Algorithm to Schedule the Specialized Transportation Vehicles

Shinya **Kikuchi** / Department of Civil and Environmental Engineering, / University of Delaware, Newark, Delaware, USA / Mikiharu **Arimura** / Muroran Institute of Technology, Muroran, Hokkaido (Japan)

Advanced Planning and Scheduling Tools in Ausias

Kenneth **Darby-Dowman** / James **Little** / Gautam **Mitra** / Department of Mathematics and Statistics, Brunel University, Uxbridge, Middlesex (United Kingdom) /
Antonio **Marques,** Vicente **Sebastian** / Manuel **Torregrosa** / ETRA I&D, Valencia (Spain)

A Branch-Fix-and-Relax Approach to Solving Crew Scheduling Problems

Kenneth **Darby-Dowman** / James **Little** / Gautam **Mitra** / Department of Mathematics and Statistics, Brunel University, Uxbridge, Middlesex (United Kingdom)

An Ant System for Bus Driver Scheduling

Paul Forsyth / Anthony Wren / Scheduling and Constraint Management Group, School of Computer Studies, University of Leeds, Leeds (United Kingdom)

CSXTs Automated Router for Unscheduled Locomotive Maintenance

Keith C. **Campbell** / Operations Research Group, US Airways, Arlington, Virginia (United States)

DSSs for generating timetables and for determining infrastructural extensions

Michael A. **Odijk** / Delft University of Technology, Department of Mathematics and Computer Science, Delft (The Netherlands) / Peter J. **Zwaneveld** / TNO INRO / Erasmus University, Rotterdam School of Management, Rotterdam (The Netherlands) **Jurjen S. Hooghiemstra** / Railned Innovation, Utrecht(The Netherlands) / Leo G. **Kroon** / Erasmus University and NS Reizigers, Department of Logistics, Utrecht (The Netherlands) / Marc **Salomon** / Rabobank Tilburg University, Department of Econometrics, Tilburg (The Netherlands)

ALLEGRO Project

Patrick **Brichard** / Jean-Charles **Urvoy** / RATP, Paris (France)

Computer Aided Passenger Information Systems in the Berlin region - Concept, Data Management, Geographical Applications, Experiences

Martin **Müller-Elschner** / IVU GmbH, Berlin (Germany) / Jürgen **Ross** / Berliner Verkehrsbetriebe (BVG), Berlin (Germany)

Connection Optimization - Method and Application

Volker **Bach** / Andreas **Knauf** / Uwe **Strubbe** / IVU Gesellschaft für Informatik,Verkers-und Unweltplanung mbH, Berlin (Germany)

Lecture Notes in Economics and Mathematical Systems

For information about Vols. 1–284
please contact your bookseller or Springer-Verlag

Vol. 285: Toward Interactive and Intelligent Decision Support Systems. Volume 1. Proceedings, 1986. Edited by Y. Sawaragi, K. Inoue and H. Nakayama. XII, 445 pages. 1987.

Vol. 286: Toward Interactive and Intelligent Decision Support Systems. Volume 2. Proceedings, 1986. Edited by Y. Sawaragi, K. Inoue and H. Nakayama. XII, 450 pages. 1987.

Vol. 287: Dynamical Systems. Proceedings, 1985. Edited by A.B. Kurzhanski and K. Sigmund. VI, 215 pages. 1987.

Vol. 288: G.D. Rudebusch, The Estimation of Macroeconomic Disequilibrium Models with Regime Classification Information. VII,128 pages. 1987.

Vol. 289: B.R. Meijboom, Planning in Decentralized Firms. X, 168 pages. 1987.

Vol. 290: D.A. Carlson, A. Haurie, Infinite Horizon Optimal Control. XI, 254 pages. 1987.

Vol. 291: N. Takahashi, Design of Adaptive Organizations. VI, 140 pages. 1987.

Vol. 292: I. Tchijov, L. Tomaszewicz (Eds.), Input-Output Modeling. Proceedings, 1985. VI, 195 pages. 1987.

Vol. 293: D. Batten, J. Casti, B. Johansson (Eds.), Economic Evolution and Structural Adjustment. Proceedings, 1985. VI, 382 pages.

Vol. 294: J. Jahn, W. Knabs (Eds.), Recent Advances and Historical Development of Vector Optimization. VII, 405 pages. 1987.

Vol. 295. H. Meister, The Purification Problem for Constrained Games with Incomplete Information. X, 127 pages. 1987.

Vol. 296: A. Börsch-Supan, Econometric Analysis of Discrete Choice. VIII, 211 pages. 1987.

Vol. 297: V. Fedorov, H. Läuter (Eds.), Model-Oriented Data Analysis. Proceedings, 1987. VI, 239 pages. 1988.

Vol. 298: S.H. Chew, Q. Zheng, Integral Global Optimization. VII, 179 pages. 1988.

Vol. 299: K. Marti, Descent Directions and Efficient Solutions in Discretely Distributed Stochastic Programs. XIV, 178 pages. 1988.

Vol. 300: U. Derigs, Programming in Networks and Graphs. XI, 315 pages. 1988.

Vol. 301: J. Kacprzyk, M. Roubens (Eds.), Non-Conventional Preference Relations in Decision Making. VII, 155 pages. 1988.

Vol. 302: H.A. Eiselt, G. Pederzoli (Eds.), Advances in Optimization and Control. Proceedings, 1986. VIII, 372 pages. 1988.

Vol. 303: F.X. Diebold, Empirical Modeling of Exchange Rate Dynamics. VII, 143 pages. 1988.

Vol. 304: A. Kurzhanski, K. Neumann, D. Pallaschke (Eds.), Optimization, Parallel Processing and Applications. Proceedings, 1987. VI, 292 pages. 1988.

Vol. 305: G.-J.C.Th. van Schijndel, Dynamic Firm and Investor Behaviour under Progressive Personal Taxation. X, 215 pages.1988.

Vol. 306: Ch. Klein, A Static Microeconomic Model of Pure Competition. VIII, 139 pages. 1988.

Vol. 307: T.K. Dijkstra (Ed.), On Model Uncertainty and its Statistical Implications. VII, 138 pages. 1988.

Vol. 308: J.R. Daduna, A. Wren (Eds.), Computer-Aided Transit Scheduling. VIII, 339 pages. 1988.

Vol. 309: G. Ricci, K. Velupillai (Eds.), Growth Cycles and Multisectoral Economics: The Goodwin Tradition. III, 126 pages. 1988.

Vol. 310: J. Kacprzyk, M. Fedrizzi (Eds.), Combining Fuzzy Imprecision with Probabilistic Uncertainty in Decision Making. IX, 399 pages. 1988.

Vol. 311: R. Färe, Fundamentals of Production Theory. IX, 163 pages. 1988.

Vol. 312: J. Krishnakumar, Estimation of Simultaneous Equation Models with Error Components Structure. X, 357 pages. 1988.

Vol. 313: W. Jammernegg, Sequential Binary Investment Decisions. VI, 156 pages. 1988.

Vol. 314: R. Tietz, W. Albers, R. Selten (Eds.), Bounded Rational Behavior in Experimental Games and Markets. VI, 368 pages. 1988.

Vol. 315: I. Orishimo, G.J.D. Hewings, P. Nijkamp (Eds), Information Technology: Social and Spatial Perspectives. Proceedings 1986. VI, 268 pages. 1988.

Vol. 316: R.L. Basmann, D.J. Slottje, K. Hayes, J.D. Johnson, D.J. Molina, The Generalized Fechner-Thurstone Direct Utility Function and Some of its Uses. VIII, 159 pages. 1988.

Vol. 317: L. Bianco, A. La Bella (Eds.), Freight Transport Planning and Logistics. Proceedings, 1987. X, 568 pages. 1988.

Vol. 318: T. Doup, Simplicial Algorithms on the Simplotope. VIII, 262 pages. 1988.

Vol. 319: D.T. Luc, Theory of Vector Optimization. VIII, 173 pages. 1989.

Vol. 320: D. van der Wijst, Financial Structure in Small Business. VII, 181 pages. 1989.

Vol. 321: M. Di Matteo, R.M. Goodwin, A. Vercelli (Eds.), Technological and Social Factors in Long Term Fluctuations. Proceedings. IX, 442 pages. 1989.

Vol. 322: T. Kollintzas (Ed.), The Rational Expectations Equilibrium Inventory Model. XI, 269 pages. 1989.

Vol. 323: M.B.M. de Koster, Capacity Oriented Analysis and Design of Production Systems. XII, 245 pages. 1989.

Vol. 324: I.M. Bomze, B.M. Pötscher, Game Theoretical Foundations of Evolutionary Stability. VI, 145 pages. 1989.

Vol. 325: P. Ferri, E. Greenberg, The Labor Market and Business Cycle Theories. X, 183 pages. 1989.

Vol. 326: Ch. Sauer, Alternative Theories of Output, Unemployment, and Inflation in Germany: 1960–1985. XIII, 206 pages. 1989.

Vol. 327: M. Tawada, Production Structure and International Trade. V, 132 pages. 1989.

Vol. 328: W. Güth, B. Kalkofen, Unique Solutions for Strategic Games. VII, 200 pages. 1989.

Vol. 329: G. Tillmann, Equity, Incentives, and Taxation. VI, 132 pages. 1989.

Vol. 330: P.M. Kort, Optimal Dynamic Investment Policies of a Value Maximizing Firm. VII, 185 pages. 1989.

Vol. 331: A. Lewandowski, A.P. Wierzbicki (Eds.), Aspiration Based Decision Support Systems. X, 400 pages. 1989.

Vol. 332: T.R. Gulledge, Jr., L.A. Litteral (Eds.), Cost Analysis Applications of Economics and Operations Research. Proceedings. VII, 422 pages. 1989.

Vol. 333: N. Dellaert, Production to Order. VII, 158 pages. 1989.

Vol. 334: H.-W. Lorenz, Nonlinear Dynamical Economics and Chaotic Motion. XI, 248 pages. 1989.

Vol. 335: A.G. Lockett, G. Islei (Eds.), Improving Decision Making in Organisations. Proceedings. IX, 606 pages. 1989.

Vol. 336: T. Puu, Nonlinear Economic Dynamics. VII, 119 pages. 1989.

Vol. 337: A. Lewandowski, I. Stanchev (Eds.), Methodology and Software for Interactive Decision Support. VIII, 309 pages. 1989.

Vol. 338: J.K. Ho, R.P. Sundarraj, DECOMP: An Implementation of Dantzig-Wolfe Decomposition for Linear Programming. VI, 206 pages.

Vol. 339: J. Terceiro Lomba, Estimation of Dynamic Econometric Models with Errors in Variables. VIII, 116 pages. 1990.

Vol. 340: T. Vasko, R. Ayres, L. Fontvieille (Eds.), Life Cycles and Long Waves. XIV, 293 pages. 1990.

Vol. 341: G.R. Uhlich, Descriptive Theories of Bargaining. IX, 165 pages. 1990.

Vol. 342: K. Okuguchi, F. Szidarovszky, The Theory of Oligopoly with Multi-Product Firms. V, 167 pages. 1990.

Vol. 343: C. Chiarella, The Elements of a Nonlinear Theory of Economic Dynamics. IX, 149 pages. 1990.

Vol. 344: K. Neumann, Stochastic Project Networks. XI, 237 pages. 1990.

Vol. 345: A. Cambini, E. Castagnoli, L. Martein, P Mazzoleni, S. Schaible (Eds.), Generalized Convexity and Fractional Programming with Economic Applications. Proceedings, 1988. VII, 361 pages. 1990.

Vol. 346: R. von Randow (Ed.), Integer Programming and Related Areas. A Classified Bibliography 1984–1987. XIII, 514 pages. 1990.

Vol. 347: D. Ríos Insua, Sensitivity Analysis in Multi-objective Decision Making. XI, 193 pages. 1990.

Vol. 348: H. Störmer, Binary Functions and their Applications. VIII, 151 pages. 1990.

Vol. 349: G.A. Pfann, Dynamic Modelling of Stochastic Demand for Manufacturing Employment. VI, 158 pages. 1990.

Vol. 350: W.-B. Zhang, Economic Dynamics. X, 232 pages. 1990.

Vol. 351: A. Lewandowski, V. Volkovich (Eds.), Multiobjective Problems of Mathematical Programming. Proceedings, 1988. VII, 315 pages. 1991.

Vol. 352: O. van Hilten, Optimal Firm Behaviour in the Context of Technological Progress and a Business Cycle. XII, 229 pages. 1991.

Vol. 353: G. Ricci (Ed.), Decision Processes in Economics. Proceedings, 1989. III, 209 pages 1991.

Vol. 354: M. Ivaldi, A Structural Analysis of Expectation Formation. XII, 230 pages. 1991.

Vol. 355: M. Salomon. Deterministic Lotsizing Models for Production Planning. VII, 158 pages. 1991.

Vol. 356: P. Korhonen, A. Lewandowski, J . Wallenius (Eds.), Multiple Criteria Decision Support. Proceedings, 1989. XII, 393 pages. 1991.

Vol. 357: P. Zörnig, Degeneracy Graphs and Simplex Cycling. XV, 194 pages. 1991.

Vol. 358: P. Knottnerus, Linear Models with Correlated Disturbances. VIII, 196 pages. 1991.

Vol. 359: E. de Jong, Exchange Rate Determination and Optimal Economic Policy Under Various Exchange Rate Regimes. VII, 270 pages. 1991.

Vol. 360: P. Stalder, Regime Translations, Spillovers and Buffer Stocks. VI, 193 pages . 1991.

Vol. 361: C. F. Daganzo, Logistics Systems Analysis. X, 321 pages. 1991.

Vol. 362: F. Gehrels, Essays in Macroeconomics of an Open Economy. VII, 183 pages. 1991.

Vol. 363: C. Puppe, Distorted Probabilities and Choice under Risk. VIII, 100 pages . 1991

Vol. 364: B. Horvath, Are Policy Variables Exogenous? XII, 162 pages. 1991.

Vol. 365: G. A. Heuer, U. Leopold-Wildburger. Balanced Silverman Games on General Discrete Sets. V, 140 pages. 1991.

Vol. 366: J. Gruber (Ed.), Econometric Decision Models. Proceedings, 1989. VIII, 636 pages. 1991.

Vol. 367: M. Grauer, D. B. Pressmar (Eds.), Parallel Computing and Mathematical Optimization. Proceedings. V, 208 pages. 1991.

Vol. 368: M. Fedrizzi, J. Kacprzyk, M. Roubens (Eds.), Interactive Fuzzy Optimization. VII, 216 pages. 1991.

Vol. 369: R. Koblo, The Visible Hand. VIII, 131 pages.1991.

Vol. 370: M. J. Beckmann, M. N. Gopalan, R. Subramanian (Eds.), Stochastic Processes and their Applications. Proceedings, 1990. XLI, 292 pages. 1991.

Vol. 371: A. Schmutzler, Flexibility and Adjustment to Information in Sequential Decision Problems. VIII, 198 pages. 1991.

Vol. 372: J. Esteban, The Social Viability of Money. X, 202 pages. 1991.

Vol. 373: A. Billot, Economic Theory of Fuzzy Equilibria. XIII, 164 pages. 1992.

Vol. 374: G. Pflug, U. Dieter (Eds.), Simulation and Optimization. Proceedings, 1990. X, 162 pages. 1992.

Vol. 375: S.-J. Chen, Ch.-L. Hwang, Fuzzy Multiple Attribute Decision Making. XII, 536 pages. 1992.

Vol. 376: K.-H. Jöckel, G. Rothe, W. Sendler (Eds.), Bootstrapping and Related Techniques. Proceedings, 1990. VIII, 247 pages. 1992.

Vol. 377: A. Villar, Operator Theorems with Applications to Distributive Problems and Equilibrium Models. XVI, 160 pages. 1992.

Vol. 378: W. Krabs, J. Zowe (Eds.), Modern Methods of Optimization. Proceedings, 1990. VIII, 348 pages. 1992.

Vol. 379: K. Marti (Ed.), Stochastic Optimization. Proceedings, 1990. VII, 182 pages. 1992.

Vol. 380: J. Odelstad, Invariance and Structural Dependence. XII, 245 pages. 1992.

Vol. 381: C. Giannini, Topics in Structural VAR Econometrics. XI, 131 pages. 1992.

Vol. 382: W. Oettli, D. Pallaschke (Eds.), Advances in Optimization. Proceedings, 1991. X, 527 pages. 1992.

Vol. 383: J. Vartiainen, Capital Accumulation in a Corporatist Economy. VII, 177 pages. 1992.

Vol. 384: A. Martina, Lectures on the Economic Theory of Taxation. XII, 313 pages. 1992.

Vol. 385: J. Gardeazabal, M. Regúlez, The Monetary Model of Exchange Rates and Cointegration. X, 194 pages. 1992.

Vol. 386: M. Desrochers, J.-M. Rousseau (Eds.), Computer-Aided Transit Scheduling. Proceedings, 1990. XIII, 432 pages. 1992.

Vol. 387: W. Gaertner, M. Klemisch-Ahlert, Social Choice and Bargaining Perspectives on Distributive Justice. VIII, 131 pages. 1992.

Vol. 388: D. Bartmann, M. J. Beckmann, Inventory Control. XV, 252 pages. 1992.

Vol. 389: B. Dutta, D. Mookherjee, T. Parthasarathy, T. Raghavan, D. Ray, S. Tijs (Eds.), Game Theory and Economic Applications. Proceedings, 1990. IX, 454 pages. 1992.

Vol. 390: G. Sorger, Minimum Impatience Theorem for Recursive Economic Models. X, 162 pages. 1992.

Vol. 391: C. Keser, Experimental Duopoly Markets with Demand Inertia. X, 150 pages. 1992.

Vol. 392: K. Frauendorfer, Stochastic Two-Stage Programming. VIII, 228 pages. 1992.

Vol. 393: B. Lucke, Price Stabilization on World Agricultural Markets. XI, 274 pages. 1992.

Vol. 394: Y.-J. Lai, C.-L. Hwang, Fuzzy Mathematical Programming. XIII, 301 pages. 1992.

Vol. 395: G. Haag, U. Mueller, K. G. Troitzsch (Eds.), Economic Evolution and Demographic Change. XVI, 409 pages. 1992.

Vol. 396: R. V. V. Vidal (Ed.), Applied Simulated Annealing. VIII, 358 pages. 1992.

Vol. 397: J. Wessels, A. P. Wierzbicki (Eds.), User-Oriented Methodology and Techniques of Decision Analysis and Support. Proceedings, 1991. XII, 295 pages. 1993.

Vol. 398: J.-P. Urbain, Exogeneity in Error Correction Models. XI, 189 pages. 1993.

Vol. 399: F. Gori, L. Geronazzo, M. Galeotti (Eds.), Nonlinear Dynamics in Economics and Social Sciences. Proceedings, 1991. VIII, 367 pages. 1993.

Vol. 400: H. Tanizaki, Nonlinear Filters. XII, 203 pages. 1993.

Vol. 401: K. Mosler, M. Scarsini, Stochastic Orders and Applications. V, 379 pages. 1993.

Vol. 402: A. van den Elzen, Adjustment Processes for Exchange Economies and Noncooperative Games. VII, 146 pages. 1993.

Vol. 403: G. Brennscheidt, Predictive Behavior. VI, 227 pages. 1993.

Vol. 404: Y.-J. Lai, Ch.-L. Hwang, Fuzzy Multiple Objective Decision Making. XIV, 475 pages. 1994.

Vol. 405: S. Komlósi, T. Rapcsák, S. Schaible (Eds.), Generalized Convexity. Proceedings, 1992. VIII, 404 pages. 1994.

Vol. 406: N. M. Hung, N. V. Quyen, Dynamic Timing Decisions Under Uncertainty. X, 194 pages. 1994.

Vol. 407: M. Ooms, Empirical Vector Autoregressive Modeling. XIII, 380 pages. 1994.

Vol. 408: K. Haase, Lotsizing and Scheduling for Production Planning. VIII, 118 pages. 1994.

Vol. 409: A. Sprecher, Resource-Constrained Project Scheduling. XII, 142 pages. 1994.

Vol. 410: R. Winkelmann, Count Data Models. XI, 213 pages. 1994.

Vol. 411: S. Dauzère-Péres, J.-B. Lasserre, An Integrated Approach in Production Planning and Scheduling. XVI, 137 pages. 1994.

Vol. 412: B. Kuon, Two-Person Bargaining Experiments with Incomplete Information. IX, 293 pages. 1994.

Vol. 413: R. Fiorito (Ed.), Inventory, Business Cycles and Monetary Transmission. VI, 287 pages. 1994.

Vol. 414: Y. Crama, A. Oerlemans, F. Spieksma, Production Planning in Automated Manufacturing. X, 210 pages. 1994.

Vol. 415: P. C. Nicola, Imperfect General Equilibrium. XI, 167 pages. 1994.

Vol. 416: H. S. J. Cesar, Control and Game Models of the Greenhouse Effect. XI, 225 pages. 1994.

Vol. 417: B. Ran, D. E. Boyce, Dynamic Urban Transportation Network Models. XV, 391 pages. 1994.

Vol. 418: P. Bogetoft, Non-Cooperative Planning Theory. XI, 309 pages. 1994.

Vol. 419: T. Maruyama, W. Takahashi (Eds.), Nonlinear and Convex Analysis in Economic Theory. VIII. 306 pages. 1995.

Vol. 420: M. Peeters, Time-To-Build. Interrelated Investment and Labour Demand Modelling. With Applications to Six OECD Countries. IX, 204 pages. 1995.

Vol. 421: C. Dang, Triangulations and Simplicial Methods. IX, 196 pages. 1995.

Vol. 422: D. S. Bridges, G. B. Mehta, Representations of Preference Orderings. X, 165 pages. 1995.

Vol. 423: K. Marti, P. Kall (Eds.), Stochastic Programming. Numerical Techniques and Engineering Applications. VIII, 351 pages. 1995.

Vol. 424: G. A. Heuer, U. Leopold-Wildburger, Silverman's Game. X, 283 pages. 1995.

Vol. 425: J. Kohlas, P.-A. Monney, A Mathematical Theory of Hints. XIII, 419 pages, 1995.

Vol. 426: B. Finkenstädt, Nonlinear Dynamics in Economics. IX, 156 pages. 1995.

Vol. 427: F. W. van Tongeren, Microsimulation Modelling of the Corporate Firm. XVII, 275 pages. 1995.

Vol. 428: A. A. Powell, Ch. W. Murphy, Inside a Modern Macroeconometric Model. XVIII, 424 pages. 1995.

Vol. 429: R. Durier, C. Michelot, Recent Developments in Optimization. VIII, 356 pages. 1995.

Vol. 430: J. R. Daduna, I. Branco, J. M. Pinto Paixão (Eds.), Computer-Aided Transit Scheduling. XIV, 374 pages. 1995.

Vol. 431: A. Aulin, Causal and Stochastic Elements in Business Cycles. XI, 116 pages. 1996.

Vol. 432: M. Tamiz (Ed.), Multi-Objective Programming and Goal Programming. VI, 359 pages. 1996.

Vol. 433: J. Menon, Exchange Rates and Prices. XIV, 313 pages. 1996.

Vol. 434: M. W. J. Blok, Dynamic Models of the Firm. VII, 193 pages. 1996.

Vol. 435: L. Chen, Interest Rate Dynamics, Derivatives Pricing, and Risk Management. XII, 149 pages. 1996.

Vol. 436: M. Klemisch-Ahlert, Bargaining in Economic and Ethical Environments. IX, 155 pages. 1996.

Vol. 437: C. Jordan, Batching and Scheduling. IX, 178 pages. 1996.

Vol. 438: A. Villar, General Equilibrium with Increasing Returns. XIII, 164 pages. 1996.

Vol. 439: M. Zenner, Learning to Become Rational. VII, 201 pages. 1996.

Vol. 440: W. Ryll, Litigation and Settlement in a Game with Incomplete Information. VIII, 174 pages. 1996.

Vol. 441: H. Dawid, Adaptive Learning by Genetic Algorithms. IX, 166 pages.1996.

Vol. 442: L. Corchón, Theories of Imperfectly Competitive Markets. XIII, 163 pages. 1996.

Vol. 443: G. Lang, On Overlapping Generations Models with Productive Capital. X, 98 pages. 1996.

Vol. 444: S. Jørgensen, G. Zaccour (Eds.), Dynamic Competitive Analysis in Marketing. X, 285 pages. 1996.

Vol. 445: A. H. Christer, S. Osaki, L. C. Thomas (Eds.), Stochastic Modelling in Innovative Manufactoring. X, 361 pages. 1997.

Vol. 446: G. Dhaene, Encompassing. X, 160 pages. 1997.

Vol. 447: A. Artale, Rings in Auctions. X, 172 pages. 1997.

Vol. 448: G. Fandel, T. Gal (Eds.), Multiple Criteria Decision Making. XII, 678 pages. 1997.

Vol. 449: F. Fang, M. Sanglier (Eds.), Complexity and Self-Organization in Social and Economic Systems. IX, 317 pages, 1997.

Vol. 450: P. M. Pardalos, D. W. Hearn, W. W. Hager, (Eds.), Network Optimization. VIII, 485 pages, 1997.

Vol. 451: M. Salge, Rational Bubbles. Theoretical Basis, Economic Relevance, and Empirical Evidence with a Special Emphasis on the German Stock Market.IX, 265 pages. 1997.

Vol. 452: P. Gritzmann, R. Horst, E. Sachs, R. Tichatschke (Eds.), Recent Advances in Optimization. VIII, 379 pages. 1997.

Vol. 453: A. S. Tangian, J. Gruber (Eds.), Constructing Scalar-Valued Objective Functions. VIII, 298 pages. 1997.

Vol. 454: H.-M. Krolzig, Markov-Switching Vector Autoregressions. XIV, 358 pages. 1997.

Vol. 455: R. Caballero, F. Ruiz, R. E. Steuer (Eds.), Advances in Multiple Objective and Goal Programming. VIII, 391 pages. 1997.

Vol. 456: R. Conte, R. Hegselmann, P. Terna (Eds.), Simulating Social Phenomena. VIII, 536 pages. 1997.

Vol. 457: C. Hsu, Volume and the Nonlinear Dynamics of Stock Returns. VIII, 133 pages. 1998.

Vol. 458: K. Marti, P. Kall (Eds.), Stochastic Programming Methods and Technical Applications. X, 437 pages. 1998.

Vol. 459: H. K. Ryu, D. J. Slottje, Measuring Trends in U.S. Income Inequality. XI, 195 pages. 1998.

Vol. 460: B. Fleischmann, J. A. E. E. van Nunen, M. G. Speranza, P. Stähly, Advances in Distribution Logistic. XI, 535 pages. 1998.

Vol. 461: U. Schmidt, Axiomatic Utility Theory under Risk. XV, 201 pages. 1998.

Vol. 462: L. von Auer, Dynamic Preferences, Choice Mechanisms, and Welfare. XII, 226 pages. 1998.

Vol. 463: G. Abraham-Frois (Ed.), Non-Linear Dynamics and Endogenous Cycles. VI, 204 pages. 1998.

Vol. 464: A. Aulin, The Impact of Science on Economic Growth and its Cycles. IX, 204 pages. 1998.

Vol. 465: T. J. Stewart, R. C. van den Honert (Eds.), Trends in Multicriteria Decision Making. X, 448 pages. 1998.

Vol. 466: A. Sadrieh, The Alternating Double Auction Market. VII, 350 pages. 1998.

Vol. 467: H. Hennig-Schmidt, Bargaining in a Video Experiment. Determinants of Boundedly Rational Behavior. XII, 221 pages. 1999.

Vol. 468: A. Ziegler, A Game Theory Analysis of Options. XIV, 145 pages. 1999.

Vol. 469: M. P. Vogel, Environmental Kuznets Curves. XIII, 197 pages. 1999.

Vol. 470: M. Ammann, Pricing Derivative Credit Risk. XII, 228 pages. 1999.

Vol. 471: N. H. M. Wilson (Ed.), Computer-Aided Transit Scheduling. XI, 444 pages. 1999.